Plunderers of the Earth

the Erosion of Civilization,
the Mad Crusade to Control the Climate,
and the Untold Stories of Soil and CO_2.

Julius Ruechel

Copyright © 2024 Julius Ruechel
www.juliusruechel.com
All rights reserved.
ISBN: 9798335219761

To all the heretics who refuse to be cowed into silence...

Contents

List of Figures 7
Introduction 13

PART ONE — DECONSTRUCTION
1. Nation Building — "The Rain Follows the Plow" 27
2. The Seeds of Pessimism and the Rise of the Doomsday Cult 67
3. "The Era of Global Boiling Has Begun" 111
4. A Continent Built for Wildfires 129
5. "Our Most Shameful Waste" 137
6. The Toxic Marriage Between Science and Politics 163
7. Local Knowledge 185
8. Re-Writing History to Warm the Present and Cool the Past 201
9. Icarus Writes an Algorithm 241
10. CO_2: A Tragically Misunderstood Villain 263
11. Fixing the Algorithm with Hypothetical Feedback Loops 289

PART TWO — RECONSTRUCTION
12. The Elusive Fingerprint of Fossil Fuels 301
13. The Blind Men and the Geological Elephant 315
14. The Boring Billion 345
15. Clouds and Cosmic Radiation 357
16. CO_2 Fertilization: The Fuel for Life 375
17. The Story of Humus & the Hidden World Beneath our Feet 389
18. Coal and Fire in an Oxygen-Fueled World 417
19. The Great Dying 425
20. Atomic Gardening 437
21. A Disturbance in the Force 447

22.	Sod, Soil, and Worms	465
23.	From Net Carbon Sink to Net Carbon Emitter	491
24.	The Slow Slide into the Icehouse	517
25.	Mining the Soil in My Father's Fields	545
26.	From Stone Age Farmers to Agri-Food Corporations	567
27.	Wards of the State — the Story of the Century	591
	Case Study # 1: The Sahel, a Systemic Unravelling	592
	Case Study # 2: Australia, a Continent on Fire	616
	Case Study # 3: Haiti, a Long Descent into Hell	624
Epilogue — Water Wars		665

List of Figures

Figure 1.	*"What if Democracy and Climate Mitigation Are Incompatible?"* — Foreign Policy magazine, 2022	22
Figure 2.	1870s railroad poster to encourage western settlement.	32
Figure 3.	*"American Progress"* — painting by John Gast, 1872.	38
Figure 4.	*"They are giving all, will you send them wheat?"* — WWI government poster	49
Figure 5.	A dust storm approaches Stratford, Texas, in 1935.	57
Figure 6.	Kansas rabbit drive during the Dust Bowl.	59
Figure 7.	A South Dakota farm buried by dust, 1936.	60
Figure 8.	*"… An Opportunity to Subvert Capitalism."* — Columbia University Climate School, 2021	80
Figure 9.	Corn yields from 1866 until today.	82
Figure 10.	Federal debt held by the public, 1900 to 2053.	86
Figure 11.	*"The Ice Age Cometh?"* — Science News, 1975	101
Figure 12.	*"Climate Change, where the weather is always "your fault", and the only solution is Communism."* — Internet Meme	110
Figure 13.	*"B.C. under state of emergency as fast-moving wildfire destroys homes near West Kelowna."* — CBC news, 2023	117
Figure 14.	A wind-driven hot ember shower during a wildfire.	123
Figure 15.	A prescribed burn in a ponderosa pine-grassland.	124
Figure 16.	The mountain pine beetle.	127
Figure 17.	Annual wildfire burn acreage in the USA since 1916.	132
Figure 18.	Historic burn acreage in the USA from the pre-industrial era until today.	133
Figure 19.	*"Our most shameful waste! Only <u>you</u> can prevent forest fires!"* — Smokey the Bear public awareness campaign.	138
Figure 20.	Annual wildfire burn acreage in the USA since 1983.	139
Figure 21.	Annual burn acreage vs the total number of fires, 1993-2022.	141

Figure 22.	The spread of mountain pine beetle outbreaks in B.C., 2000-2012.	142
Figure 23.	The decline in mineral content in food in the USA, 1914-2018.	150
Figure 24.	Iron, zinc, and boron deficiencies in Douglas Fir trees.	153
Figure 25.	Soil erosion after logging in BC forests.	154
Figure 26.	*"Chop Down Forests To Save The Planet? Maybe Not As Crazy As It Sounds."* — Forbes, 2023	156
Figure 27.	*"Forests Are Losing Their Ability to Hold Carbon."* — Scientific American, 2023	157
Figure 28.	*"Why Billionaires are Obsessed With Blocking Out the Sun."* — Time Magazine, 2023.	158
Figure 29.	*"Politicians always have lots to say about wildfires, except when arson charges laid."* – Toronto Sun, 2023.	160
Figure 30.	NASA map: arid and semi-arid landscapes make up two-thirds of the world's land surface.	165
Figure 31.	*"Cow burps are a major contributor to climate change – can scientists change that?"* — PBS, 2022.	167
Figure 32.	*"Why eating grass-fed beef isn't going to help fight climate change."* — The Conversation, 2017.	167
Figure 33.	*"Four Pests Campaign."* — Chinese poster, 1958.	168
Figure 34.	The historic range of plains and woodland bison in North America.	177
Figure 35.	Grassland/forest landscapes in British Columbia.	182
Figure 36.	Map of the North Okanagan Valley.	187
Figure 37.	Mountain pine beetle outbreaks versus Pacific sea surface temperatures (Pacific Decadal Oscillation).	191
Figure 38.	Heat island effect infographic.	196
Figure 39.	The heat island effect in Phoenix, Arizona, versus the surrounding rural Maricopa County.	197
Figure 40.	Media manipulation in 2023: reporting land surface temperatures instead of air temperatures to manufacture a heatwave.	199
Figure 41.	National Geographic climate charts from 1976.	202
Figure 42.	More propaganda headlines from 2023: *"Earth's Hottest Day In Over 100,000 Years."*	204

Figure 43.	NASA's manipulated global temperature record.	205
Figure 44.	Heat Wave Magnitude Index, Fourth National Climate Assessment, 2018.	207
Figure 45.	Satellite-based global temperature record, 1979-2023.	208
Figure 46.	Warming the present and cooling the past: Comparison of NASA's temperature records, 1999 vs 2017.	212
Figure 47.	Raw vs "adjusted" temperature data, Guira, Venezuela.	215
Figure 48.	Graphical illustration of "data adjustments" made to the official U.S. Historical Climatology Network to warm the present and cool the past.	217
Figure 49.	"Global Average Temperate Change" — Wikipedia's version of the "hockey stick graph".	218
Figure 50.	The Medieval Warm Period and the Little Ice Age according to the IPCC's First Climate Assessment Report in 1990, before they were "erased".	219
Figure 51.	*"Glaciers Are Retreating."* — Aberdeen Herald, 1910.	221
Figure 52.	Glacier National Park, 1914 vs 2001.	223
Figure 53.	The latest version of the "hockey stick graph": tree ring chronology data from the IPCC's sixth assessment report.	227
Figure 54.	The same tree ring chronology dataset re-processed using normal standardization formulas fails to re-create the "hockey stick graph".	228
Figure 55.	Ice core data from Greenland: temperature and carbon dioxide over the last 10,000 years.	230
Figure 56.	Antarctic ice core data: 450,000 years of glacial-interglacial cycles.	232
Figure 57.	5,000-year-old tree stump located 100 km north of the current treeline in what is now permanently frozen tundra.	234
Figure 58.	Changes in alpine treeline elevation over the last 3,000 years.	235
Figure 59.	Annual CO_2 emissions since 1880.	238
Figure 60.	CO_2 emissions overlaid on the "unadjusted" 1999 NASA temperature record.	239
Figure 61.	Winter highs and summer lows according to the 2018 Fourth National Climate Assessment.	245

Figure 62.	Hot days: number of days above 95 °F from 1895 to 2023 at all U.S. historical weather stations.	247
Figure 63.	Hot days: number of days above 90 °F in the contiguous U.S. since 1900.	248
Figure 64.	Hot nights: number of nights above 70 °F in the contiguous U.S. since 1900.	249
Figure 65.	Cold nights: number of nights below 0 °F in the contiguous U.S. since 1900.	250
Figure 66.	Precipitation trends in the continuous U.S. since 1901.	251
Figure 67.	Raw vs "adjusted" Global Mean Sea Level.	252
Figure 68.	*"Call for anti-greenhouse action"* (a call to yield national sovereignty to fight climate change) — Canberra Times, 1989.	254
Figure 69.	Photo of the airport on the Maldives, which is currently undergoing a $1 billion expansion.	255
Figure 70.	Obama's new house in Martha's Vineyard, right at sea level — Barron's, 2019.	256
Figure 71.	*"Now the Pentagon tells Bush: Climate change will destroy us"* and plunge Britain *"into a 'Siberian' climate by 2020"* — The Guardian, 2004.	257
Figure 72.	Sunspots vs hurricanes.	259
Figure 73.	Changes in Antarctic ice mass, 2003-2008.	260
Figure 74.	Ice core data from Vostok, Antarctica: temperature, CO_2, and dust.	264
Figure 75.	CO_2 solubility in water.	267
Figure 76.	Earth's thin atmosphere as seen from space.	271
Figure 77.	The warming effect of atmospheric carbon dioxide, by the numbers.	275
Figure 78.	Milankovitch Cycles — Obliquity, the 41,000-year cycle in Earth's tilt.	282
Figure 79.	Milankovitch Cycles — Eccentricity, the 100,000-year cycle in the shape of Earth's orbit.	284
Figure 80.	Milankovitch Cycles — Axial Precession, the 25,771-year wobble in Earth's axis (spinning top effect).	285
Figure 81.	800,000 years of CO_2 across at least eight ice ages.	303

Figure 82.	The (lack of) correlation between fossil fuel emissions vs atmospheric CO_2.	304
Figure 83.	Chart overlay: emissions vs atmospheric CO_2.	305
Figure 84.	Covid lockdowns didn't affect the rise in CO_2 levels.	307
Figure 85.	CO_2 fertilization is greening the Earth.	352
Figure 86.	The abundance of Banded Iron Formations throughout Earth's history.	355
Figure 87.	The correlation between cloud cover and cosmic rays.	361
Figure 88.	The correlation between solar activity and ocean temperatures.	362
Figure 89.	A 60,000-year temperature record preserved in evaporite deposits from the Dead Sea.	364
Figure 90.	The spiral arms of our Milky Way Galaxy.	366
Figure 91.	The correlation between cosmic ray flux and sea temperature changes.	367
Figure 92.	Our solar system's orbital path through the Milky Way Galaxy in relation to major geological events in Earth's history.	369
Figure 93.	Al Gore declares that any scientist who does not view global warming as an immediate "crisis" is "unethical." — The Indianapolis Star, 1990.	372
Figure 94.	Global temperatures and atmospheric CO_2 over geologic time.	376
Figure 95.	The effect of CO_2 concentration on plant growth.	384
Figure 96.	CO_2 fertilization effect, by the numbers.	385
Figure 97.	50 years of holistic versus continuous grazing.	400
Figure 98.	1,200 years of California's historic drought cycles.	413
Figure 99.	2,500 years of alternating wet and drought cycles in New Mexico.	414
Figure 100.	Artist's rendition of the swamp forests of the Carboniferous period.	418
Figure 101.	Reconstructed atmospheric oxygen over the last 600 million years.	420
Figure 102.	Soil organic matter improvements caused by mob grazing.	486
Figure 103.	Agricultural acreage from 7000 BCE to 2016.	496

Figure 104.	Annotated temperature record of the last 66 million years.	519
Figure 105.	Plate tectonics: the Tethys Sea 100 mya.	525
Figure 106.	Plate tectonics: Europe from 60 to 30 mya.	528
Figure 107.	Atmospheric CO_2 over the last 50 million years.	530
Figure 108.	C_3 versus C_4 plants: photosynthesis and CO_2.	532
Figure 109.	The last five million years of our gradually cooling climate.	536
Figure 110.	Atmospheric oxygen falls whenever biomass declines.	542
Figure 111.	Soil erosion in Namibia caused by continuous grazing.	563
Figure 112.	Population growth over the last 10,000 years.	572
Figure 113.	Map of the Sahel region of Africa.	593
Figure 114.	Deforestation and soil erosion in Haiti is visible from space.	625
Figure 115.	*The Course of Empire — The Savage State* (oil on canvas).	659
Figure 116.	*The Course of Empire — The Arcadian or Pastoral State* (oil on canvas).	659
Figure 117.	*The Course of Empire — The Consummation of Empire* (oil on canvas).	660
Figure 118.	*The Course of Empire — Destruction* (oil on canvas).	660
Figure 119.	*The Course of Empire — Desolation* (oil on canvas).	661
Figure 120.	Groundwater observation wells, Westwold and Lumby, B.C., 2023.	667
Figure 121.	A century of annual precipitation in Western Canada.	677
Figure 122.	Changes in seasonal rainfall in Canada, 1948 to 2014.	678
Figure 123.	Google Timelapse — deforestation in the B.C. Interior, from 1984 until 2022.	679

Introduction

"These plunderers of the world [the Romans], after exhausting the land by their devastations, are rifling the ocean. If the enemy be rich, they are rapacious; if he be poor, they lust for dominion, neither East nor West has been able to satisfy them. Alone among men they covet with equal eagerness poverty and riches. To ravage, to slaughter, to usurp under false titles, they call empire; and where they make a desert, they call it peace."

— Tacitus, 1st century AD Roman historian.

This book began as a short article intended to bring awareness to the insane policies being rolled out against farmers under the guise of "fighting climate change" and to push back against the hyperventilating climate propaganda that politicians, activist-scientists, and the media used to exploit the wildfires in my home province of British Columbia during the smoky summer of 2023, which proved to be one of the most destructive wildfire seasons in BC's recorded history. However, you cannot disarm a popular delusion if you do not expose the underlying forces driving the delusion. And you cannot untangle a distorted perception of reality without repairing the broken sense of history upon which it is built. And so, what began as a simple article morphed into a much deeper investigation into the historical origins of the climate crusade, the politics of nation-building, and the forces unravelling the post-WWII liberal international order.

As that story took shape, a number of seemingly unrelated loose threads that have rumbled about in the back of my mind for many years began to find a home within the context of that broader story —surprising results from a series of soil tests I did many years ago in the pastures and fields on my parents' farm; ice core data from Greenland that unexpectedly showed atmospheric CO_2 levels beginning to rise around 7,000 years ago against the backdrop of a cooling climate (long before the Industrial Revolution and at a time when the first large civilizations emerged); the mysterious and sudden societal collapse that occurred 3,500 years ago when the Bronze Age came to an abrupt and tumultuous end over a span of only 50 years; the unsatisfactory explanations held forth to explain the brutal ice age that turned most of our planet into a giant snowball 460 million years ago (at a time when CO_2 levels were at least 10x higher than they are today); and the strange fact that CO_2 levels continued to rise at the exact same pace during the global economic slowdown caused by Covid lockdowns. And then there's the quest for the holy grail that every farmer and gardener yearns to find: a recipe for how to re-build humus in exhausted soils — that quest also emerged to play a prominent role in the complex ecological story told within these pages.

In a sane world, the story that emerged to tie all these seemingly disconnected threads together should be both the scientific and political story of the century. And yet it isn't because, as long as crusading eyes are fixated on the distorted one-dimensional version of history that seeks to portray CO_2 as the alleged "control knob on the climate", few can see this other story unfolding right beneath our feet. And fewer still recognize its significance.

Great detective stories are not a trek towards a pre-planned destination — they are a journey into the unknown created by doggedly following a trail of cookie crumbs to see where they lead. But in the realm of science, detective stories don't start with a blank slate; they start first and foremost by colliding with the broadly accepted and oftentimes fiercely defended consensus explanations for how the world works, which stand as obstacles to new insights and the expansion of knowledge.

A lesson taught to every first-year geology student is the story of when German geologist Alfred Wegener proposed in 1912 that our continents were drifting around the surface of our planet — the theory of continental drift. He noticed that the eastern edge of South America and the western edge of Africa seem to fit together like pieces of a jigsaw puzzle, and that both continents had identical rock layers full of identical fossils of land-dwelling dinosaurs despite being separated by the vast Atlantic Ocean. The only explanation, he concluded, was that these two continents had once been joined and that the continents themselves were slowly drifting around the surface of our planet. The old guard of geologists (many of whom we deservedly revere as the Fathers of Geology for their own groundbreaking contributions to scientific knowledge) ridiculed Wegener's idea and went to their graves venomously opposing any and all evidence of continental plates floating around the planet on a gooey bed of semi-molten rock. The head of the American Association for Petroleum Geologists snarled, *"If we are to believe Wegener's hypothesis, we must forget everything that has been learned in the last 70 years and start all over again."* Their belief in a static Earth's crust was unshakable to the bitter end, evidence be damned. As the famous physicist Max Planck once observed, *"A new scientific truth does not triumph by convincing its opponents and making them see the light, but rather because its opponents eventually die, and a new generation grows up that is familiar with it ..."* In other words, Planck declared, *"science progresses one funeral at a time."*

And yet, even the idea of scientific progress as a linear progression towards ever greater knowledge is an optimistic fairy tale that disguises a much darker reality. Science is littered with countless false starts, dead ends, long episodes of groupthink that take academia deep into the weeds, and detours down bizarre sidetracks that sometimes persist for generations before they are suddenly discarded and purged from the collective consciousness as though they'd never happened.

Perhaps the most amusing example of one of these detours into scientific la-la land is the case of the Martian Canals. Today we have detailed photographs of what Mars looks like: at its poles the planet has ice caps made of water and frozen carbon dioxide, which shrink during the summer and grow larger during the winter. The Martian landscape is covered by vast plains, dry lake beds, huge craters, deep canyons, and giant volcanoes. Its surface also noticeably changes color and brightness because of seasonal dust storms that occur every spring and summer, and because a dusting of frost (made of CO_2) covers the Martian landscape at high elevations during the bitterly cold winter months.

But during the 19th and early 20th century, telescopes were significantly weaker than those we use today. Astronomers relied on hand-drawn sketches of the hazy images visible through the lenses of their feeble telescopes to show other people what they had seen and, as with any other scientific observation, scientists developed theories to try to explain the vague features and changing colors of the Martian landscape. The most popular theory to emerge, which spread rapidly through the scientific community and gripped the public imagination, was that the Martian surface was covered by a vast network of irrigation canals and engineering projects built by an advanced Martian civilization in order to transport meltwater from the ice caps to the equatorial regions to irrigate their crops. The color changes of the Martian surface were thought to indicate the seasonal growth of plants.

As bizarre as this theory seems from our perspective today, it was perfectly suited to the mood of its era. The 19th and early 20th century was a time when huge engineering projects like the Suez Canal, the Panama Canal, and the Erie Canal (along with countless smaller irrigation canals) were being built to overcome natural obstacles here on our own planet, even as adventurers explored the unknown corners of the Earth where they routinely encountered strange cultures that were wholly alien to their own Western way of life. And so, the Martian Canal Theory emerged as a neat, tidy, sensible explanation, completely logical relative to the Zeitgeist of that era. It was also completely wrong, though it did leave us with a rich legacy of Martian cartoon characters to entertain us on Saturday mornings.

By 1909, higher resolution telescopes were providing conclusive images to expose the folly of the Martian Canals Theory. And yet, as late as 1916 the editor of *Scientific American* and *Popular Science Monthly* was still ferociously denouncing anyone who dared suggest that Mars was not, in fact, home to an advanced canal-building civilization. Even as late as January of 1965, only months before NASA's Mariner 4 spacecraft did a flyby of the Martian surface and dazzled the world with high-resolution photographs of the barren Martian surface, a book published in cooperation with NASA called *Sourcebook on the Space Sciences* still left the door ajar to the possibility of Martian canals as it informed readers that *"Although there is no unanimous opinion concerning the existence of the canals, most astronomers would probably agree that there are apparently linear (or approximately linear) markings, perhaps 40 to 160 kilometers (25 to 100 miles) or more across and of considerable length."*[1] Once we become emotionally invested in an idea, that idea is nearly impossible to kill. As Carl Sagan once noted: *"One of the saddest lessons of history is this: If we've been bamboozled long enough, we tend to reject any evidence of the bamboozle. We're no longer interested in finding out the truth. The bamboozle has captured us. It's simply too painful to acknowledge, even to ourselves, that we've been taken."*

---

[1] "Martian canals" *Wikipedia*. 5 June 2024,
  https://en.wikipedia.org/wiki/Martian_canals

A less amusing and more modern example is the demonization of dietary cholesterol — that's the cholesterol found in animal-based foods, in contrast to serum cholesterol that's found in our blood. Since the 1950s, the scientific community has relentlessly warned us that eating dietary cholesterol contributes to high cholesterol in the bloodstream, which elevates the risk of heart attack. For 65 years, the USDA dietary guidelines recommended eating as little dietary cholesterol as possible. The consensus was overwhelming and indisputable. Skeptics were pilloried. The French nevertheless continued to defy scientific orthodoxy with their unusually low rate of coronary heart disease despite consuming notoriously large quantities of cheese, foie gras, and other fatty animal-based foods. Scientists shrugged their shoulders and dismissed it as the French Paradox.

Then, in 2015, the USDA shocked the world when it withdrew its long-standing warning.[2] The scientific consensus had been wrong all along. Despite decades of research establishing the link between dietary cholesterol and heart attacks, in 2015 the USDA re-evaluated the evidence and found that the evidence did not in fact support the assumed role of dietary cholesterol in the development of heart disease. It's rather sobering to think that it took 65 years to wrestle the scientific consensus back from its detour down the dietary cholesterol rabbit hole — and bear in mind that it wasn't new data that dismantled this consensus belief but rather a reappraisal of the same old data that had led to the erroneous conclusion in the first place. As the author of a research paper published by History of Geo- and Space Sciences pointed out in 2021, *"the history of science is nothing more than a long stroll through the cemetery where ideas that were overwhelmingly accepted are now resting in peace."*[3]

Our scientific institutions try to cultivate an aura of gravitas and impartial expertise, yet in truth scientists are just as fallible, biased, partisan, susceptible to groupthink, influenced by political trends, and

---

[2] Mozaffarian, D. and Ludwig, D.S. (2015). The 2015 US Dietary Guidelines – Ending the 35% Limit on Total Dietary Fat. *Journal of the American Medical Association.* 313(24). https://www.ncbi.nlm.nih.gov/pmc/articles/PMC6129189/

[3] Richet, P. (2021). The temperature-CO2 climate connection: an epistemological reappraisal of ice core messages. *Hist. Geo Space.* Sci., 12, 97-110. https://doi.org/10.5194/hgss-12-97-2021 (see also https://wattsupwiththat.com/2021/05/27/the-temperature-co2-climate-connection-an-epistemological-reappraisal-of-ice-core-messages/)

easily blinded by their emotions and opinions as anyone else — perhaps more so because of the hubris that frequently creeps into the hearts of those with knowledge, learning, influence, and credentials — perhaps even more so because of the pressure to cater to the political winds in order to open doors to funding and career promotions. New ideas that challenge scientific orthodoxy always arise from the margins as blasphemous heresies to consensus beliefs. If they could get away with it, the Inquisition would just as readily be led by scientists as by priests, with no less vicious punishments for heretics and their "dangerous" ideas.

At its most basic level, science only works as a process capable of separating fact from fiction when even the most outrageous heretics retain their freedom of speech in order to challenge consensus ideas in the face of a hostile dogmatic establishment. Although our hyper-bureaucratized consensus-seeking scientific institutions would have us believe otherwise, science only functions as a process for revealing truths as long it remains an uncensored, full-contact, evidence-fueled contest between conflicting ideas — then and only then can truths boil to the surface and bad ideas be defeated. Scientific knowledge does not and cannot progress by consensus. According to legend, when Einstein was confronted with a German newspaper headline that claimed, *"One hundred German physicists claim Einstein's theory of relativity is wrong,"* Einstein allegedly replied, *"If I were wrong, it would only take one."*

In the quest for knowledge, consensus is the path to stagnation — to echo chambers from which there is no escape. The only way to test whether the main body of scientists is advancing knowledge or losing itself down another rabbit hole is to ensure that even the most heretical voices out on the fringes retain their voice. The peer review process is heavily weighted towards like-minded institutional peers — they'll catch the little mistakes, but not the big ones. It's only out along the boiling fringes — among the lunatics, the rebels, and the free thinkers who don't easily succumb to groupthink — where ideas are truly put through the gauntlet and get their feet held to the fire. As Carl Sagan once said, *"Science is more than a body of knowledge, it's a way of thinking, a way of skeptically interrogating the universe with a fine understanding of human fallibility. If we are not able to ask skeptical questions, to interrogate those who tell us that something is true, to be skeptical of those in authority, then we're up for grabs for the next charlatan, political or religious, who comes ambling along."*

Ever since our burgeoning administrative state embraced science as an instrument of the progressive state — ever since shallow politicians began orienting towards scientific consensus to inform their policies and allocate funding — science has been degraded into an increasingly pale replica of its former self as the scientific community gradually adapted, consciously and unconsciously, to cater to the biases of its gatekeepers. A man's mind is shaped by the weight of his leash. And no academic discipline is a better example of the corrupting influence of politics than the field of climate science.

As I sat down to deconstruct the avalanche of climate propaganda with which we are bombarded every summer, it became clear that to reveal the full story behind the headlines, the whole narrative about "carbon dioxide as the control knob on the climate" had to be deconstructed in its entirety — not only to expose the fraud (and I don't use that term lightly) but also to tell the stories of how our once noble scientific institutions ended up falling into this mess and to explain where this bizarre hysterical climate ideology came from.

As I began to trace the origins of our increasingly dysfunctional scientific institutions, the full story behind the ecosystem collapse that occurred on the Great Plains during the 1930s came into view, which was predicated upon another equally misguided 19th-century climate theory — "the rains follow the plow" — a theory that was also widely endorsed by the scientific and political establishment of its era before it fell out of fashion. This earlier climate theory also confused coincidental correlations as cause and effect and, just as today's crusade against CO_2 is being used to usher in a new global social order, it too was hijacked to serve as a nation-building exercise to give birth to the cohesive, centrally directed American nation as we know it today. There are uncanny parallels between these two belief systems, a century apart yet united in the conviction that in some way Man is able to control the weather.

Buried beneath the layers of hubris, arrogance, and self-serving corruption, there's also the paternalistic impulse of the noble lie. As early as the 5th century BC, Plato wrote about the *Noble Lie* — a myth or lie knowingly spread by society's elites for the benefit of society as a whole. In his book, Plato relates the Parable of the Metals.[4] In the tale, Socrates

---

[4] "*Republic* (Plato)." *Wikipedia*. 5 June 2024, https://en.wikipedia.org/wiki/Republic_(Plato)

(Plato's teacher) tells a noble lie to convince everyone in the city to dutifully perform their social role. He tells the people that although everyone is born from the womb of their mother country so that all are siblings, their natures are different because the gods have lined each person's soul either with gold (guardians), silver (auxiliaries), or bronze or iron (producers) as befitting their role in society. By consequence, if anyone with a bronze or iron nature were to attempt to rule the city, it would violate the natural order and cause the city to be destroyed. Socrates claims that as long as the people believe this myth, it will have a good effect by making them more inclined to care for the state and for one another and to accept their role in the established social order. Thus, the Noble Lie is a falsehood told in service of some allegedly noble purpose — a myth told by those who view themselves as society's natural shepherds in order to control an otherwise uncooperative people.

The late 19th- and early 20th-century belief that "the rains follow the plow" was not only used as a pretext to launch a nation-building exercise to completely restructure society. It ultimately — unwittingly — was also directly responsible for creating the conditions for the catastrophic man-made ecosystem collapse that occurred on the Great Plains during the Dust Bowl of the 1930s. Far from being a foolish but benign scientific misadventure like the Martian Canal episode, the unintended delayed consequences of the government policies that were built on top of this erroneous 19th-century climate theory ultimately destroyed the lives of millions before it was finally abandoned and forgotten in favor of new obsessions.

History rhymes as our current climate crusade launches us headlong towards a new authoritarian global world order in which the state relies on a "scientific" priesthood to legitimize its desired institutional policies while harnessing the coercive power of the state to censor and even criminalize those who refuse to go along with the "scientific consensus". A case in point, in February of 2024, Canada's NDP party tabled a bill called the "Fossil Fuel Advertising Act", which seeks to criminalize the "promotion" of fossil fuels. As the National Post reported, if passed the bill would *"prescribe jail time even for Canadians who say scientifically true things such as how burning natural gas is cleaner than burning coal"* and the legislation is so broad that *"it could technically apply to something as simple as a Facebook post or even an 'I Love Canadian Oil and Gas' bumper sticker."*[5]

Under the guise of dealing with a series of never-ending "global" emergencies, our leaders are systematically dismantling all the self-correcting mechanisms that make both science and democracy work. Author Xi Van Fleet, who lived through Mao's murderous Cultural Revolution as a schoolgirl and was swept up as in it as a member of Mao's Red Guard before emigrating to the United States, has described the political processes at work within Western society today as a full-blown cultural revolution, which begins by robbing people of their understanding of history.[6] What is happening to our civilization goes far beyond the rot that has embedded itself in the climate sciences — the untold story of CO_2 will serve as my lens to explain the larger unravelling of our scientific institutions and the bastardization of the political architecture underpinning our civilization.

The very idea of representative democracy hangs in the balance as all the classical liberal principles underpinning western civilization — individual autonomy, freedom of speech, property rights, representative governance, limited government, free markets, independent media, impartial scientific institutions, an independent judiciary, national sovereignty, etc. — are all systematically hollowed out by the crusade to socially engineer the climate and, by extension, rewrite the fabric of Western civilization. Those advocating for these allegedly world-saving policies are no longer even making an effort to disguise the fact they increasingly view democracy as an intolerable obstacle that must be cordoned off to prevent it from getting in the way of "good governance", as shown by the media headline in figure 1.[7] In their preferred version of democracy, it's the scientific-technological elites, politicians, bureaucrats, corporate leaders, and the donor class who collectively decide the course of nations, not the voice of the people.

---

[5] Hopper, T. (2024, Feb 07). NDP bill would prescribe jail terms for speaking well of fossil fuels. *The National Post*. https://nationalpost.com/opinion/ndp-bill-jail-terms-fossil-fuels

[6] Carlson, T. (2024, Feb 26). *Xi Van Fleet (Episode 77)*. [Video]. Tucker Carlson Network. https://tuckercarlson.com/the-tucker-carlson-encounter-xi-van-fleet-2

[7] Abadi, C. (2022, Jan 07). What if Democracy and Climate Change Mitigation are Incompatible? *Foreign Policy magazine*. https://foreignpolicy.com/2022/01/07/climate-change-democracy/

> **Foreign Policy**
> https://foreignpolicy.com › 2022/01/07 › climate-chang... ⋮
>
> **What if Democracy and Climate Mitigation Are Incompatible?**
>
> Jan 7, 2022 — What if Democracy and Climate Mitigation Are Incompatible? **Elected officials work through compromise, but a warming planet waits for no one.**

Figure 1. https://foreignpolicy.com/2022/01/07/climate-change-democracy/

Although climate change hysteria is wholly unwarranted (and fraudulent) — as I will demonstrate in this book — the myopic focus on greenhouse gas emissions is disguising a completely different unfolding ecological story that has gone largely unnoticed even as our misguided climate policies accelerate that ecological crisis and block its solutions. We are reshaping the world with vast technological powers at a time when the scientific process of discovery, which is meant to give us the wisdom to understand the powers in our hands, has been struck blind by politics, leaving us completely unaware of the price that will soon be paid for our blindness. We're so busy staring at the sky that we've blinded ourselves to what is unfolding right beneath our feet.

I won't spoil the story by giving away the details — a good detective story doesn't spoil the journey of discovery by pasting the conclusions into the opening paragraphs of the book. Besides, it really does take the full journey down the cookie crumb trail to be able to see what has been hiding in plain sight and to understand the magnitude of its implications.

However, I will tease you with an innocent question with a seemingly obvious answer: what is causing the sharp rise in carbon dioxide in our atmosphere? I assure you, regardless of which side of the aisle you stand in the climate change debate, the answer is not nearly as simple as you think. What emerges over the course of these pages requires a complete rethink of our assumptions about the processes that put carbon dioxide into our atmosphere, what keeps it there, and what it says about the carbon cycle when our CO_2-starved plants fail to pull it back out of the atmosphere as fast as smokestacks can produce it.

"Once you eliminate the impossible, whatever remains, no matter how improbable, must be the truth."

— Sir Arthur Conan Doyle, The Casebook of Sherlock Holmes

The story I will tell over the coming pages has two parts. The first half of the book is the deconstruction phase where I show you how society got roped into this mad crusade to control the climate. By pulling back the curtain, I am clearing the palette of all the misconceptions that activist-scientists, media, crony capitalists, and politicians have used to bamboozle us to make us see what they want us to see. In effect, part one is the deconstruction of the "noble lie" that carbon dioxide is the "control knob" on the climate.

The second half of this book is the reconstruction phase where I build on this clean palette by walking you through the biological, chemical, and geological forces that shape the air we breathe and create the soil beneath our feet. By the time that story reaches its conclusion, you will have the background knowledge and historical context to be able to see the complex ecological story that I want to show you — once you see it, I think you will agree that this truly is the story of the century.

Because nature works on a different timescale from the speed at which impatient humans live their lives, few can see the slow but relentless forces that ultimately unravel civilizations. That is why this ecological story is so important — understanding these inescapable forces makes us confront hard realities that cannot be escaped through progressive social engineering schemes and scientific fantasies, hard realities that the neo-liberal international rules-based order has ignored for far too long during its utopian quest to reshape the world by bureaucratic decree, hard realities that force us not only to rethink our relationship with our natural ecosystems but also to reconsider the architecture of our political decision-making process.

The story that emerges over the course of these pages is ultimately a tale of how complex political and ecological systems unravel — often in tandem — whenever a society embraces centralized decision-making, empowers a meddlesome administrative state, and thereby creates perverse incentives that gradually hollow out once-thriving civilizations. As Austrian economist Friedrich Hayek once said, *"The more the state "plans", the more difficult planning becomes for the individual."* We are a zombie civilization, bribed with bread and circuses bought for us with our own money, which mindlessly repeats the slogans of democracy, science, liberty, and private property ownership yet no longer understands or even really cares what any of these ideas once meant. The climate crusade featured in this book is but one of the many faces of this dysfunctional system.

Science and technology can only provide partial solutions to our eroding civilization because the primary obstacles to reform are the perverse incentives created by our centralized political system. Everyone fanaticises that angels can be found to lead us because no-one dares dismantle the powerful all-corrupting centralized system at the heart of the dysfunction. Wall Street does not want to cede territory to Main Street; central gatekeepers do not want to cede territory to local decision makers; academia does not want to be cut off from the bounty of milk flowing from the government's teat; kleptocratic rulers, fat cats, and professional activists don't want to dismantle the system that enables their graft; and the corporate grifters are terrified of losing the leverage that gives them preferential access to subsidies, power, influence, government contracts, and regulatory advantages. Even the citizens themselves have become so dependent on the abusive system that is exploiting them at every turn that, even as they buckle under its weight, they too rally in its defense because they can imagine no other way.

Einstein once said, *"we cannot solve our problems with the same thinking we used when we created them."* It is my hope that the story I will relate to you over the coming pages will open your eyes to new possibilities. By stripping away the layers of fiction that are clouding our vision about our climate and warping our understanding of our history, and by exposing how political forces are using that distorted sense of reality to erode our civilization, the seemingly insurmountable chaos of our era become definable, tangible, and fixable. The solutions are within our grasp, but only if we are willing to face what we become when we place our lives in the hands of central planners.

PART ONE

—

Deconstruction

Chapter 1

Nation Building — "The Rain Follows the Plow"

"The most effective way to destroy people is to deny and obliterate their own understanding of their history."

— George Orwell

Ancient Greek mythology tells the story of Daedalus, the mythical master craftsman who was commissioned by King Minos of the Island of Crete to create an elaborate Labyrinth to serve as a prison for the Minotaur (the mythical bull-headed monster that feeds on the sacrifice of society's innocent youth). As soon as Daedalus finished building the Labyrinth to keep the Minotaur's destructive appetites at bay, King Minos imprisoned him in a high tower, along with his young son Icarus, to prevent them from revealing the escape route out of the Labyrinth out of fear that if the secret of the Labyrinth were to be revealed, the uncontrollable Minotaur might be set free. But crafty Daedalus used birds' feathers and beeswax to build wings for himself and for his beloved son so that together they might break free of their captivity.

According to the myth, Daedalus' hands trembled as he gave his son his wings, and tears streamed down his aged cheeks as he warned young Icarus to fly neither too low nor too high to prevent the dampness of the sea from clogging the wings' feathers and to prevent the sun's heat from melting the fragile wax that held the wings together. Like a bird leading its fledgling out of a nest, Daedalus led the way through the window in their prison tower as they leapt into the empty air to escape their captivity on their newly crafted wings, even as the plowman, the fisherman, and the shepherd, their feet planted firmly on the earth, watched their flight in awe from below.

But young Icarus was so delighted by his daring flight that he stopped following the path taken by his father and, ignoring his father's cries, he soared ever higher, reaching for the heavens until the devouring sun began to melt his wax wings. His bare arms flailed helplessly as he tumbled into the sea and drowned, destroyed by his own hubris, leaving Daedalus to curse the hands that had built his son's wings as he laid his child to rest.

The tragic fate of Icarus is a story about the impulsive bull-headedness of youth, a reflection on society's vulnerabilities as the torch is passed between generations, a warning about the fate that awaits those who try to soar among the gods, and a lesson that great power must forever be tempered by many layers of restraint to prevent those who wield it from destroying themselves or those they love. Who is to blame for Icarus' tragic death? Youthful Icarus, eager to explore the new world opened to him by his father's inventions, for falling prey to the human impulse to test the limits of the power placed in his hands? Or crafty Daedalus, striving to free his son from the chains of the past, for designing a system that failed to account for the timeless truths about human nature?

Although Icarus' fate stands as a timely warning for our technological age, we need an updated version of the myth in which Daedalus builds a political system that not only tempts Icarus to soar to new heights of hubris, but that also enables Icarus to force the plowman, the fisherman, the shepherd, and everyone else to join him on his self-destructive flight. The modern centrally controlled nation state, which concentrates unprecedented power, resources, and (most importantly) *decision-making* in the hands of a small number of easily influenced gatekeepers and technological elites, has created a uniquely dangerous platform for Icarus to impose his delusions on entire nations.

And there are few things that capture the imagination of a central planner like a convincing story that, if only society follows some plausible sounding recipe promoted by some elite lab-coat-wearing "expert", Man can control the weather.

~

In the late 19th century, another now-discredited climatological theory emerged in America to infect the minds of scientists, policymakers, and journalists all around the world. The theory emerged from the observation that the dry grasslands of western Nebraska and other parts of the Great Plains (historically known as the Great American Desert) had become greener and wetter as more people settled and cultivated the prairie. This observation became the basis for a popular scientific theory that "the rains follow the plow."[1] And, like any great fiction that is widely embraced by society as gospel truth, an entire economy was soon built on top of this myth. It took decades for the disastrous (and lethal) consequences of this climate fairy tale to become apparent, but ultimately it brought the entire nation to its knees in the worst man-made ecological disaster in American history: the Dust Bowl of the Dirty Thirties.

According to the theory, prairie soils would *"absorb the rain like a huge sponge once the sod had been broken. This moisture would then be slowly given back to the atmosphere by evaporation. Each year, as cultivation extended across the Plains…the moisture and rainfall would also increase until the region was fit for agriculture without irrigation."*[2]

The adherents of the theory believed that by peeling back the sod and by plowing deep, the soil would be able to catch and hold more rain. Plowing would loosen the soil so that rain could penetrate deeper into the earth instead of running off the hard soil and draining into the rivers. Once captured by the soil, this extra moisture would gradually be released back into the air via evaporation, which in turn would lead to more clouds and more rainfall, thus perpetuating the cycle. Furthermore, they believed that the added evapotranspiration of water from the leaves of agricultural crops and newly planted trees would inject still more moisture into the air, leading to even more cloud formation and more rainfall, much like how the Amazon rainforest creates its own weather by pumping water up through the roots to release moisture into the air through the leaves. And they believed that removing native vegetation, like sagebrush and cacti, would reduce the amount of sunlight that would be reflected back into the air, which in turn would help lower the air temperature over cultivated soils, thereby forcing moist air passing overhead to condense and fall as rain.

> *"God speed the plow…. By this wonderful provision, which is only man's mastery over nature, the clouds are dispensing copious rains … [the plow] is the instrument which separates civilization from*

---

[1] "Rain follows the plow." *Wikipedia.* Retrieved 17 June 2023, https://en.wikipedia.org/wiki/Rain_follows_the_plow

[2] Ferril, Jean M. "Rainfall follows the plow." *Encyclopedia of the Great Plains.* University of Nebraska-Lincoln. Retrieved 20 Sept 2023, http://plainshumanities.unl.edu/encyclopedia/doc/egp.ii.049

savagery; and converts a desert into a farm or garden.... To be more concise, Rain follows the plow."

— Charles Dana Wilber,
"The Great Valleys and Prairies of Nebraska and the Northwest", 1881.

The excitement of the era was palpable. It was such a simple, obvious, and plausible technological solution to an age-old problem. And yet it was wrong.

The theory gained support from the highest echelons of government. For example, after President James A. Garfield, himself a farmer from Ohio, won the election in November of 1880, he hosted two of the theory's most widely recognized advocates at his home: naturalist and geologist Professor Samuel H. Aughey from the University of Nebraska (Aughey was also a shameless promoter of settlement on the Great Plains) and journalist and author Charles Dana Wilber (Wilber coined the catchphrase that popularized the theory in the public imagination — he also happened to be a land speculator in Nebraska). President Garfield spent all of a day and most of the night in earnest conversation with his two influential advisors.[3] Science had spoken, politicians had listened, and an entire nation was being reimagined on the basis of this faulty climatological theory.

The railroads latched onto the theory and began aggressively promoting it as a means to attract settlers to the Great Plains. The U.S. Government had given enormous land grants to the railroads so that the railroads could carve up these land grants into smaller parcels to sell to homesteaders, who in turn would depend on these railroads to transport their farm produce to markets and ports in the East. This circular system of mutually reinforcing incentives was designed to achieve the government's strategic goal of securing its hold over the untamed West, guarantee a reliable domestic food supply for its population, and churn out massive surpluses of grain for overseas export. The railroads even went as far as advertising that the 20-inch rainfall line *"was moving steadily westward at the rate of about eighteen miles per annum, keeping just*

---

[3] Arrington, T. (2013). *James A. Garfield and "Rain Follows the Plow"*. National Park Service. Viewed 26 Aug 2023, https://www.nps.gov/articles/000/james-a-garfield-and-rain-follows-the-plow.htm

ahead of and propelled by the advancing population"[4] [author's note: the 20-inch rainfall line is the dividing line that separates arid from humid climates; farmland in the arid zone needs irrigation to grow crops while farmland in the humid zone can grow crops without irrigation].

Pamphlets of Professor Aughey's speeches were distributed to prospective immigrants across Europe to inspire them to emigrate to America's Great Plains to grow wheat. One Santa Fe Railroad pamphlet even cheekily asked, "*Who killed the Great American Desert?*" — the pamphlet depicted a Kansas farmer whose horses were pulling a steel plow across the prairie; the farmer proudly answers, "*I did, with my team and plow.*"[5] The message was clear: Icarus was proudly proclaiming that science and technology had finally been able to achieve what the shaman of yesteryear could not — Man had finally learned to control the weather. In time, the question asked by the pamphlet would prove prophetic, but for all the wrong reasons.

---

[4] Ferril, Jean M. "Rainfall follows the plow." *Encyclopedia of the Great Plains*. University of Nebraska-Lincoln. 20 Sept 2023, http://plainshumanities.unl.edu/encyclopedia/doc/egp.ii.049

[5] Ferril, Jean M. "Rainfall follows the plow."

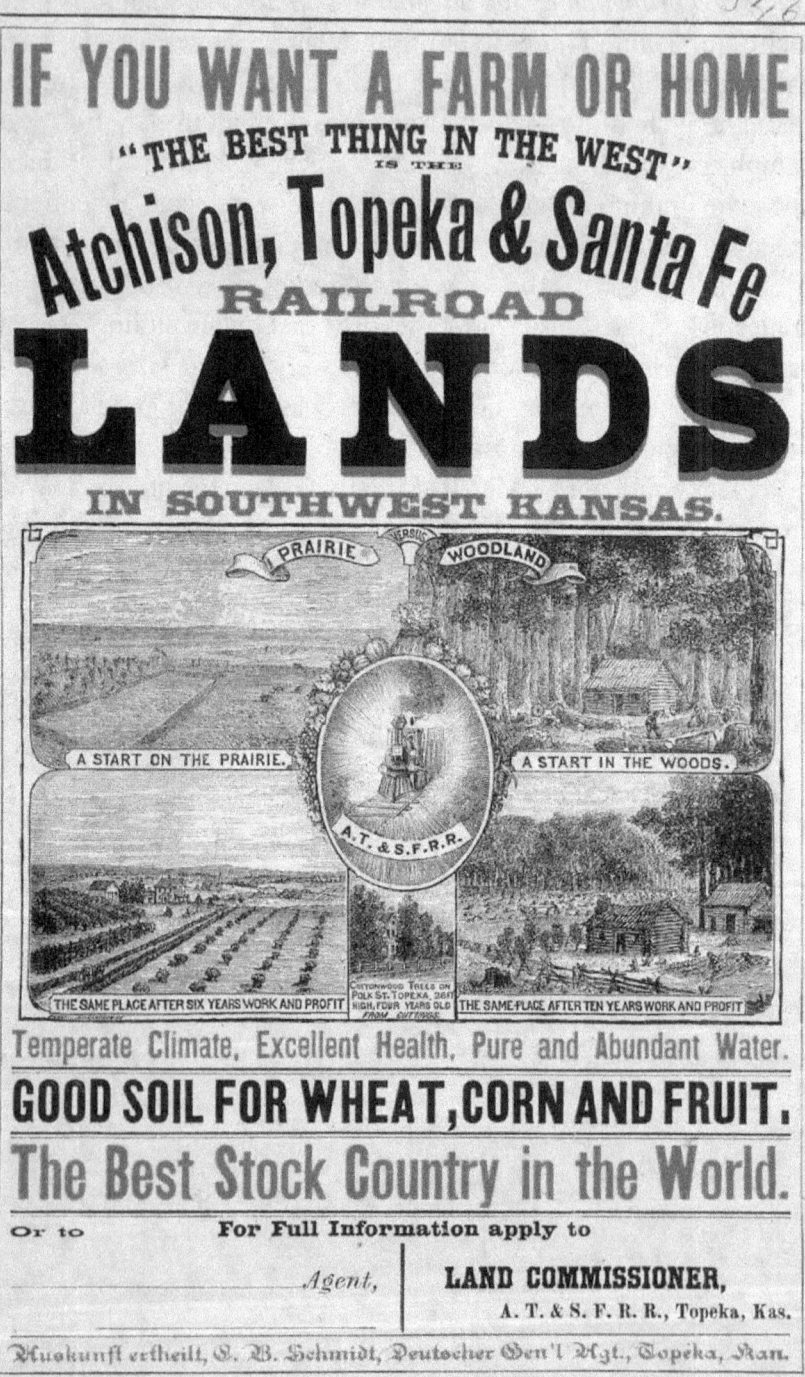

Figure 2. 1870s railroad poster. Source: Kansas Historical Society

Early explorers had a good reason to call the Great Plains "the Great American Desert". This arid, mostly treeless grassland desert, which covers almost one-third of the continental United States and stretches over 3,000 miles (4,800 km) through the center of the continent from Alberta to Texas, is created by the rain shadow of the mountain ranges that stretch all the way up and down the western edge of the North American continent. The region is so dry that almost no natural rivers or lakes emerge from within its entire expanse. Major Stephen H. Long toured over 26,000 miles of the Great Plains over five expeditions during the early 1800s; he concluded that the plains region was *"almost totally unfit for cultivation"*.[6]

Almost a century later, Long's assessment was echoed by famed geologist John Wesley Powell after Powell returned from his famous 1869 expedition down the Colorado River. Like Long, Powell declared that the Great American Desert was unfit for farming. But by then the push for settlement was well underway and Powell's assessment was *"greeted by howls of derision and indignation in Washington: an insult to progress and American know-how."*[7]

In 1878, following a report by geologist G. K. Gilbert, which demonstrated that water levels in the Great Salt Lake in Utah had risen following the introduction of farming by Mormons in the area starting in 1847, Powell once again tried to push back. Powell explained that the higher water levels of the lake had not been caused by increased rainfall triggered by agriculture (as suggested by the theory that "the rain follows the plow") but rather because of the increased runoff from the surrounding drainage basin triggered by the impact of farming, cattle production, and tree cutting, and that if there had been any increase in rainfall during recent years it was only because of cyclical weather patterns that would surely soon be followed by a corresponding cyclical decrease in rainfall. Once again, he was ridiculed and ignored thanks to the vigorous efforts of the western land development lobby, which managed to get its people appointed to key roles in the U.S. Geological Survey and on the Public Land Commission. They successfully portrayed resistance to their utopian

---

[6] "The Challenges of the Plain." *Nebraskastudies.org*. Retrieved 12 Feb 2024, https://www.nebraskastudies.org/en/1850-1874/the-challenges-of-the-plains/

[7] "Rain follows the plow." *All things environmental*. Retrieved 12 Feb 2024, https://enviropaul.wordpress.com/2013/10/15/rain-follows-the-plow/

theory as a conspiracy originating from cattlemen trying to hold back the plow in order to protect their access to open grazing lands for their cattle.[8]

Policymakers, railroad barons, bankers, investors, and developers all looked at the Great Plains as a potential Garden of Eden if only the right farming methods could be found to make the desert bloom. "The rains follow the plow" was just the ticket they needed to open the West to settlement — science, policy objectives, greed, and wishful thinking were all in alignment, as they so often are when there are bountiful rewards for those who swim with the tide of public opinion and government policy, and when there is nothing but condemnation for the Debbie Downers who dare to pour water on everyone else's parade. Besides, it was far too late to turn back. The American Indian Wars that had delayed the settlement of the American West had reached their bitter end, more than 60 million bison that had once kept the sod healthy on the arid Great Plains had been exterminated to starve the Plains Indians into submission and to make room for land surveyors and barbed-wire fences, policies had been enacted, legislation had been signed, investments had been made, loans had been granted, and the campaign was in full swing to encourage millions of farmers from all over the world to settle on the Great Plains to plow up the grasslands in order to grow wheat.

The government threw its entire weight behind this new nation-building project. This was the dawn of aggressive central planning in America when the laissez-faire attitude of the Founding Fathers gave way to active government-incentivized nation building. The consensus had spoken, science was on their side, what could possibly go wrong?

The scientific theory that the "rain follows the plow" emerged at a pivotal moment in America's coming of age story. The U.S. Civil War had transformed the hitherto voluntary union between loosely connected states into a federally directed nation organized around a centrally planned vision of its future. The idea that "the rain follows the plow" captured an entire nation not because it had any valid scientific basis but because it encapsulated all the needs, wants, desires, and dreams of a fledgling industrial nation that was struggling with existential questions. And it earned enthusiastic buy-in from its citizenry, which was swept up in the

---

[8] Smith, H. N. (1947). Rain Follows the Plow: The Notion of Increased Rainfall for the Great Plains, 1844-1880. *Huntington Library Quarterly*, 10(2), 169–193. https://doi.org/10.2307/3815643

optimism of how the world could be reshaped through science, technology, and enlightened central planning, though public enthusiasm for the theory was also clearly reinforced by the seemingly boundless financial opportunities that emerged from the blossoming economy that sprang up on the heels of westward expansion.

The latter half of the 1800s was a precarious time for the fragile young American nation. On the international front, the double-edged sword of industrialization was empowering rival nations in Europe with the technological might and economic clout to wage war on an unprecedented scale. Napoleon's rampage through Europe in the early 1800s and the extension of this war in North America (known as the War of 1812, when British and Canadian troops burned down the White House) demonstrated the force that hostile industrializing nations could unleash against weaker rivals when they optimize their entire nation to feed their military ambitions. Britain's aggressive land grabs in southern Africa in the 1860s and 1870s following the discovery of diamonds and gold in the independent Boer republics, the subsequent Boer wars (1880-1881, 1999-1902) that Britain waged against the Boers when they tried to defend their territory, and Britain's deliberate policy of using war crimes against Boer women and children to crush the Boer resistance all served as a powerful warning to weaker nations of the fate awaiting any vulnerable nation that fell under the jealous eyes of a European colonial power.

Compounding this sense of vulnerability, the Mexican-American War from 1846 to 1848 reinforced America's realization that in order to prevent rival nations from establishing a hostile presence their doorstep and in order to solidify its grip over its newly acquired Western conquests, the United States had to consolidate its hold over the North American continent — from the Pacific to the Atlantic — to ensure that no other powerful nation would ever gain a foothold in its backyard. At the conclusion of the Mexican-American War, Mexico had been forced to cede over 55% of its territory to the United States, including all of California, Nevada, Utah, New Mexico, most of Arizona and Colorado, and parts of Oklahoma and Kansas. But while the territorial gains looked impressive on paper, in reality the United States lacked the necessary population to secure its dominion over these new territories.

Western settlement was the essential ingredient that would allow America to achieve its existential geopolitical imperatives. It would allow Washington to project power across the Great American Desert, bind the

West Coast to the East, secure its hold over its new territories and over the entire continent, and enable it to build the necessary railroads and telegraphs to tie it all together as a single, cohesive, defendable unit under the central oversight of the U.S. federal government and the U.S. military.

Federal decision-makers recognized that America urgently needed a much larger population base and a much bigger economic footprint to keep up with the rapidly expanding industrial and military capacity of rival nations — western settlement would allow it to attract millions of new immigrants out onto the prairie, turn the region into an economic engine for exporting agricultural products to the rest of the world (while securing its own domestic food supply for its growing nation), and give the United States the expanded population base to be able to field (and fund) a much larger army during times of war. As grain and other agricultural products began to flow from western farms (via transcontinental railroads and Mississippi barges) to eastern cities and ports, the entire nation was essentially being bound together culturally and economically as a cohesive nation, thereby turning the page on the loose union between semi-independent states that had existed prior to the Civil War. The former economic independence of member states had given the South the economic independence to attempt to secede from the Union when it didn't like laws being imposed upon it by the North — America's nation-building efforts after the Civil War were consciously designed to squash the centrifugal forces tearing at the fabric of its new nation.

Moreover, western settlement would create new financial opportunities to help lift America out of the economic hangover left by the recent Civil War (1861-1865). Westward migration would also take pressure off overcrowded Eastern cities and help defuse social and religious tensions between new immigrants and old stock Americans as immigration shifted from coming mostly from the British Isles to include increasingly large numbers of immigrants from non-English-speaking countries in eastern and southern Europe. And, as all of America was swept up in the exciting story of the settlement of the West (and as western settlement provided Americans of all creeds with a multitude of shiny new investment opportunities in land development and railroad companies listed on the New York stock exchange), a new common cultural identity was being imprinted on a nation that had only recently been bitterly divided between north and south, between east and west, and between immigrant and old stock Americans. In short, western settlement was the nation-building exercise that would finally unite the country as a cohesive cultural entity.

Furthermore, "taming the west" by carving up the prairie into homestead plots would finally force Indians to exchange rifles and tomahawks for plows to replace their hitherto nomadic way of life, thus putting an end to the never-ending cycle of Indian Wars that had raged all along the western flank of American settlements for more than two and a half centuries, which had limited America's western expansion, restrained immigration, and thus left America vulnerable to the ambitions of rival European powers.

In other words, America's vision of its future, its unassailable grip over the North American continent, and its place on the world stage all hinged on successfully settling the dry American West. But it would only work if prospective homesteaders could be convinced that the rain would follow their plows out onto the dry prairie to make the Great American Desert bloom.

And so, with politicians, military planners, corporations, and the citizens themselves all heavily incentivized to see only what they wanted to believe, this scientifically endorsed fallacy became the mythological basis upon which to build a nation. It provided the shared belief system that would compel a divided people to act in unison towards some common vision of the future. And it served its purpose well — America's central planners brought the continent to heel on the back of this scientific myth. By the time the bottom fell out of the science to expose the fallacies underpinning the theory, there was no turning back from the centralized quasi-imperial juggernaut that America had become. The noble lie had achieved its purpose.

Figure 3. "American Progress" by John Gast, painted in 1872, represented America's vision of the modernization of the new West, led by the female personification of the United States (named "Columbia"). In the painting, Columbia leads civilization westward, bringing light to darkness via the telegraph wire, the railroad, the homesteader, the plow, and the schoolbook tucked under her right arm.

As early as 1862, the Homestead Act granted 160 acres of free land on the Great Plains to anyone willing to fence the land, build a house on it, and break the soil with the plow. In time, these farmers became known as "sodbusters", a nickname they shared with a John Deere plow from the same era. Government planners even offered free train rides to people travelling East by rail to homestead on the plains. Scores of people with big dreams and little farming experience jumped on an opportunity that seemed too good to pass up. Upon arrival, helpful government agents encouraged the homesteaders to plow deep to attract and absorb the bountiful spring rains and to harrow after each rainfall to pulverize the topsoil in order to keep it moist and loose. Early wheat harvests exceeded everyone's expectations.

But correlation does not automatically equal causation. The early years of settlement coincided with a cool wet cycle on the Great Plains with well above average rainfall. The wet cycle gave the illusion of confirming the theory — introducing the plow to the Great Plains really did seem to bring an increase in rainfall to the region. The illusion was compounded by the fact that the first crops flourished in the untapped fertile virgin soils created by thousands of years of bison migrations. By the early 1880s there was so much grain flooding the market that farmers on the Great Plains didn't even have the capacity to store it all — what could not be stored in bins and barns or loaded onto rail cars was left to rot on the ground.

But by the 1890s the initial flush of fertility that comes from plowing virgin soils was wearing out. The oversaturated grain market was collapsing. And the ever-fluctuating climate entered the hotter and drier part of its cycle. By 1893, severe drought and collapsing prices forced hundreds of thousands of bankrupt farmers to flee the Plains. They had built their dreams and staked their lives on a scientifically endorsed theory that existed only in the imaginations of scientists, developers, and policymakers. When the correlation stopped working, the farmers' dreams turned into a nightmare. Boom turned to Bust. The farmers paid the price.

The drought of the 1890s weakened confidence in the theory in the eyes of the scientific community, but with so much riding on western settlement, the theory merely evolved a new twist in order to continue to accommodate America's hopes and dreams for the West. Too much evaporation rather than too little rainfall was asserted to be the critical problem in the semi-arid regions[9] — in other words, the plow might not be able to bring rain to the region, but if the plow could be used in just the right way to ensure that rain was able to easily penetrate into the ground, then moisture from rainfall could be captured and stored underground through persistent deep plowing in both fall and spring and by harrowing the soil after every rainfall. Previous optimism that the rain would follow the plow as cultivation expanded across the plains turned into optimism that farmers who followed the correct scientifically approved farming recipe promoted by government experts (plowing often and plowing deep) would be rewarded with bountiful crops because they were doing the required work to productively harness the rain delivered by Mother Nature and bring the West's dry climate to heel. And so the theory evolved to

---

[9] Libecap, G. D., & Hansen, Z. K. (2002). "Rain Follows the Plow" and Dryfarming Doctrine: The Climate Information Problem and Homestead Failure in the Upper Great Plains, 1890-1925. *The Journal of Economic History*, 62(1), 86–120. http://www.jstor.org/stable/2697973

claim that as long as the farmer "followed the science", *"the farmer will always have a crop: in the wet years his crop will be large; in the driest year it will be sufficient to sustain him."*[10]

In other words, any farmer who failed out on the Great Plains had no-one to blame but himself because anyone willing to put in the hard work of capturing and harnessing rainfall through the government's prescribed regimen of intensive tillage was guaranteed success. Every government organization supplying information to new homesteaders in the region emphasized the same principles and farming techniques. And so, the revised theory persisted that the rains would reward the farmer and his plow as long as he followed the most up-to-date farming practices promoted by government's experts. Policymakers and land developers eagerly embraced this revised version of the theory — they forged ahead by aggressively expanding on the homesteading policies of the previous boom and bust cycle. Intensive cultivation of the prairie accelerated. The worst was yet to come.

The Progressive Era of American politics, which began in the 1890s, kicked central planning into overdrive with the creation of countless new institutions staffed by government-chosen scientific experts (the Progressive Era also pioneered the idea of using the public education system as a deliberate tool to shape society's attitudes, morals, and values — the teacher ceased being a servant to parents and became, instead, a tool of the state). As you will soon see, this new progressive political era would have a profound impact on the disaster yet to come out on the Great Plains. And the institutions and ethos that emerged during that era are essential for understanding both the liberal rules-based order under which we are still living today and the hidden ecological story unfolding beneath our feet today.

The rise of progressivism reveals a profound shift in the American psyche as society lost its distrust of big government and began instead to idolize the state as a benevolent shepherd to help solve society's problems. It is the moment when the state stopped being the servant of the people and instead became their master. As early as 1840, in his famed book *Democracy in America*, French political philosopher Alexis de

---

[10] Libecap, G. D., & Hansen, Z. K. (2002). "Rain Follows the Plow" and Dryfarming Doctrine: The Climate Information Problem and Homestead Failure in the Upper Great Plains, 1890-1925. *The Journal of Economic History*, 62(1), 86–120. http://www.jstor.org/stable/2697973

Tocqueville had already foreseen the evolution of democratic nations as citizens willingly surrendered ever more of their individual freedoms to an all-powerful central government. He predicted that *"after having thus successively taken each member of the community in its powerful grasp, and fashioned them at will, the supreme power then extends its arm over the whole community. Society will develop a new kind of servitude which covers the surface of society with a network of complicated rules, through which the most original minds and the most energetic characters cannot penetrate. It does not tyrannise but it compresses, enervates, extinguishes, and stupefies a people, till each nation is reduced to nothing better than a flock of timid and industrious animals, of which the government is the shepherd."*

Today, we associate the word "progressivism" with left-leaning social liberalism, but its roots are not partisan. It evolved from the top-down concept of government, adopted in Europe since the French Revolution, which viewed the state rather than the individual as the principal building block of society and as the agent of social change (this European perspective, borne from the idea of King and Church serving as society's shepherds, simply transferred those monarchial powers to a democratically elected government while maintaining the idea that all rights and freedoms are on loan to the individual by the state, to be granted, modified, or withdrawn as needed to achieve the collective good). Even today, the conservative who wants to use the coercive power of the state to mold society according to his vision of the future is in every way also a progressive in the 19th century understanding of the term. Each side recognizes the corruption spawned during the reign of the other, yet neither is willing to reverse the expansion of centralized power because both have come to view state power as essential to achieving whatever they perceive as the greater good of the nation.

At its heart, progressivism is the optimistic belief that a paternalistic state informed by science can serve as a benevolent and nurturing force to lift people up and help them realize their true potential.[11] And yet, in spite of the high ideals, the idolatry of the state ultimately boils down to the

---

[11] Kendall, E. M., 2012. *Diverging Wilsonianisms: Liberal internationalism, the Peace movement, and the ambiguous legacy of Woodrow Wilson.* [PhD thesis, Case Western Reserve University, History department. https://etd.ohiolink.edu/acprod/odb_etd/ws/send_file/send?accession=case1323399909&disposition=inline

worship of government force. In effect, progressivism is the polar opposite of the "laissez-faire" attitude of the prior classical liberal era, introduced by America's Founding Fathers, who fiercely distrusted strong federal authority and viewed it as a threat to individual liberty. They sought to limit government power at every turn and left society's "progress" in the hands of individual citizens, believing that the sole purpose of government was to protect the rights and freedoms of the individual. Thus, the political system they built was designed to allow society to evolve bottom-up through the sum of the free choices of its citizens rather than through top-down decrees imposed by central planners working for a coercive state. As Thomas Jefferson once wisely declared, *"the government that governs best governs least."*

Progressivism turned all of what the Founding Fathers had created upside down by subordinating individuals to a paternalistic state, effectively reducing the individual to the status of quasi-governmental property — mere chess pieces controlled by carrots and sticks in service of "national" interests (a.k.a. nation-building). By the late 19th century and early 20th century, both liberals and conservatives in America had enthusiastically turned their backs on the classical liberal restraints embedded within their Constitution and instead wholeheartedly embraced this new political philosophy of progressivism, albeit with different priorities about how this immense new government power should be used. In Canada, the conservative party even proudly called itself the Progressive Conservative Party of Canada for much of the 20th century. In the U.S., of the four Progressive Era presidents — William McKinley (R), Theodore Roosevelt (R), William Howard Taft (R), and Woodrow Wilson (D) — only the last was a Democrat, though the merger of progressivism and liberalism under Wilson would arguably have the most profound and lasting impact on America and on the world.

Wilson explicitly argued that *"Thomas Jefferson's idea that 'the government that governs best governs least' no longer fit 'the practical politics of America.'"*[12] As America threw itself behind the idea of Big Government, all that remained was to retool the political architecture to throw off the classical liberal restraints imposed by its Founding Fathers.

---

[12] Zimmerman, J. (2015, Dec 01). President Woodrow Wilson's legacy — progressive politics and racism. *WHYY PBS NPR*. https://whyy.org/articles/president-woodrow-wilsons-legacy-progressive-politics-and-racism/

At the core of that political reorganization was the desire to invert the power relationship between the federal government and the individual states.

The Constitution as it was written by the Founding Fathers had placed the majority of political power in the hands of the individual states and had structured the federal government to be subordinate to those states, effectively making the federal government a tool of the collective will of the individual states, but not their master. But at the very beginning of his first term, Wilson signed the 17th Amendment to the U.S. Constitution into law. It changed everything.

More than any other piece of legislation in U.S. history, the 17th Amendment threw open the door to the massive centralization of power in federal hands. In the original Constitution, U.S. Senators were not directly elected by voters but were instead appointed by individual state legislatures. As Wikipedia explains, *"state legislators even had the right to "instruct" their senators to vote for or against proposals, thus giving state legislatures both direct and indirect representation in the federal government."*[13] In other words, as long as Senators were appointed by state legislatures, those state-appointed Senators prevented the federal government from acting against state interests. Thus, the Senate served as a counterweight to prevent either Congress or the President from swallowing up state powers. This arrangement also prevented Senators and the federal government as a whole from being co-opted by "special interests" (like Wall Street) because state legislators held the state-appointed Senators by a leash, who in turn held the federal government by a leash. Anyone wishing to exert "undue influence" on the federal government would therefore first have to navigate a complex maze of state legislatures, which is much harder than it is in today's arrangement in which lobbyists simply have to book a flight to Washington.

State-appointed Senators were the Founding Fathers' most important constitutional check on the expansion of federal power because, through their state-appointed Senators, individual states had the power to overturn any law put forward by either Congress or by the President that robbed state legislatures of their local decision-making authority. But the 17th

---

[13] "Seventeenth Amendment to the United States Constitution." Wikipedia. 17 May 2024, https://en.wikipedia.org/wiki/Seventeenth_Amendment_to_the_United_States_Constitution

Amendment turned this relationship on its head so that Senators would henceforth be elected directly by voters from each state rather than being appointed by State legislators. As state legislatures lost control over their Senators, they lost control over the growing federal Leviathan because it untethered the link subordinating the federal government to local state legislatures, giving rise to the doublespeak that has become so common today in which elected federal representatives publicly promise their citizens the moon even as they dance to the tune of well-heeled donors and influential Washington powerbrokers behind closed doors. By the stroke of Wilson's pen, the federal government was cut free from its leash. The formerly thin crust of federal powers was thus transformed into the bottomless Washington Swamp. As President Gerald Ford once said, "*A government big enough to give you everything you want is a government big enough to take from you everything you have.*"

The 17th Amendment triggered the beginning of the massive accumulation of power in federal hands that continues to this day. The original role of the federal government had been little more than a treaty between semi-autonomous sovereign states; the 17th Amendment made the federal government their master. However, Wilson was only building on precedents set by those who had come before him. It was Abraham Lincoln who dealt the fatal blow to the idea of the United States as a union between semi-autonomous states when he refused to allow confederate states to leave the Union. Lincoln transformed the voluntary union between semi-autonomous states *in* America into the singular nation called the United States *of* America. This subtle language shift from *in* to *of* is clearly visible in the writings of the pre- and post-Civil War era.

When the option of divorce is removed from a marriage, the voluntary effort to preserve and cultivate a union that works for all participants is replaced by a power struggle over who will be the master and who will be the subordinate in the relationship. Woodrow Wilson provided a decisive answer to that outstanding question by removing the most important tether that subordinated the federal government to state legislators. All eyes shifted to the throne in Washington. Local decision making and individual autonomy has been in decline ever since. As popular blogger El Gato Malo recently wrote, "*Preventing deadlocks" was a thin gruel of pretext for what amounted to the greatest power grab in American history. This is what set up Wilson who, in turn, set up FDR who so egregiously altered the character of federal intrusiveness in America for all time.*"[14]

Wilson didn't waste any time amassing power and resources under this radically revised constitutional architecture. He was instrumental in pushing for a permanent progressive income tax to fund the expanding federal Leviathan[15] — until then, income taxes had only been imposed during wartime (before Wilson the day-to-day functions of the federal government were funded via tariffs and excise taxes on various goods; which was more than sufficient since the federal government had very little to do as long as decision-making largely rested in state hands).

Wilson also pioneered using the judiciary as an instrument for social change — he promoted the idea of a "Living Constitution" to encourage judges to engage in lawmaking by "reinterpreting" the meaning of the Constitution to suit the mood of the times rather than having to wait for elected representatives to make laws through the more tedious and difficult legislative process, which requires the consent of the governed (a.k.a. input from "voters" who might have their own ideas about how to run their lives).

Wilson massively expanded the permanently staffed administrative institutions, suggesting that these institutions could somehow be kept separate from the corrupting influence of politics under the assumption that technical expertise and scientific objectivity would prevent those working inside these institutions from falling prey to the lure of politics and publicity. Ah, the irony... since the academic field of eugenics was all the rage in those days, Wilson promptly pointed at eugenics as scientific justification for his decision to institutionalize racist Jim Crow laws in order to purge the expanding federal bureaucracy of African-Americans (who by then were working at all levels of the federal government) and prevent the hiring of any new ones, thus effectively reversing all the progress that had been made since the end of the Civil War to slowly heal America's racial divides (Wilson was also a vocal defender of the Ku Klux Klan).

---

[14] El Gato Malo (2024, Feb 20). "federalism fixes this – the free market case for repealing the 17th amendment." *bad cattitude*. https://boriquagato.substack.com/p/federalism-fixes-this

[15] Levy, Michael. "Underwood-Simmons Tariff Act". *Encyclopedia Britannica*, 3 Apr. 2023, https://www.britannica.com/event/Underwood-Simmons-Tariff-Act.

Wilson proceeded to turn the USA into a virtual police state via the Espionage Act of 1917, which enabled him to prosecute anyone who opposed U.S. involvement in WWI — that law is being used even now as the pretense to pursue charges against Julian Assange (the Australian journalist who published leaked documents and videos exposing US war crimes) and Edward Snowden (the former National Security Agency contractor who blew the whistle on America's illegal surveillance of its own citizens after his efforts to address the issue via internal channels at the NSA were ignored).[16] Wilson also passed the Trading with the Enemy Act in 1917, which Franklin D. Roosevelt (FDR) infamously used in 1933 to seize all the private gold holdings of the American people in order to extinguish the last remaining bulwark designed by the Founding Fathers to restrain a predatory government and its printing presses from robbing its citizens through a policy of deliberate inflation.

Woodrow Wilson also injected the progressive ethos into U.S. foreign policy, ushering in the era of militant liberal internationalism that seeks to impose the U.S. vision of democracy on foreign nations by "encouraging" them with various economic carrots and sticks, by staging coups and employing other tactics designed to "facilitate" regime change (the U.S. is known to have intervened in at least 81 foreign elections just in the period from 1946 to 2000),[17] and by using the bludgeon of the American war machine to spread its "enlightened" vision of democracy (the U.S. military footprint includes 750 military bases on foreign soil in at least 80 countries worldwide)[18]. In every sense, Woodrow Wilson's progressive vision of big government created the architecture for the modern-day liberal world order.

In his 1887 book, *The Study of Administration*, future president Woodrow Wilson captured the euphoric enthusiasm for progressivism that was sweeping though his country and through the entirety of the Western

---

[16] Anderson, W. L. (2020, July 2). "The Disastrous Legacy of Woodrow Wilson." *The Mises Institute (Mises Wire)*. https://mises.org/mises-wire/disastrous-legacy-woodrow-wilson

[17] "United States involvement in regime change." *Wikipedia*. 6 June 2024, https://en.wikipedia.org/wiki/United_States_involvement_in_regime_change

[18] Hussein, M. and Haddad, M. (2021, Sept 10). Infographic: US military presence around the world. *Al Jazeera News*. https://www.aljazeera.com/news/2021/9/10/infographic-us-military-presence-around-the-world-interactive

world when he wrote: "*Administration is everywhere putting its hands to new undertakings. The utility, cheapness, and success of the government's postal service, for instance, point towards the early establishment of governmental control of the telegraph system. Or, even if our government is not to follow the lead of the governments of Europe in buying or building both telegraph and railroad lines, no one can doubt that **in some way it must make itself master of masterful corporations**. ... The idea of the state and the consequent ideal of its duty are undergoing noteworthy change; ... **Seeing every day new things which the state ought to do, the next thing is to see clearly how it ought to do them*** [my emphasis]."[19] Wilson and his enthusiastic peers drove a giant wooden stake through the principles of limited government. As journalist John T. Flynn later pointed out when Wilson's ideological successor, Franklin D. Roosevelt, began rolling out his own vision of a planned economy under the direction of a strong central government, *"These two ideas — the idea of a free republic and the idea of a Planned Economy—cannot live together."*

With shiny new Progressive institutions rolling out plans for everything and everyone, this time Icarus would be able to soar even higher and take everyone along for the ride. But this time Icarus' progressive hubris would take down the entire American economy, devastate the entire ecosystem of the Great Plains, and bring the entire nation to its knees in the worst man-made disaster in American history. And so, against the backdrop of this emerging Progressive ethos, our story returns to the Great Plains.

In 1909, the enlarged Homestead Act sparked a new homesteading boom that was even bigger than the first. This time millions took up the call to try their luck busting sod on the Great Plains. Concerns about the viability of small 160-acre homesteads (so many of which had failed during the droughts of the 1890s because they had been unable to sustain themselves during the dry phase of the climate cycle) led to a minor adjustment in the size of homestead allotments — this time homesteaders were encouraged to take double parcels (320 acres) of free land for the same conditions as during the previous cycle (build fences and a house, and break the soil with the plow). Back in 1878, geologist John Wesley Powell's assessment of the region had led him to conclude that farm

---

[19] Woodrow Wilson. (1886, Nov 1). *The Study of Administration*. [Essay]. Teaching American History. https://teachingamericanhistory.org/document/the-study-of-administration/

allocations should be 2,560 acres in semi-arid regions, not 160 or 320 acres,[20] but his recommendation continued to be ignored because it didn't fit with the government's population density targets for the region and didn't fit the small-farm homesteading model being promoted by central planners and developers. 320 acres was the compromise — scientific and technological advances in farming along with good old fashioned hard work would make up the difference... or so they hoped.

Once again, this new Boom coincided with another wet cycle with above average rainfall that stretched from the early 1900s into the 1920s. Bountiful harvests driven by abundant rains and fertile virgin soils reinforced everyone's optimism that American ingenuity and hard work would continue to push the American agricultural frontier west towards the foothills of the Rocky Mountains. A number of ecologists, American Indians, cattlemen, and farmers who had lived through previous drought cycles on the open prairie and understood the region's cyclical climatic nature, made the tedious journey to Washington to testify before Congress in order to warn central planners of the consequences of plowing up the deep-rooted prairie grasses that sustained this fragile arid ecosystem. Once again, they were ignored.

The outbreak of the First World War in 1914 tripled the price of grain, and the government launched a propaganda campaign to encourage farmers to *"Grow Wheat to Win the War"*. Farmers jumped on the opportunity by taking on ever bigger loans to invest in the latest tractors, harvesters, and other modern equipment in order to expand the amount of land they could put to the plow. The number of acres under cultivation across the Great Plains jumped from around 50 million acres in the late 1890s to well over 100 million acres by the 1920s.

---

[20] Libecap, G. D., & Hansen, Z. K. (2002). "Rain Follows the Plow" and Dryfarming Doctrine: The Climate Information Problem and Homestead Failure in the Upper Great Plains, 1890-1925. *The Journal of Economic History*, 62(1), 86–120. http://www.jstor.org/stable/2697973

Figure 4. Government poster during WWI encouraging farmers to support the war effort by expanding wheat production on the Great Plains.

Then the war ended and indebted and war-torn European nations were forced to rein in their spending. Grain exports plunged and grain prices collapsed. Heavily indebted farmers were forced into a cycle of taking on even more debt to service earlier debts for land and equipment purchased during the war. Cheap loans facilitated by the loose monetary policies of the Federal Reserve encouraged farmers to invest in even more tractors and more equipment and to plow up even more grasslands in a desperate bid to avoid bankruptcy by boosting production to compensate for shrinking margins while they waited for grain prices to recover. As the old joke says, "we lose money on every sale, but we make up for it in volume."

All of this happened against the backdrop of the newly minted Federal Reserve Bank, signed into law in 1913 by newly elected President Woodrow Wilson just before the start of the war. The Founding Fathers had been strongly opposed to a national banking system — indeed, Britain's effort to place the colonies under the monetary control of the Bank of England was the "last straw" that triggered the American Revolution. Thomas Jefferson once declared: *"I sincerely believe that banking establishments are more dangerous than standing armies, and that the principle of spending money to be paid by posterity, under the name of funding, is but swindling futurity on a large scale."*

To understand how Woodrow Wilson's banking and financial reforms created the destructive incentives that plunged the Great Plains into disaster during the 1930s, I'm going to take a brief detour here to explain America's evolving financial architecture. It has important ramifications to understand the "story of the century" that will emerge in the later chapters of this book. The past is the key to the future.

The centralization of the U.S. banking system began a couple of generations before Woodrow Wilson, in 1861, upon the outbreak of the Civil War, when Abraham Lincoln brought an end to the decentralized network of state banks and free banking institutions that had controlled the monetary supply since 1837. The gold standard restrained military adventures; the printing press enabled them. Lincoln took the U.S. off the gold standard and imposed a system of national banks in order to issue a fiat paper currency, known as the "greenback", which helped Lincoln pay for the war but resulted in severe inflation. Without gold to back each dollar that was printed, Lincoln also founded a new secret federal police force that was granted Gestapo-like powers, called the Secret Service, which originated as a branch of the U.S. Treasury department in order to prevent counterfeiting of the new fiat currency (their role in protecting politicians from resentful citizens came much later).[21]

The greenback printing press wasn't enough to pay for Lincoln's war effort so, in 1862, Lincoln also founded the IRS to collect America's first ever income tax in order to help fund his war. This set the precedent for the Progressive Era a generation later when Congress passed the 16th Amendment, which granted the federal government the authority to impose direct taxes on the American people to "provide for the common Defence and general Welfare of the United States." President William Howard Taft proposed the 16th Amendment in 1909, and it was finally signed into law in 1913 by none other than Woodrow Wilson. Wilson immediately followed up with the Revenue Act of 1913, which launched the era of federal income taxes that has persisted ever since.

Critics of the 16th Amendment warned of an excessive centralization of federal power at the expense of citizens and states, but they were shouted down. By the time the U.S. entered WWI, the top marginal income tax

---

[21] Glawson, J. D. (2024, Jan 9). "America's Fiat Money Gestapo: The Untold History of the Secret Service." *The Mises Institute (Mises Wire)*. https://mises.org/miseswire/americas-fiat-money-gestapo-untold-history-secret-service

rate had surged to 73% to pay for Wilson's decision to convert the U.S. into a global military power and involve the U.S. in European infighting. During WWII the top marginal tax rate soared as high as 94%. The fiscal restraints imposed by the Founding Fathers to limit federal spending and federal policymaking were gone. Whoever controls (or influences) the power of the purse controls the trajectory of the nation. And right on cue, the pigs began lining up at the trough.

In the decades after the Civil War, the U.S. gradually returned to a gold standard and a more decentralized banking system, but just as Lincoln's precedent had opened the gates to a permanent federal income tax, his decision to take the U.S. off the gold standard and centralize the money supply had set a precedent for others to follow. Those who had benefited from the flood of centrally distributed easy money were not happy about the return to fiscal responsibility. A bitter running debate emerged between those who advocated for sound money based on the gold standard versus those who benefited from inflation and wanted to continue to inflate the monetary supply in order to flood the economy with cash, make it easier to pay off their debts, and uncork generous government spending in the name of the "common" good.

In 1880, soon-to-be president James A. Garfield, commented on a resolution put before the House of Representatives that sought to take the issuance of coins and paper money out of the hands of banks altogether and place that authority in the hands of the federal government: "[…] *never was there a measure offered to the Congress of so vast and far-reaching centralism. It would convert the Treasury of the United States into a manufactory of paper money. It makes the House of Representatives and the Senate, or the caucus of the party which happens to be in the majority, the absolute dictator of the financial and business affairs of this country. […] The government should prescribe general laws in reference to the quality and character of our paper-money but should never become the direct manufacturer and issuer of it.*" How would the Treasury pay off its debts in the future, he asked. Through inflation. "*Print it to death — that is the way to dispose of the public debt, says this resolution.*"[22] The motion was defeated.

---

[22] Brice, S. M. and Vincent, C. (1882) *Financial Catechism and History of the Financial Legislation of the United States from 1862-1896*. Chicago Franklin Print Co. https://archive.org/details/financialcatechi00bricuoft/mode/2up

But the pigs had had their first taste of the trough and were not about to let it go. Despite President Garfield's adamant resistance, bankers held on to their dreams of an elastic monetary supply under the control of a central banking cartel, which would bail them out whenever their reckless lending habits inevitably led to panics and economic contractions. Furthermore, the big Wall Street banks were not happy that they had begun losing ground to rapidly growing smaller state banks and other local banking institutions. Turning over control of the money supply to a Washington-controlled central bank would enable the politically connected Wall Street banks to regain their dominant status in the financial food chain.[23]

Likewise, the decentralized banking system was also frustrating to big businessmen looking to gain monopolistic market share, so they too embraced the idea of progressive statism as a way to access subsidies and establish cartels through a centrally controlled banking system. Both bankers and businessmen believed that once a central bank had been given an absolute monopoly over the monetary supply and over establishing reserve requirements, that central bank could be controlled by whispering in the ears of the powerbrokers in Washington, who would bail them out in times of trouble and would subsidize their operations through a smooth, controlled, permanent policy of inflation. How right they were. The path to mega-cartels, mega-corporations, and neo-feudal serfdom was paved by these centralizing financial "reforms". The sun has slowly set on local craftsmen, family farmers, and small independent businesses ever since as they are eclipsed by the financial might of the juggernauts who have priority access to the Washington Swamp.

By 1913, both political parties were fully captured by the Progressive spirit of their era, including the progressive dream of being able to issue unlimited public debt (subsidized by the power of the printing press) in order to finance their Big Government wish list. Wilson and his progressive peers saw the Federal Reserve system as the pinnacle achievement of the progressive mindset of that era, which sought to impose central planning on everything from farming, to forest management, to the economy, to the welfare of its citizens, and to the

---

[23] *"The Mystery of Banking. XV. Central Banking in the United States III: The National Banking System."* The Mises Institute. Retrieved 24 June 2024.
https://mises.org/online-book/mystery-banking/xv-central-banking-united-states-iii-national-banking-system

education of the nation's children, without recognizing (or perhaps deliberately ignoring) the distorted incentives and rampant corruption that would emerge from this centralized control. Woodrow Wilson's dream of making the administrative state the master of a masterful nation had become reality — nothing exemplifies government command-and-control more than the act of taking control of a nation's money supply. Thanks to the all-consuming administrative state that was expanding into every corner of American life, Icarus could finally force everyone to join him in his flirtations with the Sun.

The Federal Reserve's official mandate was to counter the economic boom-and-bust cycles (caused by overinvestment during the boom years), which had haunted the unregulated capitalist economy under the previous laissez-faire system. The "wisdom" of the central bankers at the Federal Reserve would finally give the administrative state the power to master the markets and put an end to all those disruptive economic booms and busts... or so they thought.

WWI broke out in 1914. Despite its official neutral stance, the newly empowered U.S. government immediately stepped up with generous helpings of grain and financing for its European friends. By 1917, the U.S. had officially entered the war. The Federal Reserve was instrumental in helping to finance the war by facilitating war bond sales and by providing loans to select banks at preferential rates.

Then, from 1921 to 1929, the Federal Reserve increased the monetary supply by 62%[24] as it showered America with a flood of dollars to kickstart the economy out of the hangover left by the First World War. America boomed. The stock market quadrupled in value in under a decade. Like everywhere else in America during the Roaring Twenties, the Federal Reserve's loose monetary policies fueled countless speculative investments out on the Great Plains. Faced with falling grain prices, local bankers encouraged farmers to take out cheap loans to service debts for land and equipment purchased during the war. The administrative state and its fancy new institutions were inadvertently luring farmers into a horrific financial trap from which there would be no escape.

By the late 1920s, the vast expanses of virgin soils that had been put to the plow since the Homestead Act of 1909 were starting to lose their fertile

---

[24] Smith, G. F. (2023, Jan 12). "The Great Depression's Patsy." *The Mises Institute (Mises Wire)*. https://mises.org/mises-wire/great-depressions-patsy

flush. And the weather gradually turned drier and hotter as the wet part of the climate cycle came to an end. Farmers responded by trying to make up for plummeting crop yields by further expanding acreage under cultivation — they had no choice if they wanted to make their mortgage payments. The easy money coming from the Federal Reserve offered the illusion of a way out by making it easy to use debt to buy machinery and expand production. But since bankers offered loans to one and all, the expanded production only drove prices down further. By the end of the 1920s farmers were on the brink — the entire Great Plains economy was hanging by a thread.

As the 1920s came to an end, the Federal Reserve finally began to recognize that its loose monetary policies had pumped so much money into the financial system that it was on the verge of triggering hyperinflation. Over in Europe, Germany had also tried to print its way to economic recovery following WWI only to see its policy of "prosperity via the printing press" spectacularly blow up in its face — the cautionary tale of the hyperinflation that devastated Weimar Germany during the early 1920s helped the Federal Reserve to recognize the dangerous inflationary game they too were playing. Prosperity cannot be faked with a printing press. As the Federal Reserve began to unwind its decade-long loose monetary policies at the end of the 1920s, interest rates began to rise, bringing an abrupt end to a decade of speculative investments, cheap loans, and easy money. What goes up, must come down.

The frothy financial bubble created by the Federal Reserve had fueled the illusion of prosperity during the Roaring Twenties, but as the bubble began to deflate the stock market began to sell off, triggering the Wall Street Crash of 1929. And then the administrative state, which thought itself as the wise "master of masterful corporations", panicked and began a series of reactionary moves (price controls, trade wars, etc.) that transformed the market crash of '29 into the decade-long Great Depression. And so, the economic crisis began to feed into the ecological crisis brewing out on the Great Plains.

As the 1920s ushered in warmer and drier weather out on the Great Plains, many farmers began to recognize the false assumptions of the government-promoted climate fairy tale underpinning their plow-dependent way of life. Climate is cyclical; the rain did not follow the plow, nor could the plow compensate for drought — indeed, the plow and the monoculture production system were clearly exacerbating the effect of dry winds evaporating moisture from prairie soils. Although they could see that their soil was struggling, they were locked into a production system and were unable to change course. Many realized that the best way to protect their fragile soils during the dry weather was to stop tilling the soil, replant the most vulnerable parts of the prairie with a protective cover of sod, and switch to grazing cattle, but they were already too indebted to raise the capital needed to completely retool their farms from growing grain to grazing livestock. Besides, just as John Wesley Powell had predicted, homesteads of 320 acres of dryland prairie were too small to financially sustain themselves via tillage agriculture during the dry part of the climate cycle and much too small to sustain a family by switching to raising cattle. The trapped farmers' only hope of avoiding foreclosure was to plow up more grassland with the tools they already owned in the desperate hope of being able to expand their crop acreage enough to make up for falling yields while they waited for their luck to turn. ("Plow like your life depends, because it does.") This downward spiral accelerated as the stock market crashed in 1929, interest rates spiked, loans were called in, and grain markets cratered.

In an effort to "protect" farmers and other highly indebted industries, the panicked government tried to stabilize falling domestic prices by imposing steep tariffs on foreign imports (the infamous Smoot-Hawley Tariff Act of 1930). This triggered a vicious escalating trade war with America's trading partners who were outraged by the tariffs and responded in kind in order to protect their own domestic markets. As bankers and government institutions watched their balance sheets erode, they began calling in loans they had made to central banks in Europe in the aftermath of WWI, which had the immediate effect of sucking all the liquidity out of the European financial system and exporting America's economic depression to all its trading partners. World trade collapsed, wheat exports dried up, and wheat prices plunged still further. Any last hope of a reprieve out on the Great Plains was extinguished.

Decades of abundant rainfall had masked the dangerous consequences of plowing up the fragile prairie and exposing the bare soil to sun and wind. But then, as the 1930s began, temperatures soared in a record-shattering heatwave that not only baked the Great Plains, but also shattered temperature records all across Europe.[25] As the mercury climbed higher, the rains began to fail in earnest, the winds picked up, and the entire Great Plains ecosystem collapsed.

The drought of the 1930s turned the Great Plains into a Dust Bowl that stretched from Alberta to Texas. Decades of plowing the fragile prairie grasslands had left the entire region wholly incapable of withstanding the hot and dry part of the natural climate variability of the American West. The heavily tilled soils were utterly devastated by the hot dry winds sweeping off the Rocky Mountains. What had once been a robust natural ecosystem uniquely adapted to the dry windy prairie climate had become a fragile, brittle, wholly artificial environment created by the farmers who had been fooled by the hubris of scientists and central planners. Lured onto the prairie by a dangerous government-promoted climate fairy tale and encouraged by central planners and central bank monetary policies to plow up as much acreage as possible to grow wheat, few had lived on the Great Plains long enough to have a memory of the region's long-term multidecadal climate cycles that alternate between decades of abundant rains followed by decades of severe drought. They trusted their government's assessment of the climate, stripped the deep-rooted native sod off the prairie according to the advice offered by government planners, strapped themselves into a debt-fueled export-dependant monoculture-focused production system encouraged by government planners and facilitated by central bankers, and dutifully followed government-recommended agricultural techniques that were ultimately responsible for triggering an entire ecosystem collapse when the next dry period in the climate cycle began. By the time Icarus became aware of the wax melting off his fragile wings, the damage was irreversible.

The dust storms began in 1931 and continued through the entire 1930s as the wind relentlessly carried away the parched and exhausted topsoil that had been exposed by the plow. Stripped of its protective sod skin, there was nothing to slow evaporation from the soil, nothing to shield the dry topsoil from the wind, and nothing to bind the delicate soil to prevent it from being carried away.

From Texas to Canada, black blizzards of blowing topsoil blocked out the sun. A particularly severe dust storm hit Oklahoma and Texas on April

---

[25] "1930s Dust Bowl led to extreme heat around Northern Hemisphere. (2022, Nov 29). *Science Daily.* https://www.sciencedaily.com/releases/2022/11/221129134521.htm

14th, 1935. Known as Black Sunday, it is estimated to have displaced over 300,000 tons of topsoil in a single storm. Avis D. Carlson described the experience of living through the storms: *"The impact is like a shovelful of fine sand flung against the face. [...] People caught in their own yards grope for the doorstep. Cars come to a standstill, for no light in the world can penetrate that swirling murk.... The nightmare is deepest during the storms. But on the occasional bright day and the usual gray day we cannot shake from it. We live with the dust, eat it, sleep with it, watch it strip us of possessions and the hope of possessions."*[26] Lungs filled with dust caused a condition known as dust pneumonia, also known as the "brown plague", which killed hundreds. By 1934, over 100 million acres of farmland had lost all or most of its topsoil to the winds.[27]

Figure 5. A dust storm approaches Stratford, Texas, in 1935.

Years of plowing up the prairie made the Great Plains uniquely vulnerable to drought conditions and so, indirectly, the plow also created the ideal conditions for yet another plague that swept across the Great

---

[26] Ganzel, B. (2003) "The Dust Bowl." *Wessels Living History Farm*. Retrieved on 8 June 2024. https://livinghistoryfarm.org/farming-in-the-1930s/water/the-dust-bowl/

[27] Corey, G. (2020, June 13). "The Great Dust Bowl of the 1930s was a Policy-Made Disaster." *Foundation for Economic Freedom*. https://fee.org/articles/the-great-dust-bowl-of-the-1930s-was-a-policy-made-disaster/

Plains during the Dust Bowl — locusts. Locusts are large grasshoppers that typically lead solitary lives until drought conditions force them to congregate in the few areas that still have some vegetation. As they come into close contact with each other they release a chemical neurotransmitter called serotonin, which makes the locusts more active, hungry, and social, causing them to swarm and begin migrating as a group in search of food. Millions upon millions of swarming locusts descended upon the Great Plains and consumed everything in their path. Entire fields were consumed within hours. They even ate the wooden handles off farm implements.[28] According to the History channel, "*[t]he July 1931 swarm was said to be so thick that it blocked out the sun and one could shovel the grasshoppers with a scoop. Cornstalks were eaten to the ground and fields left completely bare.*"[29] Some scientists from the era described the locusts as a "*metabolic wildfire*"[30], similar to the way wildfires consume vegetation as they burn across the landscape.

Laura Ingalls Wilder, the author of the famous *Little House on the Prairie* book series described the swarm that devastated their family farm: "*A cloud was over the sun. It was not like any cloud they had ever seen before. It was a cloud of something like snowflakes, but they were larger than snowflakes, and thin and glittering. Light shone through each flickering particle....Plunk! Something hit Laura's head and fell to the ground. She looked down and saw the largest grasshopper she had ever seen. Then huge brown grasshoppers were hitting the ground all around her, hitting her head and her face and her arms. They came thudding down like hail. The cloud was hailing grasshoppers. The cloud was grasshoppers.*"[31] By 1937, the Colorado National Guard was using

---

[28] "Locusts: The Bothersome Bug that Shaped America." *Arrow Exterminators Blog.* April 24, 2019. https://www.arrowexterminators.com/learning-center/blog/locusts-the-bothersome-bug-that-shaped-america

[29] "Grasshoppers devastate Midwestern crops." *History.com.* July 24, 2020. https://www.history.com/this-day-in-history/grasshoppers-bring-ruin-to-midwest

[30] Eames, W. (2021, July 23). "The Locust Plagues of the 1930's." *Without a Trace* (Issue #41). https://web.archive.org/web/20220624231252/https://news.trace.com/p/the-locust-that-plagued-the-1930s

[31] Eames, W. (2021, July 23). "The Locust Plagues of the 1930's." *Without a Trace* (Issue #41). https://web.archive.org/web/20220624231252/https://news.trace.com/p/the-locust-that-plagued-the-1930s

flamethrowers and explosives in a desperate attempt to try to stop their advance.[32]

The decreasing food supply during the drought also forced jackrabbits out of their native habitats and into the cropland. As they congregated on the crops, the close quarters caused them to multiply in astronomical numbers, triggering a jackrabbit plague of biblical proportions that likewise ate everything in its path. They even stripped the bark off trees and dug up the roots of alfalfa plants in order to consume them. People organized enormous rabbit drives to try to round up the jackrabbits to kill them. Local farmers in Lane County, Kansas, estimated that a single 8 square mile block of land contained a population of over 150,000 jackrabbits![33]

Figure 6. Kansas rabbit drive during the Dust Bowl.

By the end of the 1930s, more than half a million people were left homeless out on the Great Plains. Nearly 750,000 families lost their farms

---

[32] Morgan, A. (2014, June 9). "In 1937, Colorado Guard used flamethrowers and explosives against plague of locusts." U.S. National Guard. https://www.nationalguard.mil/news/article-view/article/575751/in-1937-colorado-guard-used-flamethrowers-and-explosives-against-plague-of-locu/

[33] Ehmke, V. (2016, May 17). "'Rabit drive' in '35 literally a hair-raising experience." *The Hutchinson News.* https://www.hutchnews.com/story/news/local/kansas-agland/2016/05/17/rabbit-drive-in-35/20717529007/

to bankruptcy or foreclosure. 2.5 million people fled the Plains, of which 200,000 fled to California. Some counties, like Baca County in Colorado, saw over 40% of their population leave.

Figure 7. A farm in South Dakota buried by dust during the Dirty Thirties. May 1936.

This time the now firmly discredited scientific theory that "the rain follows the plow" would not be resurrected. It had outlived its usefulness. The plains were settled, the railroads were complete, America's hold over the continent was secure, and no rival nation could hope to threaten America's centrally directed economic clout on the world stage. Despite the government's instrumental role in creating the ecological collapse of the Dirty Thirties and the immeasurable suffering it imposed on its unsuspecting population by sticking its nose into every corner of American life, Woodrow Wilson's cherished administrative state had successfully embedded itself as an indispensable force at the heart of American society — it has been so thoroughly normalized that few people today can even conceive how there could be another way. The rains did not follow the plow, but the administrative state was here to stay. This unholy marriage between centralized government and masterful corporations — the corporate state (the polar opposite of a free market) — has defined the trajectory of Western politics ever since.

After a decade of reactionary government policies that systematically prevented the economy from returning to a secure footing, the downward economic spiral and high unemployment of the Great Depression era was finally "solved" by the outbreak of the Second World War. America's entry into the war resolved the unemployment crisis through the draft and through the nation-wide war-time mobilization that employed millions as long-idle factories were retooled for the war effort and funded by more money-printing. After the war, the rise of consumerism facilitated by cheap debt kept the factories humming even as inflation relentlessly ate away at the savings produced by yesterday's hard work.

Those who remained on the Great Plains after the Dust Bowl were left to pick up the pieces of their lives. Through trial, error, and unrelenting stubbornness, they gradually developed the dryland farming techniques that allow agriculture to persist in the region today. But the memory of the dark days of the Dust Bowl are imprinted in their cultural DNA, passed from generation to generation as a deep enduring distrust of central planners, a deep skepticism of the grand ideas that are cooked up by academics, and a visceral distaste for the greasy entourage that makes a living by whispering in the ears of political and bureaucratic gatekeepers.

The hucksters who had so enthusiastically lined their pockets by luring homesteaders out into the Great American Desert shifted their focus to the bonanza of new opportunities created by the very thing that America's Founding Fathers had feared — a strong centralized government. Once Woodrow Wilson and his peers built the administrative state, it was only a matter of time until all of society would be co-opted into the great game of whispering in the ear of the gatekeepers, some for profit, some for power, some to save their hands from hard work, some to satisfy their ideological impulses, and some out of necessity to prevent themselves from being steamrolled by the consequences of everyone else's whispering.

Any system that concentrates enormous power in the hands of a small number of gatekeepers — any system where power and resources are centralized — is a system ripe for plucking by those who perfect the art of whispering in the ears and pulling at the heartstrings of the gatekeepers. The progressive belief that government should serve as a tool to push society towards "progress" became the defining belief system of the next century — and everyone has been fighting each other ever since in a never-ending culture war to control the agenda of this Leviathan and to secure a slice of this ill-begotten pie. Meanwhile, the old market signals of supply and demand have been completely distorted by government planners and regulators tinkering with the money supply, through their meddlesome policies, and through their never-ending debt-fueled spending. The slow organic process that once steered the direction of the nation's cultural development has been completely replaced with diktats and artificial incentives created by central planners. We live in an entirely contrived world awash with artificial carrots and sticks.

The institutions created during the Progressive era, during the New Deal era, and in the aftermath of the Second World War have only increased their power and reach in the ensuing decades until today nearly all of society has become dependent on them in some way. And the manipulative strategies to win the sympathies of the gatekeepers (and raid the public purse) have multiplied like rabbits. This last point is the key explanation behind today's growing chaos as the corruption and institutional rot reaches its paralyzing zenith. By creating a big, centralized government, America incentivised every corporation, every institution, every university, every scientist, every citizen, and every foreign government to try to find a way to manipulate the gatekeepers until today it's no longer clear who actually runs the government and for whose benefit. The same is true in every other Western country — welcome to the international rules-based liberal world order.

Anyone who refuses to play this manipulative game risks being squeezed out of business by those who do. It's not for nothing that the Founding Fathers tried to limit the size of government. The increasingly bizarre beliefs that are emerging during our era — not least of which is the bizarre crusade that has been launched against carbon dioxide in another ill-fated attempt to control the weather — are ultimately symptoms of this endless game to whisper in the ear of Leviathan. The carnival is held together by a thin veneer of democracy, which gives Leviathan an illusion of legitimacy, and by society's misplaced trust in scientific and political institutions that have long since tossed objectivity, evidence-based debate, and integrity out of the window in order to pursue self-serving political agendas and line their pockets.

~

The United States wasn't the only country to fall for the optimistic climate fairy tale that "the rain follows the plow". It was a popular myth that influenced attitudes and government policies all around the world during the late 19th and early 20th century. For example, Canada's government-sponsored railroads and the promise of free land lured millions of Eastern Canadians and European immigrants out onto the Canadian Plains as Canada rushed to secure its hold over its own western regions before these sparsely populated regions could be absorbed by its aggressively expanding southern neighbor. The race to reach the Pacific was no less existential for America's rival northern neighbor.

Once again, the plow and the railroad were the means by which to tie together a fledgling nation. Wholly oblivious to the dry cycle that would eventually catch up to them and their plowed fields, Canadian homesteaders were also doomed to experience the full range of horrors unleashed by the Dust Bowl. Canada's Natural History Society suggests that, like in the U.S., nearly 750,000 Canadian farms were lost between 1930 and 1935, the majority in southeastern Alberta and southern Saskatchewan.[34] In the rural parts of the Canadian prairies, two-thirds of the population depended on government relief during the Dirty Thirties compared to the Canadian national average of one-fifth.[35] If it hadn't been for the crops grown by farmers in the more humid eastern half of the continent, millions of Canadians and Americans would have died of starvation.

But there is a twist to the Canadian story. There is a region in southeastern Alberta and southwestern Saskatchewan called the Prairie Dry Belt. Until 1905 this region of the prairie was strictly off-limits to homesteaders because even starry-eyed government planners recognized that this area was far too dry for cultivation and was only suitable for grazing livestock. But under pressure from developers, railroads, and other special interest groups, the government lifted the ban in 1905. The region reaped enormous bumper crops during the wet weather of 1915 and 1916. But then, starting a full 13 years (!) before the rest of the Great Plains ecosystem fell apart, ten years of unrelenting drought, prairie fires, and locusts utterly devastated the Prairie Dry Belt. Entire towns were abandoned and the homesteaders within this Prairie Dry Belt region were

---

[34] Phipps, E. (2022, April 12). "Dust and Depression." *Canada's National History Society.* https://www.canadashistory.ca/education/lesson-plans/dust-and-depression

[35] "Great Depression in Canada." *Wikipedia.* May 24, 2024. https://en.wikipedia.org/wiki/Great_Depression_in_Canada

left destitute. By 1926, some counties in the Prairie Dry Belt had lost up to 80% of their inhabitants![36]

The Prairie Dry Belt Disaster is considered one of the worst social and economic disasters in Canadian history — the Dust Bowl of the 1930s was merely a pale sequel. In other words, if you include the Prairie Dry Belt Disaster in the count, it took three catastrophic ecosystem collapses — three devastating examples of the consequences of ignoring the natural multidecadal climate variability on the Great Plains — before the Canadian and American administrative states finally faced up to the fact that they were recommending farming practices that were wholly out of sync with the variable climate of the Great Plains ecosystem. People saw what they wanted to see, believed what they wanted to believe, and vilified anyone who dared say otherwise. They built polices that matched their optimistic enthusiasm for science and technology and reflected their conviction that through hard work they could push the agricultural frontier right up to the foot of the Rocky Mountains. The more cynical interpretation is that as long as parts of the government's nation-building agenda remained unfulfilled, politicians and bureaucrats simply didn't want to hear anything that contradicted the faulty science underpinning their nation-building policies, which quickly taught the few policymakers and scientists who knew better to censor themselves in order not to risk their careers by trying to swim against the tide. For every John Wesley Powell, there were dozens of others who put their careers first by keeping their mouths shut.

Australia also took the bait by trying to expand wheat farms into dry marginal lands from 1895 to 1945 based on the idea that the rains would follow the plow.[37] Tillage-based farming expanded as far north as latitude 26 degrees South, a full 6 degrees (approx. 660 km (410 miles)) further north than where crops are grown in Australia today! And, much like what happened out on the North American prairie, a series of episodic droughts led to devastating crop failures, rabbit plagues, and colossal dust storms that stripped away entire horizons of Australian topsoil.[38] The same

---

[36] Jones, D. C. (2015). "Prairie Dry Belt Disaster." *The Canadian Encyclopedia.* https://www.thecanadianencyclopedia.ca/en/article/prairie-dry-belt-disaster

[37] "Rain follows the plow." *Wikipedia.* 17 June 2024. https://en.wikipedia.org/wiki/Rain_follows_the_plow

[38] Cattle, S. R. (2016). The case for a southeastern Australian Dust Bowl, 1895-1945. *Aeolian Research*, 12, 1-20 https://doi.org/10.1016/j.aeolia.2016.02.001

wishful thinking also led the British in South Africa to try to expand tillage agriculture into the Karoo and Kalahari deserts in the late 19th and early 20th century, with similar disasterous results. Those who allow the "consensus" to serve as their compass are no wiser than lemmings following one another over the edge of a cliff.

Chapter 2

The Seeds of Pessimism and the Rise of the Doomsday Cult

"Let me make the songs of a nation, and I care not who makes its laws."

— Plato

The belief that "rain follows the plow" emerged as an idea that perfectly captured the optimistic mood of its era. The scientific and technological revolution and the seemingly magical machinery of the Industrial Age seemed poised to turn humanity into an omnipotent force for positive change. Diseases that had haunted humanity since the dawn of time were being conquered by science and technology. Pumps, irrigation canals, and mega-dams were bringing water to parched deserts. High-rises, concrete, and ambitious public works were transforming city skylines. The Suez Canal and the Panama Canal were linking oceans and shrinking the distance between continents. The slow plodding pace of horse-powered transport was being made obsolete, first by the steam engine and then by the internal combustion engine, which had the power to move literal mountains. The Wright Brothers taught us that Man could, indeed, fly. There wasn't a horizon that could not be conquered by human ingenuity and hard work thanks to scientific knowledge and technological progress. And who better to harness this technological wizardry for the collective good than Wilson's cherished administrative state. Dazzled by the power of our own creations, like Icarus, we thought anything was possible. Where once our lives were controlled by the whims of capricious gods, now science and technology were allowing us to reinvent ourselves as the Masters of our Universe.

Society's optimism knew no bounds as we contemplated the potential for science, technology, industrialization, and the administrative state to create a better world. The First World War was even optimistically viewed as "the war to end all wars", which would sweep away the era of great power politics and usher in a new liberal era of international peace and cooperation as unaligned powers were brought to heel. After the End of the First World War, Woodrow Wilson specifically argued for a "new world order" that would emphasize collective security, democracy, and self-determination. And it was Wilson who first proposed the League of Nations as the template for modern global governance (though the U.S. never joined the League). After the League of Nations failed to prevent World War II, Franklin D. Roosevelt recycled Wilson's idea via the creation of the United Nations as a replacement for the League of Nations, this time firmly under the dominant thumb of the United States.

The ecological collapse of the Dust Bowl and the misery of the Great Depression didn't dampen enthusiasm for the administrative state. On the contrary, these crises only fueled public support for Roosevelt's New Deal, which sought to massively expand the concentration of power in the executive branch — essentially Woodrow Wilson's vision of the hyperactive administrative state on steroids. Since European nations didn't have anything like the constitutional limits on government power that were built into the American Constitution, nor a Bill of Rights designed to prevent the government from trampling individual rights in pursuit of national objectives, Europe went still further by completely blurring the lines between industry and state power (fascism in Italy under Mussolini, national socialism in Germany under Hitler, and communism in Russia under Lenin and Stalin). It was an era when society's faith in science, technology, and the administrative state was so strong that when these forces failed to deliver on their promises, society simply concluded that it was only because these forces needed more power and more funding to work optimally.

When Roosevelt published his 1933 book, *Looking Forward*, in which he set out his reasoning and hopes for the major reforms of his New Deal, Mussolini wrote a glowing review of Roosevelt's book in which he pointed out the similarities between Roosevelt's "Planned Economy" and Mussolini's economic fascism: *"Reminiscent of Fascism is the principle that the state no longer leaves the economy to its own devices.... Without question, the mood accompanying this sea change resembles that of Fascism."*[1] The admiration was mutual. Roosevelt confided in a White House correspondent that *"I don't mind telling you in confidence that I am*

---

[1] Boaz, D. (2007, Sept 28). "Hitler, Mussolini, Roosevelt." *The Cato Institute*. https://www.cato.org/commentary/hitler-mussolini-roosevelt

keeping in fairly close touch with that admirable Italian gentleman." One of Roosevelt's advisors went still further in his description of Mussolini's program to modernize Italy: "*It's the cleanest ... most efficiently operating piece of social machinery I've ever seen. It makes me envious.*"[2] As late as 1940, even Winston Churchill was still describing Mussolini as "*a great man*".[3]

Not to be left out, Hitler also praised Roosevelt's efforts to build a planned economy, telling American ambassador William Dodd that he was "*in accord with the President in the view that the virtue of duty, readiness for sacrifice, and discipline should dominate the entire people. These moral demands which the President places before every individual citizen of the United States are also the quintessence of the German state philosophy, which finds its expression in the slogan 'The Public Welfare Transcends the Interest of the Individual.'*"[4]

In the aftermath of the New Deal era, journalist John T. Flynn provided a more sober reassessment of the era. In his book *The Roosevelt Myth* he wrote that "*The mere New Dealers, as that term came to be understood, comprised those wandering, vague dreamers who held to a shadowy conviction that somehow the safety of humankind depended upon the creation of some sort of ill-defined but benevolent state that would end poverty, give everybody a job and an easy old age, and who supposed that this could be done because they had discovered that money grew in government buildings.[...] There was indeed a good deal of tolerance for the idea of planning our capitalist system even in the most conservative circles. And a man could support publicly and with vehemence this system of the Planned Economy without incurring the odium of being too much of a radical for polite and practical society. There was only one trouble with it. This was what Mussolini had adopted—the Planned Capitalist State. And he gave it a name—fascism. Then came Hitler and adopted the*

---

[2] Gordon, D. (2018, Sept 13). "Three New Deals: Why the Nazis and Fascists Loved FDR. *Mises Institute.* https://mises.org/mises-daily/three-new-deals-why-nazis-and-fascists-loved-fdr

[3] DiLorenzo, T. J. (2012, Sept 3). "The Rise of Economic Fascism in America." *Mises Institute (Mises Wire).* https://mises.org/mises-wire/rise-economic-fascism-america

[4] Gordon, D. (2018, Sept 13).

same idea. His party was called the Nazi party, which was derived from the initials of its true name [national socialism]."[5]

Neither Hitler's blatant disregard for individual rights, nor the rampant corruption spawned under Mussolini's leadership dampened enthusiasm for Roosevelt's New Deal. In the end, it was only the foresight of the Founding Fathers to create a constitutional labyrinth of restrictions designed to limit government power that prevented Roosevelt's Planned Economy from reaching the extremes committed by his European peers — those limits held back his worst excesses, but only barely, even as his contemporaries in Europe descended into state-sponsored murder and genocide as a means to socially engineer a "better" society. But the appetite for economic planning in America nonetheless grew deep roots under Roosevelt (as it did in other Western allied nations), and the constitutional limits on that immense state power are now systematically falling away under the guise of fighting climate change, dealing with pandemics, fighting "misinformation", achieving "diversity, equality, and inclusivity", preventing the people from voting the "wrong" people into power, and "protecting national interests". The excuses evolve, but the fires burning in the hearts of those who hold power (and those who learn to influence them) are forever the same. Likewise, state-sponsored education, planned and fine-tuned by state planners, has left deeply corrosive imprints on the psyche of Western Civilization, softening up entire generations to the idea of ever-expanding state power.

Even the revolving door between government and industry, which spawned the colossal corruption that epitomized Mussolini's reign, is equally well entrenched in our modern "liberal democratic" hyper-regulated world order. Change the names and the dates, and the biographies of any one of Mussolini's peers could just as easily be written today about any of the movers and shakers in our modern liberal world order as their careers oscillate back and forth between industry and the regulatory bodies tasked with regulating those very same industries: *"Signor Caiano, one of Mussolini's most trusted advisers, was an officer in the Royal Navy before and during the war; when the war was over, he joined the Orlando Shipbuilding Company; in October 1922, he entered Mussolini's cabinet, and the subsidies for naval construction and the*

---

[5] Flynn, J. T. (1948). *The Roosevelt Myth*. The Devin-Adair Company, New York. https://mises.org/library/book/roosevelt-myth

merchant marine came under the control of his department. General Cavallero, at the close of the war, left the army and entered the Pirelli Rubber Company...; in 1925 he became undersecretary at the Ministry of War; in 1930 he left the Ministry of War, and entered the service of the Ansaldo armament firm. Among the directors of the big . . . companies in Italy, retired generals and generals on active service became very numerous after the advent of Fascism."[6]

The spiralling crises of the Depression Era only reinforced the public sentiment that the classical liberal principle of limited government was unfit to meet the challenges of the modern economic state and that the only solution was to further expand the authority of the administrative state to serve as an even more masterful master to masterful corporations.

The meat grinder of the Second World War ended with two atomic bombs that seemed to confirm the optimistic idea that when government focuses the minds of a unified nation on a single task (and focuses the efforts of the scientific community in particular by giving them seemingly unlimited funds to pursue state-defined objectives), even the most powerful enemies can be vanquished. By splitting the atom, we had learned to play with the same forces that power the sun. Icarus was delighted.

The rollout of the vast institutional apparatus underpinning the expanding administrative state, which began during the Progressive Era, was dramatically accelerated under Roosevelt's New Deal, and then expanded still further in the decades following WWII as the administrative state progressively extended its reach ever further into the lives of its citizens. When you're holding a hammer, every problem looks like a nail. Once the administrative state had made itself indispensable at the center of the system, new things that the state "ought to do" (and how to do them) were discovered every day. Wilson's ghost hung over it all.

The state took control of the management of public forests and fisheries. It developed herbicides and pesticides to try to make them as cheap as possible. It bombarded seeds and grains with radiation to accelerate genetic mutations in search of more productive and more frost- and drought-resistant crop varieties. It engaged in massive public

---

[6] DiLorenzo, T. J. (2012, Sept 3). "The Rise of Economic Fascism in America." *Mises Institute (Mises Wire).* https://mises.org/mises-wire/rise-economic-fascism-america

sanitation projects. It built the interstate highway system. It dammed rivers to generate power and build irrigation networks to bring water to dry areas. Visions of limitless energy harvested from the atom tantalized the imaginations of scientists and policymakers. By empowering the administrative state to oversee science and industry in pursuit of "the greater good", anything seemed possible.

The age of small scientific endeavours conducted by independent universities was increasingly replaced by state-sponsored scientific experts and by capitalists with state contracts, all toiling in service of the mega-state. JFK characterized this new hive-minded ethos best when he said, "*Ask not what your country can do for you — ask what you can do for your country.*" The Founding Fathers built a country designed to protect the individual from the state. The Progressive Era turned that idea on its head as it subordinated the individual to the needs of the state — the tribe before the individual — America First — Amerika Über Alles — or as Hitler observed in his praise of Roosevelt's Planned Economy, "*The Public Welfare Transcends the Interest of the Individual.*"

Frightened by what European nationalism had produced in the hands of strong centralized governments, the progressive vision was updated in the aftermath of WWII to refocus on the idea of a new global tribe under the watchful eye of a benevolent American-led international world order. In 1832, German philosopher Georg Hegel famously once said, "*the only thing we learn from history is that we learn nothing from history.*" Indeed.

One of the early beneficiaries of the massive bonanza of state-funded research that accompanied the rollout of ever more powerful institutions was weather modification. Weather modification efforts began in the late 19th century with attempts to trigger rainfall on demand and to prevent frost by firing canons and other explosives into clouds. By the late 1940s those efforts had matured into something much bigger as the government began sponsoring massive, sophisticated weather modification programs like cloud seeding, fog eradication, and hurricane modification to try to control the weather for both agricultural and military purposes. Weather control, like everything else, was seen as an essential state tool. As Kristine Harper documents in her eye-opening book *Make It Rain*,[7] by the mid 20th century, federal legislators had come to view weather control as

---

[7] Harper, K. (2017). *Make It Rain*. University of Chicago Press, Chicago.

equally important as atomic energy for both domestic and military purposes.

> *"The nation that first learns to plot the paths of air masses accurately and learns to control the time and place of precipitation will dominate the globe."*
>
> — General George C. Kenny, U.S. Air Force

Meteorologists tried to dissuade the weather modification programs, arguing that since meteorologists couldn't even predict the weather with any reliable accuracy and didn't even yet fully understand the complex physics driving the behaviour of clouds, rainfall, and weather patterns, it was highly unlikely that weather controllers (many of whom didn't even have a background in meteorology) would be able to harness forces that no-one fully understood, and that government funds would be better spent on funding meteorologists as the first step to understanding the forces that create our weather. But the idea of weather modification was very popular with politicians, farmers, and the general public. Furthermore, during the 1950s (under President Eisenhower), the military was deeply concerned that the Soviets might beat the U.S. in learning how to control the weather. And so, despite the cautionary words of the meteorologists, ever larger sums of money were poured into weather control programs.

Project Skyfire was run by the U.S. Forest Service to try to reduce lightning strikes and extinguish forest fires by triggering rainfall. Project Skywater was run by the Bureau of Reclamation to try to tap into atmospheric water to fill reservoirs used for irrigation and hydroelectric power generation. And Project Stormfury was run by the Weather Bureau in partnership with the Air Force and the Navy to try to snuff out and/or steer hurricanes. Classified versions of all these experiments were also conducted in secret by the U.S. government — the U.S. military used its weather modification techniques to try to break a crippling drought in India during the 1960s, and during the Vietnam War it tried to modify the weather to use it as a weapon against the North Vietnamese military (Operation Popeye), as was revealed in the explosive *Pentagon Papers* published by journalist Seymour Hersh and columnist Jack Anderson in 1971.

But the mood in America was changing. The unbridled optimism of the late 19th and early 20th century about the unlimited potential of science, technology, and industry to engineer a better world were giving way to an altogether darker and more pessimistic mood.

As early as 1961, in his farewell address, President Eisenhower warned about the double-edged sword of concentrating so much power, influence, and funding in the hands of the central government.[8] In his speech he specifically singled out both the military-industrial complex and the scientific-technological elite:

"Our military organization today bears little relation to that known by any of my predecessors in peace time, or indeed by the fighting men of World War II or Korea.

Until the latest of our world conflicts, the United States had no armaments industry. American makers of plowshares could, with time and as required, make swords as well. But now we can no longer risk emergency improvisation of national defense; we have been compelled to create a permanent armaments industry of vast proportions. Added to this, three and a half million men and women are directly engaged in the defense establishment. We annually spend on military security more than the net income of all United States corporations.

This conjunction of an immense military establishment and a large arms industry is new in the American experience. The total influence – economic, political, even spiritual – is felt in every city, every state house, every office of the Federal government. We recognize the imperative need for this development. Yet we must not fail to comprehend its grave implications. Our toil, resources and livelihood are all involved; so is the very structure of our society.

In the councils of government, we must guard against the acquisition of unwarranted influence, whether sought or unsought,

---

[8] "President Dwight D. Eisenhower's Farewell Address (1961)." (1961, Jan 17). *National Archives.* https://www.archives.gov/milestone-documents/president-dwight-d-eisenhowers-farewell-address

by the military-industrial complex. The potential for the disastrous rise of misplaced power exists and will persist.

We must never let the weight of this combination endanger our liberties or democratic processes. We should take nothing for granted. Only an alert and knowledgeable citizenry can compel the proper meshing of the huge industrial and military machinery of defense with our peaceful methods and goals, so that security and liberty may prosper together.

Akin to, and largely responsible for the sweeping changes in our industrial-military posture, has been the technological revolution during recent decades.

In this revolution, research has become central; it also becomes more formalized, complex, and costly. A steadily increasing share is conducted for, by, or at the direction of, the Federal government.

Today, the solitary inventor, tinkering in his shop, has been overshadowed by task forces of scientists in laboratories and testing fields. In the same fashion, the free university, historically the fountainhead of free ideas and scientific discovery, has experienced a revolution in the conduct of research. Partly because of the huge costs involved, a government contract becomes virtually a substitute for intellectual curiosity. For every old blackboard there are now hundreds of new electronic computers.

The prospect of domination of the nation's scholars by Federal employment, project allocations, and the power of money is ever present and is gravely to be regarded.

Yet, in holding scientific research and discovery in respect, as we should, we must also be alert to the equal and opposite danger that public policy could itself become the captive of a scientific-technological elite."

Only four months later, Eisenhower's successor, President John F. Kennedy, fired the CIA's Director, Allen Dulles, and vowed to *"splinter the CIA into a thousand pieces and scatter them to the winds."* Sadly, two years later his efforts to rein in the Deep State came to an abrupt end when

President Kennedy was assassinated by a lone gunman, who was himself assassinated by another lone gunman, who promptly died of pneumonia while in prison not long after he was arrested — and yet, for "national security reasons", government records related to the assassination remain sealed more than 60 years later as each successive president finds a fresh excuse to postpone the date of release. Make of that what you will.

As the Cold War with the Soviet Union intensified and as the Soviets began testing their own nuclear weapons as a deterrent to American nuclear bombs, the risk of a nuclear apocalypse took the proverbial bloom off the atomic rose and focused the public imagination on the dangerous unintended consequences of scientific and technological development. But there were countless other factors that soured the starry-eyed optimism of a science- and technology-obsessed American public.

DDT had once been viewed as a miracle pesticide in the fight against malaria, but following the publication of environmentalist Rachel Carson's book, *Silent Spring,* in which she laid out her belief that chemical pesticides were causing massive environmental problems, public opinion turned against DDT and led to its ban out of concern that it was causing cancer and killing wildlife. Her book is frequently cited as the catalyst for the environmental movement that began in the 1960s and the creation of the Environmental Protection Agency in the 1970s. One reviewer of her book stated that Carson had *"quite self-consciously decided to write a book calling into question the paradigm of scientific progress that defined postwar American culture."*[9]

The health problems caused by leaded gasoline (and leaded pipes) were also exposed during the 1960s. The wonderous mineral Asbestos, which had been celebrated for its fire-resistant properties and was being used almost everywhere as a building material in everything from fire-resistant coatings, in concrete, in the construction of pipes, in insulation, in drywall, in flooring, and in roofing shingles, also turned out to be a massive health risk. Court documents from the late 1970s proved that the asbestos industry had known about these dangers since the 1930s but had concealed them from the public.[10] The popular and widely used tranquilizer Thalidomide was proven to cause birth defects in the 1960s. Countless wildlife species were driven to the brink of extinction by pollution and

---

[9] "Silent Spring. *Wikipedia.* 5 May 2024. https://en.wikipedia.org/wiki/Silent_Spring
[10] "Asbestos." *Wikipedia.* 4 June 2024 https://en.wikipedia.org/wiki/Asbestos

other human activities. Many of the nation's waterways had been turned into toxic cesspools by industrial pollution. The term *acid rain* was coined to describe the acidic rain caused by sulfur dioxide and nitrogen oxide emissions spewing out of industrial smokestacks. And the list goes on and on. In the span of only a few short decades, many of the wonderous scientific and technological advances that had once captured the public imagination as symbols of progress were turning into objects of fear. The bountiful well of seemingly unlimited scientific and technological progress was beginning to look more like a poisoned chalice.

Each new scandal further undermined public optimism and eroded the euphoric belief that science, technology, and industry were forces for good. And, predictably, society turned to government to demand that it take a much more active role in regulating private industry for the good of society as a whole, which only expanded the number of government institutions that could be lobbied for power, influence, and money. Free market capitalism became regulated capitalism, and quickly degenerated into crony capitalism as the regulated soon learned how to leverage regulation as a tool to outmanoeuver their less politically connected competitors.

Likewise, environmental activism took on a decidedly more polished flavor as it transformed itself from a local volunteer initiative focused on confronting local environmental issues into a lucrative national and then international business model permanently in search of new causes and compelling stories to sustain the careers of professional activists and keep funding taps open. Local voices were completely sidelined. Nothing opens wallets (and public coffers) faster than an emotionally compelling scary story with global implications. Scientists and researchers also discovered this same basic truth about the funds divvied out by government gatekeepers — peddling in fear pays handsomely as long as you can keep the drumbeat of fear alive. As philosopher Eric Hoffer once observed, *"every good cause begins as a movement, becomes a business, and eventually degenerates into a racket."*

The other realization that crept into the public consciousness during the post-WWII era is that science, technology, and free market capitalism — through the unprecedented success of all its medical innovations, improvements in public sanitation, and higher standards of living — had triggered a massive expansion in the global population. From 1900 to 1970, the population more than doubled from 1.63 to 3.7 billion. By 1999, it had crossed the 6 billion mark. Today it stands at over 8 billion. The 19th century optimism that science, technology, and the capitalist economic model could solve all our problems turned into 20th century pessimism and then to naked fear that if mankind was allowed to continue on its present prosperous path, our rapidly expanding population would quickly exhaust our planet's resources, lead to mass starvation, and cause an environmental collapse on a scale that would make the Dust Bowl of the 1930s look like a picnic.

As early as 1867, Karl Marx questioned the sustainability of the capitalist economic model (though not for environmental reasons) when he wrote that capitalism *"cannot stand still, but must always be expanding or contracting"*. According to Marx, zero growth is not possible under capitalism.[11] Marx's skepticism of the capitalist economic model and his advocacy for a stable, centrally planned economic system with the authority to override individual free choice in service of the collective good was revived by the environmental movement and by those concerned about the implications of the planet's rapidly expanding population. This change in sentiment is captured by the following quotes and headlines:

> *"I think if we don't overthrow capitalism, we don't have a chance of saving the world ecologically. I think it is possible to have an ecologically sound society under socialism. I don't think it is possible under capitalism."*
>
> — Judith Bari, U.S. environmentalist, feminist, labor leader, and principal organizer of Earth First! anti-logging campaigns, as quoted in the Springfield State Journal Register on June 25th, 1992[12]

~

> *"Anyone who believes exponential growth can go on forever in a finite world is either a madman or an economist."*

---

[11] "Growth imperative." *Wikipedia*. Feb 23 2024. https://en.wikipedia.org/wiki/Growth_imperative

[12] "Be aware of the extremists who populate environmentalism's history." *The Athens News*. Mar 9, 2006. https://www.athensnews.com/news/local/be-aware-of-the-extremists-who-populate-environmentalisms-history/article_aded7c7f-4e30-564e-8160-beacbfe2e3c8.html

— Paul Ehrlich, biologist and author of *The Population Bomb*. Ehrlich's famous 1990 quote has been recycled on numerous occasions by famous British television presenter Sir David Attenborough. However, Ehrlich was also merely recycling the quote as the original can be traced back to a 1973 article called *"The No-Growth Society"* published in The Journal of the American Academy of Arts and Sciences (a.k.a. Daedalus).

~

"It's absurd, really, to think that there can be unlimited economic development on a planet with finite natural resources."

— Dame Jane Goodall, famed primatologist, anthropologist, and chimpanzee expert on chimpanzees, speaking at the 2019 Population Matters conference.

~

"We are in the beginning of a mass extinction, and all you can talk about is money and fairy tales of eternal economic growth."

— Greta Thunberg, environmental activist and the face of the *Fridays for Future* movement, in her famous speech at the U.N.'s 2019 Climate Action Summit in New York City.

~

"Anyone who thinks that you can have infinite growth in a finite environment is either a madman or an economist."

— Sir David Attenborough, British broadcaster and biologist, has recycled the famous quote while speaking at the 2013 Royal Geographical Society and at the 2019 Annual Meeting of the World Economic Forum.

~

And then there's this gem from Columbia University (figure 8):

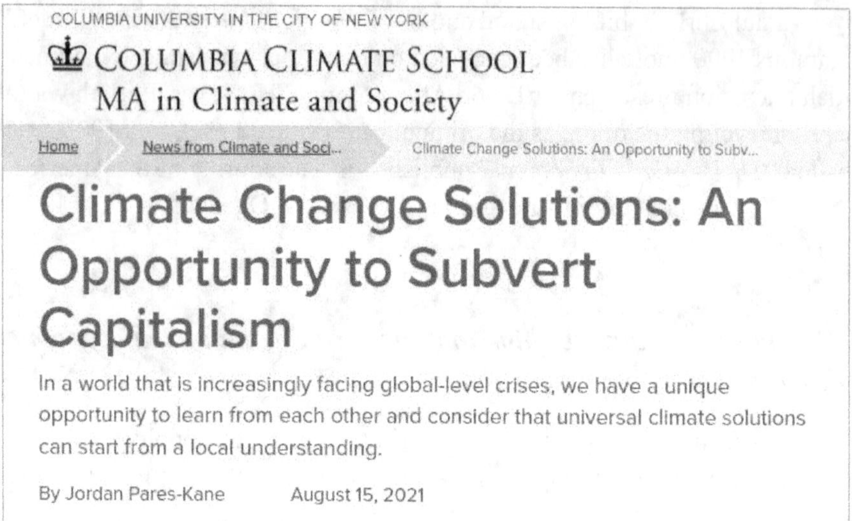

> **COLUMBIA UNIVERSITY IN THE CITY OF NEW YORK**
> **COLUMBIA CLIMATE SCHOOL**
> MA in Climate and Society
>
> Home › News from Climate and Soci... › Climate Change Solutions: An Opportunity to Subv...
>
> # Climate Change Solutions: An Opportunity to Subvert Capitalism
>
> In a world that is increasingly facing global-level crises, we have a unique opportunity to learn from each other and consider that universal climate solutions can start from a local understanding.
>
> By Jordan Pares-Kane   August 15, 2021

Figure 8. Columbia University in New York says the quiet part out loud.
Source: https://climatesociety.climate.columbia.edu/news/climate-change-solutions-opportunity-subvert-capitalism

~

The fact that individual property ownership and individual autonomy have always triumphed over the collective decision-making found in socialist and communist countries (not only to improve human happiness, encourage innovation, and generate abundance, but also to create a cleaner and more sustainable environment) has not deterred enthusiasm for socialist economic planning as the solution to the perceived problems of "infinite growth in a finite world". The Fall of the Soviet Union clearly exposed how 60 years of central planning had created vast toxic wastelands, destroyed entire ecosystems (like the entire ecosystem around the Aral Sea), and was responsible for the persistent scarcities that made life so miserable behind the Iron Curtain (not to mention the horrific human rights abuses committed in pursuit of censoring the heretics who dared criticize the system). Yet the lessons that should have been learned from that 60-year experiment in central planning fell on deaf ears.

People see only what they want to believe. "This time will be different." "If only it had been democratic socialism." "If only the right people had been in charge." Good intentions create the blindness that shields people from the fact that the centrally directed system itself is to blame for creating the incentives that inevitably spiral into corruption, authoritarianism, and economic and environmental collapse.

Few things are as poorly understood as the self-correcting forces unleashed by a true free market, popularized by Adam Smith in the 18th century via his metaphor of the "invisible hand of the market", which recognized that despite its imperfections, natural self-interest in combination with private property rights in a truly free market creates far better outcomes than stringent government control ever could. But the critics of free markets continue to carry the day. As economist John Maynard Keynes once cynically observed, *"Capitalism is the extraordinary belief that the nastiest men for the nastiest of motives will somehow work together for the benefit of all."*

~

Biologists like Paul Ehrlich were about to play an outsized role in shaping the public mood as they turned their attentions from studying animal populations to studying our own species. Unfortunately, their biological models consistently fall short as they fail to account for human ingenuity to create abundance where yesterday there was scarcity, and so they only see catastrophes ahead as they calculate how long it will take for a growing population to overwhelm its existing resources — how do you account for lifesaving and abundance-creating innovations that have yet to be invented? By contrast, the free-market economist recognizes that wherever human ingenuity is allowed to flourish free from government coercion, innovative self-serving individuals have continually produced knowledge and resources for themselves that ultimately benefit society as a whole as their innovations are mimicked by others and spread through the population (in contrast to animals that can't innovate their way out of a shortage).

The pessimistic biologist sees a world of finite resources and turns to the supposedly enlightened state planner acting in service of the alleged collective good in the hope that the angelic gatekeeper can ensure that everyone gets his "fair" share of a finite pie. The optimistic economist turns to the free market as the engine of innovation while recoiling in fear of the central planner who suffocates the free market and keeps society's most productive members in chains. Competition, not state-mandated cooperation or government subsidies is the engine that drives innovation and leads to an ever-expanding pie.

During the 1960s, 70s, and 80s, the biologists' message that the world was running out of resources thanks to overpopulation and overconsumption was driven home by media images of starving children in Africa (which drove TV ratings to new heights while spawning a whole new industry of self-serving charities and intranational organizations weaponizing empathy to enrich themselves). The suffering was real and the emergency aid was imperative, but the lessons drawn from these crises were not. Emotions ignored the statistics, which demonstrated that population growth was slowing in every country that was getting richer (people have less kids as soon as they stop depending on subsistence farming and begin to build wealth); that global poverty and starvation declined sharply throughout each of these decades despite the growing population; that environments get cleaner and pollution goes down as countries get richer (people care about their backyards as soon as they stop having to live hand-to-mouth); that starvation and poverty decrease wherever central planning is replaced by free markets, robust property rights, and respect for individual autonomy (because when the government gets out of the way, people are able to find innovative ways to improve their lives); and that thanks to the efforts of "self-serving" free market innovators, global food production was expanding faster than the world's population (as illustrated by rising corn yields in figure 9 below).

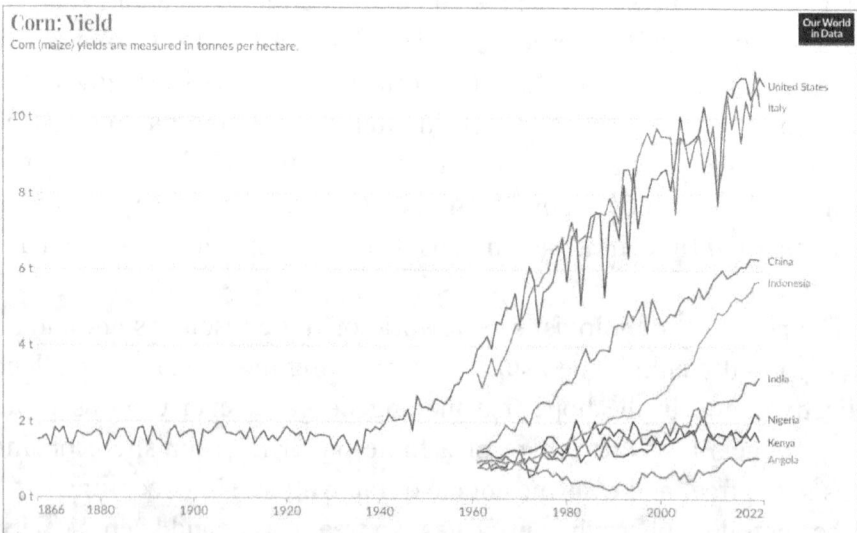

Figure 9. Corn yields per hectare hovered below 2 tons per hectare prior to the 1940s. By 2022, yields in Western countries that use chemical fertilizers have soared to between 10 and 12 tons per hectare. Meanwhile, yields in countries with limited access to fertilizer still hover somewhere between 2 and 3 tons per hectare.

Source: World in Data (https://ourworldindata.org/crop-yields)

But a single impactful media picture speaks louder than a thousand dry statistics. The warnings about overpopulation were noticed, but the lessons on how to rescue developing countries from the spiral of overpopulation by helping them embrace free markets and achieve Western prosperity were systematically drowned out by the drumbeat of fear emanating from academia and the media, all of which led directly to calls for more government intervention. The subtext was clear — if the developing world achieved Western standards of living, the planet would be toast. Emotions are shaped by powerful images and anecdotal stories, not by charts and graphs. Even today, pessimistic scientists are busy calculating the human carrying capacity of our planet, with the most common estimates *"running between two and four billion [!?!] depending on how optimistic researchers are about international cooperation to solve collective action problems."*[13]

Fears of overpopulation are nothing new. Back when the world's population was under 200 million, the city-States of Ancient Greece were deeply troubled by fears of overpopulation as they contemplated the delicate balance between their population and their nation's food supply. Earlier still in 18th-century BC, back when the world's population sat at under 50 million, the Babylonian flood myth of Atra-Hasis,[14] which closely resembles the Christian story of Noah's Ark, tells of a time when "the land was bellowing like a bull" under the stress of overpopulation. In the myth, the gods punish the people for the sin of multiplying unchecked by sending droughts, plagues, and other natural disasters to kill them. The story ends with a final cataclysmic flood that only spares a select group of people who were forewarned by the gods, but the gods condemn their offspring to forevermore be plagued with barrenness, stillbirths, high infant mortality, natural disasters, and the condition that priestesses henceforth must live a life of celibacy in order to restore a cosmic balance between the human population and the land's carrying capacity.

The Scientific Revolution, which began in the 16th century, and the Age of Enlightenment that followed in its wake had briefly filled the world with optimism that science and technology would finally enable humanity

---

[13] "Carrying capacity." *Wikipedia*. 11 May 2024.
https://en.wikipedia.org/wiki/Carrying_capacity

[14] Feen, R. H. (1996). Keeping the Balance: Ancient Greek Philosophical Concerns with Population and Environment. *Population and Environment*, 17(6), 447–458.
http://www.jstor.org/stable/27503492

to reliably feed itself without the persistent famines of bygone eras. But in 1798, economist Thomas Malthus shattered that optimism when he published *An Essay on the Principle of Population*, in which he laid out his grim theory on population growth, supported by lots of statistics and an exponential mathematical formula to calculate future population growth. In a nutshell, his theory restated the age-old fear that the supply of food cannot keep up with the growth of the human population, inevitably leading to poverty, famine, disease, war, and catastrophe unless society takes active measures to control population growth. Philosopher Thomas Carlyle nicknamed Malthus' Theory of Population "the Dismal Science".

Rather than explain how technology and innovation had made it possible for the world's population to feed itself despite growing from 200 million people during the time of the Ancient Greeks to over 1 billion by 1798, Malthus instead laid out his arguments for why every additional soul added to the planet after 1798 would simply divide the planet's resources into ever thinner slices until civilization itself collapsed under the strain. He calculated how long until England would run out of farmland. And he postulated that poor people would abuse the abundance delivered by science and technology to simply increase the size of their families, thereby cancelling out any temporary headway made by innovation.

Malthus argued that *"the aggregate mass of happiness"* would be better served without England's Poor Laws, which provided life-saving assistance to those trapped in extreme poverty.[15] Malthus also warned against allowing individuals to save too much money, arguing that *"the principles of saving, pushed to excess, would destroy the motive to production"* and so, to maximize the productive output of a nation, a nation had to balance *"the power to produce and the will to consume."*[16] In other words, if the poor were allowed to get too rich and build up too many savings, it would undermine the collective productivity of the nation because those who live in fear of starvation will work harder than those who don't. Thus, by Malthus' logic, maximizing the productive output of

---

[15] Kagan, J. (2024, Apr 16). Who Is Thomas Malthus? *Investopedia.* https://www.investopedia.com/terms/t/thomas-malthus.asp

[16] MacRae, D. G.. "Thomas Malthus." Encyclopedia Britannica, 16 May. 2024, https://www.britannica.com/money/Thomas-Malthus.

the poor by keeping them running on a treadmill is the most humane way to maximize the nation's ability to feed its growing population.

Famed economist John Maynard Keynes cited Malthus' economic writings as inspiration for his own work,[17] in which he argued that in a healthy economy citizens must be incentivised to either spend more than they save or invest more than they save, and that in order to create jobs the government should increase spending even if it means going into debt. He also was instrumental in arguing for the introduction of paper money (fiat currency) and the abolishment of the gold standard, calling it a "barbarous relic" because a commodity-backed currency limited the government's ability to inflate in the monetary supply and prevented the government from running the printing presses during economic slumps or during times of war. In short, the "barbarous relic" was an obstacle to the orgy of spending demanded by progressive thinkers and your savings account was holding back the economy.

Unsurprisingly, Keynes monetary theories were embraced by the progressive administrative state (and by war planners, wall-street investors, politicians with ambitious social programs, and so on). By the 1930s, Keynesianism had become the guiding monetary philosophy of the U.S. Federal Reserve, and it has continued to influence modern monetary theory throughout the Western World even today as Woodrow Wilson's cherished administrative state continues to impose itself as the master of the world's financial system.

In the world of banking, 1913 marks the date when the invisible hand of the marketplace was replaced by the visible hand of the government financial planner. It also marks the date when debt-financed federal spending slowly began eclipsing the private economy, creating the federal money trough that has become a magnet for an ever-greater number of pigs. Today, government spending is on the verge of eclipsing private spending in the economy. According to the IMF, government spending in France accounts for more than 58% of GDP, Germany comes in at 49%, the United Kingdom comes in at 44%, and in the United States it sits at

---

[17] Rutherford, R. P. (1987). Malthus and Keynes. *Oxford Economic Papers*, 39(1), 175–189. http://www.jstor.org/stable/2663135

36%.[18] But even these numbers fail to account for the complete picture. As Canada's MacDonald Laurier Institute pointed out, although government spending in Canada officially only accounts for 44% of GDP, by the time you add in compliance costs and the impact of tax subsidies that number rises to more than 64% of GDP![19]

And the debts have risen in lockstep. Figure 10 below shows how U.S. public debt has increased as a percentage of GDP ever since Wilson created the Federal Reserve as the keystone of his cherished hyperactive administrative state. It's an endless vicious circle: the bigger the public trough, the more pigs it attracts, the bigger the trough must get to satisfy all those pigs' demands, and the louder calls get for an increase in the size of the trough as everyone demands their "fair share" of the spoils.

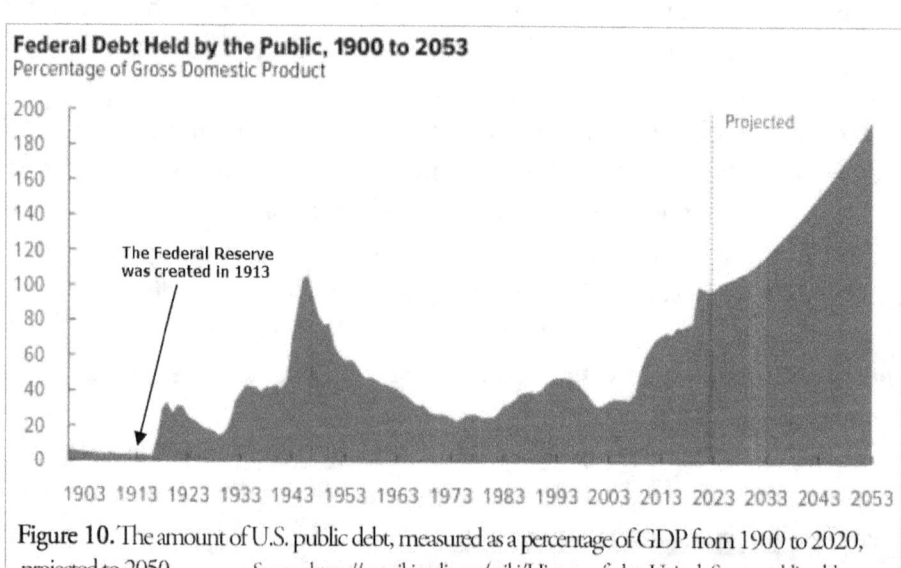

Figure 10. The amount of U.S. public debt, measured as a percentage of GDP from 1900 to 2020, projected to 2050. Source: https://en.wikipedia.org/wiki/History_of_the_United_States_public_debt

Easy money, endless debt, and relentless money printing has all but destroyed the culture of saving, while creating a policy of persistent low-level inflation (most Western nations have a target set of between 1.5 to 2% inflation per year — which they frequently exceed), which continually

---

[18] "Government expenditure, percent of GDP." *International Monetary Fund*. Accessed 9 Jun 2024.
https://www.imf.org/external/datamapper/exp@FPP/USA/FRA/JPN/GBR/SWE/ESP/ITA/ZAF/IND

[19] "Size of government in Canada." *MacDonald Laurier Institute*. Accessed 9 Jun 2024.
https://macdonaldlaurier.ca/size-of-government-in-canada/

robs people of their purchasing power, erodes their savings, and encourages risky investments to stay ahead of the inflation curve. These are all the consequences of Keynesian monetary theory and the embrace of government-controlled money, which has inflated away 97% of the value of the U.S. dollar over the 110-year life of the Federal Reserve. A dollar in 2024 only has the equivalent purchasing power of what three cents could buy in 1913. In 1995, following his tenure as Chairman of the Federal Reserve, Paul Volker acknowledged the Federal Reserve's power to destroy money through inflation: *"It is a sobering fact that the prominence of central banks in this century has coincided with a general tendency towards more inflation, not less. If the overriding objective is price stability, we did better with the 19th century gold standard and passive central banks, with currency boards, or even with 'free banking'. The truly unique power of a central bank, after all, is the power to create money and, ultimately, the power to create is [also] the power to destroy."*

With the influence of Malthus's theories on state planners and with the goal to try to optimize the economy for the collective good, free market capitalism was systematically replaced by state-managed capitalism and state-managed money. Welcome to the treadmill of the modern rat race, where inflation is used as a deliberate policy to maximize the output of the poor and erode their savings even as it inflates the assets of investors and inflates away government debts. There's a philosophical thread that leads directly from Malthus's dismal science, through Woodrow Wilson's establishment of the Federal Reserve, through Keynes' advocacy for an economic system run by government fiat and public debt, to the treadmill of the modern rat race — these are the foundations of our modern world.

Today we stand on the threshold of the next chapter of that story as central banks around the world prepare to roll out central bank digital currencies (CBDCs), with unprecedented new opportunities to use money as a means for control and coercion in service of some centrally planned vision of the "greater good". The United Nations has made no secret of the key role it sees for central bank digital currencies in achieving its Sustainable Development Goals as it seeks to reform the global financial system to create a *"society where living conditions and resources meet human needs without undermining the planetary integrity and stability of the natural system.*"[20] Even the World Bank has spilled considerable ink explaining the importance of central bank digital currencies in helping the world manage its "climate risks" and create a long-term "sustainable" global economy.[21] It's nothing less than Woodrow Wilson's utopian optimism about harnessing the coercive power of the administrative state, applied to the finances of the entire planet. Back in ancient Greece, Plato once said, *"Let me make the songs of a nation, and I care not who makes its laws"*; in 1790, banker Mayer Anselm Rothschild echoed Plato when he said, *"Permit me to issue and control the money of a nation, and I care not who makes its laws."*

Malthus' reach doesn't stop there. His theories and the overpopulation fears raised by his economic formulas were used during the 19th and early 20th century to justify many colonial policies, not least of which was the 19th-century race to secure land and resources through territorial expansion (i.e., the Scramble for Africa) in a bid by European nations to escape their Malthusian destiny as their populations outgrew their domestic agricultural capacities (and as soil degradation undermined their domestic food output — more on that later). Only later, as the bills for running a colonial empire began to stack up did these colonial powers discover that colonial empires frequently cost the home country as much or more to maintain than what they brought back in terms of resources and revenues, and that it was far cheaper to simply buy things from other countries at market prices than to try to force their resentful colonial subjects to keep selling commodities to the home country at below-market prices.

Grimmer still was Malthus' influence on the British response to the Irish Potato Famine. When blight infected European potato crops during the 1840s, Ireland was particularly hard hit. As famine set in, the British declined to send aid. Malthus' theories provided moral and intellectual cover to blame the disaster on Irish overpopulation rather than on exploitative political decisions imposed by England, which had encouraged landowners to clear the land of tenant farmers in order to

---

[20] "Sustainable development." *Wikipedia.* 7 Jun 2024.
https://en.wikipedia.org/wiki/Sustainable_development

[21] Lee, S. and Park, J. (2022). Environmental Implications of a Central Bank Digital Currency (Analytical Note). *World Bank Group.*
https://documents1.worldbank.org/curated/en/099143507042228192/pdf/IDU011 0707310d28b041b70bb560cc1f4b1f41d4.pdf

convert those lands into plantations for commodity crop production for export back to England. In essence, a cow or sheep destined for export was worth more to a landlord than a crop-sharing agreement with a tenant farmer, so the tenant farmers had to go. (A similar expulsion of tenant farmers happened in Scotland in the 18th and 19th centuries, known as the Highland Clearances.)

This left the dispossessed and destitute Irish population to fend for themselves on tiny subsistence plots on marginal lands even as landlords exported vast quantities of beef, pork, and other crops to England. The average size of these subsistence plots was less than half an acre — a dismal postage-stamp-sized allotment of land that will become all too familiar by the time this book reaches its conclusions. The potato was the only crop capable of producing enough calories to sustain a family on such a tiny plot of land, yet it was a starvation diet utterly devoid of meat protein even at the best of times and a recipe for certain death when potato blight arrived to destroy the last thin thread by which these impoverished families clung to life. Between 1845 and 1852, over 1 million Irish starved to death even as Irish exports of beef, pork, and other agricultural commodities surged over the same time period. Another 2.1 million people fled Ireland (many to America). This episode in Irish history is known as the Great Hunger. Malthus provided the lens through which 19th-century nations viewed this crisis. The hardest heart belongs to the man who has the approval of his own conscience. Beware of those who believe in a "finite world with finite resources" — there are no limits to their willingness to "help" you make the sacrifices they think you should make for the collective good.

Malthus' influence on 20th century thinking proved to be even deadlier. Germany was late to the Malthusian game of colonial expansion but then, in 1918, Allied powers forcibly stripped Germany of its recently acquired colonies following Germany's defeat in WWI. This was followed by the catastrophic hyperinflation triggered by Germany's central bank when it tried to use the printing press to outrun its towering war debts, which left large segments of the German population destitute, hungry, utterly disillusioned with liberal democracy, and in search of a strong central planner capable of rescuing Germany's sinking economy. And then, following the start of the Great Depression in 1929, the U.S. inadvertently exported the Great Depression to Europe when it called in its loans to cover losses triggered by the collapsing U.S. stock market, which sucked the last glimmers of hope out of the German economy and left a full one-third of all German households unemployed and without any means to feed their families. All this served to intensify already present Malthusian fears in Germany about running out of the necessary resources and agricultural land required to feed its growing population and secure a sustainable future for the German people.

Thus, in late 19th- and early 20th-century Germany, scientifically-endorsed Malthusian fears of an expanding population were woven together with Darwinian ideas about the evolution of societies/races/cultures competing for space and resources with other societies/races/cultures, which led to the German concept of *Lebensraum* — a concept best defined as the geographic footprint required for the survival of a nation bound together by a shared culture and shared history. Acquiring *Lebensraum* for the German people became the leading geopolitical obsession of the Nazi Regime. It was used to justify Nazi Germany's annexation of other German-speaking regions in Central and Eastern Europe, the series of decisions that led to Germany's invasion of its non-German neighbors (underpinned by the *Drang nach Osten,* a.k.a. the eastward push to acquire *Lebensraum* in the East), and ultimately to justify the genocidal death camps used to exterminate Jews, gypsies, homosexuals, political dissidents, and anyone else they deemed as an obstacle to their vision of a thriving German nation.

Antisemitism was hardly a 20th-century German invention — persecution and pogroms against Jews have been a tragic universal constant throughout European and indeed world history. But through the lens of *Lebensraum* and in combination with society's unquestioning faith in state planning, antisemitism took on a whole other dimension. Historian Niall Ferguson documents how in the 1920s (even before Hitler came to power) European intellectuals as a collective group wholeheartedly embraced extreme nationalism and racism: *"A university degree, far from inoculating Germans against Nazism, made them more likely to embrace it. [...] Non-Jewish German academia did not just follow Hitler down the path to hell. It led the way."* Even the *"'final solution of the Jewish question' began as speech—to be precise, it began as lectures and monographs and scholarly articles. It began in the songs of student fraternities. With extraordinary speed after 1933, however, it crossed into* conduct: first, systematic pseudo-legal discrimination and ultimately, a program of technocratic genocide."[22]

Once a belief takes a hold of a population, it becomes the lens through which to focus all of a nation's preoccupations. Far from being the driver of national socialism in Germany, Hitler was merely the crucible that expressed all the preoccupations of his era, a ruthless salesman catering to the ethos of his time and exploiting a public mania that began in the halls of academia. Portraying Hitler as a uniquely evil man rather than as the political expression of the Zeitgeist of his era absolves western society as a whole and public intellectuals in particular of their role in creating the horrors of that age. True evil is so much more than the beliefs of any single demented mind; true evil emerges from *any* society that endorses any system of governance that enables those in power (well intended or otherwise) to harness the power of the state to impose their beliefs onto others. State planning always begins with the belief that "if you want to make an omelette, you have to be prepared to break some eggs" and ends in the unthinkable as ends justify increasingly authoritarian means, especially once the search for scapegoats begins when the promised utopia fails to materialize according to plan. The U.S Founding Fathers had good reasons to create a labyrinth by which to restrain the hand of government — until society rediscovers those reasons and resurrects those limits in the political architecture of their nations, we will continue to stumble from one intellectually-endorsed mania to the next as state planners give themselves license to act upon the preoccupations, manias, and conspiratorial manipulations of their eras.

In a twist of bitter irony, Germany's population today is 27% larger yet its territory is only about half the size that it was back in 1933 when the Nazis came to power by preaching about *Lebensraum*. And yet, today Germany's public health officials are battling an epidemic of obesity rather than the Malthusian prediction of scarcity and famine even as Germany has become both the world's third largest *export* nation and the world's third largest food exporter (according to the World Bank). Meanwhile, Germany's birthrates have fallen far below population replacement rates simply because Germany became a wealthy nation. The

---

[22] Ferguson, N. (2023, Dec 11). Niall Ferguson: The Treason of the Intellectuals. *The Free Press*. https://www.thefp.com/p/niall-ferguson-treason-intellectuals-third-reich

incentives create the system. Prosperity turns children into a financial liability instead of a source of cheap labour. If you want to curb birthrates and increase food supplies, give people inviolable property rights and then get out of the way so they can begin accumulating wealth through their own hard work — this is the key point that Malthus missed in his dismal population calculations and the point that state planners everywhere cannot seem to grasp, as grasping it would alleviate taxpayers of the burden of having to give them a job. As novelist Upton Sinclair once wisely observed, *"it is difficult to get a man to understand something when his salary depends on his not understanding it."* Over and over again, we see that prosperity, innovation, and a strict commitment to property rights defy the apocalyptic Malthusian predictions of those who see the world as a "finite planet with finite resources".

The Nazis came and went, and intellectuals emerged as champions of a new vision of global multinational cooperation (under the stewardship of the USA and the United Nations, and based on an updated vision of Woodrow Wilson's liberal internationalism) in the hope that this new co-operative global vision would safeguard the world from backsliding into the authoritarian horrors of aggressive 20th century nationalism — henceforth, by hook or by crook, national competition would be replaced by global collaboration, and woe to anyone who fails to sign up to the collaborators' vision of that collaborative US- and UN-led future. And so, the progressive ethos evolved from aggressive nation-building to a new era of aggressively building a collaborative international community, with a network of US-led intranational organizations emerging to facilitate this new globalist vision. In essence, WWII discredited progressive nationalism but replaced it with progressive internationalism (a.k.a. globalism), which is just progressivism with an even grander purpose (and requires even grander institutions).

As progressivism evolved, the standard bearers of Malthus' eternal pessimism evolved right along with it. Enduring fears about overpopulation and running out of national resources simply reinvented themselves as a global rather than as a national problem as the global jet-setting academic intelligentsia refocused their efforts on educating the world about the countdown to the day when the planet would be overrun by humans and unravel into apocalyptic scarcity. By dissolving the boundaries between nations, the policy failings of individual nations became invisible even as the consequences of those failings became the responsibility of the planet as a whole.

1948 saw the publication of two influential books, Fairfield Osborn's *Our Plundered Planet* and William Vogt's *Road to Survival*, which portrayed human beings as the destroyers of the natural world. These books served as inspiration for the apocalyptic environmental literature that was yet to come. In 1968, ecologist Garret Hardin published an influential essay entitled *The Tragedy of the Commons*, in which he argued that *"a finite world can only support a finite population"* and that the *"freedom to breed will bring ruin to all"*.

Meanwhile, starting in the 1960s, the development of chemical fertilizers, innovations in plant breeding (propelled in part by using gamma radiation to speed up random mutations in plant genomes to create more productive, disease-resistant, and drought-resistant plant varieties), as well as innovations in irrigation and machinery led to an explosion of agricultural yields, nicknamed the Green Revolution, which turned a world of scarcity into a world of plenty in spite of our growing population. American agronomist Norman Borlaug, known as the "father of the Green Revolution" is singlehandedly credited for having prevented over one *billion* deaths because of the high-yielding, disease-resistant wheat varieties he developed while working in an agricultural research position at the non-profit International Maize and Wheat Improvement Center in Mexico (it hosts the largest maize and wheat gene bank in the world). Yet few people have ever heard of Norman Borlaug – a scientific giant among giants – and his contributions to the world. They have, however, heard plenty from Paul Ehrlich.

In 1968, American biologist Paul Ehrlich published his infamous book, *The Population Bomb*. It became a best-seller with more than three million copies sold over the next decade. It painted an apocalyptic picture that predicted worldwide famines due to overpopulation, accompanied by massive societal upheavals that could only be avoided by aggressive policies designed to severely limit population growth and curtail resource use. Ehrlich famously predicted that *"The battle to feed humanity is over. In the 1970s, the world will undergo famines. Hundreds of millions of people are going to starve to death in spite of any crash programs* embarked upon now. Population control is the only answer."²³ However, the prophesized mass starvation failed to materialize, famine and poverty continued to retreat around the world, and global birthrates began to fall quite naturally. As societies get richer, they naturally choose to have less children as savings replace children as a retirement strategy and as the financial liability of school fees and grocery bills outweigh the benefits of being able to extract free labour from one's children.

Nevertheless, Ehrlich went on to win one prestigious award after another, including the prestigious 1980 Crafoord Prize, the Royal Swedish Academy of Sciences' highest award. His book's alarmism reverberated around the globe, especially among elites and policymakers who view themselves at the natural stewards of humanity. Prince Phillip, husband of Queen Elizabeth II and founder of the World Wildlife Fund, famously declared in 1989 that, *"If I were reincarnated, I would wish to be returned to Earth as a killer virus to lower human population levels."*²⁴ Charming.

The global intelligentsia took their self-appointed role as the stewards of humanity very seriously. An influential informal organization of intellectuals and global business leaders called the Club of Rome was founded in 1968, the same year as *The Population Bomb* hit bookshelves, in order to discuss pressing global issues and overpopulation in particular. This think-tank's first report, published in 1972 to considerable media attention, was called *The Limits of Growth*. Environmental groups leveraged the fears expressed by the report to fuel their fearmongering messages and keep donations flowing. The report inspired China to impose its cruel and ultimately self-destructive one-child policy as part of Chairman Deng Xiaoping's post-Maoist reforms, which were designed to modernize China according to modern economic and scientific principles. Once again, policy oriented towards the attention-seeking doom-mongering scientists that are so beloved by newspaper editors and politicians while innovators like Norman Borlaug are forgotten even as

---

²³ Williams, W. (2010, Mar 17). Walter Williams: Global warmers apparently don't intend to go quietly. *The State Journal-Register*. https://www.sj-r.com/story/opinion/columns/2010/03/17/walter-williams-global-warmers-apparently/42946506007/

²⁴ Be aware of the extremists who populate environmentalism's history." *The Athens News*. Mar 9, 2006. https://www.athensnews.com/news/local/be-aware-of-the-extremists-who-populate-environmentalisms-history/article_aded7c7f-4e30-564e-8160-beacbfe2e3c8.html

they work tirelessly in the anonymity of their laboratories, fields, and workshops.

The Club of Rome's influence also led to the establishment of the United Nations Environmental Program (UNEP). The program was established by one of the Club of Rome's associates, a wealthy Canadian businessman named Maurice Strong. Under his leadership, UNEP launched the era of international environmental diplomacy and organized the never-ending string of annual UN Environmental Conferences hosted in exotic locations so wealthy businessmen and politicians can travel halfway around the world on their private jets to lecture us about our carbon footprint. Amusing lectures aside, it's also where our leaders are slowly pledging away ever-growing slices of our national sovereignty in increasingly binding commitments to the United Nations' "sustainability goals" for a new global development framework (i.e. Agenda 21, the Kyoto Accords, Copenhagen (COP 15), Paris (COP 21), Agenda 2030, 30 by 30, and so on). During the Rio Earth Summit in 1992, organized by UNEP under Maurice Strong's leadership, Strong commented that, *"The concept of national sovereignty has been an immutable, indeed sacred, principle of international relations. It is a principle which will yield only slowly and reluctantly to the new imperatives of global environmental cooperation. It is simply not feasible for sovereignty to be exercised unilaterally by individual nation states, however powerful. The global community must be assured of environmental security."*[25] In 2009, Strong doubled down on his 1992 criticism of national sovereignty in an essay in which he stated, *"Our concepts of ballot-box democracy may need to be modified to produce strong governments capable of making difficult decisions..."*[26]

Strong grew up during the Great Depression and later said that the Depression radicalized him to become *"a socialist in ideology, a capitalist in methodology."* Strong went on to serve as advisor to Kofi Annan, Secretary-General of the United Nations; Annan praised Strong for his "global vision" and "wise counsel". In yet another 1992 interview, Strong

---

[25] Balaam, D. N. and Dillman, B. (2015). *Introduction to International Political Economy.* Routledge Publishing.
https://books.google.ca/books?id=gvIvCgAAQBAJ&pg=PA535&redir_esc=y#v=onepage&q&f=false

[26] "Maurice Strong." *Wikipedia.* 9 Jan 2024.
https://en.wikiquote.org/wiki/Maurice_Strong

outlined the plot for a book he would like to write, stating that, *"What if a small group of world leaders were to conclude that the principal risk to the Earth comes from the actions of the rich countries? And if the world is to survive, those rich countries would have to sign an agreement reducing their impact on the environment. Will they do it? The group's conclusion is 'no'. The rich countries won't do it. They won't change. So, in order to save the planet, the group decides: Isn't the only hope for the planet that the industrialized civilizations collapse? Isn't it our responsibility to bring that about?"*[27]

Meanwhile, at the same time as Strong was organizing climate change conferences for the United Nations and fantasizing about saving the world by bringing about the collapse of rich industrialized nations, Zimbabwe gained its independence (in 1980) and immediately imposed socialist central planning under its vicious new strongman dictator, Robert Mugabe. Mugabe self-identified as a Marxist-Leninist and overtly called for Zimbabwe to be transformed into a Marxist one-party state,[28] which he mostly achieved after conducting a brutal terrorist campaign (with Soviet and Chinese support) against both black and white Rhodesians in the years before Independence, which quickly escalated into genocide (known as the 'Gukurahundi'[29]) against the rival Ndebele tribe as soon as he came into power — this well-documented genocide was conducted *"under Mugabe's explicit orders,"*[30] even as the US (under presidents Reagan, HW Bush, Clinton, and GW Bush) looked the other way and even made complimentary speeches about him,[31] [32] which raises some

---

[27] "Maurice Strong." *Wikipedia.* 9 Jan 2024. https://en.wikiquote.org/wiki/Maurice_Strong

[28] "Robert Mugabe." *Wikipedia.* 27 Jul 2024. https://en.wikipedia.org/wiki/Robert_Mugabe

[29] Nyoka, S. (2024, Jul 12). I cannot forgive Mugabe's soldiers – massacre survivor. *BBC News.* https://www.bbc.com/news/articles/c51yl354ed2o

[30] Dzirutwe, M. (2019, Sept 6). Mugabe's legacy: thousands killed in 'rain that washes away the chaff'. *Reuters.* https://www.reuters.com/article/us-zimbabwe-mugabe-violence-idUSKCN1VR18H/

[31] "How the United States Supported White Genocide in Southern Africa." (2023, Aug 9). *The American Tribune.* https://www.theamericantribune.news/p/how-the-united-states-supported-white

[32] Reagan Library. (2017, Jul 11). *President Reagan with Prime Minister Mugabe of Zimbabwe on September 13, 1983.* [Video]. YouTube. https://youtu.be/EjYmnK6FhJE?si=wIrqNVinsPdqhAs_

uncomfortable questions about the priorities, objectives, and ideological predispositions among our political classes and the bureaucratic institutions advising them. And so, a country that had long been known as the breadbasket of Africa because of its deep, rich soils and its vast surpluses of corn, wheat, and other agricultural crops was rapidly turned into an economic basket case, utterly dependent on food aid grown in the very same rich industrialized nations that Maurice Strong had singled out as the principal threat to the survival of the planet.

As Mugabe's central planners strangled Zimbabwe's economy, gutted its agricultural production with socialist diktats, plundered the nation to the benefit of their own offshore bank accounts, adopted land management policies that turned grasslands into deserts, triggered hyperinflation that left the entire country utterly impoverished, and waged genocide against its own peoples, Mugabe turned to blaming global warming as the reason for his country's demise and denounced capitalism as the root cause of that global warming. In 2009, the United Nations invited Mugabe (along with fellow Venezuelan dictator and self-described Marxist, Hugo Chávez[33]) to speak at UNEP's 2009 Climate Change Conference in Copenhagen (COP 15) where they used the podium to attack capitalism and ranted against capitalism as the root cause of climate change — Chávez even gave the opening speech.[34] In 2017, the United Nations even appointed Mugabe to serve as a World Health Organization goodwill ambassador. The WHO's head, Dr. Tedros A. Ghebreyesus, praised Mugabe for his commitment to universal health coverage even as critics pointed out that despite Zimbabwe's utopian commitments, its entire healthcare system had collapsed with staff going unpaid and medicines in critically short supply.[35] It is also worth noting that before the United Nations appointed Dr. Tedros to serve as the head of the WHO, Tedros was a member of Ethiopia's Marxist-Leninist terrorist organization (the Tigray People's Liberation Front (TPLF)), which violently overthrew the Ethiopian government in 1991, and that he then served as one of three officials in charge of the TPLF's security forces from 2013 to 2015 and as

---

[33] "Hugo Chavez." *Wikipedia*. 29 Jul 2024. https://en.wikipedia.org/wiki/Hugo_Chávez

[34] "Mugabe blames capitalism for climate woes." *Independent Online (IOL)*. Dec 16, 2009. https://www.iol.co.za/mercury/world/mugabe-blames-capitalism-for-climate-woes-467979

[35] "Robert Mugabe's WHO appointment condemned as 'an insult'". *BBC News*. Oct 21, 2017. https://www.bbc.com/news/world-africa-41702662

Ethiopia's foreign minister until 2016, during which the TPLF's security forces were accused of widespread ethnically motivated killings, arbitrary detentions, and the torture of the Amhara people of Ethiopia.[36] [37] The Amhara Professionals Union considers him to be a genocide suspect[38] and Nobel peace prize nominee David Steinman has lodged an official complaint with the International Criminal Court in the Hague calling for Dr. Tedros to be prosecuted for genocide, as reported by the Daily Mail in 2020.[39]

~

Back in the 1970s, the eternal optimism that scientists would eventually crack the secret of how to control the weather ("designer weather on demand") gradually sputtered out. Bloated U.S. weather modification budgets were slashed because, other than some limited success with cloud seeding and eliminating fog, they were largely ineffective while creating a nightmare of unintended consequences and legal liabilities. What worked in the controlled conditions of the lab didn't work so well in the uncontrolled conditions of nature (artificially produced rain only fell if it was already raining naturally in the immediate vicinity). There were simply too many variables at work in a natural ecosystem — you can't control what you don't fully understand. If only politicians had listened to the meteorologists back in the 1940s when they pointed out that it's unrealistic to think weather modification schemes could control the weather before meteorologists fully understood what makes and controls the behaviour of clouds. However, just in case someone else figured out a reliable weather modification recipe, in 1977 an international treaty was

---

[36] Banyan, R. (2020, Dec. 14). WHO chief Tedros Ghebreyesus is accused of aiding genocide in Ethiopia: Nobel peace prize nominee lodges complaint with International Criminal Court. *Daily Mail.* https://www.dailymail.co.uk/news/article-9052247/WHO-chief-Tedros-Ghebreyseus-accused-aiding-genocide-Ethiopia-nobel-peace-prize-nominee.html

[37] Josh Walkos. (2024, Jun 17). *The Troubled History of Tedros.* Twitter. https://x.com/JoshWalkos/status/1802743890899783923

[38] "International Organizations Leadership Recruitment Policies: the Failed Experiment of Dr. Tedros A. Ghebreyesus Candidacy for WHO Director General Position." (2017, Apr.). *Amhara Professionals Union Research Department.* https://www.moreshwegenie.org/sites/default/files/multimedia/APU_opposes_candidacy_of_Dr_TAG.pdf

[39] Banyan, R. (2020, Dec. 14).

signed that banned weather warfare and all hostile uses of weather/climate modification techniques.

Where weather optimism had failed, climate pessimism was already charging in to soak up the bonanza of state funding. Climatology emerged in the 1960s and 1970s as a separate subdiscipline of the Earth Sciences, distinct yet overlapping with other Earth Sciences like geology, astronomy, metrology, and oceanography, all of which study Earth at a planetary level. And, right on cue, a bitter debate began as to (1) whether our industrial greenhouse gases would tip our climate into a civilization-threatening Hothouse Earth scenario, (2) whether natural climate cycles were about to tip us into the next civilization-threatening Ice Age, or (3) whether our industrial air pollution would dim the sun and trigger a new civilization-threatening Ice Age. Optimistic dreams of modifying the local weather for our benefit were replaced by the pessimistic belief that the global climate was teetering on the edge of the abyss — hot or cold was to be decided later. And so, dreams of controlling the local weather seamlessly transitioned into dreams about global climate control. To keep our global climate in the Goldilocks Zone (not too hot and not too cold), we allegedly had no choice but to find ways to limit our industrial impact on the planetary ecosystem. The Malthusian doomsday cult and the environmentalists finally found a suitable partner to try to get the world to pay attention to their warnings about humanity's growing unsustainable impact on the planet — and so, the climate crusade was born even as other environmental concerns were pushed into the background.

It became apparent from weather station records that from 1940 to 1970 the Earth had experienced massive cooling. In 1969, environmentalist Nigel Calder warned that, *"The threat of a new ice age must now stand alongside nuclear war as a likely source of wholesale death and misery for mankind."*[40] In 1972, the Department of Geological Sciences at Brown University wrote a letter to President Nixon warning that, *"Dear Mr. President: ...We feel obliged to inform you on the results of the scientific conference held here recently. ... The main conclusion of the meeting was that a global deterioration of climate, by order of magnitude larger than any hitherto experienced by civilized mankind, is a very real possibility*

---

[40] Bratby, P. (2008, May 15). Memorandum by Dr. Phillip Bratby. *UK government* (www.parliament.uk). https://publications.parliament.uk/pa/ld200708/ldselect/ldeconaf/195/195we07.htm

and indeed may be due very soon. *The cooling has natural causes and falls within the rank of processes which produced the last ice age. ... The present rate of the cooling seems fast enough to bring glacial temperatures in about a century...*"[41] In the July 1975 issue of National Wildlife, C.C. Wallen of the World Meteorological Organization proclaimed, *"The cooling since 1940 has been large enough and consistent enough that it will not soon be reversed."*[42] In 1978, the New York Times wrote, *"An international team of specialists has concluded from eight indexes of climate that there is no end in sight to the cooling trend of the last 30 years, at least in the Northern Hemisphere."*[43]

The media amplified the message. For example, the front cover of the March 1975 issue of Science News asked whether, *"The Ice Age Cometh?"* as it showed glaciers plowing through the skyscrapers of Manhattan (see figure 11).

---

[41] Maximus, F. (2009, Oct 21). An Important Letter Sent to the President about the Danger of Climate Change. *The Street*. https://www.thestreet.com/economonitor/us/an-important-letter-sent-to-the-president-about-the-danger-of-climate-change

[42] Williams, W. (2010, Mar 17). Walter Williams: Global warmers apparently don't intend to go quietly. *The State Journal-Register*. https://www.sj-r.com/story/opinion/columns/2010/03/17/walter-williams-global-warmers-apparently/42946506007/

[43] Sullivan, W. (1978, Jan 5). International Team of Specialists Finds No End in Sight to 30-Year Cooling Trend in Northern Hemisphere. *New York Times*. https://www.nytimes.com/1978/01/05/archives/international-team-of-specialists-finds-no-end-in-sight-to-30year.html?_r=1

Figure 11. From the cover of *Science News* on March 1st, 1975.

Source: www.sciencenews.org/sn-magazine/march-1-1975

The Department of Geological Sciences at Brown University, which blamed natural forces, were drowned out by those blaming the cooling on pollution, soot, and aerosols.[44] Just as the environmental movement had raised awareness about humanity's ability to destroy local ecosystems through polluting industrial activities, now that idea was expanded to the climate of the entire planet. In 1970, the Boston Globe wrote, *"Scientist Predicts a New Ice Age by 21st Century: Air pollution may obliterate the sun and cause a new ice age in the first third of the next century. ... If the current rate of increase in electric power generation continues, the demands for cooling water [for operating the generators] will boil dry the entire flow of the rivers and streams of the continental United States. ... By the next century 'the consumption of oxygen in combustion processes, world-wide, will surpass all of the processes which return oxygen to the atmosphere."*[45] The unholy alliance between climate science, Malthusian fears about overpopulation, and doom-mongering environmental activism was complete.

Meanwhile, in the early 1960s, scientists set about to measure atmospheric carbon dioxide levels and discovered that they were rising sharply. Scientists took notice and began to speculate whether CO_2 emissions produced by our industrial activity could produce a strong enough greenhouse gas effect to raise global temperatures — the Hothouse Earth scenario. Meanwhile, global temperatures were falling. Undeterred, in 1972 a collection of 31 essays by American scientists was published, called *Global Ecology, Readings Toward a Rational Strategy for Man,* in which the competing scenarios were reconciled.[46] The aforementioned eternal pessimist Paul Ehrlich and environmental scientist John Holdren served as both contributors and editors of the publication. In one of the essays, Holdren wrote that *"it seems however that a competing effect has dominated the situation since the 1940s. This is the reduced transparency of the atmosphere to incoming light as a result of*

---

[44] Struck, D. (2014, Jan 10). How the "Global Cooling" Story Came to Be. *Scientific American.* https://www.scientificamerican.com/article/how-the-global-cooling-story-came-to-be/

[45] Ayres, J. (1970, Apr 16). Scientist predicts a new ice age by 21st century. *The Boston Globe.* https://www.newspapers.com/article/the-boston-globe-new-ice-age-forecast/36613964/?locale=en-CA

[46] Holdren, J. P. (Ed.). (1971). *Global Ecology: Readings Toward a Rational Strategy for Man.* Harcourt Brace Jovanovich.

urban air pollution (smoke, aerosols), agricultural air pollution (dust), and volcanic ash. This screening phenomenon is said to be responsible for the present world cooling trend – a total of .2C over the past quarter century."* Holdren went on to say that we are compounding this cooling effect by increasing cloud cover by adding man-made condensation nuclei to the atmosphere through jet exhausts and other pollutants and by increasing the reflectivity of the earth's surface (albedo) through urbanization, deforestation, and the enlargement of deserts. In other words, Holdren was pointing at the sum of all our industrial activities as the primary driver for the global cooling that had occurred from the 1940s to the 1970s while simultaneously pointing at our CO_2 emissions as having a competing heating effect. It's a wonderfully safe bet when, no matter what happens next, it's our fault and, in typical human fashion, whatever happens next is projected indefinitely out into the future to create a terrifying doomsday scenario that can only be prevented by following the orders of the guy holding the microphone and wearing the lab coat.

Holdren's article was building upon another scientific article co-authored by Holdren and Ehrlich a year earlier, in 1971, and published in the prestigious peer-reviewed academic journal *Science*. In this article the gloomy pair discussed remedies to the world's supposed problems, stating that *"population control is obviously not a panacea – it is necessary but not alone sufficient to see us through the crisis. [...] Precisely because population is the most difficult and slowest to yield among the components of environmental deterioration, we must start on it at once. To ignore population today because the problem is a tough one is to commit ourselves to even gloomier prospects 20 years hence, when most of the "easy" means to reduce per capita impact on the environment will have been exhausted. The desperate and repressive measures for population control which might be contemplated then are reason in themselves to proceed with foresight, alacrity, and compassion today."*[47]

The final article featured in Holdren and Ehrlich's aforementioned 1972 *Global Ecology* publication was authored by Ehrlich. It was called "*Looking backward from 2000 AD*". It painted a picture of the world as he believed it would look by the year 2000 if drastic steps to deindustrialize and depopulate were not taken immediately to prevent famine and

---

[47] Ehrlich, P. R., & Holdren, J. P. (1971). Impact of Population Growth. *Science*, 171(3977), 1212–1217. http://www.jstor.org/stable/1731166

ecological collapse. The article predicted that if the world failed to implement their recommendations to prevent their predicted ecological changes, by the year 2000 the population of the United States would crash by 90% to 20 to 25 million (the population in the USA in 1970 was 203 million).[48]

I've already discussed how influential Ehrlich's pessimistic world view was, but it is worthwhile noting that Holdren later became President Obama's senior science advisor on science and technology issues. His beliefs directly influenced policy.

Likewise, despite Ehrlich's track record of predictions that never come true, he also remains deeply influential even today. In a CBS article published in 2023 called *"Scientists say planet in midst of sixth mass extinction, Earth's wildlife running out of places to live"*, an unrelenting 90-year-old Paul Ehrlich is quoted declaring that *"humanity is not sustainable. To maintain our lifestyle (yours and mine, basically) for the entire planet, you'd need five more Earths. Not clear where they're gonna come from."*[49]

Back in the 1970s, policymakers and scientists discussed outlandish Earth-saving proposals to save the world from the deep freeze of the impending Ice Age. One such proposal envisioned using aircraft to dump black soot on the ice pack in order to try to melt sea ice in the Arctic Ocean.[50] Another proposal envisioned warming the Arctic Ocean by building a 90-kilometer-wide dam across the shallow Bering Strait in order to block cold Pacific water from entering the Arctic Ocean, and then using nuclear power to run giant pumps to suck cold water from the Arctic Ocean into the Pacific, which would effectively increase the amount of warm water flowing into the Arctic from the Atlantic Ocean via the Gulf Stream.[51]

---

[48] Ehrlich, P. R., & Holdren, J. P. (1971). Impact of Population Growth. *Science*, 171(3977), 1212–1217. http://www.jstor.org/stable/1731166

[49] Pelley, S. (2023, Jan 1). Scientists say planet in midst of sixth mass extinction, Earth's wildlife running out of places to live. *CBS News (60 minutes)*. https://www.cbsnews.com/news/earth-mass-extinction-60-minutes-2023-01-01/

[50] Richard, K. (2021, Aug 2). In the 1970s Climate Modification Proposals Included Purposely Melting Arctic Sea Ice With Black Soot. *No Tricks Zone*. https://notrickszone.com/2021/08/02/in-the-1970s-climate-modification-proposals-included-purposely-melting-arctic-sea-ice-with-black-soot/

[51] "Bering Strait." *Wikipedia*. 30 May 2024. https://en.m.wikipedia.org/wiki/Bering_Strait

By 1979 temperatures were rising again, but fears about a climate destabilized by human activities didn't go away. Instead, they merely refocused on the warming trend and projected that warming trend indefinitely into the future with dire predictions of civilizational collapse in the absence of de-industrialization. The media did a complete 180-degree reversal in a matter of months. On February 14th, 1979, only 13 months after the New York Times had published its ominous warning about global cooling as the existential threat to humanity ("*An international team of specialists has concluded from eight indexes of climate that there is no end in sight to the cooling trend of the last 30 years...*"),[52] the same New York Times was raising the alarm about global warming as it proclaimed that, "*Climatologists are Warned North Pole Might Melt: There is a real possibility that some people now in their infancy will live to a time when the ice at the North Pole will have melted...*"[53] Both articles were written by the same journalist (Walter Sullivan). [Incidentally, by 2010 even the idea of building a dam across the Bering Sea had been recycled by one enterprising Dutch scientist as a way to block warm Pacific waters from entering the Arctic Ocean in order to prevent the icepack from melting.[54]]

Another lasting impact on the ethos of the era was the 1970s' oil crisis, triggered by the Yom Kippur War in Israel in 1973 and by the Iranian Revolution in 1979. Oil prices increased by over 250%, triggered a crippling recession during a period of massive price inflation (stagflation). It exposed the West's dangerous reliance on oil exporters from the volatile Middle East. Then, in the 1980s, mine workers who worked for the government-run coal industry in the U.K. went on strike in the bitter (and often violent) 1984-1985 Miners' Strike that brought the country to a standstill. Prime Minister Margaret Thatcher unleashed the police in a violent repression designed to teach the miners a lesson, with countless

---

[52] Sullivan, W. (1978, Jan 5). International Team of Specialists Finds No End in Sight to 30-Year Cooling Trend in Northern Hemisphere. *New York Times*. https://www.nytimes.com/1978/01/05/archives/international-team-of-specialists-finds-no-end-in-sight-to-30year.html?_r=1

[53] Sullivan, W. (1979, Feb 14)). Climatologists Are Warned North Pole Might Melt. *New York Times*. https://www.nytimes.com/1979/02/14/archives/climatologists-are-warned-north-pole-might-melt-another-projection.html

[54] Burke, J. (2010, Nov 26). Could a massive dam between Alaska and Russia save the Arctic? *Anchorage Daily News*. https://www.adn.com/arctic/article/could-massive-dam-between-alaska-and-russia-save-arctic/2010/11/26/

false imprisonments and malicious prosecution tactics used to try to intimidate the miners' union into capitulation. One of her chief advisors envisioned the Miners' Strike as a communist revolution.

Documents later revealed that Thatcher decided during the Strikes that she would try to destroy Britain's nationalized coal industry altogether (the much-reduced coal industry was ultimately privatized in the U.K. in 1994) and to push the country towards adopting nuclear power instead. The move was seen as a way to deal with the country's energy vulnerabilities while simultaneously eliminating the threat of socialist labour unions once and for all. Another of her advisors, Sir Crispin Tickell, a British diplomat and environmentalist who had studied climatology at Harvard, advised her to explore climate change as a pretext to push the country away from coal and oil and towards nuclear energy.[55] Nuclear energy would not only eliminate the unions, which had held the country hostage and had stoked up communist and socialist sympathies across the British Isles, but it would also provide the U.K. with a "clean" domestically-controlled energy source (with uranium sourced from friendly uranium-rich countries like Canada), thus relieving the country's reliance on "dirty" Middle Eastern oil.

The Miners' Strike ended in 1985, but Thatcher's work to eliminate the coal industry had only begun. She poured massive funds into climate research and helped push the world towards launching the Intergovernmental Panel on Climate Change. In 1989, in an address to the UN General Assembly, she warned of *"vast increases in carbon dioxide"* as she advocated for the world to embrace nuclear power. Armies of climate scientists suddenly found themselves the beneficiaries of generous government grants to investigate the link between carbon dioxide and climate change. As the cheeky internet meme so succinctly points out, *"97% of scientists agree with whoever is funding them."*

Thatcher's conservative peers in other western countries, most especially U.S. President George Bush Sr., Germany's Chancellor Helmut Kohl, and Canadian Prime Minister Brian Mulroney, took up her banner and also started pouring vast funds into climate research. Their efforts to tarnish the hydrocarbon industry were wildly successful, but their efforts

---

[55] Kay, W. W. (2022, Jul 20). Margaret Thatcher and the Rise of the Climate Ruse. *Friends of Science.* https://blog.friendsofscience.org/2022/07/20/margaret-thatcher-and-the-rise-of-the-climate-ruse/

to promote nuclear energy faltered as public opinion failed to rally around nuclear power in no small part because of the bitter aftertaste left by the Three Mile Island and Chernobyl nuclear disasters, but also because a "green" alternative power generating industry soon emerged to take advantage of the windfall of academic funding and government subsidies that were being dished out to anyone who could find a positive spin to link their activities to the climate change narrative. The grift can only be sustained as long as the sense of crisis can be maintained — the carnival will continue as long as governments keep showering the doomsday prophets with money and subsidies.

The climate change movement has taken on a life of its own. Just as society once latched on to the idea that the "rain follows the plow" because it encapsulated all the optimism, hopes, dreams, and geopolitical challenges of the late 19th and early 20th century, "anthropogenic climate change" (a.k.a. "carbon dioxide is the control knob on the climate change") is perfectly suited to the gloomy mood of our own era — pessimism about the future, fears about overpopulation, fears about pollution, deep mistrust of capitalist corporations, and our neurotic love-hate relationship with science and technology fueled by deep skepticism about the unintended consequences of certain "bad" industries coupled with a blind trust that other "good" technologies are the panacea that can save us from impending doom. It also feeds on the sense of vulnerability that comes with being swept along by opaque forces beyond our control and the tribal sense that "other" people whose choices we disapprove of are to blame for our misery. At the same time, it recycles the optimistic belief that if only the "benevolent" nanny state confiscates enough of our tax dollars and intervenes aggressively enough in the choices of our unruly fellow human beings through a combination of top-down incentives and state-imposed coercion, society can somehow tame the climate in the same way that our forefathers mastered the forests, the rivers, and the oceans using science and technology.

The warning of what happened on the Great Plains when society enthusiastically embraced the alleged god-like wisdom of central planners and their hand-picked "climate experts" is, once again, being ignored. The multitude of other failed experiments with central planning, like prohibition, eugenics, and the New Deal, along with all the horrors unleashed by 20th century European experiments with state planning, should have taught us everything we need to know about where central planning inevitably leads. As money and power is concentrated in the hands of a small number of gatekeepers, the distorted incentives created by central planning guarantee that even the best of intentions are doomed to fail. But, as always, the allure of control and the temptations created by so much money sloshing around in the government trough are just too strong to resist.

Just as "the rains follow the plow" became the catalyst to usher in a new world order during the nation-building exercises of the late 19th and early 20th century, "anthropogenic climate change" is the belief system that is being hijacked by one and all to sell stuff, to accumulate power, to outmanoeuver competitors, to raid the public treasury, to signal virtue, to clamp down on things other people are doing that we don't like, and to impose a new global world order on the 21st century. It is the "noble lie" of our age, used to usher in a new era of "sustainable" international cooperation… and to thoroughly pick our pockets and strip us of our liberties.

In a sense, the mad crusade against carbon dioxide has something to appeal to almost everyone. It's a compelling story. It promises simple solutions. It has lots of correlations that are easy to confuse as cause and effect. It's a convenient rallying cry for virtue-signalling politicians, activists, scientists, journalists, corporations, and attention-seeking citizens. It combines the thrill of a scary story with the emotionally compelling urge to control our unruly fellow citizens. And it expresses the existential angst about the future that is felt by so many people trapped in the rat race, toiling in meaningless bullshit jobs, with few meaningful social connections and community bonds, and anxious about a financial future that is hanging under a cloud of inflation and increasingly unaffordable living conditions. As philosopher Eric Hoffer once said, "*A man is likely to mind his own business when it is worth minding. When it is not, he takes his mind off his own meaningless affairs by minding other people's business.*"

The advertised solutions are equally seductive. It promises to turbocharge the ethos of the New Deal era that is so appealing to socialist sympathizers by exponentially expanding government responsibilities and control. It promises to unite society around a common purpose and give society a common enemy around which to build a global rather than a national identity. As such, it promises to dissolve what remains of the nationalism that is blamed for the horrors of the Second World War by imprinting a new global multicultural climate-fighting identity on the world — "we're all in this together" — (while ignoring the fact that central planning is what gave birth to 19th century nationalism in the first place and then gave central planners the green light to commit unthinkable atrocities both against their own and against other peoples).

This new centrally planned global "stakeholder" economy (as it is called by the World Economic Forum) is envisioned to give hard-nosed capitalism a softer touch as a triumvirate of politicians, activists, and scientific-technological elites is installed as the updated "master of masterful corporations". Whether by accident or by design, the climate change narrative has been adapted for political purposes to serve as the foundational myth for our new "cooperative" multicultural world order.

In 1998, Canadian Liberal Member of Parliament and Minister of the Environment, Christine Stewart, told editors and reporters at the Calgary Herald, *"No matter if the science of global warming is all phony...climate change [provides] the greatest opportunity to bring about justice and equality in the world."*[56] She headed the Canadian delegation to the Kyoto climate change negotiations and signed the Kyoto Accord on behalf of Canada.

Well-intended believers have become increasingly desperate and increasing willing to embrace authoritarian measures in order to try to frighten and browbeat humanity into getting onboard. And a horde of not so well-intended con artists and thieves have piled onto the lies and doubled down on data manipulation in order to coerce society into accepting their desired authoritarian policies and to keep the taps flowing on a global Green grift that has already grown close to $5 trillion dollars per year[57] — and that's just the value of "green" corporations listed on Western stock exchanges. The "Green" narrative has become a veritable Golden Goose that keeps on laying golden eggs for anyone who learns to leverage the right buzzwords to put a "green" spin on their activities. Nothing degrades into vulgarity faster than the hubris of a "noble lie", especially when that noble lie opens a back door into the nation's treasury.

---

[56] "Christine Stewart." *Wikipedia*. 5 May 2024.
https://en.wikipedia.org/wiki/Christine_Stewart

[57] "Investing in the green economy 2023 – Entering the next phase of growth." *London Stock Exchange Group*. Sep 19, 2023. https://www.lseg.com/en/ftse-russell/research/investing-green-economy-2023-entering-next-phase-growth

The problem with building an economy on a lie is that once the lie is embraced by society as gospel truth, almost everyone becomes dependent on keeping the lie intact. Whether you sell solar panels, have a "green" stock in your retirement portfolio, depend on "green" government grants or "green" subsidies, put your reputation at stake by publicly expressing a "green" opinion, or need to virtue signal the right "green" mantras in the workplace to protect your career, everyone (including those who can see through the farce) has an incentive to keep their mouths shut and go with the flow. As 18th century French philosopher Voltaire once said, *"It is dangerous to be right in matters on which the established authorities are wrong."* The idea has become self-reinforcing for reasons that have nothing to do with science. Trust the Scienz™.

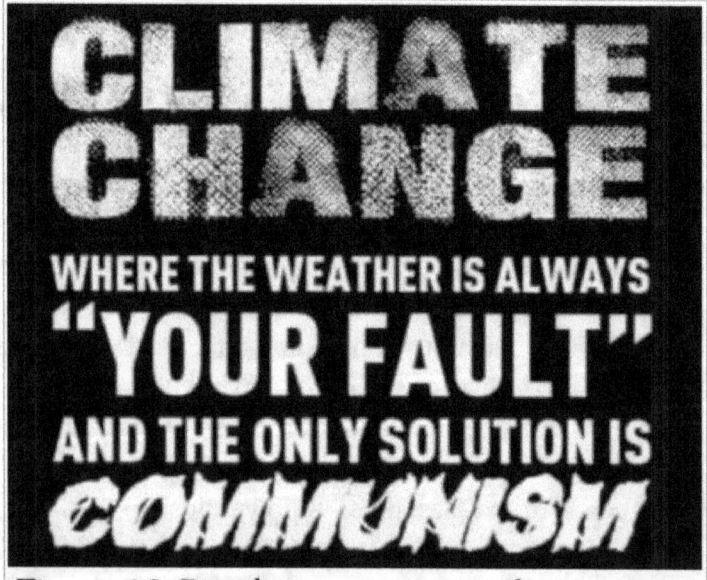

Figure 12. Popular internet meme that captures the spirit of our age.

Chapter 3

"The Era of Global Boiling Has Begun"

"A half truth is the worst of all lies because it can be defended in partiality."

— Solon (630-560 BC),
known as one of the Seven Wise Men of Greece in part because of his efforts to end the exclusive aristocratic control of government in ancient Athens.

Slowly but surely, we are inching towards a censorious, centrally planned, authoritarian surveillance state that seeks to shield itself from criticism and transparency even as it works to micromanage every single individual's carbon use right down to how much meat we eat, what kind of car we drive, how we heat our homes, and even how much fertilizer farmers are allowed to use in their fields and how they manage their lands. Just like when society believed that "the rains follow the plow", our government-endorsed "experts" are foisting a barrage of ill-advised policies on us with sweeping consequences to our civil liberties, our property rights, our economy, our food and energy production systems, and our local ecosystems. Once again, the hubris of central planning is creating a perilously brittle world that is ill prepared to weather the dynamic forces that shape our world. The law of unintended consequences was surely written with central planners in mind.

A magic trick loses its power once you see what's going on behind the curtain. But the problem we face today is the same problem faced by John Wesley Powell in the 1870s as he tried to debunk the belief that the plow could somehow turn the fragile ecosystem of the Great American Desert into a fruitful Garden of Eden: how do you get people to see what they don't want to see? How do you get people to even consider ideas that challenge the established dogma when the very act of looking has been branded a subversive act — a heresy committed by knuckle-dragging science deniers whose malodorous moral shortcomings make them reject all that is good and righteous about the world? Do we have any chance of breaking through the noise before society is forced, once again, to find out the hard way that they are being led around by the nose hairs?

The climate fearmongering has reached a fever pitch. The UN has even declared that *"the impact of climate change on health if carbon emissions remain high, could be up to twice as deadly as cancer in some parts of the world."*[1] Those pushing for this mad crusade against carbon dioxide seem to sense they are on the cusp of getting their desired regulatory changes. Their goals are within reach, all facilitated by the centralization of authority within each nation (one node controls the whole country), and increasingly by entangling each of those sovereign nodes in a labyrinth of treaties that bind them to the central node of global authority — the United Nations. (J.R.R. Tolkien's Lord of the Rings tells the story of the "one ring to rule them all, and in the darkness bind them" —the trilogy was ultimately a warning about how centralized power is, by its very nature, too dangerous to exist because it corrupts all those who lay their hands upon it (as well as all those who seek to lay their hands upon it), including those who wish to use it for a good purpose. Even the lowly and humble hobbit Frodo ultimately succumbed to the One Ring's power and refused to let it go — it was only destroyed when Gollum greedily snatched it from the hobbits' hands as he accidentally stumbled and fell into the Fires of Mordor.)[2]

Carbon taxes are being rolled out to make everything more unaffordable (remember in 2008 when President Obama proudly declared during an interview that, *"under my plan of a cap-and-trade system, electricity rates would necessarily skyrocket…"*[3]). The mandates for a forced transition to electric vehicles are here. Oil and gas development is being systematically strangled by a host of regulatory hurdles. The autonomy of family farms is being systematically undermined by "green" legislative hurdles. Wall Street is lobbying the government to invoke

---

[1] United Nations. (2022, Nov 4). *Climate change much deadlier than cancer in some places, UNDP data shows.* [Press release]. https://news.un.org/en/story/2022/11/1130202

[2] Yost, Z. (2018, Aug 1). JRR Tolkien on the Danger of Centralized Political Power. *Mises Institute.* https://mises.org/power-market/jrr-tolkien-danger-centralized-political-power

[3] Martinson, E. (2012, Apr 5). Uttered in 2008, still haunting Obama. *Politico.* https://www.politico.com/story/2012/04/uttered-in-2008-still-haunting-obama-in-2012-074892

eminent domain to seize private farmland to transfer into the hands of Wall Street investors so they can build wind and solar energy projects.[4] Restrictions on fertilizer are being rolled out (Canada, Netherlands, Germany). Mandatory livestock culls are being written into law to reduce the number of burping cows (Ireland). Denmark is the first nation to roll out an emissions tax that will tax farmers for every single cow and pig that they own in what amounts to a kind of head tax on livestock.(New Zealand also passed a similar law, but it was repealed after hefty public pushback).[5] The disgraced Dutch Prime Minister, Mark Rutte, darling of the World Economic Forum, even tried to expropriate over 3,000 farms during his tenure in a fanatic bid to reduce the agricultural footprint of his country[6] (the rationale is truly baffling — Dutch farms are among the most productive farms in the world — but hysteria knows no reason). And as I write this in the winter of 2023/2024, German farmers are protesting all across their country in a desperate bid to roll back a series of fertilizer restrictions, tax hikes, and land-use restrictions designed to meet the new "biodiversity" targets of the UN's new global *30 by 30* initiative, even as the media vigorously denounces the desperate farmers as "extremists" and condemns their protests as "anti-democratic". And throughout it all, government is clamping down on freedom of speech through a stifling censorship regime enforced on behalf of the government by media and tech companies in the ultimate dystopian corporate-government partnership. In March of 2024, in *Murthy vs Missouri* (a trial challenging the federal government's widespread pressure tactics to coerce social media companies to censor conservative views and silence critics of the Biden Administration), US Supreme Court Justice Ketanji Brown went as far as stating that in her view freedom of speech "*hamstring[s] the government in significant ways*" and that freedom of speech is an obstacle

---

[4] Lee, A. (2023, Apr 6). Wall Street titan Jamie Dimon says seize private land for wind and solar builds. *Recharge News.* https://www.rechargenews.com/energy-transition/wall-street-titan-jamie-dimon-says-seize-private-land-for-wind-and-solar-builds/2-1-1431685

[5] Associated Press. (2024, Jun 27). Gassy cows and pigs will face a carbon tax in Denmark, the first country to do so. *NPR.* https://www.npr.org/2024/06/27/nx-s1-5021147/denmark-carbon-tax-cows-pigs-farms-worlds-first

[6] Boztas, S. (2022, Nov 30). Up to 3,000 'peak polluters' given last chance to close by Dutch government. *The Guardian.* https://www.theguardian.com/environment/2022/nov/30/peak-polluters-last-chance-close-dutch-government

to the government being able to control the narrative *"in an environment of threatening circumstances"*.[7]

Perhaps the vilification of independent farmers, especially livestock farmers, shouldn't surprise us. After all, White House climate envoy, John Kerry, recently declared that *"we can't get to net zero unless agriculture is front and center as part of the solution [...] You just can't continue to both warm the planet, while also expecting to feed it. It doesn't work. So, we have to reduce emissions from the food system."*[8]

Even the World Bank has gotten into the action with a new report called *Recipe for a Liveable Planet: Achieving Net Zero Emissions in Global Agrifood Systems,*[9] which calls for central planners to completely redesign the global agriculture system (including calls for high income countries to use government planning to *"decrease consumer demand for emissions-intensive foods by fully pricing animal-source foods through repurposed subsidies and by promoting sustainable food options"* (a.k.a. make beef and other animal proteins more expensive to coerce people to move towards a "plant-based diet"). They aren't even subtle about it — they overtly state that *"Consumption of animal-source foods has damaged the planet"* and that they want high-income countries to reduce their citizens' meat and dairy consumption. They even recycle the false claim, which I already debunked in the introduction to this book, that reducing red meat consumption will also have the added benefit of reducing heart attacks — I guess they didn't get the USDA's 2015 memo. I guess when you're on a mission, what's true matters far less than what's useful to achieving your objectives.

The net effect of all these initiatives and policy changes is to systematically hollow out independent multigenerational family farms and push food production into the hands of global agricultural

---

[7] Cleveland, M. (2024, Mar 20). Justic Jackson's Comment About Free Speech 'Hamstringing' The Government Wasn't Her Worst. *The Federalist.* https://thefederalist.com/2024/03/20/justice-jacksons-comment-about-free-speech-hamstringing-the-government-wasnt-her-worst/

[8] Heller, T. (2023, Oct 15). *John Kerry: We Must Reduce Farming To Feed The World.* [Video]. YouTube. https://www.youtube.com/watch?v=y4yU1v5MCx4&t=18s&ab_channel=TonyHeller

[9] Sutton, W. R., Lotsch, A., and Prasann, A. (2024). *Recipe for a Livable Planet: Achieving Net Zero Emissions in the Agrifood System. Agriculture and Food Series. Conference Edition.* World Bank. http://hdl.handle.net/10986/41468

conglomerates that take their marching orders from distant shareholders and central planners — the perfect chess move for an administrative state seeking to expand its powerful grasp over the whole of society.

In 2022, the United Nations announced it had teamed up with Google to ensure that UN-verified climate messaging is prioritized in search results.[10] Later that year at the World Economic Forum, the UN's Under-Secretary-General for Global Communications proudly proclaimed, *"We own the science, and we think that the world should know it."*[11] The message to journalists, media organizations, and academics was clear — if you want news reports, research studies, and editorial opinions to be visible at the top of search results (which is the key to making money from advertisers), you have to toe the UN party line. Remember the new concept pioneered by activists on university campuses in recent years to rationalize their censorship of an ever-expanding list of forbidden topics: *"freedom of speech is not freedom of reach."* Those who don't toe the line become invisible in the public forum. Censorship has become more sophisticated since the days when the Inquisition tortured heretics in the public square to "educate" onlookers, but the chilling effect on the pursuit of truth is the same.

In the end, irregardless of how censorship justifies itself when it first begins, it always ends up in the same place. Idi Amin, Uganda's former brutal dictator, allegedly once said, *"There is freedom of speech, but I cannot guarantee freedom after speech."* Echoing Idi Amin, prominent Canadian environmental activist Dr. David Suzuki has called for political leaders to be thrown in jail for *"ignoring the science behind climate change"*.[12] The United Nations has floated the idea of criminalizing "climate denial."[13] Canada's NDP party recently tabled a bill criminalizing

---

[10] United Nations. (2022, Apr 22). *Google teams up with UN for verified climate information.* [Press release]. https://www.un.org/en/climatechange/google-search-information

[11] Bernstein, B. (2022, Oct 4). U.N. Communications Official Touts Google Search Partnership: 'We Own the Science'. *National Review.* https://www.nationalreview.com/news/u-n-communications-official-touts-google-search-partnership-we-own-the-science/

[12] Offman, C. (2008, Feb 7). Jail politicians who ignore climate science: Suzuki. *National Post.* https://nationalpost.com/news/jail-politicians-who-ignore-climate-science-suzuki

[13] "Climate crimes must be brought to justice." *UNESCO.* Jun 6, 2023. https://www.unesco.org/en/articles/climate-crimes-must-be-brought-justice-0

the "promotion of fossil fuels" (as I reported in the introductory paragraphs of this book). And efforts are underway to list climate change denial as a crime against humanity so it can be prosecuted via the International Criminal Court in the Hague.[14] The false empathy of the environmental movement disguises the ruthless iron fist of tyrants — the emerging green dystopia is right on our doorstep. As presidential hopefuls Vivek Ramaswamy and Robert F. Kennedy Jr. have both recently reminded us, there is no time in history when the people who were censoring speech were the good guys.

But this is also the point at which the globalists are most vulnerable to alienating the public as their final push to ram their desired policies down our throats exposes the utter lunacy of their propositions and betrays the authoritarian face behind their "green" mask. As George Orwell once said, *"All tyrannies rule through fraud and force, but once the fraud is exposed, they must rely exclusively on force."*

~

One of the most effective propaganda tools used by politicians, activists, and their partners in mainstream media is to co-opt natural disasters and link them to climate change in the public imagination. The dramatic images produced by the 2023 wildfire season here in western Canada (figure 13) were shamelessly exploited as a tool to drive fear into the hearts of citizens. (In 2016, a wildfire swept through the city of Fort McMurray in Alberta. In 2021, the entire town of Lytton, BC, was burned to the ground. And as I put the finishing touches on this book in the summer of 2024, a wildfire has just swept through the town of Jasper, Alberta, destroying 30-50% of all structures,[15] even as the entire village of Silverton, BC, has been evacuated as yet another fire bears down on it.)

---

[14] "ASP21 Side Event: Climate Change as a Crime Against Humanity: Article 15 Submission." *PILPG A Global Pro Bono Law Firm.* Dec 9, 2022.
https://www.publicinternationallawandpolicygroup.org/lawyering-justice-blog/2022/12/9/asp21-side-event-climate-change-as-a-crime-against-humanity-article-15-submission

[15] Snowdon, W. and Frew, N. (2024, Jul 25). Buildings in Jasper in ashes as 'monster' wildfire spans 36,000 hectares. *CBC News.*
https://www.cbc.ca/news/canada/edmonton/wildfire-that-roared-into-jasper-was-a-wall-of-fast-moving-flame-says-fire-official-1.7274825

Figure 13. www.cbc.ca/news/canada/british-columbia/what-you-need-to-know-about-bc-wildfires-aug-18-2023-1.6940311

Heartbreaking scenes of the 2023 wildfire that swept through the city of Kelowna and the burnt-out homes and businesses left behind by the nearby 2023 North Shuswap wildfire became fodder for a flurry of doom-mongering headlines with a thinly concealed political agenda. In July of 2023, UN Secretary-General António Guterres went as far as announcing that *"the era of global boiling has begun."*[16] In September of 2023 he

---

[16] Besheer, M. (2023, Jul 27). UN Chief: Planet Is Boiling; Time Running Out to Stop Climate Crisis. *VOA News.* https://www.voanews.com/a/un-chief-planet-is-boiling-time-running-out-to-stop-climate-crisis/7200507.html

declared that *"climate breakdown has begun"*[17] and that by failing to take action soon enough *"humanity has opened the gates to hell."*[18] At the United Nations COP28 Climate Conference in December of 2023 he warned the 10,000+ delegates who flew to Dubai on private jets to hear him speak that *"the 1.5-degree limit is only possible if we ultimately stop burning all fossil fuels. Not reduce. Not abate. Phaseout."*[19]

Here's a brief sampling of some of the other propaganda headlines that flooded the airwaves in the summer of 2023:

Associated Press: "Climate change keeps making wildfires and smoke worse. Scientists call it the 'new abnormal.'"[20]

The Globe and Mail: "The summer the sun turned red: Canadians wake up to the reality of climate change under a veil of smoke."[21]

New York Times: "Canadian Wildfires Twice as Likely Because of Climate Change, Study Finds"[22]

---

[17] United Nations. (2023, Sep 6). *Climate Breakdown Has Begun with Hottest Summer on Record, Secretary-General Warns, Calling on Leaders to 'Turn Up the Heat Now' for Climate Solutions.* [Press release]. https://press.un.org/en/2023/sgsm21926.doc.htm

[18] Milman, O. (2023, Sep 20). Humanity has 'opened gates to hell' by letting climate crisis worsen, UN secretary warns. *The Guardian.* https://www.theguardian.com/world/2023/sep/20/antonio-guterres-un-climate-summit-gates-hell

[19] Doshi, T. (2023, Dec 8). COP28 And Fossil Fuels: Showdown Between Alarmists And Pragmatists. *Forbes.* https://www.forbes.com/sites/tilakdoshi/2023/12/08/cop28-and-fossil-fuels-showdown-between-alarmists-and-pragmatists/

[20] Borenstein, S. and Walling, M. (2023, Jun 30). *Associated Press.* https://apnews.com/article/wildfire-smoke-canada-climate-change-new-normal-f22a68e7df9688ef8eccd970efde3baf

[21] Andrew-Gee, Jones, and Skrypnek. (2023, Aug 19). *The Globe and Mail.* https://www.theglobeandmail.com/canada/article-canada-wildfires-smoke-climate-change/

[22] Zhong, R. (2023, Aug 22). *The New York Times.* https://www.nytimes.com/2023/08/22/climate/canada-wildfires-climate-change.html

CBC: "'Something's Changed': Summer 2023 is screaming climate change, scientists say"[23]

MSN: "Earth could become "hell": Scientists predict consequences of uncontrolled greenhouse effect."[24] (this absurd article paints an apocalyptic picture of our oceans completely evaporating as temperatures soar above 2732 °F (1500 °C), which is more than three times hotter than the surface of Venus.)

I watched Kelowna burning with my own eyes in the summer of 2023 — it's not far from my hometown where I grew up on my parent's cattle farm. But I have also spent a lifetime watching Kelowna and other cities in BC's dry interior relentlessly expand into the surrounding woodlands, and I have had a front row seat into the dysfunctional politics of how BC's forests and watersheds are (mis)managed because my family once had a government grazing lease for our cattle on the forested mountain plateau to the east of Kelowna, which gave us plenty of opportunities to interact with the bureaucratic gears of the BC Forest Service. And so, I have witnessed firsthand how the management of these forests and watersheds has changed over my lifespan and how these policies have impacted the forests and the soils that sustain them, all of which have made our cities and the surrounding forest ecosystem more vulnerable to wildfires and more vulnerable to the drought and flood cycles that are part of the natural variations in our dry western climate.

Once again, just like out on the Great Plains during the lead up to the Dust Bowl, even as CO_2-obsessed central planners delude themselves about their ability to control the global climate, government-dictated forest management practices are inadvertently sowing the seeds for local ecological crises, like the wildfire that swept through Kelowna — crises that have nothing to do with our CO_2 emissions but have everything to do with how central planners are mismanaging our local ecosystems. John Wesley Powell must be rolling in his grave.

---

[23] Weber, B. (2023, Aug 5). *CBC News*.
https://www.cbc.ca/news/canada/calgary/climate-change-something-changed-summer-2023-canada-1.6929271

[24] KMO. (2023, Dec 21). *MSN News (ESSA News)*. https://www.msn.com/en-us/news/technology/earth-could-become-hell-scientists-predict-consequences-of-uncontrolled-greenhouse-effect/ar-AA1lRNB7

As I will demonstrate over the next three chapters, the wildfires sweeping through our western cities are not a consequence of anthropomorphic climate change. Rather, they are a symptom of the systematic failure of complex centrally planned systems — a perfect storm of failed policies in urban planning, forest management, logging practices, water management, forest soil management, building practices, and the dysfunctional intertwined relationship between local, provincial, and federal responsibilities. There is a lot to untangle here but as I unravel the lies and half-truths being told about the wildfires and their alleged link to climate change, I will also be laying the initial groundwork to be able to show you the story of the century that I alluded to at the beginning of this book.

~

The simplistic "consensus" narrative about wildfires runs something like this:
1. Forest fires are bad.
2. Since the 1980s, the number of acres burned by wildfires has increased.
3. Over this time period, the climate in western Canada has become hotter and drier.
4. The number of houses and other infrastructure destroyed by wildfires has increased over this same time period.
5. And when your house is in the path of a wall of flames, you should run and leave fire fighting to government authorities.

These five statements create a tidy self-supporting circle of half-truths. To deny them is to deny the evidence before our own eyes. But to accept them without further context turns us into a sucker for a vicious lie about our climate. As the Yiddish proverb says, *"a half truth is a whole lie."*

On one level, the story I am about to tell is about wildfires, urban mismanagement, failed forest management policies, the consequences of decades of fire suppression, the depletion of our forest soils, and the harebrained crusade launched against carbon dioxide (the boogeyman of the climate brigade and, ironically, probably the only character in this pathetic tragicomedy that doesn't deserve any of the blame).

And yet, as the various threads driving this chaos become visible, this story becomes the perfect miniature replica of how a century of increasingly centralized decision-making in the hands of increasingly powerful government institutions has created a political architecture that is wholly incapable of dealing with the complexities of our natural ecosystems (or anything else for that matter). By placing so much power and resources in the hands of a small number of gatekeepers, we have distorted the incentives towards rampant corruption and profiteering. In the process, we have all but destroyed the ability for science to discover objective truths. And the proliferating red tape, the once-size-fits-all approach of centralized government, and the misguided and deeply politicized policies that are being imposed from above have all but destroyed the autonomy of local foresters, farmers, scientists, landowners, and local governments by stripping them of the ability to opt out of the madness that has gripped society. As Elon Musk recently pointed out on Twitter, "*our civilization is being slowly strangled to death one regulation at a time.*"

By the time you finish reading this section of the book, I hope you will be able to drive out into the forests to verify for yourself what I have written about, be armed with a deeper understanding of the complex ecological processes that create healthy forests and healthy forest soils, and have a broader knowledge of our Earth's history to provide the missing context by which to judge the doomsday scenarios being painted by our media, politicians, and institutions.

~

In 1789, Alexander Mackenzie made his historic trip up the Mackenzie River from Lake Athabasca to the Arctic Ocean. What I remember most from reading his diaries was how often he mentions wildfires consuming forests on either side of the river along his journey north. One hundred and twenty-five years later, arctic explorer Vilhjalmur Stefansson described a similar scene in his book, *The Friendly Arctic*, about his expedition to the Canadian Arctic from 1913 to 1916: "*I have seen forest fires in the Canadian Northwest that were burning over an area of hundreds of square miles. The smoke from these fires sometimes fills the air for hundreds of miles around and makes the sun look like a red ball. Sometimes the smoke is so thick that it is hard to breathe.*"[25]

---

[25] Stefansson, Vilhjalmur (1922). *The Friendly Arctic: The Story of Five Years in Polar Regions.* New York: Macmillan.

The western side of North America was already a very, very smoky place long before Europeans established themselves on the continent and built an economy centered around the internal combustion engine. Indeed, wildfires were such a common, widespread, and natural part of the western North American landscape that countless species of trees, flowers, mushrooms, and other plants, along with countless birds, insects, and other animals have evolved fire-resistant characteristics and have become entirely dependent upon the fire cycle for its regenerative properties. The evolutionary pressure required to reshape so many independent parts of an ecosystem to become dependent on wildfires did not happen because of a few bad wildfire seasons; it takes persistent and widespread wildfires, repeated year after year over many thousands of generations, to reshape the evolutionary traits of so many plants and animals. Wildfires are quite literally embedded in the DNA of the dry western North American landscape — yet few people grasp the full implications of this fact because we have been lured into a false sense of reality by a century of modern wildfire suppression.

The lodgepole pine is the perfect example of one of these unique fire-adapted species that thrives in the dry fire-prone interior of British Columbia. Its thick heavy bark has evolved to become especially fire tolerant, allowing the trunk to survive low-intensity wildfires. And yet, their needles and sap simultaneously evolved to become especially flammable — they are the equivalent of hanging gasoline-soaked rags in trees with the added "bonus" that when these needles catch fire, they shower the forest floor with a rain of glowing embers that ignite grass and brush wherever the wind blows them (see figure 14). Their highly flammable nature is an evolutionary feature, not a bug.

Figure 14. A wind-driven hot ember shower during a wildfire.
Source: Kaibab National Forest, CC BY-SA 2.0, https://commons.wikimedia.org/wiki/File:Boundary_Fire_2017 _(27406794889).jpg

In the dry climate of western North America, Mother Nature evolved trees that not only tolerate frequent low-intensity fires but even actively amplify the spread of wildfires. At first glace, this seems counter-intuitive. In a dry fire-prone landscape, wouldn't it be advantageous to evolve to become <u>less</u> flammable (more like the maples and oak trees that thrive on the eastern side of the continent)?

By combining these two seemingly opposite evolutionary adaptations, these highly flammable trees with their fire-tolerant trunks have not only achieved a competitive advantage over less fire-tolerant competitors but, by virtue of those highly flammable needles and the shower of embers they release as they burn, these tree species make it easy for wildfires to ignite the dangerous fuel that continually builds up over time on the forest floor. In essence, it is a clever evolutionary strategy that is perfectly designed to reduce the risk of big high-intensity wildfires that climb up into the canopy by promoting lots of reoccurring low-intensity wildfires that remain confined to the forest floor, while also knocking out competition from less fire-tolerant tree species. Win-win. It's also part of the puzzle that allows us to understand why, after over a century of modern fire suppression, fires over the past few decades are getting bigger, hotter, and more dangerous — we have systematically suppressed the frequent low-intensity wildfires that once regularly swept through western forests to clean away fuel accumulating on the forest floor.

Figure 15. A ponderosa pine-grassland during a prescribed burn. Source: U.S. Geological Survey

The lodgepole pine is just one among countless western tree species with similar fire-tolerant characteristics, like jack pines, ponderosa pines, larch, and Douglas fir trees, all of which combine to make Western forests orders of magnitude more flammable than the deciduous forests that thrive in the far more humid climate on the eastern side of the continent. From an evolutionary perspective, if you live in a fire prone climate and you can't stop the lightning and dry conditions that cause wildfires, you adapt, you embrace these conditions and even bend them to your advantage to minimize your own risk while simultaneously using those wildfires to take out other less fire-tolerant rivals that are competing for access to the same soil, nutrients, groundwater, and sunlight. Nature is ruthless — in the western forests, wildfires have become an existential part of the never-ending competitive struggle for survival of the fittest.

But the evolutionary adaptations go far beyond being able to tolerate low-intensity wildfires and actively spreading burning embers whenever there's a spark. Many species that populate the most fire-prone microclimates, such as jack pines and lodgepole pines, have evolved to become entirely dependent on frequent low-intensity fires in order to open their seed cones — they rely on the heat of a fire to melt the waxy coating of their cones in order to release the seeds inside. The hard waxy coating protects the seeds inside from being eaten by insects and birds, allowing the seeds to remain intact and unharmed for decades. But when a fire finally melts the waxy coating, these seeds gain a head start to repopulate burned ground faster than other competing plant and tree species. This advantage comes with its own risks — without a wildfire to regularly sweep through the forest floor to open their seed cones, they cannot regenerate. By adapting to survive the frequent low-intensity wildfires of the western North American landscape, these trees have become wholly dependent on those low-intensity wildfires in order to reproduce.

Because fire-prone ecosystems depend on frequent low-intensity fires to regularly *clean* the forest floor to prevent too much fuel from building up, if the interval between wildfires is too long, too much fuel accumulates. Any wildfire that gets too big and too hot risks burning through the fire-resistant trunks or risks reaching the lower branches of the trees and climbing up into the canopy to cause a dangerous crown fire. High-intensity wildfires also risk burning through the wax-coated cones (and the seeds within them) and they risk scorching the soil by burning down into the delicate humus layer, which renders the soil sterile and nutrient deficient. The unique characteristics of trees that shed highly flammable needles and shower the forest floor with burning embers have, in effect, harnessed one of nature's most destructive forces by turning it into a beneficial regenerative force at the center of a self-sustaining ecosystem. Frequent fires prevent big destructive fires. The fire cycle is how nature adapted to the dry climate and lightning-prone geography of the American West. Just as the Great Plains depended on deep-rooted sod and migrating herds of grazing bison to sustain the fragile prairie ecosystem, wildfires are integral to the health of the vast forests of the American West. When these forests are denied frequent low-intensity wildfires, the ecosystem begins to break down.

This endless fire cycle, repeated over thousands of years, created a unique environmental niche that allows countless other fire-adapted species to thrive (such as pioneer species like morel mushrooms and fireweed, which are first to recolonize freshly burned areas). Wildfires are also beneficial to deer and other grazing species, which grow fat and raise their young in the lush grass that grows in the early years after a wildfire before the trees get big enough to rob sunlight from the grass growing on the forest floor. Grass won't grow in a dense forest because insufficient sunlight reaches the forest floor. To feed large numbers of grazing wildlife, you need fire (or logging) to open up clearings in the forest canopy. Walk through any dense forest — you'll find very little for a deer to eat. The endless cycle of wildfires that opened up gaps in the forest canopy gave life to the entire historic western landscape.

At least it did until we started interfering by trying to suppress all wildfires.

~

Human interference in the delicate fire cycle has had countless entirely unintended consequences, many of which have only become visible after decades of fire suppression (sound familiar?). The most obvious unintended consequence of decades of fire suppression is that it has encouraged the build-up of vast quantities of fuel on the forest floor, which has led to larger and hotter wildfires that scorch the earth, burn the seed cones, and spread into younger forest stands that would otherwise have been resistant to the spread of low-intensity wildfires.

But decades of fire suppression have also had far more subtle consequences. When plants and insects become dependent on frequent low-intensity wildfires for their reproduction, fire suppression can undermine the reproductive success of some of the predators that are required to keep our forests healthy and free from dangerous pests. As an example, let's take the case of one of the most well-known and most destructive forest pests of the North American West — the mountain pine beetle.

Figure 16. Mountain pine beetles are approximately 3.7 mm to 7.5 mm long.

Source: Natural Resources Canada

Mountain pine beetle outbreaks have infected over 18 million acres of forests in the BC Interior since the 1990s.[26] These dead and dying forests, filled with dry red needles, create millions of acres of highly flammable fuel, which makes them a leading contributor to the massive high intensity wildfires that have plagued the West in recent years. Mountain pine beetles kill pine trees by boring through the bark and depositing their eggs in the tree's vascular tissue. As the larvae hatch, they feed on the tree's vascular tissue, which disrupts the movement of water through the tree and ultimately kills the tree. When the beetles attack in small numbers, the trees defend themselves by leaking sap — the resin contains chemicals that are toxic to the beetles, causing them to become immobilized and get stuck in the sticky sap. But if there are too many mountain pine beetles, the beetles are able to overwhelm the trees' defenses, destroy the tree's vascular tissue, and kill the tree. Furthermore, as the beetles bore into the tree, they also infect the tree with a fungus called the blue stain fungus. This fungus disrupts the tree's ability to produce its natural defensive resin, which helps the beetles but accelerates the death of the tree. Nature is a cruel never-ending war with unexpected alliances.

---

[26] Natural Resources Canada. (n.d.). *Mountain Pine Beetle.* Government of Canada. Viewed 8 Jun 2024. https://natural-resources.canada.ca/our-natural-resources/forests/insects-disturbances/top-forest-insects-and-diseases-canada/mountain-pine-beetle/13381

One of the mountain pine beetle's natural predators is a pyrophytic (fire-loving) beetle called the fire beetle (genus *Melanophila*). They get their name because they are attracted to burning forest fires — the fire beetle has infrared sensors that allow it to detect active wildfires from up to 130 km away. They fly towards the wildfires in order to mate and then lay their eggs in the freshly burned wood because the wildfires cleanse the wood of the aphids that normally prey on fire beetle eggs. By suppressing the low-intensity wildfires that leave behind lots of charred debris and by reducing the burn acreage of low-intensity wildfires, we have inadvertently suppressed one of the mountain pine beetle's natural predators.

Frequent low-intensity wildfires play another important but frequently overlooked role in protecting trees from mountain pine beetle assaults, which has also been undermined by our effort to suppress wildfires: although the fire-resistant tree trunks can withstand a fairly high degree of heat and charring before the fire becomes a threat to the tree, the beetles boring into those trunks and the larvae growing beneath the bark cannot. A study in the Tweedsmuir Park in central British Columbia used a series of controlled burns to demonstrate that low-intensity wildfires reduced the population of mountain pine beetles in the burned areas by up to 50%.[27] Frequent low-intensity wildfires reduced beetle populations by "cooking" the beetles crawling underneath the trees' bark. Decades of fire suppression have given the mountain pine beetles a competitive advantage, and now we're paying for it with massive mountain pine beetle outbreaks and massive high-intensity wildfires that are caused by these beetle infestations. When you mess with complex ecological systems, prepare for the unexpected.

---

[27] Safranyik, Linton, Shore, and Hawkes. (2001). *The effects of prescribed burning on mountain pine beetle in lodgepole pine.* Information Report BC-X-391. Natural Resources Canada. https://www.for.gov.bc.ca/hfd/library/documents/bib48463.pdf

Chapter 3

A Continent Built for Wildfires

"People talk glibly about science. What is science? People come out of the university with a master's degree, or a PhD, and you take them into the field, and they literally don't believe anything unless its a peer reviewed paper. It's the only thing they'll accept. And you say to them, 'Let's observe, let's think, let's discuss.' They don't do it! It's just 'is it in a peer-reviewed paper or not?' That's their view of science. [...] Gone into the universities as bright young people and come out of them brain dead, not even knowing what science means. They think it means peer-reviewed papers, etc. No, that's academia. If a paper is peer-reviewed it means everybody thought the same, therefore they approved it. An unintended consequence is that when new knowledge emerges, new scientific insights, they can never ever be peer-reviewed, so we're blocking all new advances in science that are big advances. If you look at the breakthroughs in science, almost always they don't come from the center of that profession, they come from the fringe. The finest candlemakers in the world couldn't even think of electric lights."

— Allan Savory, founder of Holistic Management[1]

The dry, lightning-prone, fire-dependent ecology of western North America is the unavoidable consequence of its geography, in particular the Coast Mountains and the Rocky Mountains, which run perpendicular to the trade winds bringing moisture from the Pacific Ocean onto the continent. In essence, it's the same story that creates the climate of the Great American Desert out on the prairie to the east of the Rocky Mountains. As the jet stream pushes water-filled clouds up and over the mountain ranges, they dump their rainfall on the western side of each mountain range, leaving a dry arid rain shadow on the eastern side. Tofino

---

[1] Marijn Poels. (2020, Oct 14). *What is science – Return to Eden* [Video]. YouTube. https://youtu.be/rt0m4PltGHA?si=GC2CQ50SlTIBrghw

on the windward side of the Insular Mountains that run the length of Vancouver Island is very wet, Victoria on the leeward side is dry and fire prone. Vancouver on the windward side of the Coast Mountains is very wet, while Kelowna on the Interior Plateau on the leeward side of the Coast Mountains is extremely dry and extremely fire prone. Revelstoke on the windward side of the Rocky Mountains has a wet climate, while Rocky Mountain House on the leeward side of the Rockies is bone dry and very fire prone. Furthermore, as these mountain ranges force air up into higher elevations, strong electrostatic charges build in the clouds and shower the mountains with frequent cloud-to-ground lightning storms, making wildfires inevitable.

With lots of water up in the mountains to tap into for irrigation but with a hot dry climate in the valleys below, the Interior Plateau of British Columbia is a wonderful place to live, grow crops, raise cattle, harvest timber, play on the lakes, and retire. In recent years, Metropolitan Kelowna has emerged as the fastest growing city in Canada, growing at over 13.5% *per year*. Its urban sprawl is gradually supplanting the region's agricultural heritage and spreading out into the surrounding pine and Douglas fir forests. But nearly a century of suppressing wildfires on the dry Interior Plateau has lured a lot of people (and especially city planners) into forgetting that forest fires are a simple fact of life in this region that people need to live with and plan for. No area goes unburnt for long. Suppressing wildfires over decades simply created a massive build-up of dry brush, dry pine needles, dry pinecones, and dry grass on the forest floor in the periphery of the cities, which are all just waiting for a spark. Decades of aggressive fire suppression have created ticking time bombs everywhere. A short-term solution (fire suppression) became a long-term liability because the underlying ecological processes were misunderstood.

The difference between a normal fire cycle and a disaster that burns up entire cities comes down to how well a city has learned to co-exist with wildfires burning at their margins. Fire fighting crews are merely a tool of last resort — they are a means to manage inevitable fires if preparations were done right in the months and years before the outbreak of a wildfire, but these fire fighting crews are a largely futile line of defense against an unstoppable firestorm if forest management and building policies have fallen short in the decades preceding the wildfire. When fire-prone houses are built inside coniferous forests and when fuel has been allowed to build up on the forest floor in the periphery of towns and cities, it's only a matter of time until the bill comes due, as it did in Kelowna in the summer of 2023. Blaming "climate change" is nothing more than a convenient scapegoat to avoid grappling with a far more complicated reality. Increasing the budget for water bombers and picking the pockets of taxpayers with various carbon taxes utterly fails to address the root causes. Fire is normal here. Adapt.

Much like decades of plowing on the Great Plains gave rise to the conditions that created the Dust Bowl, decades of fire suppression and poor forest management along the city's periphery created conditions that turned a normal fire-prone ecosystem into something dangerous. The temporary success of a few short decades of modern fire suppression using rakes, shovels, water pumps, and water bombers, have given urban planners the completely unrealistic expectation that these tools could continue to protect neighborhoods expanding out into the forests even as decades of unburnt fuel continues to accumulate and as those forests are denied the regular cleansing effects of frequent low-intensity fires. These unrealistic expectations created massive failures in urban planning that have put countless homes in harm's way as developers and homeowners are encouraged to build neighborhoods in the midst of these fire-prone forests. Fire suppression created this mess, more fire suppression is not going to get us out of it.

Once again, the hubris of central planners has lured an unsuspecting public into the crossfire as residents move into "safe" neighborhoods along the forest interface. It's the same old story from the Great Plains at the turn of the century playing out on a whole new venue. And, once again, rather than face up to the failures of central planning, bureaucrats and politicians are quick to blame climate change to avoid accountability (and lawsuits) — it's much easier to engage in victim blaming by shaming society for its carbon footprint. The message is clear: only the government can save you from the evil consequences of your neighbour's SUV. Deflect, divide and conquer… and reward the institutional perpetrators with more power and more money as they are tasked with solving the very problems they helped create.

~

Few people are aware of just how common and widespread wildfires once were on the North American continent long before carbon dioxide levels began to rise and long before the Industrial Revolution gave rise to the internal combustion engine. The burn acreage prior to modern fire suppression methods absolutely dwarfs the burn acreage of the modern era. The 2010 National Report on Sustainable Forests put out by the U.S. Department of Agriculture calculates that during the scorching hot and terrifyingly dry 1930s, over 50 million acres of forests and grasslands burned every year in the United States[2] — and yes, we were already fighting wildfires during the 1930s as part of the government's effort to control nature. During the cooler decades of the 1950s, 60s, and 70s (when scientists and the media were fretting about a new Ice Age), burn acreage dropped to between 2 and 5 million acres per year. Since the 1980s it has climbed back up to between 5 and 10 million acres per year, which is still an 80% reduction compared to the vast expanses of forests that burned during the 1930s, as shown in figure 17 (the chart is from the 2010 National Report on Sustainable Forests).

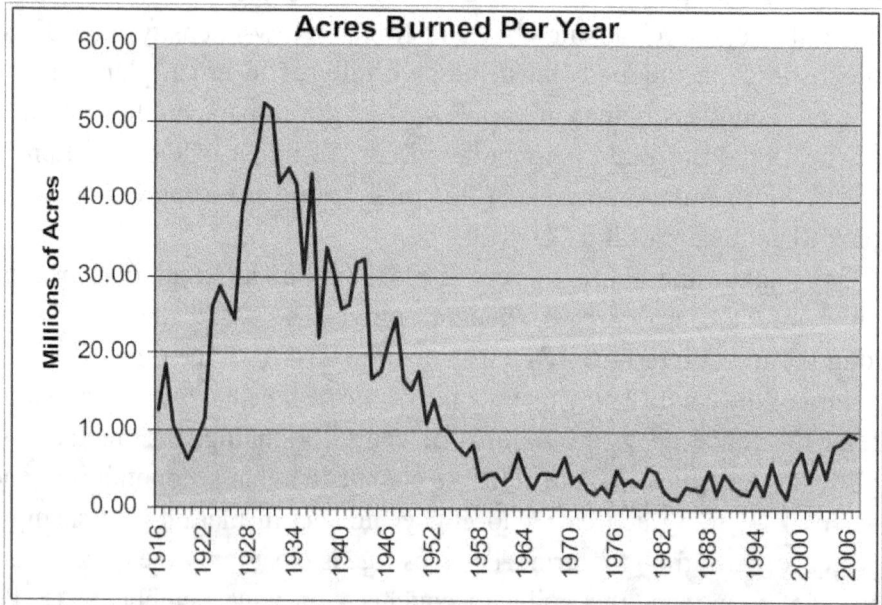

Figure 17. Total burn acreage in the United States, by year, since 1916.
Source: United States Department of Agriculture, National Report on Sustainable Forests - 2010.
https://www.fs.usda.gov/Internet/FSE_DOCUMENTS/stelprdb5293555.pdf

---

[2] USDA Forest Service. (2008, Dec 8). *National Report on Sustainable Forests – 2010.* U.S. Dept of Agriculture.
https://www.fs.usda.gov/Internet/FSE_DOCUMENTS/stelprdb5293555.pdf

The next chart (figure 18), also published by the U.S. government, is even more sobering because it shows burn acreage in the United States extending back into pre-industrial times at over 145 million acres per year! To put that in perspective, this means that approximately 7% of the entire land area of the continental United States burned every single year! This means that, on average, every single square inch of the continental United States burned once every 14 years. But since the eastern edge of the continent is much more humid and since the rainforests along the outermost western edge of the continent rarely get dry enough to burn, this means that most of North America's forests and grasslands burned far more often than every 14 years. The smoldering historic western landscape described by Alexander Mackenzie and Vilhjalmur Stefansson was the norm; today's low burn acreage is not.

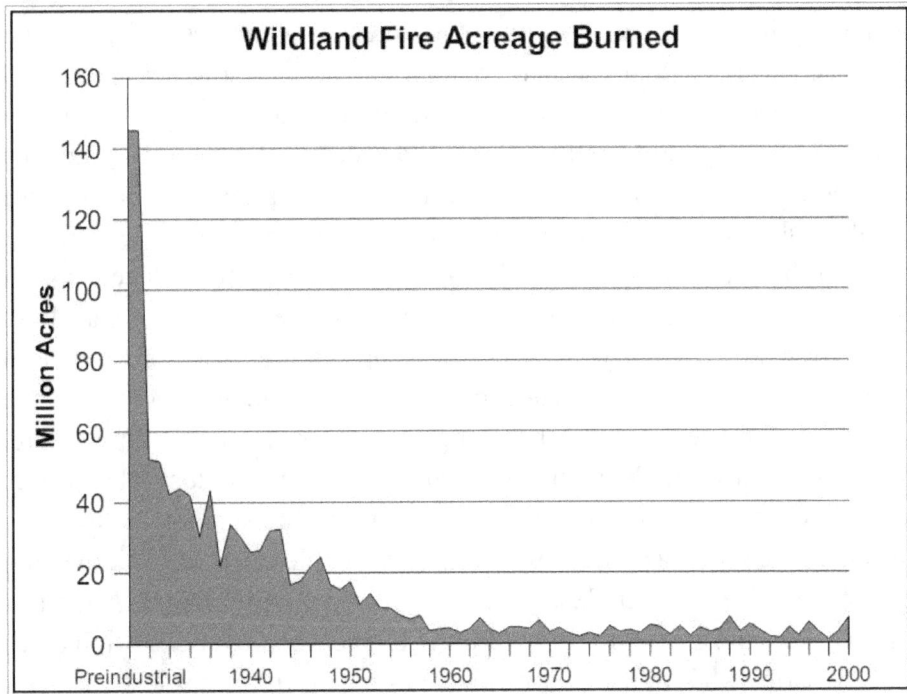

Figure 18. Historic burn acreage in the United States, including during the Pre-Industrial Era.
Source: Review and Update of the 1995 Federal Wildland Fire Management Policy
https://www.nifc.gov/PIO_bb/Policy/FederalWildlandFireManagementPolicy_2001.pdf

It's not an accident that, after decades of modern fire suppression, of the 146 plant species that are considered threatened or endangered in the

United States, 135 of those plant species benefit from wildland fire or are found in fire-adapted ecosystems.[3] Even as our forests are plagued by high-intensity wildfires, they are being starved of the regenerative benefits of the vast low-intensity wildfires that once regularly swept through our pre-industrial forests.

Many of these frequent pre-industrial era wildfires were low-intensity fires that are part of the natural fire-dependent ecology of North America caused naturally by the frequent lightning storms fueled by the mountainous topography. However, Native Americans also routinely lit wildfires to maintain forested grasslands in order to attract wild game for hunting and to reduce the risk of high-intensity fires near their villages. Those deliberate fires also served to remove dead grass to prevent it from choking the regrowth at the base of the grass sod. In dry climates where there is insufficient humidity to rot dead leftover vegetation, if dead grass is not regularly removed by either low-intensity wildfires or by grazing animals, the grass becomes uncompetitive and deteriorates, being replaced first by weeds and then ultimately deteriorating to bare ground. In dry climates, grazing herds and low-intensity wildfires both play a vital role in removing dead grass in order to keep the sod healthy. As grazing herd numbers shrink, more frequent low-intensity wildfires are needed in order to keep the grasslands healthy. Remove both, via political decisions to reduce grazing cattle and via wildfire management that prevents low-intensity wildfires, and you slowly build up the fuel load for ever more destructive high-intensity wildfires while simultaneously creating the conditions for weeds and brush to supplant lush grasses on the forest floor.

However, it would be a mistake to think that all pre-industrial wildfires were low-intensity wildfires creeping along the forest floor. Many were also massive all-consuming firestorms that completely destroyed existing tree stands. The three single largest firestorms in recorded North American history were the 1825 Miramichi Fire, which burned over 3 million acres in New Brunswick in a span of only eight hours; the Great Fire of 1910, which burned over 3 million acres in Idaho, Montana, and Washington in a span of just 6 hours; and the Great Fire of 1919 in the forests of northern Alberta, which laid waste to 5 million acres in just 9 days! These massive

---

[3] U.S. Government. (January 2001). *Review and Update of the 1995 Federal Wildland Fire Management Policy.*
https://web.archive.org/web/20201101021334/https://www.nifc.gov/PIO_bb/Policy/FederalWildlandFireManagementPolicy_2001.pdf

fast-moving firestorms absolutely dwarf the wildfires we see today. As you will soon see when I get back to discussing the massive outbreaks of mountain pine beetle, in addition to nature's many small low-intensity wildfires, nature also routinely rebalances an unhealthy ecosystem by burning everything to the ground.

Chapter 5

"Our Most Shameful Waste"

"I also expected science to banish the evils of human thought—prejudice and superstition, irrational beliefs and false fears. I expected science to be, in Carl Sagan's memorable phrase, "a candle in a demon haunted world." And here I am, not so pleased with the impact of science. Rather than serving as a cleansing force, science has in some instances been seduced by the more ancient lures of politics and publicity. [...] Once you abandon strict adherence to what science tells us, once you start arranging the truth in a press conference, then anything is possible. In one context, maybe you will get some mobilization against nuclear war. But in another context, you get Lysenkoism. In another, you get Nazi euthanasia. The danger is always there if you subvert science to political ends."

— Michael Crichton, from his cheeky Caltech lecture called "Aliens Cause Global Warming"[1]

The U.S. Forest Service was established in 1905 during the height of the Progressive buildout of government institutions. Its primary mission was to suppress ALL wildfires in order to protect marketable timber and defend civilian infrastructure. In the early years, the Forest Service focused on suppressing fires near cities, roads, and other infrastructure but then, as it built out its capabilities, it created a nation-wide network of fire watchtowers manned by fire rangers and rolled out new fire-fighting technologies like water bombers. By 1960, its efforts had become so effective that it succeeded in reducing forest fires by over 80% from 1930 levels (they were also helped along by the significant cooling that occurred in the global climate from the 1940s to the 1970s).

---

[1] Crichton, M. (2003, Jan 17). *Aliens Cause Global Warming*. Caltech Michelin Lecture.
https://stephenschneider.stanford.edu/Publications/PDF_Papers/Crichton2003.pdf

Beginning in the early 1900s and continuing until the 1970s, the Forest Service tried to completely replace the entire natural fire cycle with timber harvesting and fire suppression. In essence, we began farming our forests. The nature of our forests was changed forever through monoculture plantations, large blocks of similar aged trees, and forest management strategies designed to maximize the harvest of marketable timber. Meanwhile, decades of fire suppression led to decades of unburnt fuel accumulating on the forest floor.

Figure 19. Smokey the Bear was rolled out as the firefighting mascot in 1944 to enlist the public's help in preventing forest fires.

In 1974, the U.S. Forest Service finally recognized fire as a natural, necessary, and beneficial ecological process and slowly began to shift from away from trying to suppress all wildfires and towards managing wildfires to create a healthier forest ecology. By 2020, that strategy had evolved to where the U.S. intentionally ignites approximately 6,000 separate controlled burns that burn approximately 3.5 million acres each year — a drop in the bucket compared to historic burn acreages but a big step in the right direction nonetheless to restore a natural fire cycle to our forests. Of course, being a little more relaxed about wildfires happening way out in the mountains is one thing, but it's not so easy to light a controlled burn on the outskirts of a city. But more on that in a moment...

You're probably familiar with some version of the chart shown in figure 20, showing the increase in burn acreage in the United States since the 1980s. Politicians, media, climate activists, and dishonest climate scientists routinely use this chart to terrify the unsuspecting public by misrepresenting it as evidence of "climate change", which hides the fact that this chart begins just as our weather was beginning to recover from the chilly lows of the 1970s and ignores the fact that the Forest Service had just begun the gradual transition from a policy of complete fire suppression to fire management as an ecological process, a transition that was phased in very gradually beginning in 1974 and faced lots of criticism from politicians and media (especially after a few controlled burns spiralled out of control, like the 2000 Cerro Grande Fire in New Mexico that started as a controlled burn but ultimately turned into a 43,000-acre firestorm that destroyed over 400 homes in Los Alamos).[2] The fact that media, politicians, activists, and so many prominent "climate scientists" use the 1980s as the starting point for their scary-looking chart is the first warning sign that the half-truth, stripped of context, is being used as a propaganda tool to tell a whole lie.

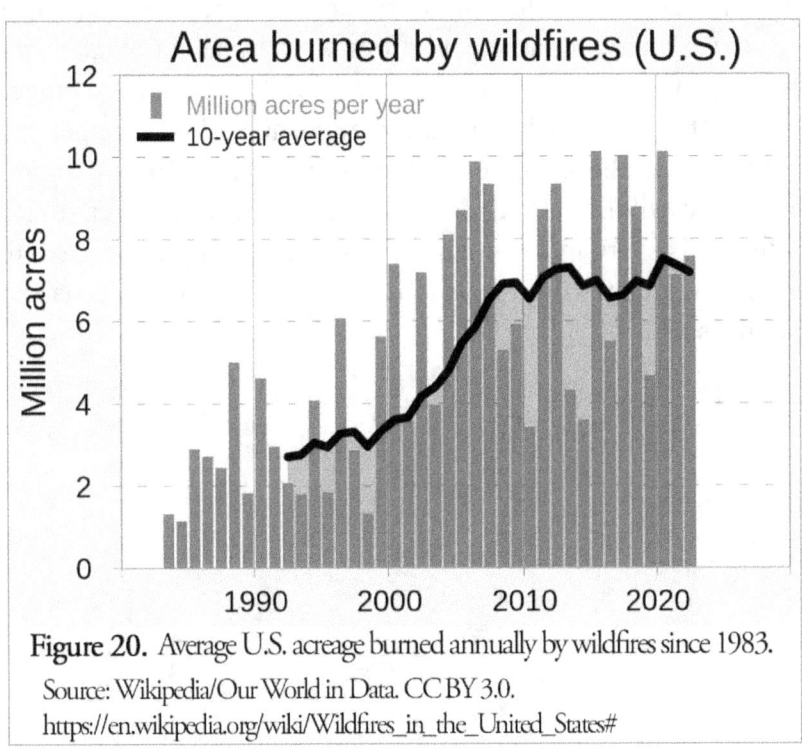

Figure 20. Average U.S. acreage burned annually by wildfires since 1983. Source: Wikipedia/Our World in Data. CC BY 3.0. https://en.wikipedia.org/wiki/Wildfires_in_the_United_States#

---

[2] "Cerro Grande Fire." *Wikipedia*. 20 May 2024. https://en.wikipedia.org/wiki/Cerro_Grande_Fire

Before I dive further into the gritty details of our forest management, it is important to note that the 3.5 million acres that are being burned annually in recent years as part of controlled burns are not included in the wildfire burn acreage shown in figure 20. However — and this point is key — prior to 1974 the goal was to extinguish all fires but after 1974 an increasing number of wildfires *were allowed to burn unopposed (even if they didn't start as controlled burns) as long as they were not a threat to cities and infrastructure*. And that acreage is included in this chart! Thus, the increase in burn acreage since 1974 does nevertheless reflect the gradual policy shift from complete fire suppression to managing a fire-dependent forest ecosystem that allows some wildfires to burn unopposed in order to restore the natural fire cycle to our forests. Charts stripped of context are convenient tools for promoting lies.

As the next chart (figure 21) shows, the total number of individual wildfires has decreased since the 1990s even as the burn acreage per wildfire has increased. This chart is also from the U.S., but the trend towards fewer but more intense wildfires is the same in Canada.[3] In other words, there are less fires now than in the 1980s, but they are bigger and burning hotter, no doubt partially reflecting this change in fire management that allows some fires to burn instead of trying to actively put out every wildfire. In an ideal world, we would prefer to see the opposite of this chart: more, smaller, low-intensity fires rather than fewer, larger, high-intensity wildfires. But that's not happening. As usual, there's more to the story...

---

[3] Natural Resources Canada. *Canadian National Fire Database (CNFDB)*. Accessed 30 Sept 2023. https://cwfis.cfs.nrcan.gc.ca/ha/nfdb

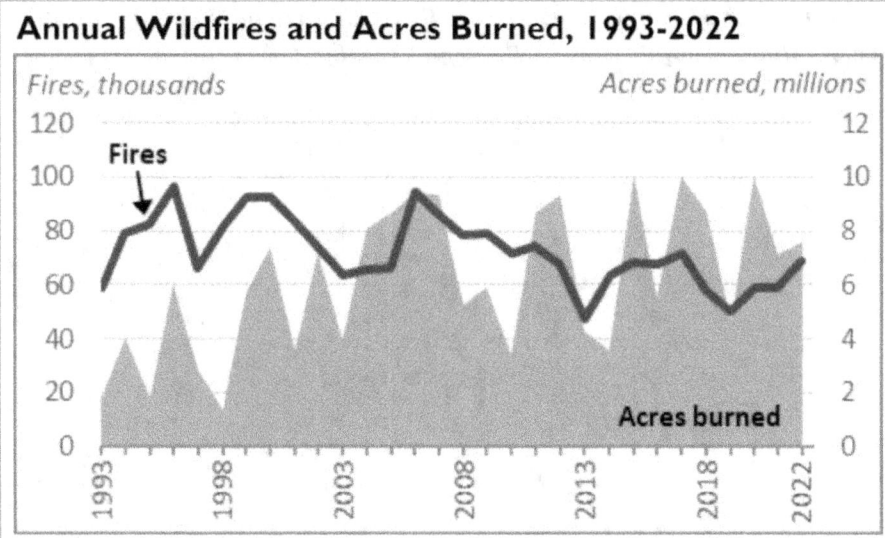

Figure 21. Burn acreage vs total number of fires each year according to the NICC Wildland Fire Summary and Statistics annual report. As you can see, wildfires are getting bigger (more intense) even as the total number of wildfires falls.

Source: Congressional Research Service Wildfire Statistics
https://crsreports.congress.gov/product/pdf/IF/IF10244

~

In 1987, Canada passed the Forest Act, which made it mandatory for timber companies to replant 100% of the trees that they harvest. Until it became law, only around 50% of forests were replanted, the rest were left to reseed themselves naturally. In 1988, the United States followed suit with the National Forest Management Act, prior to which only 30% of U.S. forests were replanted. In the early years after these new laws were passed, many timber companies replanted cutblocks with monocultures in order to meet their tree planting requirements, especially pine trees because they were cheap, fast growing, and matured into high value timber. That is, until a cycle of ever-expanding outbreaks of mountain pine beetles started turning millions of acres of forest red in the 1990s, as shown in figure 22 below. *By 2017, the mountain pine beetle had killed 58% of all marketable pine in British Columbia.*[4]

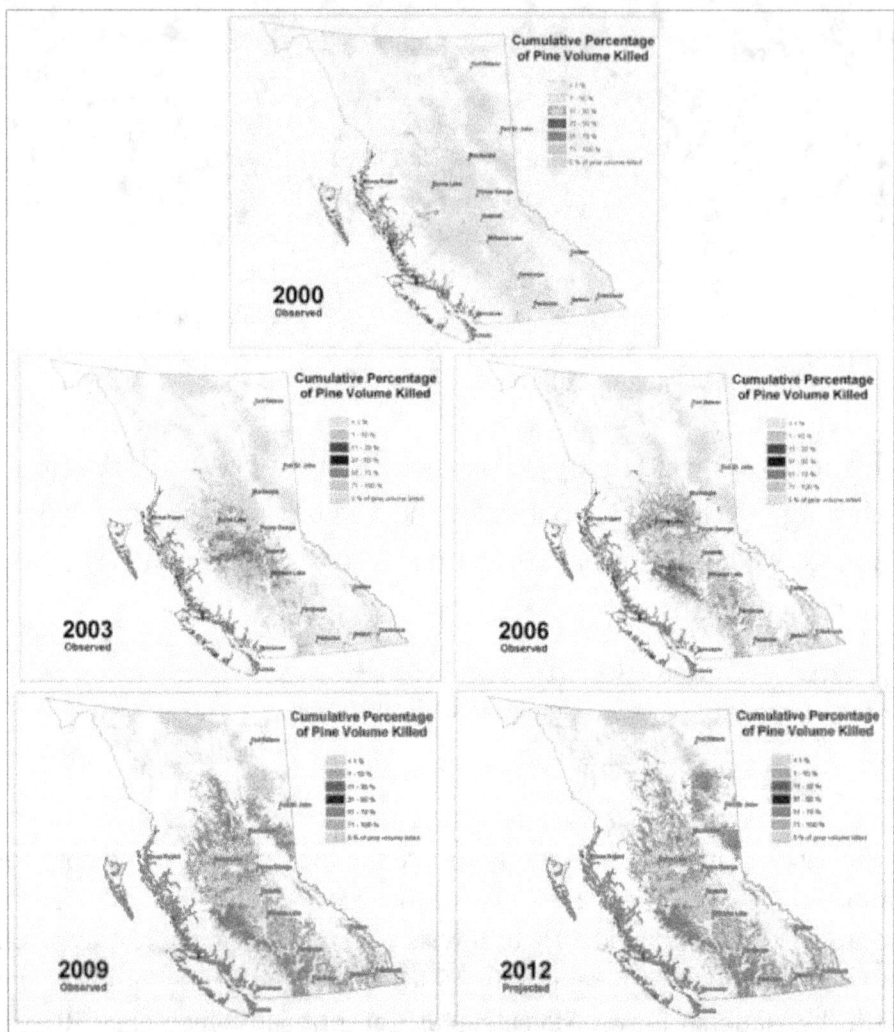

Figure 22. Mountain pine beetle outbreaks spreading in British Columbia from 2000 to 2012. By 2012, 53% of merchantable pine had been killed.
Source: BC Ministry of Forests, Lands, & Natural Resources
https://www2.gov.bc.ca/assets/gov/farming-natural-resources-and-industry/forestry/forest-health/bark-beetles/history_of_the_mountain_pine_beetle_infestation.pdf

---

[4] Natural Resources Canada. (n.d.). *Mountain Pine Beetle.* Government of Canada. Viewed 8 Jun 2024. https://natural-resources.canada.ca/our-natural-resources/forests/insects-disturbances/top-forest-insects-and-diseases-canada/mountain-pine-beetle/13381

Biodiversity creates disease resistance because it limits disease and parasite outbreaks to a smaller area. As every farmer knows, whenever you plant a monoculture crop, you dramatically increase its vulnerability to disease, which is why large-scale agriculture is virtually impossible without pesticides and herbicides. With such huge monoculture food sources, parasites, bacteria, fungi, and viruses can multiply so fast that they can easily overwhelm their natural predators, thus allowing them to spread unchecked through the whole crop. The spread of the mountain pine beetle in our northern forests and the locust plagues that sometimes tear through crops on the prairie both have to be seen in the exact same light — these are complex outbreaks fueled by large uniform food sources that, when combined with temporary weather conditions that favor their spread, allow parasites to multiply very rapidly, faster than their natural predators can eat them. Big intense forest fires are nature's way of cleansing an unhealthy forest ecosystem of a plague of pests. With the massive outbreak of pine beetle infestations that have turned vast swathes of our northern forests from green to red, it shouldn't surprise anyone that wildfires have been getting bigger and hotter.

As mountain pine beetle outbreaks began to spread in the 1990s, forest companies moved away from monoculture re-plantings as they recognized the vulnerability they were creating in our forests. However, since the average turnaround time between planting and harvesting is typically on the order of 60 to 90 years, we are nonetheless left with a legacy of those early monoculture re-plantings that will follow us for many years to come.

Furthermore, as logging equipment got bigger and as timber companies sought to optimize their harvesting efficiency, clear-cut cutblocks grew in size, leading to ever larger blocks of similar-aged trees that make it easy for fire, disease, and parasites to spread across huge tracts of mature forests. Moving from monoculture to multi-species plantings helps reduce the risk of large disease outbreaks and large fires, but it's still not ideal (nor is it possible everywhere due to local growing conditions). Far more important is having a wide range of ages in the tree stand by shifting to smaller block sizes and/or by relying on select logging to create much more fire-resistant multi-aged forests rather than creating large blocks of similar-aged trees that all reach maturity at the same time. But as the mountain pine beetle spread, efforts to fight the beetle only led to even larger cutblock sizes, out of necessity, in order to remove infected trees.

In recent years this trend has finally begun to reverse (outside of beetle-killed areas) as forest managers and regulators began to recognize the risks posed by such large cutblocks, which increased the risk of disease outbreaks, accelerated soil erosion, increased the risk of mudslides, and encouraged the spread of large wildfires. But again, the legacy of earlier forest practices will be with us for some time until these large blocks mature and can be split into smaller blocks.

Another (and perhaps the most important) unintended consequence of the tree replanting law is that by replanting all the coniferous trees immediately after logging or wildfires, foresters inadvertently disrupted the natural forest succession cycle in which brush, and then leafy trees (like aspen), repopulate the barren ground before the coniferous forest finally reclaims that ground. These economically unviable bushes and deciduous trees nevertheless play an important role in the health of a forest because their leaves (and their decaying roots and limbs as they die) help rebuild a fertile, nutrient-rich, moisture-absorbent, carbon-rich forest soil, which then provides a nutrient-rich base to feed healthy bug-resistant coniferous trees and creates the spongey soil texture capable of absorbing and holding on to vast quantities of groundwater, which in turn helps make the coniferous forest more drought resistant (the link between soil organic carbon and soil moisture absorption is well documented[5]). By skipping over these earlier phases of the natural forest succession cycle through our tree planting efforts, our forests are left increasingly nutrient deficient and increasingly vulnerable to drought. Everything is connected —nutrient-deficient trees and trees growing on humus-deficient soils are significantly more vulnerable to drought, pests, and parasites like the mountain pine beetles that bore beneath the bark to feed on their vascular tissue, which thrive whenever the trees are less capable of mounting an aggressive defense as the beetles begin to bore into their trunks. But blaming all these vulnerabilities on climate change helps everyone, from forest companies to public officials, shift the blame away from themselves.

~

---

[5] Zhao, X., et al. (2022). Soil organic carbon primarily control the soil moisture characteristic during forest restoration in subtropical China. *Frontiers in Ecology and Evolution.* 10. https://doi.org/10.3389/fevo.2022.1003532

A managed forest is, in essence, an agricultural crop. Any farmer can tell you what happens to a crop if it is starved of nutrients — in addition to growth being stunted and weeds invading the crop (for example, the link between calcium-deficiency and weed outbreaks in orchards is well known[6] — grass outcompetes weeds only as long as the grass isn't being put at a disadvantage by a nutrient deficiency that gives weeds the upper hand), nutrient deficiencies also make forests more vulnerable to drought, pests, and disease. Each successive harvest takes its toll as nutrients are transported away by each harvest — no less in a managed forest as from a farmer's field.

In chapter one, I discussed how wheat planted into the virgin soils of the freshly plowed Great Plains thrived because the first crop planted in any virgin soil (soil that has not been cultivated before) benefits from centuries of accumulated soil nutrients. The second crop does okay but not as well as the first, the third does more poorly, and so on, year after year, because repeated harvests are transporting nutrients out of the field, which leaves the soil increasingly depleted of nutrients as the years go by unless there is some system to replace/rebuild fertilizer elements that have been removed from soil.

The slash-and-burn farming in the Amazon basin is perhaps the most extreme example of the soil nutrient depletion caused by repeated harvests as farmers in the Amazon burn the forest to clear it (the ash left behind provides a boost of fertilizer to the notoriously nutrient-depleted soil), then grow a series of crops on the cleared land until crop productivity dwindles away (indicating that the soil nutrient reservoir has been exhausted), and then allow the forest (or pastures) to reclaim the exhausted land, out of necessity, because the soil is no longer productive.

Many pre-industrial farming methods (and some modern regenerative farming techniques) developed complex techniques designed to continually replace nutrients in the soil by relying on crop rotations, animal dung, a delicate livestock-to-crop balance, by including fallow fields in the crop rotation (a fallow field is a field that is allowed to remain uncultivated for a growing season in order to allow the soil to rest), and by planting cover crops grown exclusively for the purpose of plowing

---

[6] Soil Works LLC. (2024, May 13). *Calcium vs. Weeds (like Morning Glory) with Daniel Unruh.* [Video]. YouTube Shorts. https://www.youtube.com/shorts/J5B1prW0z2o

them back into the soil to replenish soil nutrients. But none of those soil management strategies are being used to replenish our forest soils after repeated harvests.

Entire civilizations have been brought to their knees because, despite their best efforts to sustain the fertility of their fields, they slowly depleted the soils upon which their civilization depended. The Maya, Easter Island, and the Sumerians are three of the most well-known examples of civilizations that collapsed as a consequence of exhausted soils, but there are countless more. In his book *Dirt: The Erosion of Civilizations*,[7] geologist and author David Montgomery explains how the mountainsides of ancient Greece were once covered by thick woodlands, but as agriculture expanded from the flat valley bottoms up onto the sloped hillsides to support their growing population, the hillside soils gradually eroded away to expose the bare bedrock below. In time, olive trees came to dominate the exhausted rocky Greek hillsides (and became a staple of the Greek diet) because olive trees are among the few crops that still thrive on these heavily eroded rocky hillslopes.

In his book, David Montgomery quotes Plato, writing in the 4th century BC, who observed that *"The rich, soft soil has all run away leaving the land nothing but skin and bone. But in those days the damage had not taken place, the hills had high crests, the rocky plain of Phelleus was covered with rich soil, and the mountains were covered by thick woods, of which there are some traces today."* As early as 590 BC, the hills around Athens were already stripped bare of soil, prompting Athenian statesman and philosopher Solon to try to ban plowing on steep slopes. By the 5th century BC, Greek soils had deteriorated to the point that Greece was importing between a third and three-quarters of its food from Sicily and Egypt. Montgomery also points out that this cycle of soil erosion and nutrient depletion played a pivotal role in triggering the population collapse at the end of Bronze Age, and another population collapse after the Classical Greek Age a thousand years later (I'll be digging deeper into those collapses and the role that climate played as their trigger in later chapters). We take our soils for granted, especially in our forests.

---

[7] Montgomery, D. R. (2012). *Dirt: the Erosion of Civilizations*. University of California Press.

> *"Here in a nutshell, so to speak, we have the underlying hazard of civilization. By clearing and cultivating sloping lands—for most of our lands are more or less sloping—we expose soils to accelerated erosion by water or by wind and sometimes by both water and wind. In doing this we enter upon a regime of self-destructive agriculture."*
>
> — W. C. Lowdermilk, Conquest of the Land[8]

The modern agricultural industry has learned to make chemical fertilizers to compensate for soil nutrients removed by each harvest — this is the modern chemistry cookbook approach to adding nutrients to grow things ("feed the plant"). This cookbook approach is increasingly being combined with low-till and no-till cultivation practices to reduce the rate at which soil is lost to erosion (no-till farming techniques, aided by herbicides, were pioneered on the Great Plains in the aftermath of the Dust Bowl). The other advantage of no-till farming (something that we will be exploring in greater detail in later chapters as part of the untold story of CO_2) is that no-till farming not only reduces soil erosion but also keeps the soil covered with plant debris, which significantly reduces *soil degassing*. Whenever the sod cover is removed to expose bare soil directly to the oxygen-rich atmosphere, oxygen in the air begins to oxidize the carbon particles in the soil. As this organic matter (a.k.a. humus) is oxidized, the carbon bonds with oxygen to form carbon dioxide, which then degasses into the air. The carbon component of bare soil is quite literally floating away, ever so slowly robbing farmers (and foresters) of their fertile soils and of their soil's moisture absorption capacity. It's an incalculable loss because humus plays a pivotal role in storing soil nutrients (thus reducing the dependency on fertilizer inputs). And humus also serves as the soil's sponge to give soil its ability to soak up and store rainwater to prevent it from draining away between the particles of inorganic sand, silt, and clay.

The cookbook approach to soil fertility stands in stark contrast to the older organic approach to managing soil fertility, inherited from the era before chemical fertilizers, which relies on harnessing biological and

---

[8] Lowdermilk, W. C. (1953). *Conquest of the Land Through Seven Thousand Years*. Natural Resources Conservation Service. http://soilquality.org/history/files/conquest_of_the_land.pdf

chemical properties in the soil to create a rich soil-microbial ecosystem that feeds crops with nutrients from below ("feed the soil") while crop residues and manure are continually added back to the soil to replace the carbon that is continually lost to degassing during tillage and to wind and water erosion. These two entirely different approaches to managing soil fertility are not mutually exclusive — each have their place, each have their positives and negatives, and as the example of no-till farming shows, a merger of the two philosophical approaches is paying dividends.

However, the modern forest industry typically uses neither method to manage forest soils. Trees get harvested, new seedlings are planted, and then the trees are left to fend for themselves on increasingly nutrient-depleted, carbon-depleted, and eroded soils.

Despite the fact that modern forests are, in essence, agricultural crops, very few managed forests ever see a speck of fertilizer to replace the nutrients that are exported out of the forest during the timber harvest. Furthermore, by suppressing wildfires we have interrupted the low-intensity wildfires that historically recycled nutrients back to the soil from all the accumulating debris on the forest floor. To make matters worse, clear-cut logging in large cutblocks has accelerated the erosion of the thin mountain soils upon which our forests grow. And while historic forest soils were kept fertile by the manure and grazing impact of vast herds of bison, elk, and deer, and then later by vast herds of cattle and sheep after the west was settled by ranchers, those historic herds are only a fraction of the size they once were (more on this complex issue in a moment), so grazing animals are contributing far less than they once did to help build and enrich our forest soils. We are ignoring the fertility of our forest soils at our peril.

The World Wildlife Fund estimates that half the topsoil on the planet has been lost over the last 150 years.[9] The aforementioned geologist, David Montgomery, estimates that we are losing approximately 1% of our agricultural soils every year due to erosion[10] — and that estimate doesn't include the loss of soil in our forests, which typically grow on much steeper land. Equally alarming as the loss of humus in our farm and forest

---

[9] "Threats: Soil Erosion and Degradation" *World Wildlife Fund.* Accessed 8 Jun 2024. https://www.worldwildlife.org/threats/soil-erosion-and-degradation

[10] Verso, E. (2015, Dec 9). Topsoil Erosion. *Stanford University Course Work.* http://large.stanford.edu/courses/2015/ph240/verso2/

soils is the depletion of many of the minerals and trace minerals that keep soils and plants healthy. The farming industry understandably prioritizes fertilizer elements that have the biggest impact on yield (nitrogen, phosphorus, and potassium), but calcium, magnesium, iron, and other trace minerals are often neglected. To take magnesium as an example, magnesium deficiencies in humans play a significant role in diseases like hypertension, diabetes, and neurological disorders. In trees and other plants, magnesium deficiencies result in wide range of negative consequences that include stunted growth and increased vulnerability to drought.

The graph in figure 23 is reproduced from a peer-reviewed research study entitled *Challenges in the Diagnosis of Magnesium Status*; the graph shows how the mineral content of our food, in particular calcium, magnesium, and iron, has dropped a staggering 80-90% between 1914 and 2018.[11] And that's in food grown on "well-managed" farm soils that are regularly fertilized. Now consider what is happening to our forests after many repeated harvests, especially considering that soil fertility management is extremely neglected or altogether absent in most of our forests.

---

[11] Workinger JL, Doyle RP, Bortz J. Challenges in the Diagnosis of Magnesium Status. *Nutrients*. 2018 Sep 1;10(9):1202. doi: 10.3390/nu10091202.

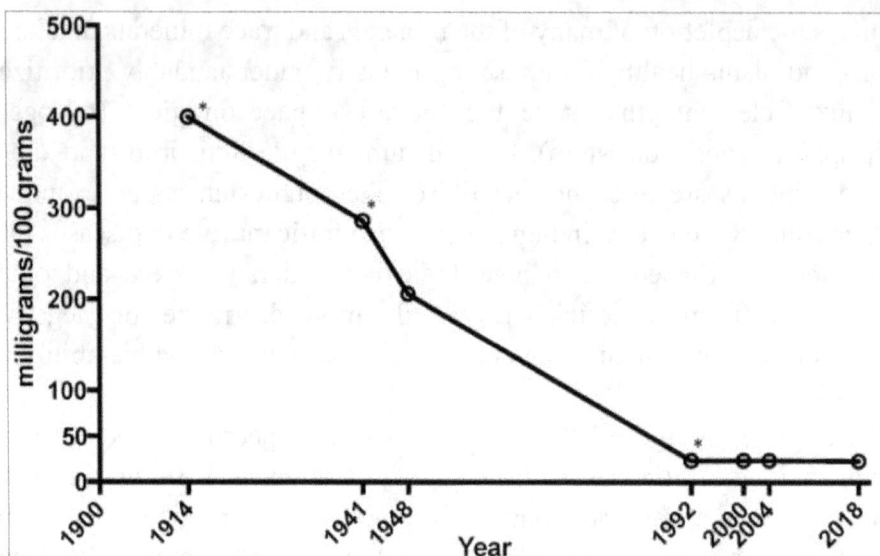

Figure 23. The average mineral content of calcium, magnesium, and iron in cabbage, lettuce, tomatoes, and spinach has dropped 80–90% between 1914 and 2018.

Source: Workinger et al, 2018.
https://www.ncbi.nlm.nih.gov/pmc/articles/PMC6163803/

As the many threads of this book coalesce towards the end this book, it will become clear how soil erosion, the mineral depletion of the remaining soil, and the destruction of the rich microbial life that lives in humus are all playing pivotal roles in making our local ecosystems (including our forests) increasingly vulnerable to natural climate variations, similar to what happened on the Great Plains in the late 19th and early 20th century. A local problem repeated all around the world by people using the same or similar land management methods gives the appearance of a global problem when, in reality, these are local management failures that are only fixable through local management changes. It merely looks like a global phenomenon because everyone is following the same recommended "best practices" promoted by our government-funded scientific-technological elites.

The forest soils in the mountains and on the plateaus of the BC interior are not like the deep, rich soils found on the Great Plains or those found in farmer's fields on BC's rich valley bottoms, which can be farmed for generations before the consequences of nutrient depletion can no longer be ignored. These fragile forest soils are little more than a thin crust over sloped bedrock, or, in the case of BC's Interior Plateau, a paper-thin skin over ancient volcanic basalt flows and glacial rubble. These soils don't have a deep reservoir of soil nutrients to begin with, and that thin reservoir is continually being depleted by each successive tree harvest and by the erosion caused by clear-cut logging practices. And so, year by year, harvest by harvest, raindrop by eroding raindrop falling on bare exposed soil, our forests are becoming increasingly nutrient deficient and more vulnerable to parasites and drought. And few recognize the real causes of this increased vulnerability because every change in our environment is automatically blamed on the burning of fossil fuels.

The proof of the relentless erosion of our forest soils is found in our rivers. I've seen it firsthand on my parents' farm as mud, sand, and gravel from the heavily logged Aberdeen Plateau to the East of Kelowna is washed downstream and is now filling up the bed of the creek that runs through their farm, causing ever more frequent spring floods because the silt-choked streambed makes it easier and easier for spring floods to jump the streambanks and flood across dozens of acres of surrounding farmland — it's a constant source of friction between local farmers and the government, which refuses to allow farmers to dredge the creek bed out of fear that they might damage salmon spawning beds but also refuses to compensate the farmers for flood-damaged crops and as their fields revert back into swamps as a consequence of government-managed logging practices in the surrounding hills. [Ironically, while farmers are prohibited from disturbing a single grain of sand in the creek beds that cut across their lands, as soon as those same floods impact townships, the government commandeers local earth-moving equipment to dredge out the swollen creeks to protect houses, and even asks local farmers to bring their excavators into town to help clear the flooded creek bed.]

On my parent's farm, we witnessed firsthand as the amount of soil and sediment washing down from the uplands increased dramatically (and the creek turned noticeable browner) after a massive increase in logging activity on the mountainous Plateau above our valley during the early 2000s. But government officials prefer to blame the floods on "climate change," and promote taxes and electric vehicles as the solution.

The impact of declining soil fertility in our forests is also becoming increasingly noticeable by the growing number of spindly, sickly, disease-prone trees in our forests, especially in areas where the soil cover is thinnest, where soil nutrients were most scarce to begin with, and where soil erosion has been the fastest. The size of the trees being harvested by loggers is getting smaller with each crop not only because we're impatient and harvest them younger, but because overcrowded stands of trees growing on nutrient-deficient mountain soils will not grow as large and are more vulnerable to disease and drought, which forces timber companies to harvest them sooner to prevent them from falling prey to fires and beetles.

In the B.C. Interior, virtually every single patch of forest has been harvested, on average, between two and four times. In Washington State, it's between three and five times. In California is between two and four. Most of these forests have never received a single drop of fertilizer to help regrow the next crop of trees. A 70-year-old tree has extracted 70 years' worth of nutrients from the soil — in other words, the impact of harvesting a tree on soil fertility is somewhat akin to harvesting 70 consecutive years of agricultural crops without putting anything back to replenish the soil.

Figure 24 shows a series of photos of normal versus nutrient-deficient seedlings published by the B.C. Ministry of Forests to illustrate the dramatic impact that nutrient deficiencies have on tree growth. From their pale appearance and stunted size, it's easy to imagine how much more vulnerable these seedlings will be to disease and wildfires as they continue to mature if these nutrient deficiencies are not addressed:

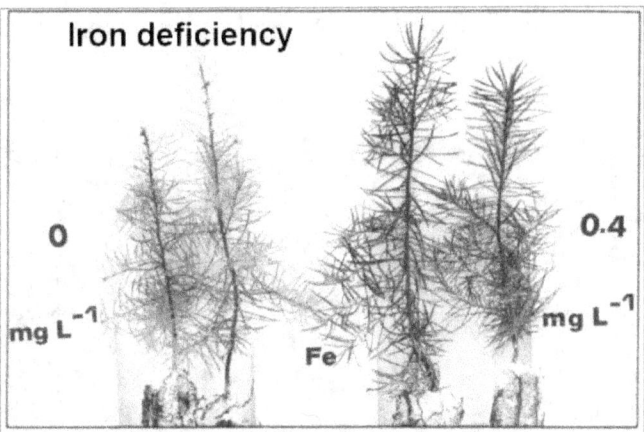

Iron deficient (0) and normal (0.4 mg iron L^{-1}) Douglas-fir photographed in October.

Normal (2.0 mg zinc L^{-1}) and zinc-deficient Douglas-fir (0) photographed in October.

Normal (0.02 mg boron L^{-1}) and boron-deficient (0) Douglas-fir photographed in October.

Figure 24. Iron, zinc, and boron deficiencies in Douglas Fir trees. Source: B.C. Ministry of Forests. https://www.for.gov.bc.ca/hfd/pubs/Docs/Frr/Frr100part2.pdf

And the photo in figure 25 shows soil erosion in a cutblock in the aftermath of logging:

Figure 25. Soil erosion in a cutblock continues to wash away sediment into the nearby stream.

Source: Government of British Columbia
https://www2.gov.bc.ca/gov/content/industry/forestry/managing-our-forest-resources/integrated-resource-monitoring/forest-range-evaluation-program/frep-training/riparian-protocol-training/riparian-protocol-training-lesson-3

Many countries with intensely managed, privately owned forests regularly fertilize their forests with nitrogen, potassium, and phosphorus, and in some cases also with key micronutrients like sulfur, boron, and zinc. Tree fertilization is most common in Sweden, Finland, and in privately-owned pine plantations in the southern United States. But only about 10% of public U.S. forests get regular fertilization.

A tree growing in fertile soil is not only more resistant to drought and beetle attacks, there is also a clear benefit in terms of timber values: a Douglas Fir forest will grow 20-30% more volume following nitrogen fertilization and returns on fertilization can be in excess of 15% when comparing the cost of fertilization to the increased value of timber.[12] As erosion and nutrient depletion continue to degrade forest soils, chemical fertilization will become an increasingly essential tool to keep our forests healthy and resistant to drought. But in vast tracts of publicly managed forests that are harvested by private timber companies, who pays for the fertilizer? How many more harvests until the problem can no longer be ignored? Will the problem ever be addressed as long as politicians, bureaucrats, "experts", and the media keep misdiagnosing the problem as "climate change" while diverting our tax dollars to subsidize solar farms instead of using those dollars to replace nutrients in our forest soils?

British Columbia has begun experimenting with forest fertilization using aerial applications of nitrogen, in particular in lodgepole pine forests that have been affected by mountain pine beetle outbreaks. But as of 2019, British Columbia had only fertilized around 150,000 hectares of forest land, or about 0.2% of its total forest area, and adding fertilizer after a disease outbreak is like fighting a rear-guard war — not ideal. An ounce of prevention would have saved a pound of cure. Likewise, in the publicly owned forests of the northwestern United States, forest fertilization is being used on a tiny area but again it is a mere drop in the bucket in terms of overall acreage. It's not that our forest managers don't understand the benefits of fertilization (they do), the problem is that government budgets are controlled by politicians who don't understand, budgets are already strained, there are a million other more publicly-visible issues competing for government money, timber companies want to keep costs and royalties low, and environmentalists scream bloody murder about using fertilizer in public forests and near watersheds. And bringing more cattle herds back into the forests to boost soil fertility the natural way is also a political non-starter as long as professional environmental activists continue their crusade against livestock. When politics and science collide, politics always win. And with climate change hysterics like John Kerry having recently added nitrogen fertilizer to their list of scapegoats for "climate change", it's only going to get even more difficult for government scientists to convince policymakers that fertilizing a public forest is a good idea.

---

[12] Fox et al., (2007). Forest Fertilization and Water Quality in the United States. *Better Crops,* 91(1).
https://www.for.gov.bc.ca/ftp/hfp/external/!publish/Locke/Fertilization%20Information/environmental/Water/Forest-Fertilize-Water-Qual-US%202007.pdf

A similar issue arises with the funding required to properly space regrowing trees or clean the forest floor — who pays? It's difficult to force timber companies to pay to manage forests they don't own and to spend money managing timber stands that might be consumed by wildfires before they can be harvested. As for the government, they are already drowning our country in debt by spending far more than they take in through taxes. Sadly, something as mundane as cleaning up the forest floor in your city's backyard isn't going to attract nearly as much attention (or as much public funding) as some flashy high-tech billionaire claiming they can fix all our problems by dimming the sun.

For example, in the latest climate boondoggle, Forbes recently reported that Bill Gates and other investors are pushing an idea to cut down millions of trees in order to bury them as carbon sinks, and then sell carbon offsets based on how much these buried trees will allegedly reduce carbon dioxide in the atmosphere (figure 26).[13] And — surprise, surprise — they are lobbying the government to give them tax credits for doing so. Various companies are already lining up tracts of forests to rake in their share of this hare-brained, ecology-destroying, money-making scheme.

Forbes

Chop Down Forests To Save The Planet? Maybe Not As Crazy As It Sounds

Kodama Systems is betting it can reduce carbon dioxide in the air by chopping down and burying trees. Now if only Uncle Sam would get on...

Jul 28, 2023

Figure 26.
https://www.forbes.com/sites/christopherhelman/2023/07/28/chop-down-forests-to-save-the-planet-maybe-not-as-crazy-as-it-sounds/

The formerly prestigious popular science magazine *Scientific American* even published a recent article[14] (based on a report published by

---

[13] Helman, C. (2023, Jul 28). Cut Down Forests To Save The Planet? Maybe Not As Crazy As It Sounds. *Forbes*. https://www.forbes.com/sites/christopherhelman/2023/07/28/chop-down-forests-to-save-the-planet-maybe-not-as-crazy-as-it-sounds/

[14] Kim, M. and E&E News. (2023, Jul 26). Forests Are Losing Their Ability to Hold Carbon. *Scientific American*. https://www.scientificamerican.com/article/forests-are-losing-their-ability-to-hold-carbon/

the USDA[15]) claiming that because rotting wood emits CO_2, aging forests are losing their ability to absorb carbon so that forests are allegedly set to become "substantial" carbon emitters as "decaying trees exceed their carbon footprint" (figure 27). It seems that in the span of just a few years, we have gone from protecting old growth forests to cutting down mature trees to prevent them from getting old. Either the entire basis of what creates and sustains forest soils and forest ecosystems appears to be completely beyond their grasp or perhaps the opportunity to milk the nanny state was simply too tempting to ignore.

Figure 27. On July 26th, 2023, Scientific American published an article rationalizing why U.S. forests are becoming major carbon emitters.
https://www.scientificamerican.com/article/forests-are-losing-their-ability-to-hold-carbon/

And the madness doesn't stop with cutting down trees only to bury them. Time Magazine reports that Bill Gates, along with George Soros and Jeff Bezos (among others), are working on "solar engineering" schemes to seed the upper atmosphere with chemicals like sulfur dioxide in an effort to try to dim the amount of sunlight reaching the surface of the Earth (figure 28).[16] Even the White House has begun pushing this research since 2022.[17] These schemes will, in all likelihood, be carried out by corporations owned by these same investors and be paid for by the government using our taxpayer dollars. It's incredible how many predatory money-making schemes are spawned once a society can be deceived into

---

[15] Forest Service U.S. Dept. of Agriculture. (2023). *Future of America's Forest and Rangelands: Forest Service 2020 Resources Planning Act Assessment.* https://www.fs.usda.gov/research/treesearch/66413

[16] De La Garza, A. (2023, Feb 24). Why Billionaires are Obsessed With Blocking Out the Sun. *Time Magazine.* https://time.com/6258126/solar-geoengineering-billionaires-george-soros/

[17] Clifford, C. (2022, Oct 13). White House is pushing ahead research to cool Earth by reflecting back sunlight. *CNBC News.* https://www.cnbc.com/2022/10/13/what-is-solar-geoengineering-sunlight-reflection-risks-and-benefits.html

believing that carbon dioxide is some kind of scary control knob on atmospheric temperatures.

Time Magazine
https://time.com › Climate › adaptation

Why Billionaires are Obsessed With Blocking Out the Sun

Feb 24, 2023 — George Soros is interested in using geoengineering to address climate change. He isn't the first billionaire to give support to the idea.

Feb 24, 2023

Figure 28. https://time.com/6258126/solar-geoengineering-billionaires-george-soros/

In essence, this story has become a classic example of how central planning fails at every level as a million competing interests line up to influence policy to their advantage. It's a never-ending vicious cycle. The same institutions that cause the problems are also tasked with fixing them. No-one is fully responsible for anything. The cost of neglect takes decades to become visible, if it is recognized at all. The local voices of those with practical experience who see the problems firsthand (like farmers and loggers and even many of the scientists working at ground level in our institutions) are routinely ignored because they have neither the right "credentials" nor the "right" political connections to get their voices heard. Those who do have connections within government are more likely to divert attention to flashy solutions that accomplish little other than to make Wall Street rich. The scientists working in government who understand what needs to be done are subordinate to clueless politicians and self-serving department heads — career promotions in these byzantine hierarchical institutions tend to favor consensus-seekers and Yes Men over independent thinkers. Meanwhile, the voting public has been bamboozled to believe there is high-tech solution to every problem. A lot of technological solutions are being developed by clever people who lack wisdom and practical experience but excel at marketing and at lobbying the gatekeepers. Self-serving academics who cater to the biases and beliefs of politicians and journalists are far more likely to get funding than the honorable scientists who don't. And the decision-makers who ultimately decide on the policies that everyone else needs to live by pay no price for being wrong — if anything, as long as they can find convenient scapegoats for why their policies failed (i.e., climate change), they're likely to be promoted (fail upwards) and be given even more taxpayer funds to fight the problems they themselves created.

As the size of government outgrew the original limits imagined by our Founding Fathers, democracy has degenerated into a kind of vulgar public spectacle optimized for pilfering the public purse, an elaborate kleptocracy where truth is secondary to whatever plausible-sounding slogans can energize a voting base, secure promotions, impress decision-makers, uncork funding, get politicians re-elected, and shower investors with tax credits. But who is ultimately to blame for this lunacy? Does the fault lie with the billionaire investors with bad ideas who have perfected the art of leveraging their connections in government in order to create money-making schemes? Does the fault lie with the Yes-Men advising and egging them on? Does the fault lie with the politicians whose careers rise and fall depending on how effectively they rub shoulders with influential people? Does the fault lie with the academics and bureaucrats defending their careers by pandering to whatever hysteria is currently in vogue as a matter of survival in politicized institutions? Does the fault like with an attention-hungry media that sells its soul to the highest bidder? Or does the fault lie with the credulous public, which habitually uses the ballot box to expand the size of government in their own effort to turn the power and resources of the nanny state to their advantage? As 19th century free market economist and property rights advocate Frederic Bastiat once wrote, *"Government is the great fiction, through which everybody endeavors to live at the expense of everybody else."*

Perhaps it is the system itself that is ultimately to blame because, by centralizing power and resources in the hands of a small number of gatekeepers, we created a political machine that incentivises an endless stream of corruption, conflicts of interests, special interest politics, and grift. What if we created this monstrosity ourselves because we thought that by empowering a vast institutional network of government "experts" to watch over us, we thought we could solve all our problems? Only now, just like the creation that Dr. Frankenstein stitched together in his laboratory, our monster has outgrown our ability to control it, much less reform it.

~

There's yet another factor that's driving an increase in forest fires that is routinely left out of the debate. People. In the U.S., human-caused fires account for a whopping 87% of all wildfires[18] (in California, it's over 90%[19]). Even more disturbingly, human-caused fires account for 97% of wildfires that threaten homes.[20] In Canada as a whole, humans ignite around 50% of wildfires. In British Columbia, it's about 40%.[21] And those numbers have only gotten worse over time. In BC, only about 25% of wildfires were caused by humans in the 1950s compared to 40% today.[22] In the United States, human-caused wildfires increased from 78% to 87% from 1984 to 2016.[23]

> Toronto Sun
>
> **WARMINGTON: Politicians always have lots to say about wildfires, except when arson charges laid**
>
> The CBC reports the accused "allegedly set fire to a fishing cabin on May 31 and to forests in the area between July 8 and Sept. 5" in an area...
>
> Sep 9, 2023

Figure 29. https://torontosun.com/opinion/columnists/warmington-politicians-always-have-lots-to-say-about-wildfires-except-when-arson-charges-laid

Of all human-caused wildfires, 29% are ignited by people burning debris, 21% are caused by arson (including arson started by would-be

---

[18] National Interagency Fire Center. (n.d.). *Wildland fire investigation: common wildfire causes*. Viewed 17 Oct 2023. https://www.nifc.gov/fire-information/fire-prevention-education-mitigation/wildfire-investigation

[19] Palm, I. (2023, Jul 23). Humans to blame for about 90% of wildfire ignitions, report finds. *KTLA News*. https://ktla.com/news/california/humans-to-blame-for-about-90-of-wildfire-ignitions-report-finds/

[20] Joosse, T. (2020, Dec 8). Human-sparked wildfires are more destructive than those caused by nature. *Science*. https://www.science.org/content/article/human-sparked-wildfires-are-more-destructive-those-caused-nature

[21] Government of British Columbia. (2023, May 18). *What causes wildfire*. https://www2.gov.bc.ca/gov/content/safety/wildfire-status/wildfire-response/what-causes-wildfire

[22] Gelineau, J. (2022, Sept 1). Human-caused wildfires dip to lowest rate since 1950 but B.C.'s hot, dry fall a concern. *Langley Advance Times*. https://www.langleyadvancetimes.com/news/human-caused-wildfires-dip-to-lowest-rate-since-1950-but-b-c-s-hot-dry-fall-a-concern-2568084

[23] Cattau et al. (2020). Anthropogenic and lightning-started fires are becoming larger and more frequent over a longer season length in the U.S.A. *Global Ecology and Biogeography*, 29: 668–681. https://doi.org/10.1111/geb.13058

firefighters in areas where there are few other sources of employment[24]), equipment use causes another 11% (i.e. hot mufflers, sparks, etc.), and campfires and kids playing with matches or fireworks account for another 5%.[25] Additionally, there are controlled burns that get out of control, cigarette butts that get thrown out of cars, hot engine exhausts that ignite dry grass, power line transformers that spark when they are overloaded, and flat tires that go unnoticed on trailers, which cause their rims to spark as they're dragged along the asphalt.[26]

In California, the explosive failure of power lines and other electrical equipment has regularly ranked among the top three sources of California wildfires.[27] Outdated electricity grids bear some of the blame, but solar mandates imposed by California's government are playing havoc with the ability of power companies to create smooth transitions between solar and conventional electricity sources when the sun goes down, especially during heat waves when everyone's air conditioning is putting pressure on power grids.[28] During the daily switch from solar to conventional power, there is a momentary demand-supply imbalance called the "duck curve",[29] which overloads grids and risks causing transformers to spark or explode, which happens most often during heat waves when everyone's air conditioner is running at full tilt. Fear of "climate change" has led the public to adopt a kind of wilful blindness to the terrible price that is being

---

[24] Some wildfires started by would-be firefighters, investigator says. (2013, May 17) *CBC News* https://www.cbc.ca/news/canada/saskatchewan/some-wildfires-started-by-would-be-firefighters-investigator-says-1.1364016

[25] Daley, J. (2017, Feb 28). Study Shows 84% of Wildfires Caused by Humans. *Smithsonian Magazine.* https://www.smithsonianmag.com/smart-news/study-shows-84-wildfires-caused-humans-180962315/

[26] Karimi, F. and Mossburg, C. (2018, Aug 5). A flat tire started the deadly Carr Fire and days of devastation in California. *CNN.* https://www.cnn.com/2018/08/04/us/carr-fire-week-wrap/index.html

[27] Penn, I. (2017, Oct 17). Power lines and electrical equipment are a leading cause of California wildfires. *Los Angeles Times.* https://www.latimes.com/business/la-fi-utility-wildfires-20171017-story.html

[28] Rotter, C. (2019, Mar 4). Solar energy may have caused California's wildfires. *Watts Up With That.* https://wattsupwiththat.com/2019/03/09/solar-energy-may-have-caused-californias-wildfires/

[29] Department of Energy. (2017, Oct 12). Confronting the Duck Curve: How to Address Over-Generation of Solar Energy. *Energy.gov.* https://www.energy.gov/eere/articles/confronting-duck-curve-how-address-over-generation-solar-energy

paid as we allow our politicians to forcibly impose their climate mandates upon us and on our power companies, consequences be damned, thus creating the very crises that are being blamed on climate change. Even the devastating fires that swept through Maui in 2023, which destroyed the city of Lahaina and claimed the lives of 101 victims, were started by power grid malfunctions despite all the headlines attempting to link these devastating fires to "climate change".[30]

In 1976, the city of Kelowna had a population of 52,000. By 2021, Kelowna's population had tripled to over 145,000 residents even as the greater metropolitan area had surged to over 222,000. Nearby Vernon has grown from 17,000 people in 1976 to over 40,000 in 2021, with a ballooning metropolitan area of over 67,000 people! With such rapidly expanding populations, with urban sprawl penetrating deep into surrounding forests, and with so many more people recreating in the forests, it's hardly surprising that humans are causing an ever-growing percentage of wildfires, all of which contribute to the increase in wildfires since the 1980s that politicians prefer to blame on "climate change".

As I mentioned earlier, when I was young my parents had a grazing lease for their cattle in the public forests on the Aberdeen Plateau to the east of Kelowna. We spent a good portion of our summers playing hide-and-seek with our cows across tens of thousands of acres of dense forests, overgrown cutblocks, and alpine meadows. Other than the occasional weekend fisherman, these forests were fairly quiet on weekends during the late 70s and early 80s when the logging trucks were not hauling timber. But as Kelowna, Winfield, Vernon, and other surrounding cities grew larger, as camping and hiking became more popular, and especially following the beginning of mass production of dirt bikes and ATVs in the 1980s, the number of people in the forests skyrocketed. If you go up onto that same plateau on any weekend now, it is absolutely buzzing with human activity. It's a story that repeats itself in the forests in the vicinity of every single town and city in North America. And it all contributes to more accidental wildfires and in particular to more dangerous wildfires that ignite in the vicinity of towns and cities that have done a poor job of managing the fire risk in their periphery.

---

[30] Blair, A. (2023, Aug 17). New data reveals dozens of power grid malfunctions as Maui wildfires began. *Hawaii News Now*.
https://www.hawaiinewsnow.com/2023/08/18/new-data-reveals-dozens-power-grid-malfunctions-leading-up-maui-wildfires/

Chapter 6

The Toxic Marriage Between Science and Politics

"When plunder becomes a way of life for a group of men in a society, over the course of time they create for themselves a legal system that authorizes it and a moral code that glorifies it."

— Frédéric Bastiat (1801-1850), French statesman, writing in the era of the Revolutions of 1848 when France and many of its European peers were rapidly turning towards socialism.

My parents' former grazing lease in the public forests in the mountains to the east of Kelowna is part of a very old tradition in the North American West that grants grazing leases to farmers so they can graze their cattle on public lands. The cows provide a valuable service — they get fat on the grass in the mountains in exchange for removing the masses of grass and brush that normally accumulate on the forest floor. Their mouths remove the grass before it turns into a tinder-dry fire hazard, their manure recycles grass and brush to serve as fertilizer for the trees (thus helping build more fertile forest soils and counteracting the soil-nutrient depletion caused by repeated timber harvests), their urine adds much-needed nitrogen to the soil, and their feet trample brush and debris so it can be composted in the soil before it becomes fuel for a wildfire (contrary to the fears being stoked by *Scientific American* and the USDA about the allegedly dangerous carbon footprint caused by decaying trees, you can't build a rich forest soil without rotting plant material).

As the cattle plow through the forests, they break off dry branches at the base of trees, which makes it harder for grassfires to reach the lower branches and climb up into the forest canopy. They also knock over small trees and brush that would otherwise compete with the timber crop for sunlight and nutrients, effectively spacing the trees in the forest without mechanical intervention. And the cows do it all without asking for a penny of taxpayer funds and without burning a drop of diesel to power their march through the forests. All combined, this has given cattle the reputation of being the poor man's bulldozer.

In essence, herds of grazing cattle accomplish many of the same goals that low-intensity wildfires accomplish, except without the smoke, without the fire risk, and by recycling plant matter accumulating on the forest floor. And they do it without requiring anyone to cut down trees just to bury them back in the ground to give tax credits to investors — Bill Gates and his friends should really spend an afternoon in the woods with a cow. In arid and semi-arid regions, you only have two tools to choose from to manage the fuel loads in the forests and grasslands — livestock or wildfires — if both are excluded, the dead grass will soon choke out the regrowth even as the accumulating dead plant debris becomes a dangerous fire hazard.

Grass co-evolved with grazing herds and with fire. In dry climates (below the 20-inch rainfall line), there isn't enough moisture to efficiently rot leftover dead grass. Without regular grazing, grass ecosystems in dry climates quickly begin to deteriorate — understanding this concept is key to understanding the hare-brained folly of the environmental crusade against grazing livestock. Perennial grasses regrow from the root base *below the level at which grazing animals rip it off while grazing,* so if grass is not regularly removed by grazing or by fire as it reaches maturity, the leftover dead grass will build up and begin to choke out the regrowth. Without regular grazing or low-intensity wildfires, semi-arid landscape like those in western North America, Africa, southern South America, the Asian steppe, etc., rapidly deteriorate as the grass becomes uncompetitive and chokes itself out, giving way first to a takeover by weeds and ultimately to bare soils. Without regular grazing (and low intensity wildfires) to keep these dry ecosystems healthy, the grassland ecosystem quickly begins to fall apart. That, in a nutshell, is the leading cause of desertification.

As will become increasingly clear in the upcoming chapters, desertification is primarily a local land management problem. And it's worth remembering that arid and semi-arid regions (which are completely unsuitable for crop production without irrigation) make up two-thirds of the world's land surface, as shown by the light-colored areas in figure 30 below. Without grazing animals to keep the grass healthy in these dry regions, they rapidly deteriorate into desert once the grazing animals are removed (they also rapidly deteriorate into desert if the grazing animals are not managed to mimic the behaviour of the wild herds that once roams across the plains, but more on that debacle in a moment).

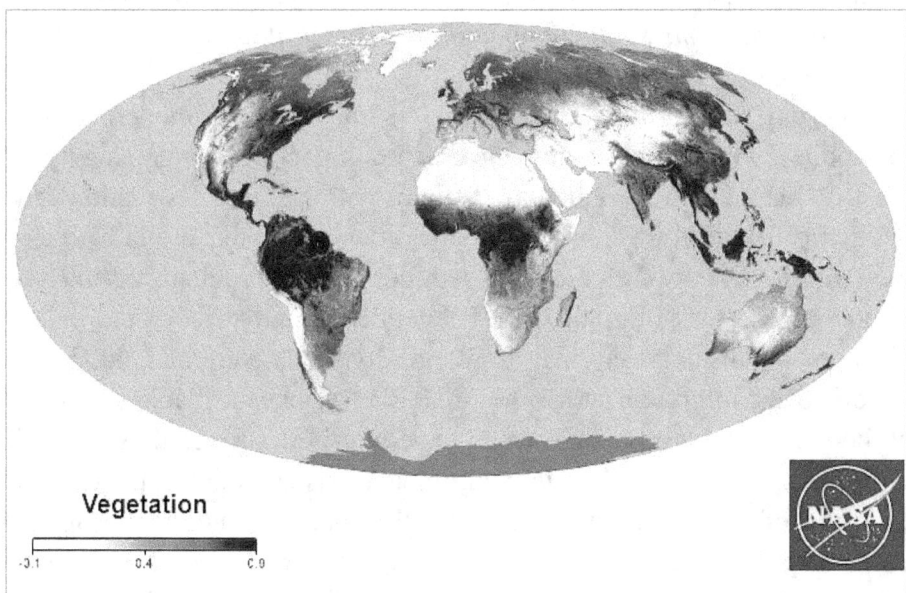

Figure 30. Arid and semi-arid landscapes make up two-thirds of the world's land surface.

Source: https://earthobservatory.nasa.gov/global-maps/MOD_NDVI_M

Like wildfires, livestock have become yet another misunderstood scapegoat for climate hysterics, who are promoting deeply flawed research that blames cows for up to 40% of methane emissions.[1] And if you thought the attack on cattle was about going after feedlots, think again… by ignoring the fact that cattle are part of a complete ecosystem and that different livestock management systems and different grazing techniques have completely different impacts on the land, the anti-livestock climate zealots triumphantly parade around "peer-reviewed research," which alleged "proves" that <u>grazed</u> beef accounts for between a quarter and a third of all greenhouse emissions from ruminant

---

[1] Booker, C. and Weber, S. (2022, Mar 6). Cow burps are a major contributor to climate change – can scientists change that? *PBS News.*
https://www.pbs.org/newshour/show/cow-burps-are-a-major-contributor-to-climate-change-can-scientists-change-that

livestock[2] and that grass-fed cattle produce 20% more methane in their lifetime than grain-fed cattle... because they live longer.[3] According to environmentalist George Monbiot, "... *above all else [...] the demand for grass-fed beef, that's the thing that is the greatest driver of habitat destruction on Earth today.*"[4]

As usual, a half-truth is a whole lie. It is true that many modern grazing techniques, which fail to mimic the grazing impact of wild grazing herds, are to blame for a colossal amount of habitat destruction and desertification. But it is not the livestock themselves that are to blame — the fault lies in how they are managed, nor can the problems be fixed by removing livestock from the land (I'll get deeper into these issues in later chapters — suffice to say that, if grazing livestock were bad, 60 million historic bison migrating across the Great Plains since the glaciers retreated at the end of the last ice age would have utterly destroyed those historic grasslands, but instead their grazing impact was essential to the creation of some of the most fertile soils on the planet, which supported an uninterrupted sea of grass that stretched from Texas to northern Alberta, including many dry regions that are capable of sustaining little more than sagebrush today now that the great herds are gone). Science has become a perversion of its former self — blind, hysterical, completely out of touch with reality, and completely incapable of understanding complex ecosystems and the dynamic interconnected forces that sustain them.

---

[2] Dunne, D. (2017, Oct 3). Grass-fed beef will not help tackle climate change, report finds. *Carbon Brief.* https://www.carbonbrief.org/grass-fed-beef-will-not-help-tackle-climate-change/

[3] Marshall, J. (2010, January 27). Grass-Fed Beef Has Bigger Carbon Footprint. *Discovery News.* https://web.archive.org/web/20120203073611/http://news.discovery.com/earth/grass-fed-beef-grain.html

[4] Oxford University Museum of Natural History (2023, Jul 13). *Allan Savory v George Monbiot debate | Is livestock grazing essential to mitigating climate change?* [Video]. YouTube. https://www.youtube.com/watch?v=-FihlOvsVkY&t=4089s&ab_channel=OxfordUniversityMuseumofNaturalHistory

>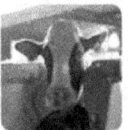
> **PBS**
> **Cow burps are a major contributor to climate change -- can scientists change that?**
> Livestock production—primarily cows—produce 14.5 percent of global greenhouse gas emissions. The majority of that is in the form of methane,...
> Mar 6, 2022

Figure 31.
https://www.pbs.org/newshour/show/cow-burps-are-a-major-contributor-to-climate-change-can-scientists-change-that

>
> **The Conversation**
> **Why eating grass-fed beef isn't going to help fight climate change**
> This is because commercial feeds tend to be less fibrous than grass, and so cows that eat them produce less methane (through belching and...
> Oct 3, 2017

Figure 32.
https://theconversation.com/why-eating-grass-fed-beef-isnt-going-to-help-fight-climate-change-84237

~

    This fanatical and escalating witch hunt for simplistic scapegoats that produce carbon dioxide and methane (which completely overlooks the essential role that livestock play in building fertile farm soils and completely overlooks the pivotal role that ruminant grazing animals play in creating fertile, drought-resistant ecosystems in dry climates) is not the first time that a witch hunt has been triggered by the reductionist thinking of academics who are unable to understand nature's full complexity. As it continues to escalate in tone and volume, the crusade against cattle and other livestock under the guise of "fighting climate change" is beginning to echo the fanaticism that led Chairman Mao in Communist China to declare war against the sparrow in 1958 as part of his Four Pests Campaign (see figure 33 below) as Mao and his advisers failed to understand the pivotal role played by the humble sparrow in keeping its local ecosystems healthy.[5] I don't make this comparison lightly, as will become increasingly apparent as this book reaches its conclusion. And so,

---

[5] "Four Pests campaign." *Wikipedia.* 28 May 2024.
   https://en.wikipedia.org/wiki/Four_Pests_campaign

I'm going to take a brief detour to briefly explore this tragic episode in China's history because the parallels to our own growing atmosphere of censorship in service of grand utopian social engineering schemes and our own embrace of land management by bureaucratic decree should serve as a warning of what is to come if we allow this hysterical climate crusading movement to continue unchecked on its present course.

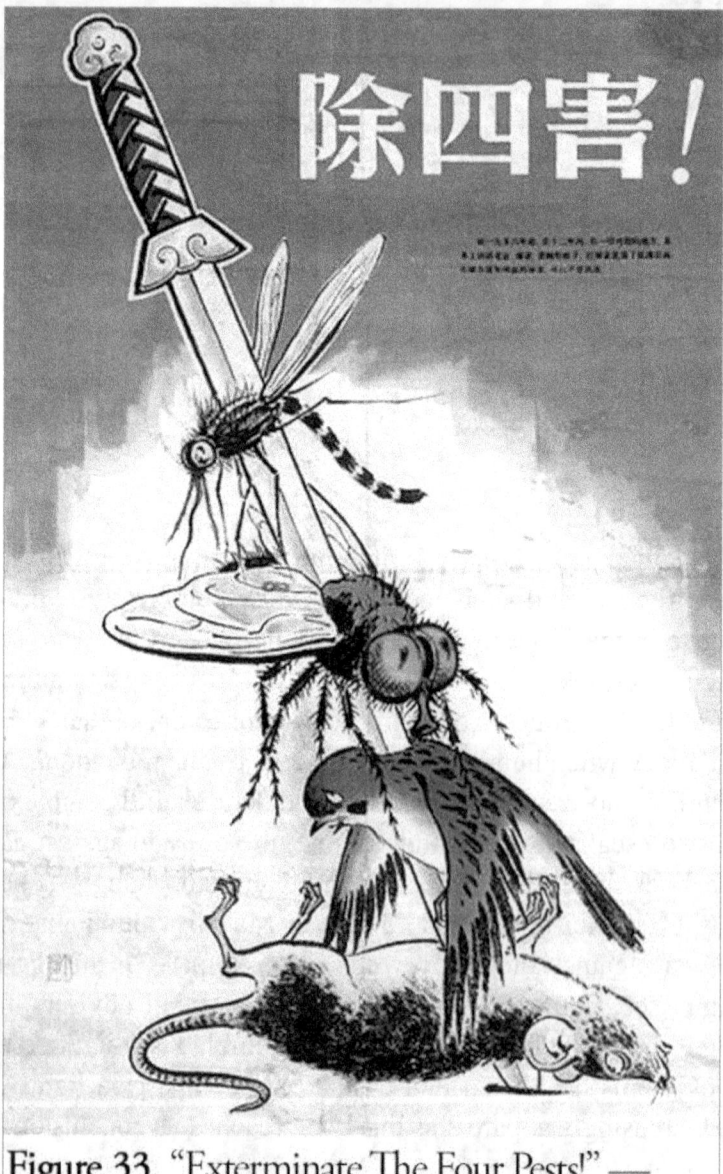

Figure 33. "Exterminate The Four Pests!" — Chinese government campaign poster from 1958.

Through a combination of his own ideological beliefs and bad advice from his scientific advisors, Mao came to believe that sparrows were eating too much valuable grain, fruit, and seeds so he organized millions of people to destroy eggs, kill chicks in their nests, shoot the birds wherever they could find them, and to bang pots and pans for days and nights on end to frighten them into taking flight until the exhausted birds literally dropped dead from the sky. The Four Pests Campaign was a "resounding success" — the poor sparrow was very nearly driven to extinction within China. With no sparrows left to eat them, locust populations boomed, swarmed across the country, and consumed the crops. As the locusts ate the crops and famine began to set in, Mao quietly ended the crusade against the sparrow and switched his focus to exterminating bed bugs instead. But it gets worse.

Mao's campaign against the sparrow was part of his "Great Leap Forward" (doesn't that sound like the UN's optimistic plan to "Build Back Better" according to the UN's sustainable development goals for 2030).[6] It was an ambitious utopian engineering campaign with the goal of modernizing the country's agricultural sector according to communist economic principles. In a series of moves reminiscent of the ecosystem collapse triggered by plowing up America's Great Plains during the late 19th and early 20th century, Mao's Great Leap Forward intentionally deforested over 25% of China's forests (approximately 247 million acres over 29 years) in "Three Great Cuttings" in order to make room for more farms. He also encouraged peasants to smelt iron and steel in backyard furnaces fueled by trees cut from the forests – this was going to help transform China from an agrarian into an industrial society. Furthermore, in a move reminiscent of the farming practices out on the Great Plains when America believed that "the rain follows the plow", Mao launched an ambitious plan to "reclaim the borderlands" and "wastelands" by deep-plowing vast expanses of grasslands and pastures in order to turn them into cropland. He also extended his new-found obsession for deep plowing to existing farms as he empowered a veritable army of bureaucrats to relieve farmers of their former independence to make day-to-day decisions on their farms so that cutting-edge scientific ideas spilling out of Soviet and Chinese universities could be imposed by diktat to modernize China's

---

[6] United Nations (2023, Mar 22-24). *Transforming our world: the 2030 Agenda for Sustainable Development*. [Press release]. https://sdgs.un.org/2030agenda

"backward" agricultural practices — practices developed by autonomous farmers over centuries by trial and error.

Farmers were ordered to *"turn up thousand-year-old soil and strive for output of 1,500 jin per mu"* [a *jin* is a unit of measurement for weighing grain equivalent to around 1.3 pounds and a *mu* is an area measurement equivalent to around 1/15th of a hectare].[7] One author describes how, as the idea of deep plowing took hold, the communist party continued to adjust yield targets ever higher to keep pace with escalating utopian ideals (even as yields began to plunge in response to the new centrally-planned farming methods). Party leaders ordered farmers to plow 4 to 5 feet deep to bring virgin soils to the surface.[8] In the 'satellite' fields of Shouzhang County, things reached such absurd extremes that leadership demanded farmers bring soil to the surface from a depth of 3.3 meters (10 feet) — which required three-man crews to dig up the soil by hand.[9] In 1958 in Liaoning province, 5 million people spent more than a month deep-plowing 3 million hectares of land.[10]

Farmers were also ordered to increase the planting density of seeds according to the "politically correct" belief that plants of the same species would cooperate rather than compete with one another and thus would support each other's growth. Before 1958 a Chinese farmer might plant 1.5 million seedlings per 2.5 acres, by 1959 that had increased tenfold to 12-15 million seedlings per 2.5 acres.[11] With too many seedlings competing for limited soil nutrients, few seedlings survived.

Even as people began to starve, the fear of being purged ensured that no one dared utter a critical word to acknowledge that the new farming practices were a disaster, much less point out that the collectivization of farming was proving to be an abject failure. Instead, local officials falsified agricultural yield numbers while party leaders, shielded from reality, horded grain in storage facilities and continued to export large

---

[7] Orlando, P. (2018, Oct 23). Food from Thought. *Unintended Consequences Blog*. https://unintendedconsequenc.es/food-from-thought/

[8] Chinese Famine of 1958-1961. (2005, Oct 30). *Adiotos Blog*. https://adiotos.wordpress.com/2005/10/30/chinese-famine-of-1958-1961/

[9] Orlando, P. (2018, Oct 23).

[10] Adiotos. (2005, Oct 30).

[11] Adiotos. (2005, Oct 30).

volumes of grain to world markets even as their people starved out in the provinces.

Mao's agricultural reforms relied on advice from the Soviet Union's most prominent scientists. The problem was that Stalin had fallen under the influence of a Soviet agronomist by the name of Trofim Lysenko who rejected genetics and natural selection and embraced the idea that environment alone can shape the traits of plants and animals. *"Put them in the proper setting and expose them to the right stimuli,"* Lysenko declared, *"and you can remake them to an almost infinite degree.*[12] This is the core essence of the Nature versus Nurture debate that has raged since Darwin proposed his theory of natural selection (in biology the debate was known as Darwinism vs Lamarckism). For example: Did random genetic mutations among the short-necked ancestors of giraffes give some giraffes a slightly longer neck than others, thus giving a slight competitive survival advantage to those offspring who inherit this longer-necked genetic mutation because they are able to browse taller trees during droughts and are therefore able get enough nutrition to have more offspring? (This is the Darwinian explanation.) Or did the necks of the short-necked ancestors of the giraffe get longer because of generations of stretching to reach the highest leaves on tall trees as each generation passes on its acquired stretches to the next generation? (This was the Lamarckian explanation.)

Lysenko was a hard-core believer that environmental challenges (nurture), not genetics (nature), determined plant characteristics and thus that these characteristics could be changed, at will, by changing the environmental stimuli. As an example of his beliefs, Lysenko argued that plants could be made to adapt to a colder climate by planting a little further north each year, thus using exposure to frost as an environmental stimulus to "harden" the plants' tolerance to frost so that, according to Lysenko, in time oranges could be made to grow in the Arctic.[13] He also soaked seeds in freezing water to "educate" Soviet crops to be able to sprout at colder soil temperatures under the belief that future generations of seeds would remember these environmental cues and inherit beneficial cold-tolerant

---

[12] Kean, S. (2017, Dec 19). The Soviet Era's Deadliest Scientist Is Regaining Popularity in Russia. *The Atlantic*. https://web.archive.org/web/20210220032906/https:/www.theatlantic.com/science/archive/2017/12/trofim-lysenko-soviet-union-russia/548786/

[13] Watkins, T. (n.d.). *The Great Leap Forward Period in China, 1958-1960*. San José State University, Department of Economics.

traits without further treatment. He promised that by applying his methods, the Soviet Union could boost crop yields and convert the empty Russian interior into vast productive farms.[14]

Lysenko's rejection of genetics and his belief that he could remake plants and animals through environmental stimuli resonated with the Marxist belief that individual humans and, indeed, society as a whole, are not bound by cultural or genetic characteristics but can be molded like clay into whatever social engineers want them to become in order to serve the collective good. Lamarckism (and its derivative Lysenkoism) was the perfect scientific theory to legitimize the Marxist belief that collective action under the directive of a centrally planned state can free us from the limiting social and genetic inheritances of our ancestors. If Nature itself could be re-shaped at will for the collective benefit of society, then so could humanity. Soviet science and Soviet politics were a match made in Heaven, and a fusion recipe destined to recreate Hell on Earth.

From Lysenko's humble beginnings as the son of a desperately poor peasant, Lysenko went on to study agronomy. By the 1920s, he was influencing state agricultural policy and was garnering praise from the state-run newspaper because of the impressive results produced by his experimental farm. Few knew at the time that he was overstating his successes, faking his experimental results, and omitting any mention of his experimental failures.[15] Trust the science.

In 1930, at the age of only 32, Stalin placed Lysenko in charge of agriculture for the whole of the vast Soviet Union. Lysenko's personal story of his humble beginnings and his Marxist-inspired agricultural philosophies all perfectly matched the doctrines of the Communist Party — who better to translate Stalin's social engineering dreams into reality in the agricultural sphere. In his 1954 book, *Agrobiology,* Lysenko summed up his vision of biology — you can see how masterfully he has blended political ideas with his scientific ideas when he wrote, "*bourgeois scientists found it necessary to invent the intraspecific struggle. They say that a fierce struggle for food, of which there is an insufficiency, goes on in nature, within the species, among its individual members – a struggle*

---

[14] Kean, S. (2017, Dec 19). The Soviet Era's Deadliest Scientist Is Regaining Popularity in Russia. *The Atlantic.*
https://web.archive.org/web/20210220032906/https://www.theatlantic.com/science/archive/2017/12/trofim-lysenko-soviet-union-russia/548786/

[15] "Lysenkoism" *Wikipedia.* 23 May 2024. https://en.wikipedia.org/wiki/Lysenkoism

for the conditions of life. The stronger, fitter individuals win. The same thing, they aver, goes on among human beings: the capitalists, you see, are brainier, are more capable by nature and heredity. We Soviet people know full well that the oppression of the working people, the domination of the capitalist class and imperialist war have nothing in common with the laws of biology. These phenomena are all governed by the laws of decaying bourgeois, capitalist society, which has outlived its day. Nor is there any intraspecific competition in nature itself."

Criticisms of Lysenko's theories were denounced as "bourgeois constructs" and the critics were accused of being "agents of international fascism." Research that contradicted Marxist beliefs was systematically destroyed. In the state-run newspapers, it became common to read public letters of apology from scientists as they confessed the error of their ways in the hope of escaping reprisals for wrongthink.[16] Nevertheless, over 3,000 biologists and geneticists were fired, imprisoned, incarcerated in psychiatric hospitals, sent to work and/or die in the gulag, or were executed outright as enemies of the state for attempting to oppose Lysenkoism. As the old saying goes, "punish one, educate a hundred". Genetic research in the Soviet Union was effectively destroyed until the death of Stalin in 1953.[17]

Outside of the halls of academia, Lysenko's influence was even more dangerous because central planners and their enforcers made it impossible to ignore directives that farmers knew from experience were nothing more than sheer lunacy. Whenever decision-making is separated from the hands that have to put those decisions into practice, things quickly turn into a nightmare. For example, Soviet farmers were forbidden to use any fertilizers or pesticides on their farms. Instead, Lysenko claimed that deep ploughing the soil and planting seeds deeper would encourage faster and deeper root growth capable of accessing nutrients deeper in the soil. If that sounds familiar, you know how this tragic episode ends.

During Lysenko's tenure, over a million Soviets citizens were forcibly displaced onto collective, state-run farms to put Lysenko's ideas into practice. As their crops began to fail and as famine began to spread, Stalin

---

[16] The Disastrous Effects Of Lysenkoism On Soviet Agriculture." *Encyclopedia.com*. Accessed 3 June 2024. https://www.encyclopedia.com/science/encyclopedias-almanacs-transcripts-and-maps/disastrous-effects-lysenkoism-soviet-agriculture

[17] "Lysenkoism" *Wikipedia*. 23 May 2024. https://en.wikipedia.org/wiki/Lysenkoism

empowered Lysenko to double down on his ideas. In a clear nod to the Marxist notion of class struggle, Lysenko advised planting crops even closer together because he believed that *"plants of the same species do not compete with each other but instead help each other survive"*.[18] His directives backfired spectacularly as plants starved one another for nutrients on overtaxed soils. Almost everything grown using Lysenko's methods withered and died. But you have to give Lysenko credit for one thing, he certainly knew how to promote himself by catering to the political winds of his era.

In the Ukraine alone, 7 to 10 million people died of starvation under Lysenko's oversight in an event known today as the *Holodomor* (1932-1933), yet throughout the Holodomor Stalin continued to export Ukrainian grain to the rest of the world to demonstrate the superiority of communist ideology. Similar horrors unfolded in Kazakhstan (another 1.5 million deaths) and elsewhere in Russia (another 2 to 3 million deaths are recorded in the North Caucasus and Lower Volga regions). As the American Institute for Economic Research reports, "Soviet officials themselves confirmed a population deficit of 15 million people, but only after these figures were revealed to the world in 1990 after the collapse of the Soviet Union."[19] A friend born under the Soviet shadow once wryly said to me, "compared to Stalin, Hitler looks likes a boy scout."

As nightmarish stories of starvation and cannibalism spread around the Soviet Union, Stalin sealed the Ukrainian borders to prevent Ukrainians from fleeing the famine-struck region. It was a deliberate decision to condemn them to their grisly fate — a kind of collective punishment for having failed to deliver on the promise of Soviet ideology. Furthermore, Stalin feared that Ukrainian opposition to his policies might spark a secession movement, so he deliberately ramped up procurement quotas to higher and higher levels so that party officials effectively vacuumed every last trace of grain out of Ukraine's granaries as a calculated strategy to use famine to exterminate political dissent. Though few who were educated in government-run Western schools have heard of the Holodomor, it stands alongside the Holocaust as one of the two deadliest examples of

---

[18] Watkins, T. (n.d.). *The Great Leap Forward Period in China, 1958-1960*. San José State University, Department of Economics.

[19] Peterson, M. N. (2024, Jan 24). Mr. Jones and the Soviet Lie. *American Institute for Economic Research*. https://www.aier.org/article/mr-jones-and-the-soviet-lie/

governments that believed the end justifies the means and thus felt morally justified to deliberately execute millions of their own citizens to serve some alleged collective social-engineering purpose.

After Stalin's death in 1953, efforts were made to try to disempower Lysenko, but Stalin's successor, Nikita Khrushchev, returned him to power where he remained until after Khrushchev's death in 1964. Lysenko's deadly reign finally came to an end in 1964 when physicist Andrei Sakharov daringly denounced Lysenko in front of the General Assembly of the Academy of Sciences of the USSR, stating that *"He is responsible for the shameful backwardness of Soviet biology and of genetics in particular, for the dissemination of pseudo-scientific views, for adventurism, for the degradation of learning, and for the defamation, firing, arrest, even death, of many genuine scientists."*[20]

But Lysenko's fall from grace was still many years away when Mao launched China's Great Leap Forward (1958-1962). Mao looked to scientists in the Soviet Union to help him lift his country out of backwardness and so Lysenko's beliefs also became the driving philosophy underpinning Mao's agricultural reforms. The ecosystem collapse that followed Mao's agricultural reforms not only repeated all the famine, suffering, cannibalism, and death experienced under Lysenko's watch in the Soviet Union but eclipsed them by an order of magnitude. Much like what happened out on the Great Plains during the 1930s, Mao's war on grasslands, forests, and marginal lands, along with his adoption of Lysenko's ideologically-motivated agricultural principles, led to catastrophic environmental degradation as the fragile topsoil in these marginal lands was laid bare to sun, wind, and rain, which triggering massive soil erosion, dust storms, salination, and droughts during a dry period, followed by devastating floods when the rains finally came and were unable to be absorbed by the bare degraded soil. What resulted was the deadliest famine and deadliest man-made disaster in human history — the Great Chinese Famine — which claimed the lives of up to 55 million people.[21]

---

[20] "Trofim Lysenko." *Wikipedia.* 4 Jun 2024.
    https://en.wikipedia.org/wiki/Trofim_Lysenko

[21] "The Great Chinese Famine." *Wikipedia.* 27 April 2024.
    https://en.wikipedia.org/wiki/Great_Chinese_Famine

Though these tragic chapters of history are rarely taught in liberal schools in the West (few people have even heard of Trofin Lysenko), they are grim lessons that as individual autonomy is increasingly overridden by bureaucratic diktat, as decision-making becomes increasingly centrally directed at the expense of local autonomy, and as the State accumulates the political authority to roll out ever harsher punishments for heretics who dare question the central planners' utopian directives, it becomes harder and harder to change course when bad ideas take root. Pay attention — today's escalating climate hysteria is slowly accumulating legal teeth. This isn't some game that will fade on its own. Speak out while you still can — the price of dissent will only get higher with each passing day as our judicial systems are increasingly weaponized as tools for suppressing wrongthink and punishing thoughtcrimes. Without freedom of speech, both science and democracy become mere illusions as our once liberal societies are slowly bent into tyrannical illiberal parodies of their former selves.

~

But let's get back to the main ecological story unfolding on our present-day farms, in our forests, and in our semi-arid deserts because of land management policies that are being imposed by our own central planners.

Most people are familiar with the role that herds of plains bison once played in creating the rich soils of the North American prairie. But where the plains gave way to trees in northeastern BC, Alberta, the Yukon, the Northwest Territories, and Alaska, in historic times vast herds of woodland bison played the same role in creating fertile woodland soils, even as they also helped recycle the flammable debris that built up on the forest floor (figure 34). These large grazing herds were once integral to the dense flourishing forests of the American West, just as plains bison (and antelope) were integral to the thriving grasslands of the open prairie and the semi-arid deserts of the southwest. The thick grass sod created beneath the canopy of trees by frequent grazing helped stabilize sloped soils and slowed the steady erosion of soil. Their manure added nutrients to the soil. Their feet helped trampled grass, twigs, and brush back down into the dirt so it could be digested easily by soil microbes to be turned into humus. And the microbes in their manure helped stimulate the soil biology that otherwise goes to sleep during the dry season.

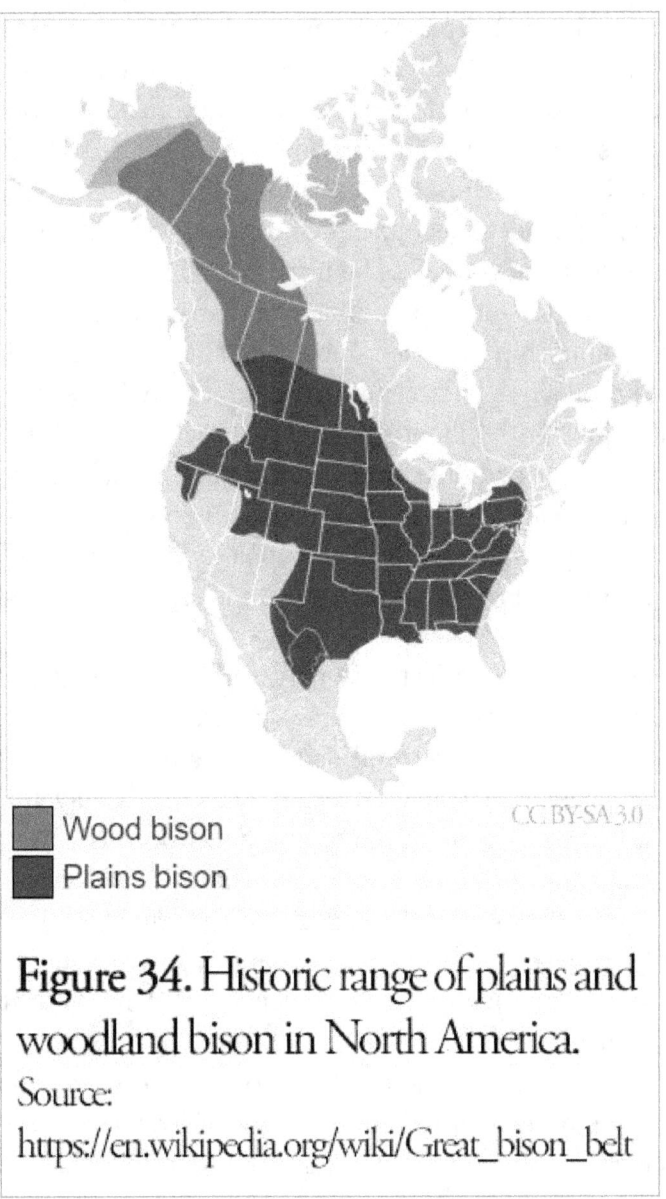

Figure 34. Historic range of plains and woodland bison in North America.
Source: https://en.wikipedia.org/wiki/Great_bison_belt

This symbiotic relationship between ruminant grazing herds and the soil biology is a cornerstone of both the carbon cycle and the nutrient cycle in the grasslands, northern forests, and semi-arid deserts of the world. In these arid and semi-arid climates, the soil's ability to store and recycle carbon, supply nutrients to plants, and build and retain humus is severely impaired without the beneficial impact of migrating ruminant animals. In any climate where annual rainfall drops below the aforementioned 20-inch rainfall line (the rainfall line that unscrupulous railroad barons used to lure homesteaders out into the plains during the 19th century), this lack of soil moisture during the dry season essentially shuts down the soil microbiology responsible for digesting plant material, building and storing humus, and recycling carbon in and out of the atmosphere. However, as long as there are ruminants that continue to graze the dry grass, microbes inside their stomachs continue to "digest" all this dry plant debris so that it can nevertheless be turned into fertilizer and transformed into soil despite the dry conditions. Nature evolved this remarkable symbiotic relationship between grass, ruminants, and soil to create thriving and productive landscapes in areas that are far too dry to flourish with plants alone — without grazing livestock, these dry ecosystems collapse into deserts and their soils become barren. I will get even deeper into the inner workings of this remarkable symbiotic relationship between grass and grazing animals later in the book as the surprising ecological story behind our rising atmospheric CO_2 levels (and its implications) become clear.

Few people understand just how powerful this symbiotic relationship is. Consider the deep rich soils of the prairie, which were created by migrating bison. Most agricultural soils around the world have a topsoil layer no more than 5 to 10 inches deep — that's the thin layer that sustains civilization. But some parts of the prairie have topsoils that are up to 15 feet deep! And they were created in less than 10,000 years (since the retreat of the ice sheets after the end of the last Ice Age) by the repeated high-density herds of bison who grazed, pooped on, and trampled the grass prairie before migrating onwards to leave the land to rest until the migration returned.

This cycle of intense high-density grazing followed by complete rest until fresh rains bring a fresh flush of regrowth, repeated year after year over millennia, created an extraordinary buildup of rich soil brimming with high concentrations of humus, soil nutrients, and carbon, which is unparalleled anywhere else in the world. The only other place in the world with similarly rich natural soils are the Chernozem soils of the Ukraine and Russia, likewise created by ancient migrating herds of ruminant animals grazing the steppe, year after year, for millennia. (It is therefore quite a feat that Lysenko and Stalin managed to trigger a famine amidst some of the richest soils on the planet, just as President Mugabe later did when he managed to turn Zimbabwe's fertile breadbasket into a food desert through his socialist planning policies before lecturing the world about how capitalism creates climate change).

Neither the slow natural composting of plant debris in forests and fields in high rainfall areas nor the slow work of the farmer laboriously adding soil amendments is capable of building rich deep soils as quickly (or as cheaply) as a migrating herd of ruminants. Cow poop is the golden currency of farming. This symbiotic relationship between ruminants, grass, and soil microbes, and the incredibly fertile soils they create is one of nature's most remarkable and least appreciated miracles. I have introduced this integral link between grazing animals and soil formation in this chapter in the context of understanding the forest fires of the American West, but we'll be coming back to this idea in greater detail later in the story as the cookie crumb trail in this unfolding detective story takes us to its unexpected destination.

Nor were the woodland bison alone in their work to improve forest soils and reduce the buildup of dry fire-prone fuel on the forest floor. In the woodlands of southern BC, Washington, and Oregon, massive herds of deer, elk, and moose once played a similar role. The area around my hometown of Lumby, 50 km to the northeast of Kelowna, was historically known as Bull Meadows because of the abundant moose that once grazed local swamps and meadows before cattle and plows took over the valley bottom.

As western North America was opened up for farming, cattle stepped into the role once played by these wild grazing species. But cattle farming in BC is declining (cattle numbers are down 25% since the 1970s, 80s, and 90s), especially in the periphery of the big cities where changing economic conditions and soaring production costs have made it increasingly difficult for BC ranchers to compete with Alberta's farm-friendly economy and because BC's political climate has created an increasingly hostile regulatory environment for cattle farmers.

The local slaughterhouse a few miles up the road from my parents' farm closed permanently in the 1990s after it burned down — it was prevented from rebuilding because of changes to local zoning laws. The local cattle auction yard — Valley Auction — closed its doors in 2021. There are still lots of cattle in the province, but their numbers are rapidly shrinking and their use as a forest and soil management tool is grossly under-appreciated and grossly under-utilized.

Like the unintended consequences of decades of fire suppression, as the presence of large herds of grazing cattle declines in our forests, this too contributes greatly to the increased fuel accumulating on the forest floor, reduces the natural fertilizer that keeps forest soils healthy and makes trees more pest-resistant, and increases the vulnerability of neighborhoods expanding into fire-prone coniferous forests surrounding so many of our western cities. In the early years, most of BC's cities were located on the valley bottoms. To a large degree, these cities were protected from forest fires by a ring of farmland that grew up around each town. Furthermore, the massive temperature difference between the valley bottoms and the surrounding mountainous plateaus meant that the valley sides were often covered by open grasslands or partially forested grasslands, too hot and dry for dense coniferous forests to grow. Historically, these hillside grasslands were prime grazing areas for cattle. Today, many of these hillside grasslands are neglected and sparsely grazed, if they get grazed at all. By the end of summer, at the peak of fire season, they are quite literally brimming with dead dry grass and tinder-dry weeds that are just waiting for a spark.

In a sense, it's a milder repeat of the much more extreme situation that has arisen in Hawaii — exemplified by the deadly wildfire that swept through the town of Lahaina, in Maui, in 2023 — where high land prices and hostile regulations have gutted the agricultural industry on the Island. Since the 1980s, crop production on Hawaii has plunged by over 200,000 acres (a decrease of over 60%) as pineapple and sugar cane plantations have shuttered their doors.[22] Likewise, Hawaii once had extensive tracts of land solely dedicated to raising livestock on pasture, but pasture use has likewise decreased by over 340,000 acres (a decrease of over 30%) since the 1980s.[23] These abandoned plantations and untended grasslands become huge flammable liabilities that grow right up to the edge of

---

[22] State of Hawaii Department of Agriculture. (2016, Feb 17). *Hawaii Agricultural Land Use Study Released*. [Press release].
https://hdoa.hawaii.gov/blog/main/nrsalus2015/

[23] State of Hawaii Department of Agriculture. (2016, Feb 17).

communities.[24] The risk is further compounded by dense stands of invasive grass species that take over neglected land in Hawaii and crowd out native plant species, like an ornamental grass called pampas grass, which grows between nine and ten feet tall and thrives in neglected pastures, along roadsides, and in both residential and remote areas. In 2023, strong winds knocked over some powerlines in the midst of the dry season, which sparked and ignited the dry grass. These same winds then pushed the grassfire into Lahaina, destroying the town and killing at least 106 people. Politicians and the media shamelessly cited climate change as a compounding factor when in reality the fault lies in their own disastrous policies that have slowly turned Hawaii into a tinderbox.

The photograph in figure 35, which I took near Vernon during a thunderstorm a few years ago, illustrates the grass hillsides that separate many of BC's valley bottoms from the dense forests on the mountainous plateaus in the North American West. Until the city began to grow up the valley sides, Vernon (barely visible in the distance on the left) was separated from the dense fire-prone coniferous forests on top of the mountainous plateaus by irrigated farmland on the valley bottom and by dry grasslands on the hillsides that were routinely grazed by cattle. (The thunderstorm over the mountain in the photo demonstrates how the topography fuels massive thunderstorms on the forested mountain plateaus, which hammer the mountaintops and hillsides with lightning and ignite wildfires but fail to find much to burn in the farmland below.)

---

[24] Broder Van Dyke, M. (2023, Aug 11). UH wildfire expert: Invasive grasses growing in the abandoned plantations fueled wildfires on Maui and Hawaii Island. *Spectrum News*. https://spectrumlocalnews.com/hi/hawaii/news/2023/08/11/uh-wildfire-expert--invasive-grasses-growing-in-the-abandoned-plantations-fueled-wildfires-on-maui-and-hawaii-island

Figure 35. Looking west up the valley towards Vernon, B.C., you can see a classic example of the grass hillsides that separate many valley bottoms from B.C.'s forested mountain plateaus. Author's photo.

Where once these valley-bottom towns were surrounded by farmland and then forests full of grazing cattle, today there are new neighborhoods growing up the sides of the valleys and spilling out into the surrounding forests of the upland plateaus. The large tracts of unbroken forests and grasslands surrounding our cities, which used to be prime grazing land for cattle, are now being chopped up into housing developments and tiny acreages, crisscrossed by fences and devoid of grazing cattle.

I can give several examples right here in my local neighborhood — a school and several houses now sit on land that was once routinely grazed by my parents' cattle. Likewise, behind our neighbors' farms (all former cattle ranches, many of which now only grow crops but no longer own cattle), there are countless new houses and small homesteads in places that were formerly forested pasturelands, but which no longer see any grazing cattle today. The school and all of these new houses and homesteads butt right up against forests that spill off the mountain plateaus. Today the brush and dry grass is building up under the trees in their backyard, un-grazed, a fire hazard growing right outside their back door. If a wildfire spills off the mountain plateau, as they inevitably do, these properties are quite literally surrounded by a steadily accumulating tinderbox of accumulating fuel that will carry that firestorm right to their back door. And when it burns, the media and the government will flood the airwaves with climate change hysteria as those burned buildings are leveraged to push a political narrative.

People can't fix a problem if the cause is misdiagnosed. As long as we're bamboozled to keep looking up at the sky, we can't see the problems accumulating at our feet. And the worst part is, by putting the blame on carbon dioxide, people look to firefighters and to climate change policies as a way to protect themselves from wildfires when, in reality, they should be focused on fire hardening their homes[25] and they should be removing the fuel building up in their neighborhoods long before fire season begins. Once the problem is understood and the propaganda is ignored, the power to protect ourselves is entirely within our hands, and it becomes clear that we don't need to control our fellow citizens with liberty-destroying climate change policies in order to protect ourselves.

For local cattle ranchers, it is no longer profitable to build and maintain all the fences required to graze this checkerboard of peripheral forest properties. On the smaller properties and in the forested neighborhoods, there is no way practical cost-effective way to even bring cattle into these neighborhoods. And as for the farmers whose grazing leases in the public forests butt up against the back of these forest neighborhoods, they cannot afford the extra manhours required to monitor cattle grazing on the edge of the cities — it's completely unaffordable to build and maintain that many fences, it's an uphill battle to deal with the wrath of environmentalists who view every cow poop as an environmental risk and every trampled tree as a crime against humanity, and it's a public relations nightmare to continually have to extract cows from the lawns and flowerbeds of angry homeowners (been there, done that!). It's easier to simply push the cows deeper into the mountains where they won't bother anyone. Meanwhile, the fuel just keeps accumulating in the periphery of the cities.

The unintended and unnoticed consequence of these subtle changes in land use is that people newly arriving in the area do not realize how much the forest in their backyard is changing because of the retreat of the cattle ranches, and that they now need to use weedwhackers, mowers, and chainsaws to replace the impact that cattle once had in managing the

---

[25] National Fire Protection Association. (2011, May 18). *Wildfire: Prevent Home Ignition, Part 1*. [Video]. YouTube. https://www.youtube.com/watch?v=zx9-pZvKW2U&ab_channel=NationalFireProtectionAssociation

forests in their backyard. They believe the illusion created by politicians and climate activists who claim that their vulnerability to wildfires is increasing because of the carbon dioxide emitted by their neighbors' gas-guzzling SUVs and by the natural gas being burned to heat homes and supply houses with electricity. Consequently, they are being bamboozled into believing that if only coercive central planners could throw enough money, people, resources, and red tape at the problem, only then will they be safer from wildfires. The net result is that the climate hysterics are conditioning voters to harden their hearts towards the poverty and suffering that are being imposed by climate policies and even leading many well-meaning citizens to demand the authoritarianism that is growing up in the midst of our once-liberal democracies. And the great irony is that the more that these central planners are given power and resources, the bigger the problems get as our institutions turn ever further away from science in order to play the political games required to secure funding.

95% of fire protection isn't what you do once the forest begins to burn, it's what you do to prepare your property for a wildfire in the months and years before a bad fire season. Like the lodgepole pines before them, if people want to live on the edges of these forests, they need to adapt, embrace the wildfires, build accordingly, fire harden their houses, eliminate fuel accumulating around their yards, and learn to live with the fires. The best way to understand what's causing the wildfires burning on the edges of our cities is to turn off the 6 o'clock news, go for a walk in the local forests, and watch what a herd of cows can do to a forest full of grass and dry brush during an afternoon of grazing.

Chapter 7

Local Knowledge

"The sanity of the individual lies in the continuity of his memory; the sanity of a group lies in the continuity of its traditions."

— Will Durant, best known for his 11-volume work, The Story of Civilization.

Exposing rampant government mismanagement of our forests doesn't automatically rule out whether a changing climate is also playing some compounding role in making our forests more vulnerable to wildfires. Whenever our professional propagandists in politics, media, and activist-science are confronted by all the local factors triggering these wildfires, they always rely on the perfect unverifiable comeback that settles everything yet proves nothing: "Sure, it was arson / lightning / pine beetles / neglected pastureland / unhealthy forests / etc., but climate change made it all so much worse." So, let's have a closer look at the "climate" to try to answer three important questions:

1. Is the local climate changing?
2. Are those climate changes playing any meaningful role in making our forests more vulnerable to wildfire?
3. And are those changes driven by global man-made CO_2 emissions or by natural forces beyond our control — in other words, is carbon dioxide acting as a control knob on our climate?

Before I dive into the long answers to these questions, I want to give you a brief overview of some of the changing local weather patterns that my family has observed in the North Okanagan over the course of the last five decades, specifically on my parent's cattle farm in Lumby. These observations provide context for the subtle changes that are happening to local weather patterns everywhere and help expose the cyclical forces that have long shaped the climate all along the western edge of the North American continent — cycles that are systematically ignored by the propaganda that is flooding the airwaves.

If there's one thing farmers tend to notice, it's anything that affects their crops and livestock. And there have indeed been a lot of them. At first glace, these changing local weather patterns seem to support the scary headlines about a rapidly changing climate. But only at first glance because, as we drill into the gritty details, a far more complicated picture emerges that has little if anything to do with global changes in the weather.

Summers in our area during the 1970s and early 80s were very hot and dry. It was great for making hay. By mid summer the surrounding hillsides were always brown and parched, as were any fields or pastures that weren't irrigated. By late summer the creeks slowed to a trickle. Without irrigation, hayfields essentially stopped growing by mid summer and ungrazed pastures turned into tinderboxes of dry grass. My parents' farm is about 20 minutes east of Vernon near the little town of Lumby (figure 36) at the bottom of our deep east-west valley that intersects the main north-south-oriented Okanagan Valley in Vernon. During the 70s, 80s, and early 90s, it was not uncommon to see my father's min-max thermometer record a temperature above 40 °C despite the fact that the temperature in nearby Vernon never quite reached 40 °C during those years.[1] Thanks to the moderating effect of the lakes in the main Okanagan Valley, Vernon's winters and nights were always a little warmer and Vernon's summer daytime highs were always a little cooler than those recorded in Lumby.

---

[1] "Highest Temperatures in Vernon by Year." *Extreme Weather Watch*. Accessed 9 Jun 2024. https://www.extremeweatherwatch.com/cities/vernon-bc/highest-temperatures-by-year

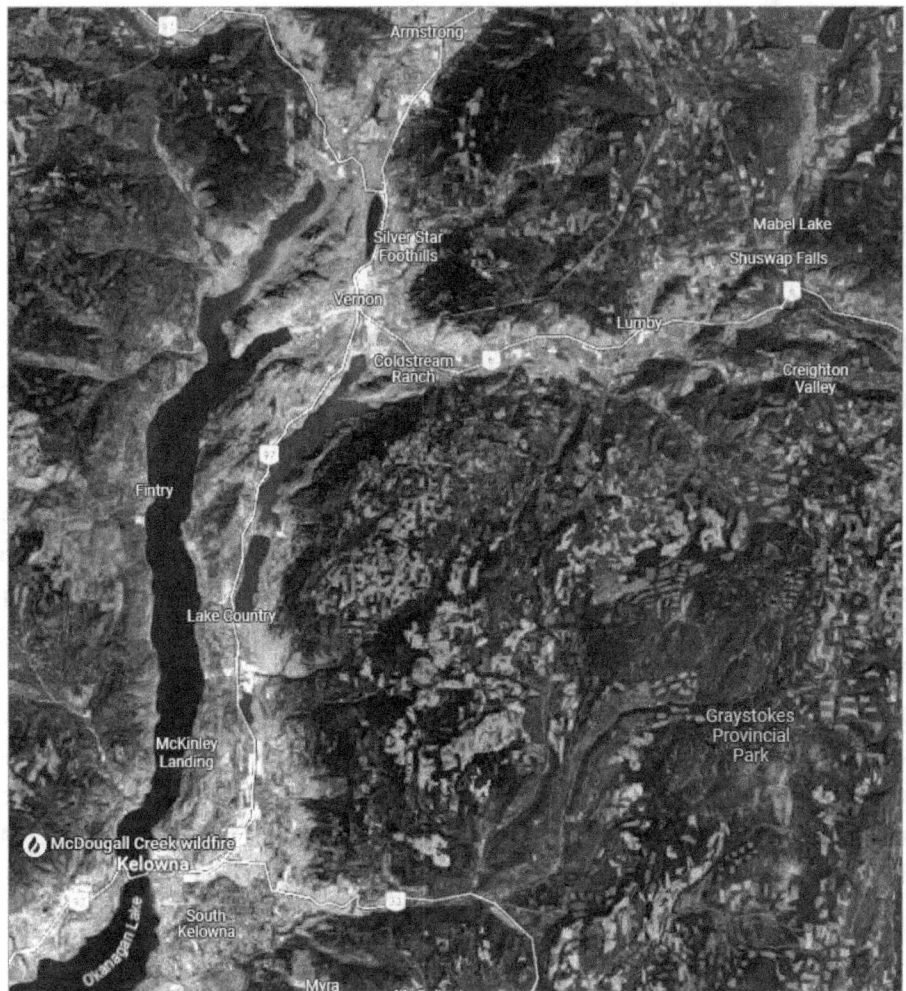

Figure 36. The lake-filled north-south oriented Okanagan Valley intersects with the east-west oriented Lumby Valley at the city of Vernon (metro population 67,000). Google Maps

During those years, our first frost reliably arrived in Lumby by the end of August (as a teenager I was in charge of moving irrigation pipes until school started in September; by late August I had to brush ice off the aluminum irrigation pipes each morning before moving the sprinklers). The killing frost came by end of September (that's the frost that kills all plant growth for the year). During the winter we routinely saw -30 °C on the farm and during a few rare years the mercury even dipped to -40 °C. Winter temperatures in Lumby are typically 10 °C colder than in nearby

Vernon because of the lake effect in the main Okanagan Valley. The cold winters months were hard on cattle, but they did give the cattle a mid-winter break from the diseases that stalk cattle during wet muddy weather.

Reminiscent of the cyclical back-and-forth between wet and dry periods that shapes the weather out on the Great Plains, as described in chapter one, all these weather patterns changed abruptly during the late 90s and early 2000s, which ushered in noticeably wetter and milder weather. That's also when mountain pine beetle outbreaks began ravaging BC's forests. This change in weather patterns was so pronounced that it prompted many local farmers to invest in bale wrapping equipment in the 2000s to give them the option to make haylage instead of dry hay because increased humidity and increased summer rains made it more difficult to dry hay crops (silage and haylage use fermentation to preserve wet forage crops by pressing the air out of the crop and covering/wrapping the compressed forage to prevent oxygen from rotting the crop). Summer nights became warmer starting in the late 90s (which created noticeably less dew at night, which sometimes allowed us to keep baling hay right through the night, whereas during the 70s and 80s we almost always were forced to stop baling hay shortly after sundown because the heavy dew made the hay too wet to bale). Likewise, winters in the 90s became noticeably milder and wetter, with mud and even rain becoming common in December and sometimes even in January, which forced us to completely overhaul our winter cattle management strategies to keep the cows dry and healthy.

During the late 90s, -40 °C became a thing of the past and -30 °C became increasingly rare down in the valley bottom in Lumby. These warmer winter temperatures undoubtedly played a role in the massive mountain pine beetle outbreaks that turned so many BC forests red during the 2000s (and led to so many big forest fires in subsequent years). Cold winters had helped keep the beetle numbers in check — as winters got warmer, more beetles survived the winter and were able to expand their territory into pine forests situated at higher elevations and at higher latitudes, which were previously too cold for pine beetles to survive during the winter.

Mountain pine beetle outbreaks are not new to BC. The outbreak that started during the late 1990s and early 2000s mirrors prior beetle outbreaks that occurred in the first decade of the 1900s, again during the 1930s and 40s, and again during the 60s. There's a multi-decadal cyclical pattern to beetle outbreaks. However, years of fire suppression, degradation of forest soils, and monoculture forest management all exacerbated the 2000s outbreak by creating vulnerable forests and ideal conditions for mountain pine beetles to multiply in record numbers.

And so, another half-truth emerges from the propaganda. Local changes in weather patterns during the 90s and early 2000s undoubtedly played a role in triggering the mountain pine beetle outbreak of the 2000s. But these weather changes are driven by multidecadal ocean current cycles, not rising carbon dioxide.

The most significant of driver of cyclical changes in sea surface temperatures off the west coast of North America is called the Pacific Decadal Oscillation (PDO) — robust research has demonstrated that these cyclical changes in ocean currents are the key driver of mountain pine beetles outbreaks in British Columbia.[2] Likewise, this same Pacific Decadal Oscillation is also the most important driver of the cyclical weather patterns out on the Great Plains, which fueled the 19th century myth that "the rain follows the plow". Everything is connected.

The Pacific Decadal Oscillation runs on a 40- to 50-year cycle (20 to 30 years of warmer and drier weather followed by 20 to 30 years of cooler and wetter weather). Much like the shorter and more well known El Niño cycle that runs on a 2- to 7-year cycle, during the warm phase of the Pacific Decadal Oscillation, ocean currents transport warm water north from the tropics, which trigger abrupt shifts between warmer and cooler temperatures off the coast of North America, which are then pushed inland by trade winds. Sometimes the Pacific Decadal Oscillation and El Niño amplify each other, sometimes they cancel each other out. (There is a similar 60-year cycle operating in the Atlantic Ocean, called the Atlantic Multidecadal Oscillation.)

These ocean currents act somewhat like a conveyer belt to transfer heat accumulating in tropical waters up towards the poles where the heat from these warm ocean currents is then radiated back out into space. These currents essentially function as a thermostat to regulate Earth's temperature; without them heat accumulating in the tropics would build

---

[2] Alfaro, R., Campbell, E., and Hawkes, B. (2010). *Historical frequency, intensity and extent of mountain pine beetle disturbance in British Columbia.* Natural Resources Canada, Canadian Forest Service.
https://publications.gc.ca/collections/collection_2011/rncan-nrcan/Fo143-3-2009-30-eng.pdf

up near the equator and make the tropics unliveable. This is the reason why, whenever the Earth's climate warms, there is little to no change in temperature near the Equator — these ocean currents continually transfer most of the warming from the tropics towards the polar regions. The UN's warnings that a warming climate is going to create hordes of "climate migrants" fleeing tropical regions is completely fictitious — even if the planet experienced a massive increase in temperature (as it has many times in geological history), there would be very little warming in the tropics because these ocean currents would export most or all of that warming to Alaska, Northern Canada, Northern Europe, and Northern Russia — places that benefit greatly from a longer growing season and milder winters. People fleeing southern countries (the UN's supposed "climate migrants") are fleeing broken political systems and ecological disasters caused by local government mismanagement — they are most certainly *not* fleeing unprecedented heat accumulating in the tropics.

In figure 37 I have overlaid two charts so that their time axes line up on the same years. In the upper chart, researchers from the Canadian Forest Service have plotted pine beetle scars in lodgepole pines in central BC based on the year in which the beetles scarred the trees — these waves of scars correspond with outbreak years. Because the Canadian Forest Service chart ends in 1995, I added a rough sketch of the big beetle outbreak that occurred during the 2000s, which peaked around 2004. The lower chart (stretched to fit the same timeline as the pine beetle outbreaks shown in the upper chart) is a screengrab from the National Oceanic and Atmospheric Administration (NOAA) showing how Pacific sea surface temperatures alternate between warm and cold phases as part of this multi-decade Pacific Decadal Oscillation.[3] You can see the strong correlation between the two cycles — the beetles thrive after warm periods of the Pacific Decadal Oscillation. Then, once the beetles finish devastating their favorite food source and as cooler temperatures return, their numbers collapse. But then, after a few decades of forest recovery to create a fresh food source, followed by a new period of warmer winters, beetle numbers surge once again. Rinse and repeat.

---

[3] "Pacific Decadal Oscillation (PDO)." *NOAA Physical Sciences Laboratory.* Accessed 5 Oct 2024. https://psl.noaa.gov/pdo/

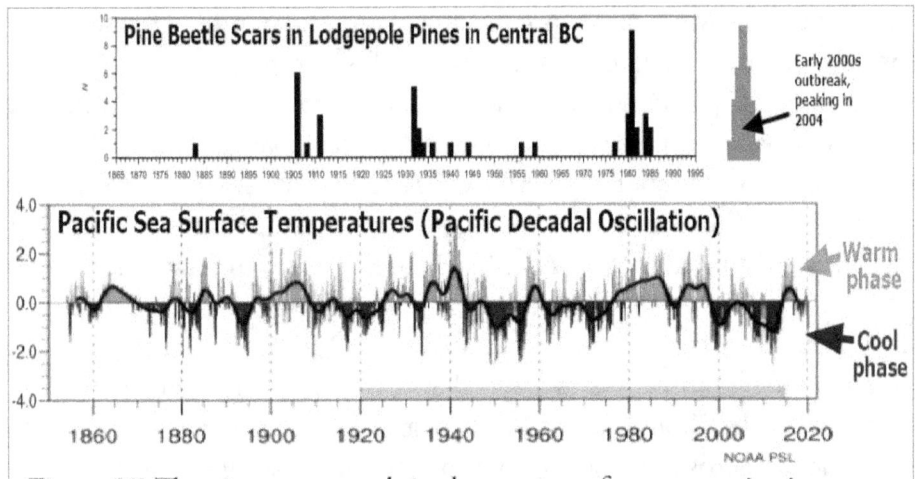

Figure 37. There is a strong correlation between scars from mountain pine beetles in lodgepole pines in central BC versus cyclical changes in sea surface temperature in the Pacific Ocean caused by the Pacific Decadal Oscillation.

Sources:
- Upper chart is from the Canadian Forest Service (https://publications.gc.ca/collections/collection_2011/rncan-nrcan/Fo143-3-2009-30-eng.pdf).
- Lower chart is a screengrab from the National Oceanic and Atmospheric Administration (https://psl.noaa.gov/pdo/).

These cyclical multi-decadal ocean currents also cause significant changes in global-average cloudiness.[4] In its most simplistic form, an increase of warm water pushing north into a cold region triggers an increase in global cloud formation, in a similar way to how steam and fog can be seen forming over the warm water of a lake as the weather turns cold in the autumn. As cloud cover increases, sunlight reflects off the clouds and back into space, which reduces how much direct sunlight reaches the Earth's surface and, in turn, affects how hot the surface gets.

Physicist John Clauser, who won the 2022 Nobel Prize in Physics for his contributions to quantum mechanics, recently co-signed a letter along with over 1,600 other scientists, which declares that there is no "climate

---

[4] Chen, Y., Hwang, Y., Zenlinka, M., and Zhou, C. (2019). Distinct Patterns of Cloud Changes Associated with Decadal Variability and Their Contribution to Observed Cloud Cover Trends. *American Meteorological Society — Journal of Climate*, 32(21). 7281-7301. https://doi.org/10.1175/JCLI-D-18-0443.1

emergency" — Mr. Clauser points outs in an article in the Epoch Times[5] that cloud cover will drop the output of solar panels by over 50%, representing a profound difference in the amount of heat that hits the Earth's surface as the amount of cloud cover changes, but notes that climate models completely overlook these cloud dynamics — the models assume that the amount of sunlight reaching the Earth's surface is unaffected by cloud cover. As Mr. Clauser points out, satellite images consistently show wide variances in cloud cover, which can shade anywhere from five to ninety-five percent of the Earth's surface at any given moment in time (the average cloud cover has averaged around 67% in recent years, though as you will soon see that average has changed dramatically across the long span of our Earth's history — our deep dive into the inner workings of our planet's climate is going to take us through some very surprising twists and turns over the coming chapters).

By omitting long-term changes in cloud cover from their climate models, the models fall short of being able to explain the full range of changes in the Earth's climate. By leaving clouds out of their energy budget calculations, climate scientists are able claim that the only way to explain the Earth's climate history on both long and short time scales is because of changes in CO_2 levels. It's ironic considering that Steven Koonin, who served as President Obama's Undersecretary for Science even pointed out in his book that: *"Just a 5 percent increase in cloud cover can largely counterbalance the temperature effect of doubling atmospheric CO_2."*[6] By deliberately ignoring cloud cover, climate modelers have been able to generate hysteria about atmospheric carbon dioxide.

Back on my parent's farm, the most noticeable change brought on by the milder weather that began in the 1990s — triggered by the Pacific Decadal Oscillation — was the increased length of the growing season. The first frost began arriving later each year and the last killing frost came later still. In recent years, pastures have continued growing through the end of October, which represents a massive and extremely beneficial extension to the growing season. A couple of years ago my parents' farm

---

[5] Jekielek, J. and Nguyen Ly, M. (2023, Sept 9). Nobel Winner Refutes Climate Change Narrative, Points Out Ignored Factor. *The Epoch Times*. https://www.theepochtimes.com/us/nobel-winner-refutes-climate-change-narrative-points-out-ignored-factor-5486267

[6] Jekielek, J. and Nguyen Ly, M. (2023, Sept 9).

didn't get its killing frost until early November! During the 70s and 80s, you could rely on early winter snows to stay on the ground in Lumby after late November, but after the mid 90s it became increasingly uncertain whether there would be any snow on the ground at Christmas. Longer growing seasons have enabled local farmers to plant crop varieties that require longer growing seasons (i.e. higher-yielding corn varieties that mature later in the fall) and temperature-sensitive cherry and apple orchards have slowly expanded their territory from the main Okanagan Valley up into our frost-prone side valley, which had previously been considered much too cold for cherries and apples because it was too far from the temperature-moderating effect of the big Okanagan lakes. Although the orchards haven't reached as far as my parent's farm yet, they are now close enough that we can sometimes hear the helicopters and wind turbines that these orchards use on cold summer nights to prevent cold air from settling over and damaging their crops — these orchards are using every technological hack in the book to grow fruit right on the margins of what is possible in our local climate. Even slight changes in weather patterns can completely change what crops you can grow. But what will happen to these marginal orchards when the Pacific Decadal Oscillation flips cooler again after they were lured into our valley by the false belief that the local warming was allegedly caused by permanent CO_2-induced climate change?

Contrary to the irrational fears being promoted on television about a warmer future, longer growing seasons are a farmer's best friend. During the 70s and 80s, we could count on getting two cuts of hay each summer, plus a third cut that was grazed by our cattle in the late fall. In recent years, three cuts of hay have become the norm in addition to a final cut that's grazed by the cattle in the late fall.

In recent years the summers have become drier again with less humidity to interfere with drying hay. The long growing season is still a thing, though we did get a proper winter again in 2022/23 with the mercury hitting -37 °C in Lumby just before Christmas of 2022 (closer to the lakes, Vernon hit a low of -29 °C on the same day).

In 2023, as the UN went into overdrive to spread their climate propaganda, the media went to great lengths to make a big deal of every single hot day in the valley. The headlines even promoted the idea that it was the hottest weather in over 120,000 years! But the mechanical min-max thermometer I have hanging in the shade on the north side of my shed did not record a single day with an air temperature above 34 °C in 2023, despite the fact that the hottest temperatures recorded in nearby Kelowna and Vernon both exceeded 38 °C on August 17th, 2023 (38.3 °C in Kelowna, 38.1 °C in Vernon).[7]

In other words, despite the drought conditions of 2023, it was not exceptionally hot here in 2023 despite what you may have heard on the news — by the time we reach the Epilogue of this book, the true significance of the fact that we had drought without exceptionally hot temperatures will become clear. What is noteworthy about the story told by the min-max thermometer on my shed is that, in contrast with the 70s and 80s when summer daytime highs were *always* consistently hotter on our farm in Lumby than in the nearby city of Vernon as a consequence of the lake effect, during the summer of 2023 the opposite happened on the hottest day as Vernon's August 17th record high comfortably beat the maximum temperature recorded by the thermometer on my woodshed... by a full four degrees! The most likely explanation for this temperature inversion, which saw hotter temperatures near the Okanagan lakes, is that it was caused by the heat island effect from all the concrete, asphalt, and construction that have been added to Vernon's growing metropolitan footprint as its population has quadrupled since the 70s. I'm also left to wonder whether Vernon's heat island effect is playing a big role in the expanding territory of the cherry and apple orchards and the rather dramatic increase in frost-free growing days in nearby Lumby.

As you'll soon see when I show you national and global climate trends in the next segment of this book, the data doesn't back up the media narrative that we are seeing unprecedented hot weather — the warming since the 1970s (when scientists feared the onset of a new Ice Age) has been rather modest and doesn't even come close to the scorching heat of the 1920s and 1930s. Furthermore, a study released in 2012 calculated that up to half of the modest warming since the 1970s is attributable to the heat island effect due to weather stations being located within or in the immediate vicinity of rapidly growing cities — the study compared how much warming was recorded by weather stations located in high quality locations away from cities versus how much warming was recorded by weather stations located near and inside cities and found very little

---

[7] "Vernon Weather in 2023." *Extreme Weather Watch*. Accessed 10 Jan 2024. https://www.extremeweatherwatch.com/cities/vernon-bc/year-2023

warming away from the urban heat islands.[8] Like termites in their mounds, we are surrounding ourselves with local bubbles of warmth of our own creation.

The US EPA states that buildings, roads, and infrastructure absorb and re-emit the sun's heat more than natural landscapes so that *"daytime temperatures in urban areas are about 1–7 °F higher than temperatures in outlying areas and nighttime temperatures are about 2-5 °F higher."*[9] A study conducted by NOAA showed that the heat island effect could increase minimum temperatures in urban areas by up to 13°F (7 °C) compared to their rural surroundings[10] — something to bear in mind the next time politicians, climate activists, and the media start hyperventilating about "record-breaking" temperatures. 40 °C recorded in 2023 in a rapidly growing city like Kelowna, Vernon, or Phoenix does not quite mean the same thing as 40 °C recorded by the same urban weather station in 1970.

The infographic in figure 38 from the EPA illustrates how the heat island effect increases both daytime and nighttime temperatures in cities. Also, note the moderating effect of the pond in the infographic, which dramatically reduces daytime temperatures and dramatically increases nighttime temperatures near the water, just as the lakes in the Okanagan Valley moderate temperature close to the lakes.

---

[8] Watts, A. (2012, Jul 29). New study shows half of the global warming in the USA is artificial. *Watts Up With That.* https://wattsupwiththat.com/2012/07/29/press-release-2/

[9] "Learn About Heat Islands." *United States Environmental Protection Agency.* Accessed 7 Nov 2023. https://www.epa.gov/heatislands/learn-about-heat-islands

[10] "Does the Urban Heat Island Affect Rainfall Variability Across the Phoenix, AZ Metropolitan Area During the Monsoon Season. *National Oceanic and Atmospheric Administration.* https://web.archive.org/web/20210508231213/https://www.wrh.noaa.gov/psr/general/monsoon/MonsoonPMA.php

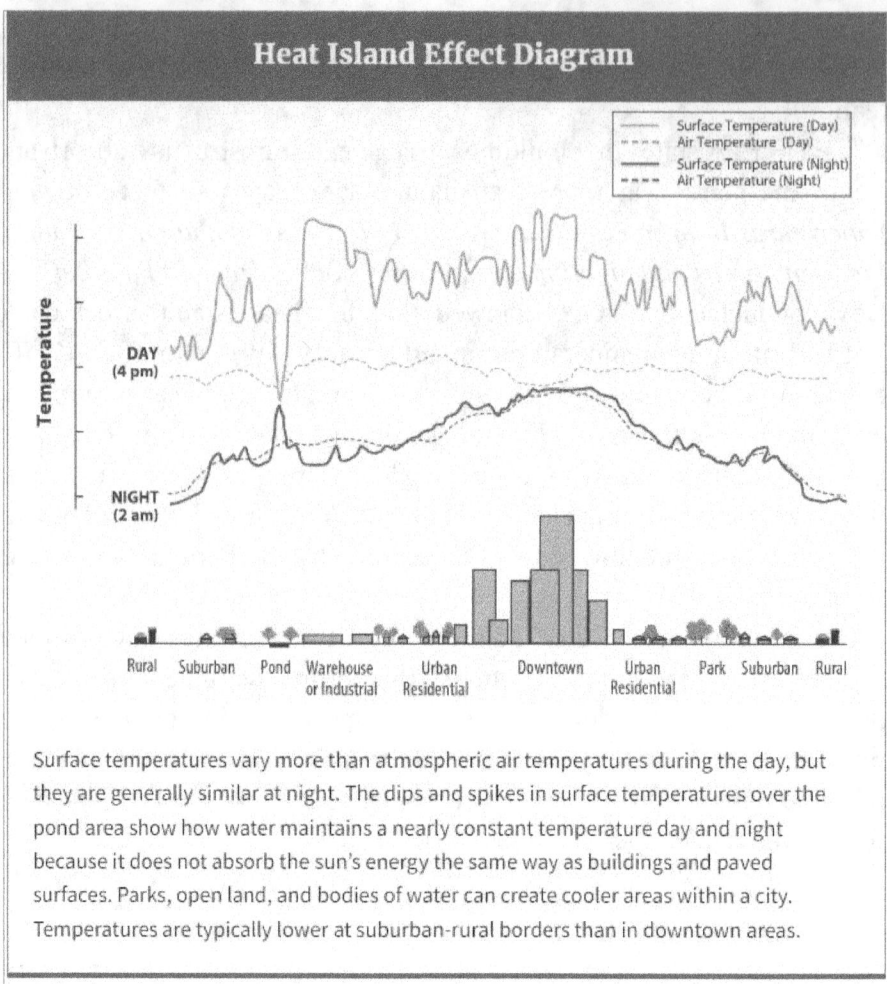

Figure 38. Heat island effect. Infographic from the EPA
https://www.epa.gov/heatislands/learn-about-heat-islands

A specific example of the magnitude of this heat island effect comes from a study published by NOAA in 2008, which studied the heat island effect in Phoenix, Arizona, located at the heart of the much larger Maricopa County. Phoenix is one of the fastest growing big cities in the USA (apparently people prefer moving to hot rather than cold climates when given the freedom to vote with their own feet). The chart in figure 39 shows how average summer nighttime temperatures in Phoenix (in light grey) have become increasingly hotter than the summer nighttime temperatures in the rural parts of Maricopa County (in dark grey). On the left-hand side of the chart, you can see that temperatures in the 1960s were almost identical in Phoenix versus the surrounding countryside. But as Phoenix grew larger, the heat island effect grew stronger so that by 2008 the city had become six to eight degrees Fahrenheit hotter than the surrounding countryside!

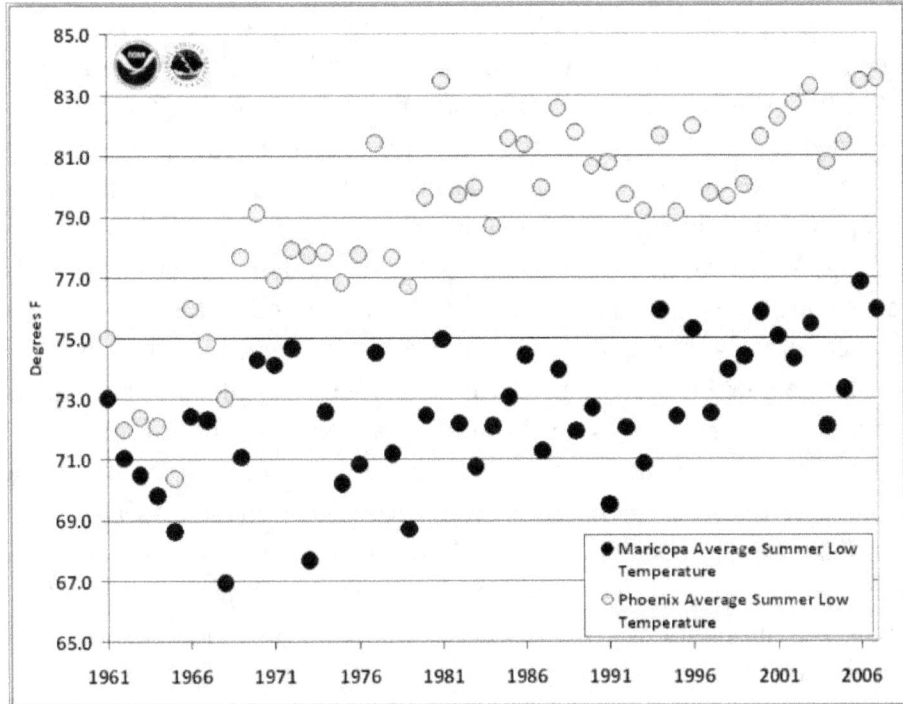

Figure 39. Average summer (June-July-August) low temperatures for Phoenix (light grey) versus rural Maricopa County (dark grey). Phoenix grew considerably warmer than Maricopa County during the late 1960s through late 1980s.

Source: National Oceanic and Atmospheric Administration.
https://web.archive.org/web/20210508231213/https://www.wrh.noaa.gov/psr/general/mons

And yet, despite the fact that the desert in Phoenix was essentially uninhabitable before air conditioning and despite the fact that the surrounding countryside isn't warming at nearly the same rate, the media has declared that "climate change" is making Phoenix *"uninhabitable without air-con"*.[11]

---

[11] Mandenberg, Y. (2023, Jul 21). Residents react after Phoenix is deemed 'uninhabitable' without air conditioning. *The Mirror*. https://www.mirror.co.uk/news/us-news/residents-react-after-phoenix-deemed-30516022

The heat island effect is only one of many ways that headline-breaking news about dramatic daily temperature records can disguise reality. Remember when the media and the European Space Agency (ESA) breathlessly reported that a European heat wave in the early summer of 2023 had caused temperatures to soar above 45 °C in many European cities, including Rome, Naples, and Taranto, and above 50 °C on the slopes of Mount Etna in Sicily? It was yet another half-truth designed to tell a whopper of a lie — it turns out they were reporting temperatures recorded right at the soil surface rather than the standard air temperature measured two meters above the Earth's surface. In essence, they were leveraging the old trick of impressing people by frying an egg on black asphalt. When the media and the ESA were called out on this grossly misleading way of reporting data, instead of replacing surface temperatures with proper air temperatures they continued to use the surface temperatures to push scary headlines but added "clarification phrases" to news stories to "inform" readers that these were surface temperatures rather than air temperatures (notice my highlights on the screengrab from a CTV news article shown in figure 40 below[12]). But how many citizens understand the difference between surface and air temperatures as they casually browse news headlines? This isn't an innocent manipulation of public perceptions; this is a malicious type of blatant and deliberate fraud. It turned out that, despite the dramatic headlines about surface temperatures in Sicily reaching 45 °C, *the air temperature measurement in Sicily only reached 32 °C*! They used a half-truth to tell a monstrous lie; it was a deliberate ploy to terrify an unsuspecting public. You can read the full story of the ESA's fraud on wattsupwithhat.com.[13]

---

[12] "A heat wave named Cerberus has southern Europe in its jaws, and it's only going to get worse" (2023, Jul 13). *CTV News*. https://www.ctvnews.ca/climate-and-environment/a-heat-wave-named-cerberus-has-southern-europe-in-its-jaws-and-it-s-only-going-to-get-worse-1.6478127

[13] Gosselin, P. (2023, Jul 19). Europe's "48°C Horror That Never Was"...ESA, Media Sharply Criticized For Manipulative Reporting. *Watts Up With That*. https://wattsupwiththat.com/2023/07/19/europes-48c-horror-that-never-wasesa-media-sharply-criticized-for-manipulative-reporting/

Figure 40. Media began reporting land surface temperatures (recorded at ground level) instead of official air temperatures (recorded 2 m above the surface) in order to create terrifying headlines. This screengrab from a CTV article published on July 13th, 2023 shows the slight-of-hand used to terrify the public (my annotations).

CTV article:
https://www.ctvnews.ca/climate-and-environment/a-heat-wave-named-cerberus-has-southern-europe-in-its-jaws-and-it-s-only-going-to-get-worse-1.6478127

Chapter 8

Re-Writing History
to Warm the Present and Cool the Past

"If you torture the data long enough, it will confess to anything."

— Ronald Coase, British economist

It's time to step back to look at the bigger picture of Earth's global climate. By the time I'm done, it will be clear that neither the temperature charts paraded around in the media, nor the narrative about carbon dioxide acting as the control knob on the climate, nor the scary stories about unprecedented changes in our climate are true. As well-known farmer and author Joel Salatin once wrote, *"Realize that agendas drive data, not the other way round."*

Even though the story of this blatant re-writing of our history is remarkable in its own right, as it unfolds some nagging questions will begin to bubble up through the noise to force us to also question some of our assumptions about the origins of the carbon dioxide accumulating in our atmosphere. That story has nothing to do with a changing global climate but everything to do with the soils on our farms and in our forests and their ability to withstand the droughts and floods that are part of the natural cyclical variations that create our planet's climate.

During the depth of the global cooling scare of the 1970s and almost a decade before Margaret Thatcher decided to use climate change as a pretext to try to end the United Kingdom's dependence on coal in favor of nuclear energy, National Geographic published a lengthy feature about climate change in its November 1976 edition[1] — it was published right at the height of the 1970s global cooling scare. The article had a series of four charts (reproduced in figure 41) showing how climate has changed over time. I want to walk you through these charts before I show you how

---

[1] Mattews, S. (1976, Nov). What's Happening To Our Climate? *National Geographic*. Nov 1976 edition. https://www.sealevel.info/NatGeo_1976-11_whats_happening_to_our_climate/

more recent charts featured in the media are re-writing the history of our planet's climate.

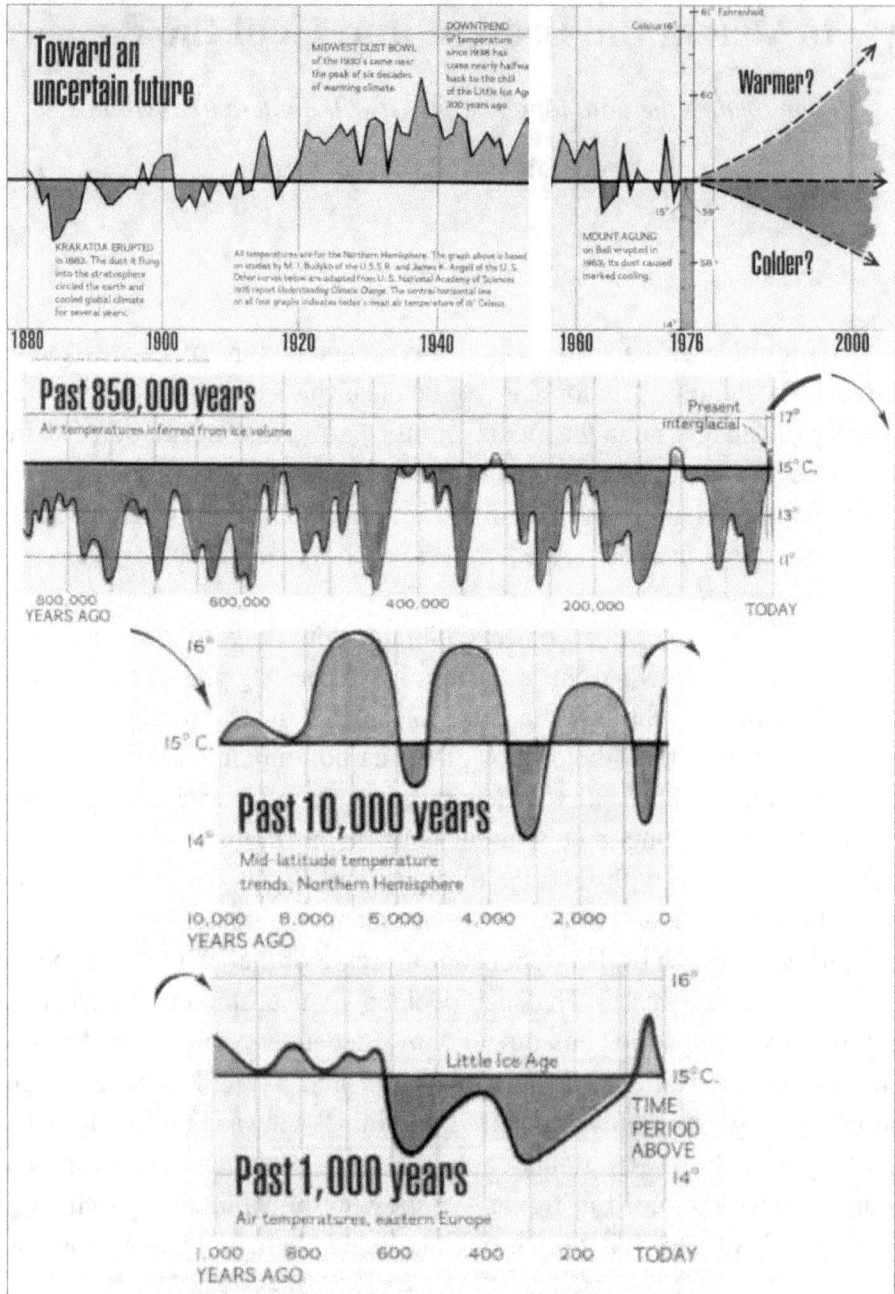

Figure 41. Published by National Geographic in its November 1976 issue in an article entitled "*What's Happening to Our Climate?*"
Source: https://www.sealevel.info/NatGeo_1976-11_whats_happening_to_our_climate/

The first of the four charts in the National Geographic series shows how climate warmed from the lows of the 1880s (just after the end of the Little Ice Age) to the extreme heat and drought of the 1930s Dust Bowl, and then subsequently cooled nearly halfway back down to the lows of the 1880s, ending with a question mark in 1976 as to whether the future would be warmer or cooler. The second chart in the National Geographic series shows how Earth's climate has undergone at least 8 irregular cyclical ice ages over the last 850,000 years and that today we are merely in a warm interglacial period between ice ages. Although we are unlikely to see glaciers in Manhattan in our lifetime, if history is our guide, then the next ice age is coming; it's just a matter of time. It's also worth noting that our current ice age cycles extend back even further than what is shown in this National Geographic chart. In total there have been over 20 glacial advances and interglacial retreats just over the last 2.58 million years, collectively known as the Quaternary Glaciation — our current warm period is merely the latest warm interglacial period within this longer Quaternary Glaciation.

The third chart in the National Geographic series zooms in on our current warm interglacial period to show only the last 10,000 years (a blink of an eye from a geological perspective); it reveals that our warm interglacial period has not been a single, stable warm period, but rather a series of alternating warm and cool periods, with each warm phase being slightly cooler than the last (the Roman Warm Period never reached the warmest temperatures experienced during earlier Minoan Warm Period, the subsequent Medieval Warm Period was even cooler than the Roman Warm Period, and our own Modern Warm Period is once again even cooler than the Medieval Warm Period. Each warming episode fails to reach the warmth of the one that preceded it.

The final National Geographic chart zooms in on the last 1,000 years to show the Little Ice Age that separated our own Modern Warm Period from the Medieval Warm Period. The unpredictable frosts and short growing seasons of the Little Ice Age caused widespread crop failures and famines all around the world for over 500 years. It also drove the Vikings out of Greenland, caused the River Thames and the Dutch canals to regularly freeze over during the winter, contributed to the extinction of the Dorset Culture in the Canadian High Arctic, contributed to the extinction of woodland bison in Canada's Yukon Territory 400 years ago, weakened the monsoon rains in northwest India leading to brutal droughts that killed millions, and contributed to the fall of the Ming Dynasty in China as crop failures, floods, and droughts triggered widespread social unrest. The Little Ice Age only ended around 1850 and was followed by the short blip of heat that peaked during the 1930s. The improving weather conditions that followed the end of the Little Ice Age undoubtedly also played a role in improving growing conditions on the Great Plains during the 1800s, which fooled so many into believing that the rain was following the plow into the Great American Desert.

There are two take-home messages from National Geographic's 1976 charts, which are important in order to understand the fraudulent climate charts that I will be showing you momentarily:

1. Climate is not stable, but rather it is driven by a multitude of powerful cyclical forces acting over both long- and shorter-time scales, which are constantly tugging climate in one direction or another.
2. The heat of the 1930s was <u>much</u> warmer than both the cold climate of 1880s (just after the world came out of the Little Ice Age) and the cool climate of the 1970s (when scientists were warning us about the next ice age).

And yet, in 2023 we were told that global temperatures were allegedly the hottest in more than 100,000 years (figure 42).

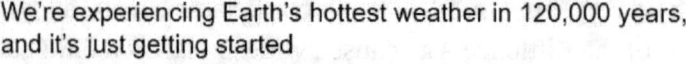

Figure 42. A sample of news headlines from July, 2023, claiming that 2023 exceeded temperature records of the last 120,000 years.

Let's start by picking apart the infamous propaganda chart that is commonly portrayed as "proof" that temperatures are spiralling out of control, and which is held forth as part of the body of "evidence" for their laughable claim that we're experiencing "the hottest weather in 120,000 years". The fraudulent graph shown in figure 43 is published on NASA's website, with my annotations added:[2]

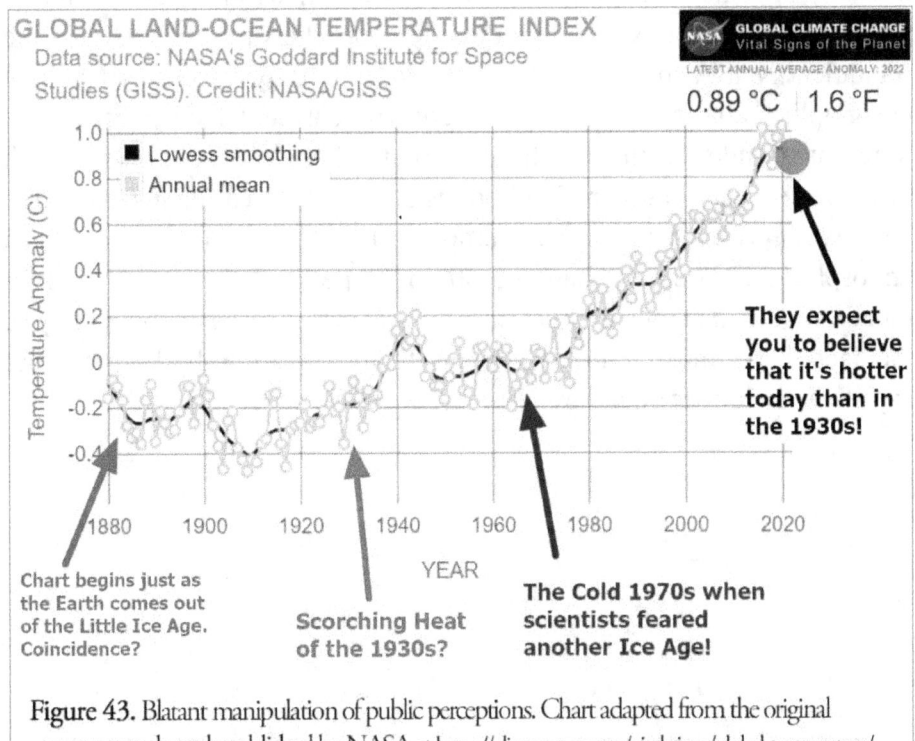

Figure 43. Blatant manipulation of public perceptions. Chart adapted from the original un-annotated graph published by NASA at https://climate.nasa.gov/vital-signs/global-temperature/

In contrast with the earlier National Geographic charts, NASA's fraudulent chart expects us to believe that the scorching heat of the 1930s was as cool (even slightly cooler) than the climate of the 1880s. Bizarrely, it also shows the scorching heat of the 1930s as being colder than the 1970s when scientists feared we were sliding into another Ice Age. Then, as the chart curves into what looks like some kind of scary runaway temperature spike, we're expected to believe that temperatures today are considerably hotter than the scorching heat of the 1930s that turned the

---

[2] "Global Temperature". *NASA.* https://climate.nasa.gov/vital-signs/global-temperature/

plowed fields of the Great Plains into a dustbowl. This is what fraud looks like when it's committed by scientific institutions.

Compare the NASA chart above with the much less well-known Heat Wave Magnitude Index shown in figure 44 below, which was published by the U.S. government in its *Fourth National Climate Assessment* in 2018.[3] The Heat Wave Magnitude Index combines the duration and intensity of extreme temperature events from all weather stations into a single numerical index. While it directly contradicts the story told by the fraudulent NASA chart, it reaffirms the story told by the earlier National Geographic charts. This official government-published Heat Wave Magnitude Index confirms that the brutal heat of the 1930s, which destroyed so many farmers' livelihoods out on the Great Plains and led to heat extremes as far away as Europe and East Asia[4] (at a time when atmospheric CO_2 hovered around 307 parts per million (ppm)), was far more extreme than the summer temperatures we experience today despite the fact that atmospheric CO_2 levels today (at around 417 ppm) are significantly higher than they were in the 1930s.

---

[3] U.S. Global Change Research Program. (2018). *Fourth National Climate Assessment*. U.S. Government. https://science2017.globalchange.gov/chapter/6/

[4] "1930s Dust Bowl led to extreme heat around Northern Hemisphere. (2022, Nov 29). *Science Daily*. https://www.sciencedaily.com/releases/2022/11/221129134521.htm

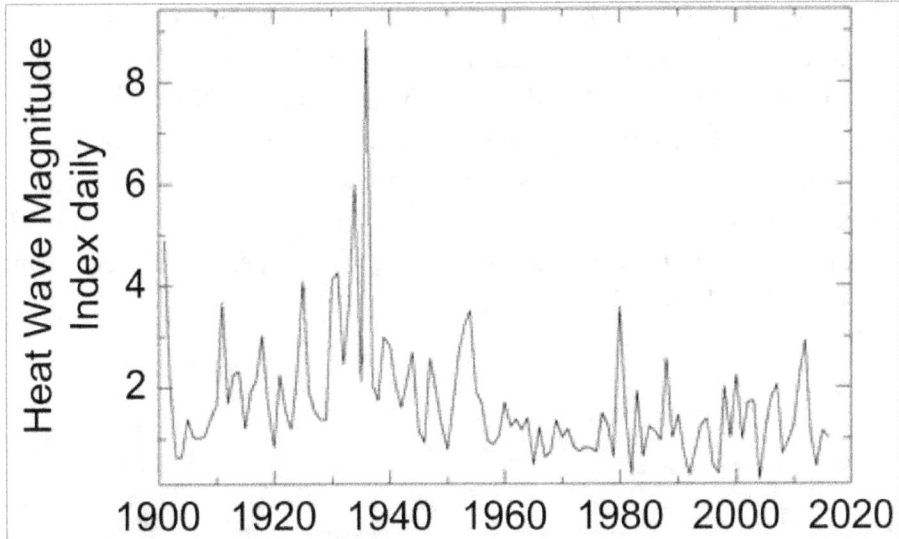

Figure 44. Heat Wave Magnitude Index for all weather stations in the contiguous United States, published in 2018 by the U.S. federal government as part of its Fourth National Climate Assessment.
Source: https://science2017.globalchange.gov/chapter/6/

This Heat Wave Magnitude Index chart also completely obliterates any illusion that rising carbon dioxide levels are causing an increase in heat waves — the chart clearly demonstrates that there's no correlation whatsoever between heat waves and rising atmospheric carbon dioxide — the worst heat waves clearly occurred when carbon dioxide was much lower, which might help explain why so few members of the public have seen this inconvenient government chart. Yet the *Fourth National Climate Assessment* is a vital government publication that every politician, climate scientist, and journalist writing about climate change is well aware of it (or should be if they are doing their job) — the story told by the Heat Wave Magnitude Index should be front-page news; the fact that it is not tells you everything you need to know about the toxic polygamous marriage between climate science, mainstream media, and politics.

The aforementioned fraudulent NASA chart in figure 43 shows 0.89 °C of warming since the 1970s. However, since 1979, the University of Alabama in Huntsville has been using satellites to measure temperatures in the lower atmosphere (shown in figure 45). By using satellites to measure temperatures in the lower atmosphere, their data doesn't suffer all the biases of poorly located ground-based weather stations, heat island effects, and so on. The University of Alabama chart only shows 0.69 °C of warming since 1979. And it doesn't show anything resembling the near parabolic rise in temperature shown in the NASA chart. Notice that on the fraudulent NASA chart, temperatures look far hotter today than during the hot summer of 1999, whereas the University of Alabama satellite-measured temperature record shows that the gradually warming trend (13-month average) still hasn't climbed above the temperatures seen during temporary heat blip of 1999.

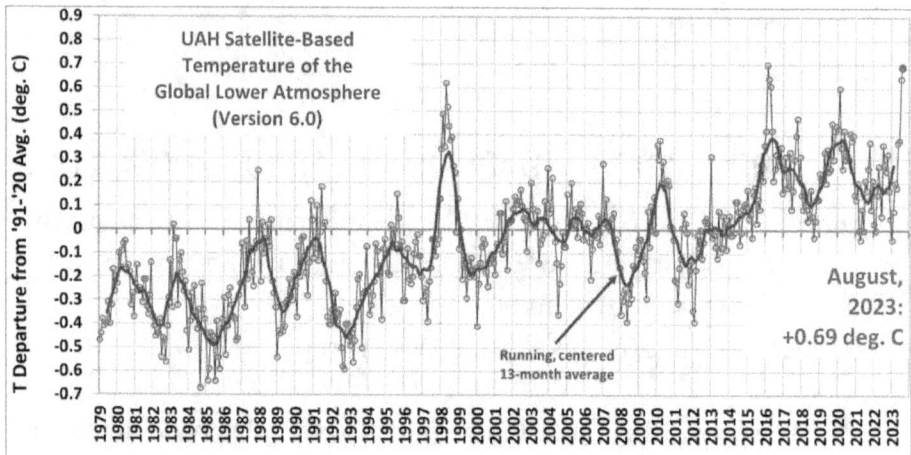

Figure 45. UAH satellite-based temperature of the global lower atmosphere, updated monthly by the University of Alabama in Huntsville (https://www.nsstc.uah.edu/climate/). Chart published by climatologist Dr. Roy Spencer, principal research scientist at UAH on his website (https://www.drroyspencer.com/).

The contrast between this combination of charts — the NASA chart, the Heat Wave Magnitude Index, and the University of Alabama satellite temperatures — illustrates the way the public is fed deliberately fraudulent temperature charts that are completely out of line with the raw unaltered temperature records. In their efforts to pander to the climate change narrative, many climate scientists and government institutions are cooling the past and warming the present by making "data adjustments" to the raw historical data in order to tell a fictitious story. In the competition for funding, fear sells. And the end justifies the means for those wanting to tell a compelling story about the climate to corral citizens into accepting a preferred political agenda, just as "the rain follows the plow" was conveniently leveraged as a nation-building exercise in the 19th century.

In November 2009, hackers broke into the servers of the Climate Research Unit at the University of East Anglia and published thousands of internal emails between some of the world's most prominent climate scientists. A second set of approximately 5,000 emails was released in 2011. These emails appeared to show a systematic and deliberate attempt to manipulate the climate record in order to match the hysterical climate narrative that has emerged in recent decades. I have reproduced a small sampling of quotes below from these "Climategate" emails, sourced from an article published in the Guardian in 2010[5] and from an article published on wattsupwiththat.com in 2023 on the 12-year anniversary of the Climategate scandal,[6] in which various climate scientists discuss how to manipulate temperature data, censor data, circumvent freedom of information requests, subvert the peer-review process, and suppress critical papers.

Here are a few excerpts of the leaked emails [my emphasis in bold]:

"I've just completed Mike's [Mann] Nature trick of adding in the real temps to each series for the last 20 years (i.e. from 1981 onwards) and from 1961 for Keith's [Briffa] **to hide the decline***."*

—Dr. Phil Jones, Director of the Climatic Research Unit, disclosed Climategate e-mail, Nov. 16, 1999.

"Also **we have applied a completely artificial adjustment to the data after 1960***, so they look closer to observed temperatures than the tree-ring data actually were...."*

— Dr. Tim Osborn, Climatic Research Unit, disclosed Climategate e-mail, Dec. 20, 2006.

"We have 25 or so years invested in the work. Why should I make the data available to you, when your aim is to try to find something wrong with it."

---

[5] Pearce, F. (2010, Jul 7). The five key leaked emails from UEA's Climatic Research Unit. *The Guardian.* https://www.theguardian.com/environment/2010/jul/07/hacked-climate-emails-analysis

[6] "Climategate: Never Forget (12th anniversary)." *Watts Up With That.* https://wattsupwiththat.com/2022/11/22/climategate-never-forget-12th-anniversary/

— Dr. Phil Jones, Director of the Climate Research Unit at East Anglia University, email to Warwick Hughes, 2004.

"I'm getting hassled by a couple of people to release the CRU station temperature data. **Don't any of you three tell anybody that the U.K. has a Freedom of Information Act.***"*

— Dr. Phil Jones, Director of the Climatic Research Unit, disclosed Climategate e-mail, Feb. 21, 2005.

"Keith's [Briffa] series...differs in large part in exactly the opposite direction that Phil's [Jones] does from ours. This is the problem we all picked up on (everyone in the room at IPCC was in agreement that **this was a problem and a potential distraction/detraction from the reasonably consensus viewpoint we'd like to show** *w/ the Jones et al and Mann et al series)."*

— Dr. Michael Mann, IPCC Lead Author, disclosed Climategate e-mail, Sep. 22, 1999.

"You might want to check with the IPCC Bureau. I've been told that IPCC is above national FOI Acts [FOI = Freedom of Information]. One way to cover yourself and all those working in AR5 [the upcoming IPCC Fifth Assessment Report] **would be to delete all e-mails at the end of the process***. Hard to do, as not everybody will remember it."*

— Dr. Phil Jones, Director of the Climatic Research Unit, on avoiding Freedom of Information requirements, disclosed Climategate e-mail, May 12, 2009.

"... I can't see either of these papers being in the next IPCC report. Kevin and I will keep them out somehow - even if we have to redefine what the peer-review literature is !"

—Dr. Phil Jones, Director of the Climatic Research Unit, disclosed Climategate e-mail, July 8th, 2004.

Unsurprisingly, given the decrepit state of the government-aligned mainstream media, journalists largely ignored the Climategate scandal and went out of their way to pretend that nothing was wrong, and many have since attempted to downplay or whitewash the emails as having been "misunderstood out of context". It's hardly a mystery why the media chose to turn a blind eye — as the science editor at Time Magazine declared on September 16th, 1989, at a global warming conference at the Smithsonian Institute (quoted by the Wall Street Journal), *"As the science editor at Time, I would freely admit that on this issue we have crossed the boundary from news reporting to advocacy."*[7]

If the Climategate emails weren't damning enough, evidence of data tampering is plain to see in many of the publicly available climate records. Geologist and climate historian Tony Heller has documented how NASA tampered with the raw temperature data via a series of "data adjustments" in order to cool the past and warm the present. I have included two screengrabs from one of Tony's videos, reproduced in the graphic in figure 46 below.[8] As Tony shows, the upper chart was original published by NASA in 1999 before the climate hysteria reached current levels. It clearly shows temperatures in 1999 were cooler than the peak temperature from 1934. But the same updated chart published by NASA in 2017 (shown in the lower chart and truncated to only cover the same time period) shows that the temperature peak in 1999 has inexplicably become hotter than 1934.

---

[7] Noyes, R. (2010, Apr 23). Twenty Years of Advocacy, Not Journalism, on Global Warming – the media has formed a consensus around climate change. *Wall Street Journal*.
https://web.archive.org/web/20210621052158/https://www.wsj.com/articles/SB10001424052748703709804575202554026555656

[8] Heller, T. (2017, Jul 20). *NASA And NOAA : Erasing The Record Heat Of 1934*. [Video]. YouTube.
https://www.youtube.com/watch?v=mt14zqcghXo&t=103s&ab_channel=TonyHeller

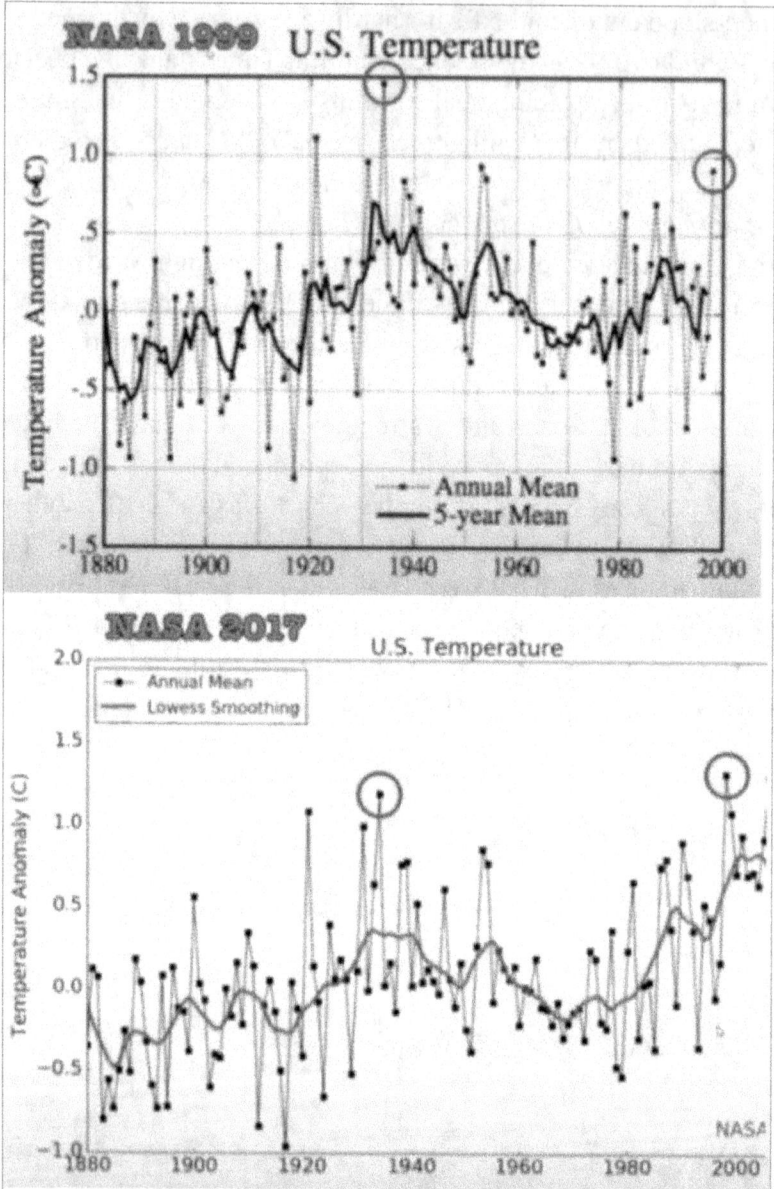

Figure 46. NASA added data adjustments sometime after 1999 to cool the past and warm the present. Note how the chart from 1999 shows the 1934 peak as being noticeably hotter than the 1999 peak, but on the "adjusted" 2017 chart, the 1999 peak is now hotter than 1934 peak.

Charts sourced from climate historian Tony Heller's YouTube video: "NASA and NOAA: Erasing the Record Heat of 1934"
https://www.youtube.com/watch?v=mt14zqcghXo&ab_channel=TonyHeller

Likewise, look at the lows reached in 1979 on both charts. In the chart from 1999, the lows of 1979 are as low as the lows of the 1880s, 1890s, and 1910s (hence the global cooling scare of the 1970s) and significantly lower than the lowest lows of the 1920s. By contrast, in the chart from 2017, the lows of 1979 are significantly warmer than lowest lows of the 1880s, 1890s, and 1910s and even slightly warmer than the lowest lows of the 1920s. The following quote from Tony Heller in the second edition of *Evidence-based Climate Science* sums up the data tampering in our historical temperature record as follows: *"Historical data has been systematically altered over the past 15 years to cool past temperatures and increase more recent temperatures. The amount of warming from 1880 to 2000 is now shown by NASA as double what was shown in 2001. Going back further to the 1975 National Academy of Sciences report, we see a completely different story—where all 1900–40 warming was lost by 1970."*[9]

The extraordinary pair of charts in figure 46 reveal the extent to which many of our government agencies are corrupting the data shown to the public in order to push a clear narrative that matches political expectations and to generate the fear that gives these institutions priority access to funding. I strongly encourage you to explore Tony Heller's full video library on YouTube *(@TonyHeller)*, check out his website at *realclimatescience.com*, and follow him on Twitter *(@TonyClimate)*. In some of his recent videos, he has zoomed in on individual weather stations to demonstrate how government agencies are systematically tampering with raw temperature data to re-write history via "data adjustments" that create dramatic warming trends that are not reflected in the raw temperature data. In his videos, Tony Heller demonstrates that one of the strategies used to create these illusions appears to be a "data homogenization" technique that effectively smears the urban heat island effect onto surrounding rural stations to create warming trends where there are none. As Tony says, they are systematically cooling the past and warming the present.

---

[9] Heller, T. (2016). *Chapter 3 – Is the NASA Surface Temperature Record and Accurate Representation?* Evidence-Based Climate Science (Second Edition). Easterbrook, D. (Ed.) 49-59 Elsevier Publishing.
https://www.sciencedirect.com/science/article/abs/pii/B9780128045886000033?via%3Dihub

The two charts in figure 47 are also from one of Tony Heller's recent videos. They illustrate the dramatic warming that is created through the process of "data homogenization". In this example (one of many examples that Tony shows from North America and from other "official temperature records" from all around the world) comes from NOAA's official temperature records for the city of Guira, in Venezuela — the first chart shows the raw NOAA temperature record, the second shows the final "adjusted" temperature record for the exact same weather station, which illustrates how a slight cooling trend has been transformed into a warming trend by tampering with the raw temperature record.[10]

---

[10] Heller, T. (2024, Feb 12). *Poisoning the Climate Soup*. [Video]. YouTube. https://www.youtube.com/watch?v=hqgRtaWKulY&ab_channel=TonyHeller

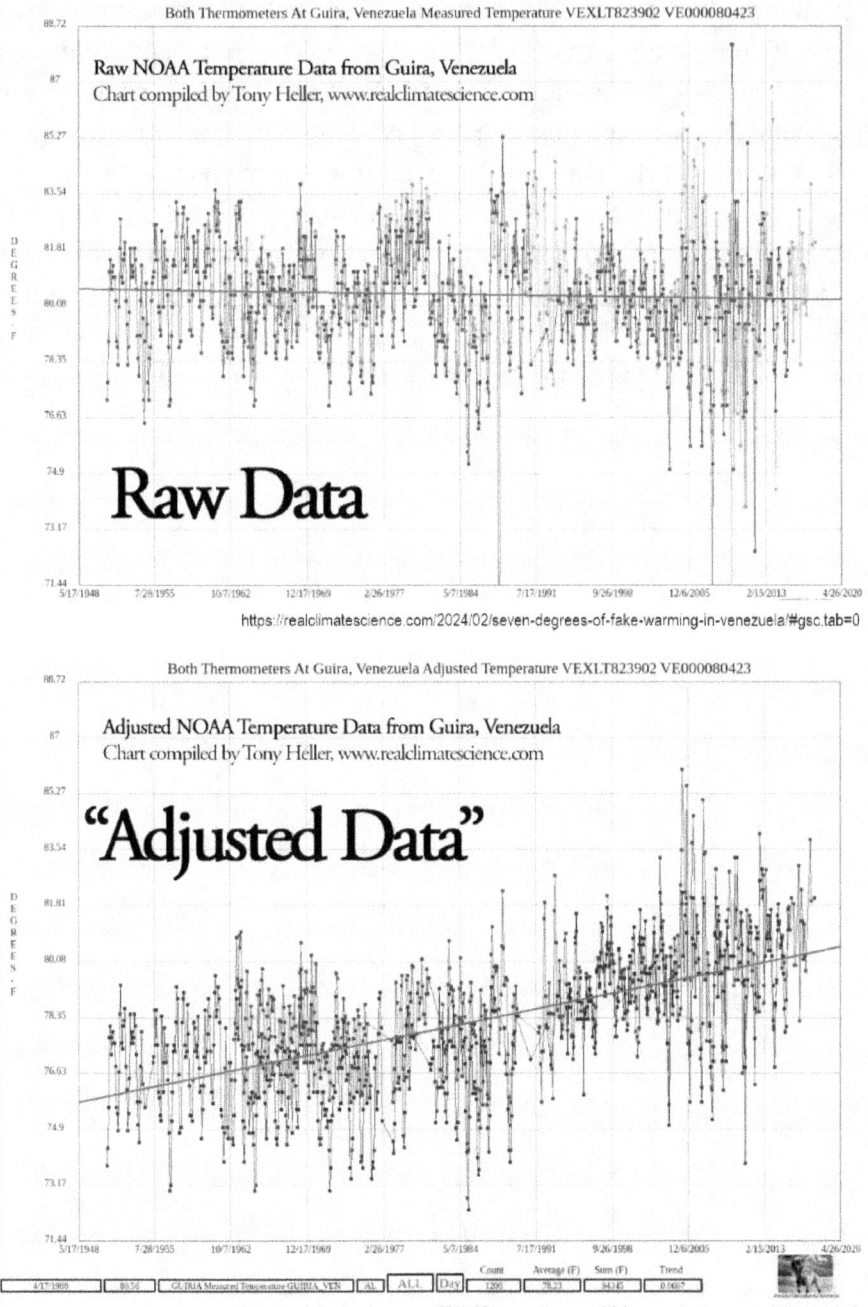

Figure 47. Raw vs "adjusted" temperature data from Guira, Venezuela

Source: https://realclimatescience.com/2024/02/seven-degrees-of-fake-warming-in-venezuela/

In figure 48 below, Tony plotted the difference between the final "adjusted" temperatures versus the raw underlying temperature data for all U.S. historical climatology network stations. It tells a damning story — as you can see, temperature "adjustments" consistently cool older measurements while consistently warming more recent temperature records, effectively projecting a strong artificial warming trend onto the final data.[11] The net effect is that, by cooling the past and warming the present, the temperature record is "bent" to match the rise in atmospheric CO_2. In keeping with the tradition of the noble lie, the temperature record needed a little help to tell the "right" story.

---

[11] Heller, T. (2024, Feb 15). *The 50 PPM Rule For Data Tampering*. [Video]. YouTube, https://www.youtube.com/watch?v=ai4j6zc2OPo&t=124s&ab_channel=TonyHeller

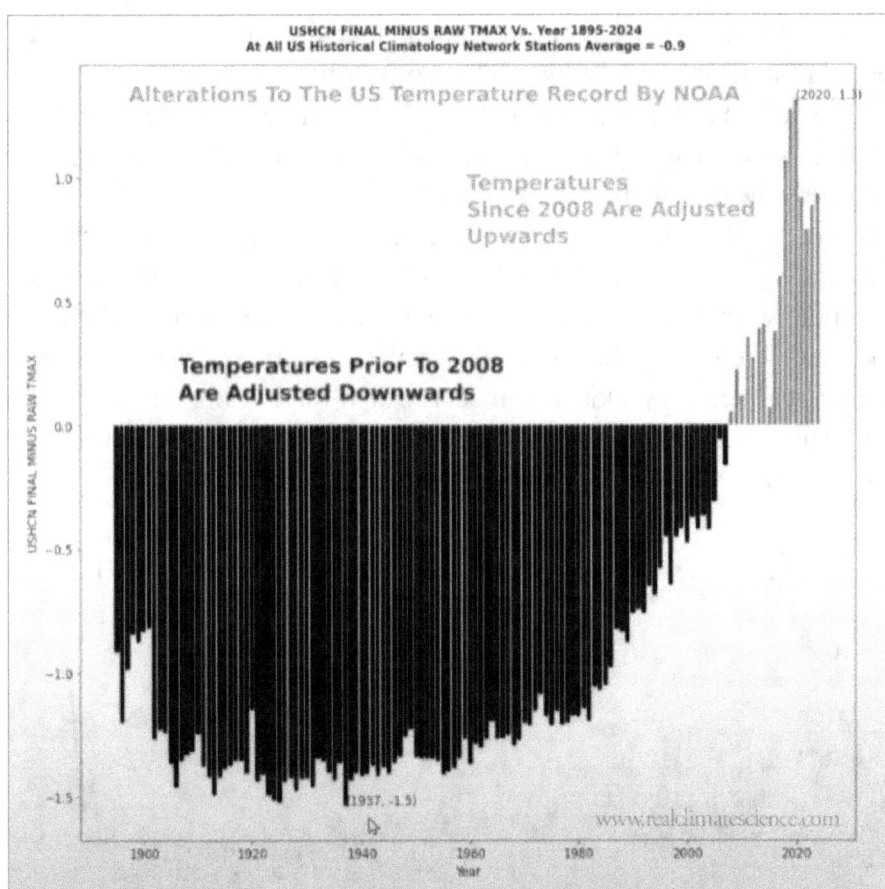

Figure 48. US Historical Climatology Network (all stations) showing the magnitude of NOAA's data adjustments to produce their official charts (chart bars are calculated by subtracting the raw value from the final value).

Chart compiled by Tony Heller:
https://realclimatescience.com/2024/02/the-50-ppm-rule-for-data-tampering/

It's not only the recent temperature record of the last century that is being manipulated. On December 6th, 2006, geologist and geophysicist Dr. David Demming testified to Congress about how he received an email from a major researcher in the climate sciences who told him, *"We need to get rid of the Medieval Warm Period."*[12] In the years since Dr.

---

[12] Heller, T. (2022, Sep 13). *"We have to get rid of the Medieval Warm Period!"* [Video]. YouTube.
https://www.youtube.com/watch?v=6sX31KEyucI&ab_channel=TonyHeller

Demming's testimony, it would appear that the long-term temperature record has indeed been "adjusted" to do just that.

As you can see in the "consensus" global temperature record in figure 49, published on Wikipedia and popularly referred to as the "hockey stick graph" by the media, the Medieval Warm Period from roughly 450 to 1300 AD along with the subsequent Little Ice Age from 1300 to 1850 have essentially been erased. Instead, the chart shows what appears to be a remarkably stable and gradually cooling climate over the past 2,000 years until, not long after the invention of the modern internal combustion engine, temperatures suddenly appear to soar (and again, the chart shows the heat of the 1930s as having been significantly cooler than the cold years of the 1970s).

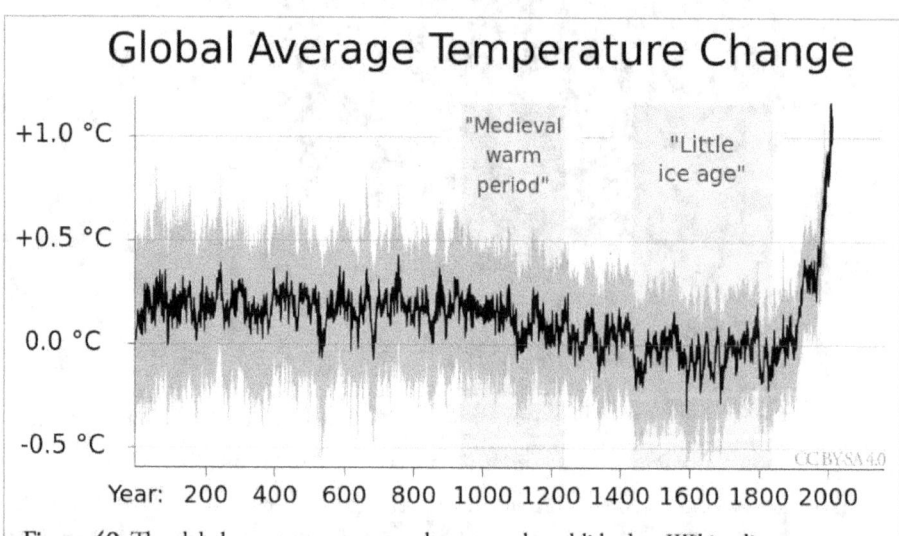

Figure 49. The global average temperature change graph, published on Wikipedia (https://en.wikipedia.org/wiki/Global_temperature_record), claims to show the temperature record of the last 2,000 years. It has erased the Medieval Warm Period and the subsequent Little Ice Age.

Compare this "hockey stick" chart to the temperature record documented by National Geographic back in 1976 (figure 41), which showed dramatic temperature swings from hot to cold over this same time period — these two versions of history would seem to be from two completely different planets!

Let's also compare the "hockey stick" reconstruction to the temperature reconstruction published back in 1990 as part of the IPCC's First Assessment Report (before climate hysteria went completely off the rails), reproduced in figure 50 below.[13] It looks nothing like the hockey stick reconstruction — there's nothing stable about the climate shown in the 1990 IPCC chart. You can see that the Little Ice Age, which drove the Vikings out of Greenland and caused advancing glaciers to destroy entire villages in the French and Swiss Alps as well as in Alaska,[14] was meaningfully colder than the heat of the 1910s through the 1930s (when those same glaciers were in rapid retreat). It also shows that the Medieval Warm Period was considerably warmer than temperatures today. But on the infamous hockey stick graph, that's not the case. The hockey stick lives in another reality.

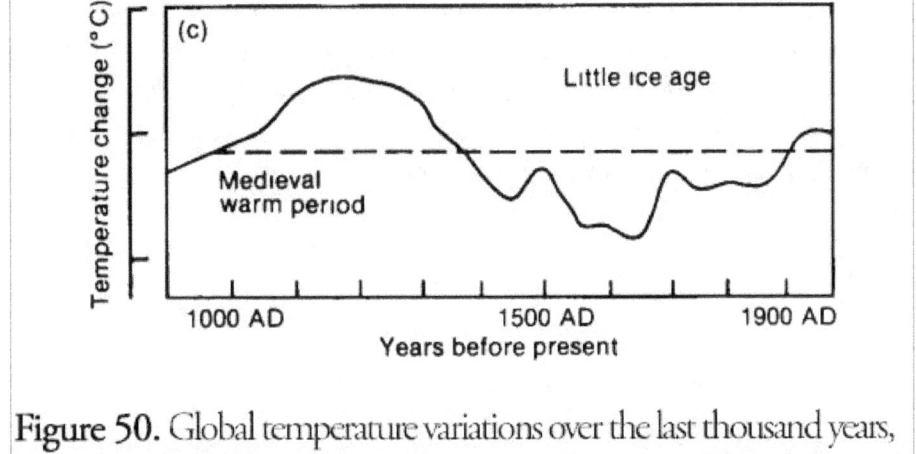

Figure 50. Global temperature variations over the last thousand years, as per the IPCC's First Climate Assessment Report, published in 1990. https://www.ipcc.ch/site/assets/uploads/2018/03/ipcc_far_wg_I_full_report.pdf

On October 13th, 1910, the Aberdeen Herald published a newspaper article entitled *Glaciers are Retreating — Withdraw from Lowlands to Mountains, Scientists Find*,[15] reproduced in figure 51, which provides further context to the rapidly retreating glaciers during the late 1800s and

---

[13] Intergovernmental Panel on Climate Change. (1990) *First Assessment Report*. World Meteorological Organization / United Nations Environment Programme. https://www.ipcc.ch/site/assets/uploads/2018/03/ipcc_far_wg_I_full_report.pdf

[14] Gedney, L. (1985, May 6). *Don't Build On A Glacier's Right-Of-Way*. Geophysical Institute, University of Alaska Fairbanks. https://www.gi.alaska.edu/alaska-science-forum/dont-build-glaciers-right-way

[15] "Glaciers are Retreating." (1910, Oct 13). *Aberdeen Herald*. https://www.newspapers.com/image/71208161/

early 1900s — the article describes glaciers rapidly retreating all around the world at a time when, according to the infamous hockey stick graph, temperatures were still near their lowest levels after more than 1,000 years of declining temperatures. The article from 1910 notes that, "*Scientists aver that save over a small area the glaciers of the world are retreating to the mountains. The glacier of Mount Sarmiento, in South America, which descended to the sea when Darwin found it in 1836, is now separated from the shore by a vigorous growth of timber. The Jacobshaven glacier, in Greenland, has retreated four miles since 1860, and the East glacier, in Spitzenbergen, is more than a mile away from its old terminal moraine. In Scandinavia the snow line is farther up the mountains, and the glaciers have withdrawn 3,000 feet from the lowlands in a century. The Arapahoe glacier, in the Rocky mountains, with characteristic American enterprise has been melting at a rapid rate for several years. [...]*" Once again, eyewitness accounts from 1910 appear to be from a completely different planet than the history shown on the hockey stick graph.

> **Aberdeen Herald**
> 13 Oct 1910, Thu · Page 7
>
> ## GLACIERS ARE RETREATING.
>
> ### Withdraw From Lowlands to Mountains, Scientists Find.
>
> Scientists aver that save over a small area the glaciers of the world are retreating to the mountains. The glacier of Mount Sarmiento, in South America, which descended to the sea when Darwin found it in 1836, is now separated from the shore by a vigorous growth of timber.
>
> The Jacobshaven glacier, in Greenland, has retreated four miles since 1860, and the East glacier, in Spitzenbergen, is more than a mile away from its old terminal moraine. In Scandinavia the snow line is farther up the mountains, and the glaciers have withdrawn 3,000 feet from the lowlands in a century.
>
> The Arapahoe glacier, in the Rocky mountains, with characteristic American enterprise has been melting at a rapid rate for several years. In the eastern Alps and one or two other small districts the glaciers are growing. In view of these facts we should not be too skeptical when old men assure us that winters nowadays are not to be compared with the winters of their boyhood.

Figure 51. "Glaciers are Retreating — Withdraw from Lowlands to Mountains, Scientists Find." Source: Aberdeen Herald, Page 7, October 13th, 1910. https://www.newspapers.com/image/71208161/

The 1976 National Geographic chart showed that temperatures cooled sharply after the heat of the 1930s and 1940s, and bottomed out during the 1970s at approximately the same temperature level as the late 1800s and early 1900s. This mirrors the story told by the Greenland Ice Sheet during the 20th century — the ice sheet shrunk rapidly from the mid-1850s until 1964 but then stabilized, neither shrinking nor growing from the 1970s until the mid 1990s, as discussed by NASA here[16] and here.[17] Yet according to the "hockey stick" graph, temperatures during the 1960s, 1970s, and 1980s were significantly warmer than during the late 1800s and early 1900s. How is it possible for the Greenland Ice Sheet to stop melting during the 60s and 70s if, according to the hockey stick graph, the 60s and 70s were supposedly significantly warmer than the late 1800s and early 1900s when those same glaciers were all melting rapidly!?! Are we meant to believe that glaciers melt when its colder and stop melting when its warmer? Or maybe the authors of the 1976 National Geographic chart got it right, but the "hockey stick" graph does not.

~

This brings up another dirty trick used by the activist-scientists. It is common to show two side-by-side photographs showing how much various glaciers have melted over the last century as evidence of global warming, much like the two photographs from Glacier National Park shown in figure 52 below, one from 1914, the other from 2001, which illustrate that glaciers retreated substantially between 1914 and 2001.

---

[16] "Kjer Glacier, Then and Now" (n.d.). *NASA Earth Observatory.* Accessed 4 Dec 2023. https://earthobservatory.nasa.gov/images/149050/kjer-glacier-then-and-now

[17] Starr, C. (2007, Jan 5). Jakobshaven Glacier Calving Front Recession from 1850 to 2006. *NASA Scientific Visualization Studio.* https://svs.gsfc.nasa.gov/3395/

Jackson and Blackfoot Glaciers in Glacier National Park, USA, in 1914 vs 2001.

Figure 52. Glacial melt between 1914 and 2001 at Glacier National Park.

https://commons.wikimedia.org/wiki/File:Jackson_and_Blackfoot_Glaciers_1914_to_2001.jpg

So, here's a trick question: ignoring everything we've been discussing about the climate history of our planet over the past pages of this book and based only on the photographic evidence of these two pictures of the glaciers in Glacier National Park in 1914 and 2001, which of the three following statements about what happened to temperatures between these two dates is true:

    A. The climate warmed up between 1914 and 2001.
    B. The climate cooled down between 1914 and 2001.
    C. There is no way of knowing based only on these two photographs.

It's a brilliant photographic deception — here's the solution to the fake story that's superimposed on these side-by-side photographs. Imagine I get up at 6:00 am and pull a giant block of ice out of the deep freezer and put it on my kitchen counter. At 8:00 am my wife gets up and observes the slowly melting block of ice on the counter, then goes away for the day and comes back home at 8:00 pm to see that the block of ice has melted significantly. Has the temperature of the room changed between <u>8:00 am</u> and 8:00 pm? In reality, the main temperature change happened at 6:00 am when I removed the block of ice from the deep freezer, just like when global temperatures finally came out of the 500-year deep freeze at the end of the Little Ice Age around the year 1850, but that tells you nothing about what temperatures have been doing since then. The melting is only evidence that temperatures are warmer than when the block was removed from the deepfreeze, but unless you can track the year-to-year melting of the ice, you have no idea whether temperatures have warmed, cooled, or stayed the same between those two later snapshots in time. If you change the dial on the thermostat, you can speed up or slow down the rate at which the ice melts but the fact that the ice block melted between 8:00 am and 8:00 pm (just like the fact that the glaciers melted from 1914 and 2001) does not tell you whether temperatures have warmed, cooled, or stayed the same between these two snapshots in time — it merely tells you that it is warmer in both 1914 and 2001 than it was during the Little Ice Age when those same glaciers were growing.

Further complicating the story are humidity and snowfall. There are many periods in history when glaciers advanced despite a warming climate because the climate got wetter, which caused more winter snowfall — I'll describe one such episode in chapter 24. Falling humidity can have the opposite effect, causing glaciers to shrink even if temperatures remain stable or are falling because less snowfall or even drier air will tip the scales towards decreasing ice — ice doesn't just melt, it also sublimates, turning directly from a solid to a gas when air gets really dry. For example, Al Gore frequently likes to tell the story of Mt. Kilimanjaro's shrinking glaciers, citing it as evidence of global warming, yet researchers have since demonstrated that the primary culprit responsible for Mt. Kilimanjaro's shrinking glaciers was less snowfall and drier air over the mountaintop because of rampant local deforestation all around the perimeter and flanks of the mountain. This local deforestation dramatically reduced the amount of humidity in the air above the mountain, which reduced snowfall and (even more importantly) the dry air sucked moisture out of the ice, which dramatically increased sublimation — more on that story in chapter 23. The main point I am trying to make here is that when activists show dramatic side-by-side photos of retreating glaciers, they're leaving out important parts of the story to create a simple illusion to manipulate heartstrings — beware!

The nuanced story about global temperatures told by the 1990 IPCC chart and the charts published by National Geographic in 1976 makes sense. These charts also fit with story told by archaeological data, while the hockey stick chart does not. During the Medieval Warm Period, grapes grew in England in regions that are still too cold to grow grapes today. In France and Germany archaeologists discovered the remains of medieval vineyards 500 km north of their northernmost range today.[18] Vikings colonized and farmed Greenland during the Medieval Warm Period; at that time they were able to grow barley,[19] which is still not possible on those same fields in Greenland today because it's still too cold (today there are not enough warm days for the barley kernels to reach maturity).[20] Likewise, during the Medieval Warm Period wheat and oats were successfully grown around Trondheim, Norway, suggesting that the climate at that time was about 1 °C warmer than it is today.[21] And the northern boundary of citrus crops in medieval China demonstrates that 13th century China was 0.9 °C to 1.0 °C warmer than today.[22] Nor was this medieval warming merely a northern hemisphere phenomenon — ice core data, sediment cores, and radiocarbon ages of glacier moraines and of elephant seal colonies all demonstrate that many regions of Antarctica also experienced this same medieval warming while also clearly demonstrating that temperatures today are still cooler than at the peak of the Medieval Warm Period.[23]

---

[18] Easterbrook, D. (2011). *Medieval Warm Period.* Evidence-Based Climate Science (First Edition). Easterbrook, D. (Ed.) Elsevier Publishing. https://est.ufba.br/sites/est.ufba.br/files/kim/medievalwarmperiod.pdf

[19] "Vikings Grew Barley in Greenland." (2012, Jan 11). *Iceland Review.* https://www.icelandreview.com/news/vikings-grew-barley-greenland/

[20] Hildebrant, S. (2012, Feb 3). Vikings grew barley in Greenland. *ScienceNordic.* https://www.sciencenordic.com/agriculture-archaeology-denmark/vikings-grew-barley-in-greenland/1447746

[21] Easterbrook, D. (2011).

[22] Zhang, D. Evidence for the existence of the medieval warm period in China. *Climatic Change* 26, 289–297 (1994). https://doi.org/10.1007/BF01092419

[23] Lüning, S., Mariusz Gałka, M., and Vahrenholt. F. (2019) The Medieval Climate Anomaly in Antarctica. *Palaeogeography, Palaeoclimatology, Palaeoecology.* 532. https://doi.org/10.1016/j.palaeo.2019.109251

The original "hockey stick" graph used tree-ring studies to reconstruct a model of past temperatures — the method relies on the idea that the width of tree rings can be correlated with past temperature changes. But the results of the hockey stick reconstruction are at odds with thousands of published papers, including ice core data from Greenland.[24] Nonetheless, some version of the "hockey stick graph" is still widely used by Wikipedia, by the media, and by activist-scientists. The two researchers (Stephen McIntyre and Ross McKitrick) who exposed the flaws in how the hockey stick chart was constructed found that the hockey stick chart was invalid *"due to collation errors, unjustifiable truncation or extrapolation of source data, obsolete data, geographical location errors, incorrect calculation of principal components and other quality control defects."*[25] On January 28, 2005, at an event hosted by the American Enterprise Institute, author and filmmaker Michael Crighton provided a brief summary all the technical problems with how the infamous "hockey stick graph" was constructed — it's a very worthwhile 5 minutes of your time (available on YouTube).[26]

The IPCC dropped the hockey stick graph from its climate assessments for a number of years after the initial problems were uncovered, but a renewed version based on more tree ring chronologies has once again reared its ugly head in the IPCC's sixth assessment report, published in 2023. Climatologist Dr. Judith Curry recently published a lengthy critique of the IPCC's latest hockey stick graph,[27] including fresh research by Stephen McIntyre into the underlying tree ring chronology used to construct this latest version of the hockey stick — the data in question is known as the "PAGES2K Asia_207" tree ring chronology shown in figure 53 below, as published on Stephen McIntyre's website

---

[24] Easterbrook, D. (2011). *Medieval Warm Period.* Evidence-Based Climate Science (First Edition). Easterbrook, D. (Ed.) Elsevier Publishing. https://est.ufba.br/sites/est.ufba.br/files/kim/medievalwarmperiod.pdf

[25] Easterbrook, D. (2011). *Medieval Warm Period.*

[26] Heller, T. (2023, Mar 20). *Michael Crichton Explains The Hockey Stick.* [Video]. YouTube. https://www.youtube.com/watch?v=3CSrOvAMMKI&ab_channel=TonyHeller

[27] Curry, J. (2024, Mar 1). IPCC's New "Hockey Stick" Temperature Graph. *Climate Etc.* https://judithcurry.com/2024/03/01/ipccs-new-hockey-stick-temperature-graph/

(climateaudit.org).[28] Once again, the tree ring reconstruction appears to show an unequivocal story about unprecedented global warming. Once again, the devil is in the details…

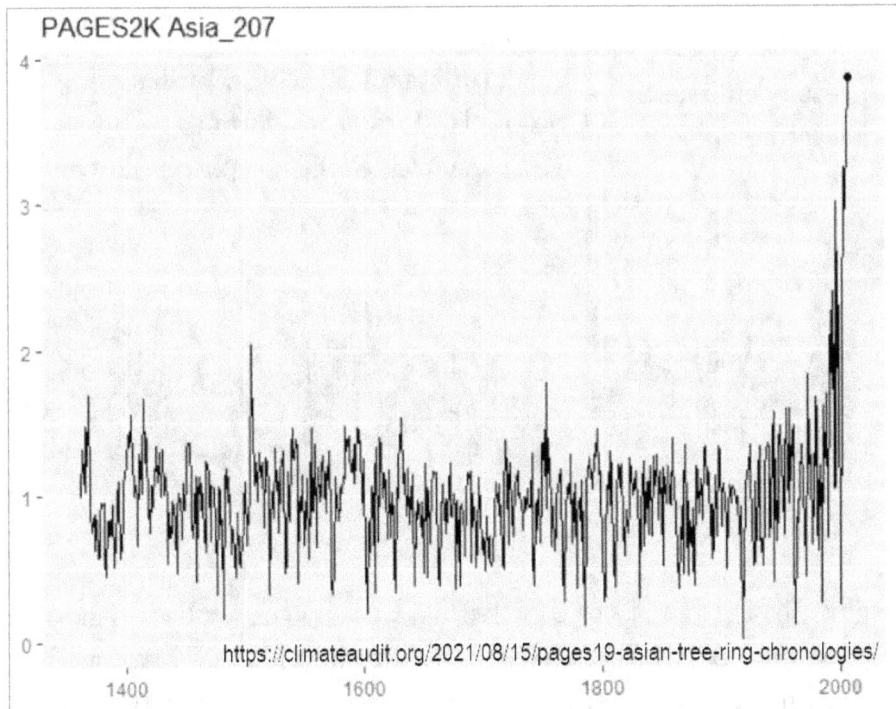

Figure 53. The IPCC's sixth assessment report published a new version of the "hockey stick graph" based on a climate reconstruction using a tree ring chronology data set known as "PAGES2K Asia_207".

Source: https://climateaudit.org/2021/08/15/pages19-asian-tree-ring-chronologies/

The issue is that tree rings get narrower as a tree gets older, so the data has to be "detrended" using complex standardization formulas to compensate for this effect. Stephen McIntyre recalculated the underlying tree ring widths using two normal standardization procedures rather than the standardization formula used by the IPCC and found there was no hockey stick effect at the end of the time period — quite the opposite, when using the normal standardized detrending formulas the temperatures appear to be *declining* in the 20th century, as shown in figure 54 below. So,

---

[28] McIntyre, S. (2021, Aug. 15). Pages19 Asian Tree Ring Chronologies. *Climate Audit*. https://climateaudit.org/2021/08/15/pages19-asian-tree-ring-chronologies/

different statistical algorithms to "detrend" the data create completely different results. But the story gets even worse.

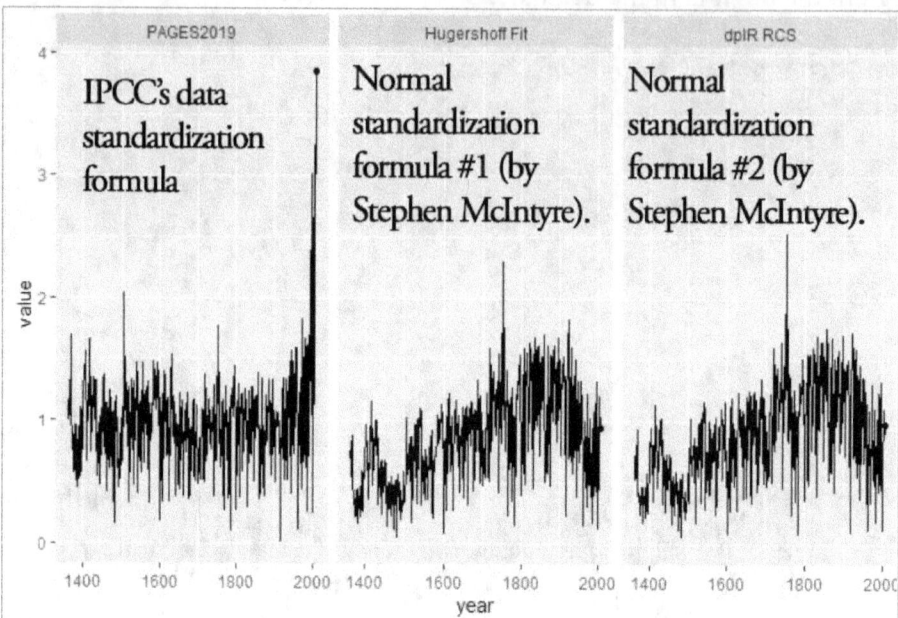

Figure 54. The formulas used for data standardization matter. Each of these three graphs is based on the same raw data from the PAGES2K Asian tree data set. The first reconstruction using the IPCC's data standardization formulas produces an alarming "hockey stick chart." But statistical analysis of the exact same data set by Stephen McIntyre (www.climateaudit.com) processed with normal standardization formulas does not produce a hockey stick — on the contrary, it shows temperatures *declining* since the early 1900s.

Source: https://climateaudit.org/2021/08/15/pages19-asian-tree-ring-chronologies/

As Dr. Curry reported in her critique, in 2023 Hampus Söderqvist managed to reverse-engineer the IPCC's "detrending" algorithm that was used to calculate the original PAGES2K Asia_207 tree ring chronology and, together with Stephen McIntyre, they then used that reverse-engineered algorithm to re-process the raw underlying data, only they left out the last 50 years of data (the runaway hockey stick portion that allegedly shows the 50 warmest years in recorded history). But instead of reducing the size of the hockey stick blade when the data from the last 50 years was left out, the "blade" got even bigger! They went through all the other tree ring data sets and found that the same thing happened to most of the other data sets as well, though not all. The moral of the story is that an algorithm that produces a hockey stick regardless of the time period fed into the algorithm clearly cannot be trusted and should be tossed out.

Furthermore, Hampus Söderqvist has verified that this same "detrending" algorithm was also used by the IPCC to process multiple other datasets, which were then held forth as "further confirmation" of their hockey stick chart. In other words, this same faulty algorithm is being used to produce other IPCC charts that are then shown to the public as "confirmation" that multiple data sets from other parts of the world tell the same story — the data may be different, but the black box that makes sausages out of the data is always the same and (surprise, surprise) the sausages that come out the other end all look the same. I've barely scratched the surface of all the issues with the IPCC's hockey stick chart, so I recommend reading Dr. Curry's full blogpost at https://judithcurry.com/2024/03/01/ipccs-new-hockey-stick-temperature-graph/. She concludes her post with a poignant question: *"Did the IPCC want to present a hockey stick temperature graph? The Clintel Foundation and McIntyre believe so, and I wouldn't be surprised, either. But it's hard to prove."*

~

Now let's turn to some ice core data, which tells an altogether different story than the narrative presented by the hockey stick charts. The graphic in figure 55 shows a climate reconstruction that uses ice core data from Greenland (the upper half of the chart shows the temperature reconstruction, the lower portion of the chart shows carbon dioxide levels over the same time period)[29] — temperatures and atmospheric CO_2 were reconstructed by measuring the gas composition inside air bubbles trapped within the layers of ice. Note that the ice core data covers the last 11,000 years but does not include data for the Industrial Era. Once again, the ice core data shows large temperature swings across those 11,000 years, which are completely absent on the IPCC-approved hockey stick charts.

---

[29] Scafetta, N. (2016). Problems in Modeling and Forecasting Climate Change: CMIP5 General Circulation Models versus a Semi-Empirical Model Based on Natural Oscillations. *International Journal of Heat and Technology* 34(Special Issue 2). http://dx.doi.org/10.18280/ijht.34S235

But the story told by the ice core data does look like a more detailed version of the 10,000-year climate history published by National Geographic in 1976, which also showed that the Medieval Warm Period was cooler than the preceding Roman Warm Period, which in turn was cooler than the preceding Minoan Warm Period.

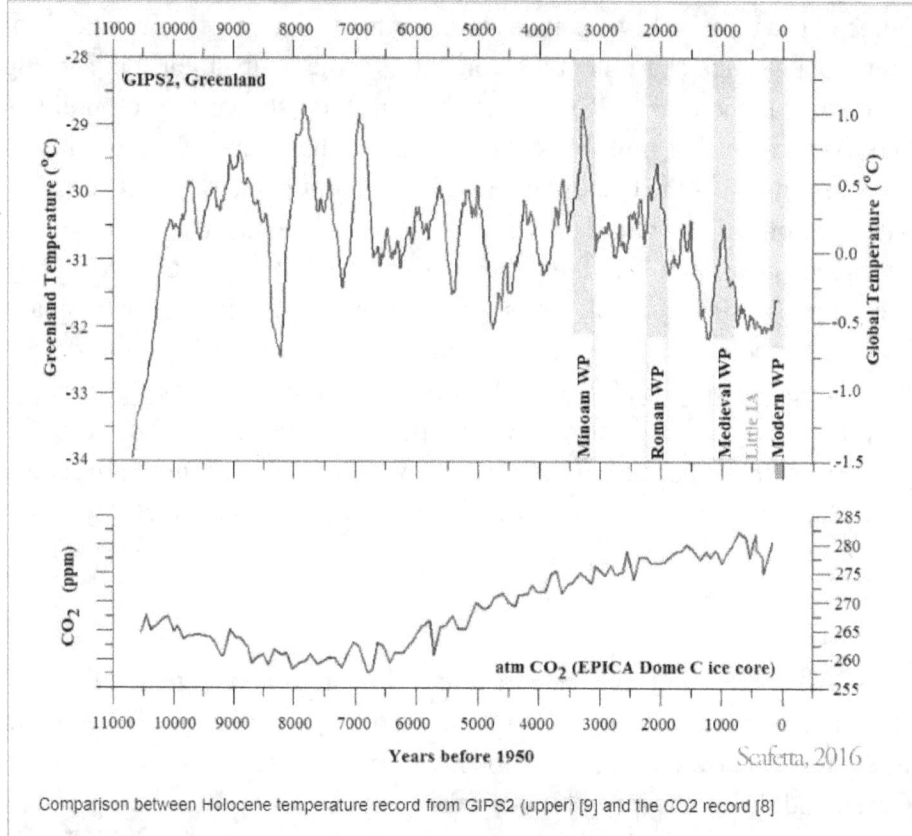

Comparison between Holocene temperature record from GIPS2 (upper) [9] and the CO_2 record [8]

Figure 55. Temperature and atmospheric carbon dioxide reconstructions from ice core data from Greenland.

Source: Problems in Modeling and Forecasting Climate Change, Oct 2016
https://www.researchgate.net/figure/Comparison-between-Holocene-temperature-record-from-GIPS2-upper-9-and-the-CO2-record_fig5_309880579

Each of these earlier warm periods coincided with flourishing civilizations and expanding populations, while the cool periods coincide with population collapses, crop failures, droughts, and famines. The Bronze Age Collapse (between the Minoan and Roman Warm Periods), the Dark Ages that followed the collapse of the Roman Empire, and the population declines and famines of the Little Ice Age that brought an end to the flourishing culture of the Middle Ages all happened during cool periods. Cooler weather leads to shorter growing seasons and crop failures. And cooler temperatures lead to less moisture evaporating from the oceans, which leads to less rainfall and less humidity over land.

The Greenland ice core data clearly demonstrates that the Minoan Warm Period 3,500 years ago, the Roman Warm Period 2,000 years ago, and the Medieval Warm Period 1,000 years ago were all part of a long-term cooling trend as each successive warm period failed to reach the highs of the previous warm period. This explains why scientists during the 1970s were so alarmed about the possibility of another Ice Age as the climate cooled from the 1940s to the 1970s — the short-term trend since the 1970s may be up, but the ice core data is telling us that on a geological time scale the long-term trend since around 8,000 years ago appears to be down. This makes sense based on the long-term geological record of glacial advances and retreats over the 800,000 years.

The chart in figure 56, published by the Utah Geological Survey, shows a more detailed version of the longest view of climate history published by National Geographic in 1976.[30] It shows how global temperatures bounced up and down over the past 450,000 years as the climate went through five bitter ice ages followed by five brief, warm, interglacial periods. As each glacial period comes to an end, temperatures tend to spike rapidly but then, after the initial spike, temperatures gradually begin to deteriorate as the warm interglacial period slowly loses its strength. These warm interglacial periods last, on average, from 10,000 to 15,000 years before the cooling accelerates to kick off the next ice age. We have been living in a warm interglacial period for around 10,000 years. Tick tock. The near-term future may be warmer but, in the long-term, winter is coming. Wrapping our minds around the vast expanses of geologic time is a difficult thing for a species with a lifespan of less than a century. It's easy to scare people by projecting short-term trends indefinitely into the future; it's not so easy if people can put those short-term trends into a long-term context.

---

[30] Eldredge, S. and Biek, B. (2019, Sept). Glad You Asked: Ice Ages – What Are They and What Causes Them? *Utah Geological Survey.* https://geology.utah.gov/map-pub/survey-notes/glad-you-asked/ice-ages-what-are-they-and-what-causes-them/

It's also worth noting on the chart that our current interglacial period has been cooler than the four previous interglacial periods despite the inconvenient fact that CO_2 levels are higher today than they were at any time over those previous warmer interglacial periods — more on that in a moment.

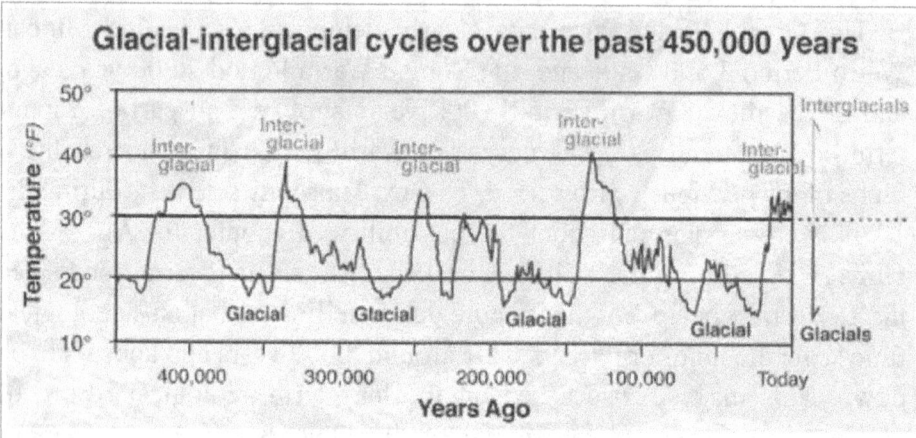

Figure 56. Glacial and interglacial periods over the last 450,000 years using climate reconstructions from several ice-core records from Antarctica. Sharp temperature rises are invariably followed by slow temperature declines as temperature gradually slides back into another glacial period.

Source: Utah Geological Survey
https://geology.utah.gov/map-pub/survey-notes/glad-you-asked/ice-ages-what-are-they-and-what-causes-them/

According to the Greenland ice core data shown back in figure 55, the warmest temperatures of the last 10,000 years happened shortly after the end of the last Ice Age around 8,000 years ago. This warm period, known as the Holocene Climate Optimum, was at least 4 °C [31] to 7 °C [32] warmer than today (at that time Siberia was 3 to 9 °C warmer in the winter and 2 to 6 °C warmer during the summer![33]). During the Holocene Climate Optimum (7,000 to 6,000 years ago), the Arctic Ocean was periodically ice free during the summer.[34] So much for the media's hyperventilating

---

[31] "Holocene Climate Optimum." *Scholarly Community Encyclopedia.* 1 Nov 2022. https://encyclopedia.pub/entry/32197

[32] van den Bilt et al. (2019). Early Holocene Temperature Oscillations Exceed Amplitude of Observed and Projected Warming in Svalbard Lakes. *Geophysical Research Letters.* 46(24) 14732-14741. https://doi.org/10.1029/2019GL084384

[33] "Holocene Climate Optimum." *Scholarly Community Encyclopedia.*

[34] "Less Ice in Arctic Ocean 6000-7000 Years Ago." (2008, Oct 20). *Science Daily.* https://www.sciencedaily.com/releases/2008/10/081020095850.htm

headlines that this past summer was the hottest in over 120,000 years and so much for the UN's claim of a climate breakdown! Despite Al Gore's theatrical never-ending failed predictions, we're still nowhere near an ice-free Arctic Ocean today. Nor should we be surprised that the hype never delivers on its alarmist predictions — in a 2006 interview with Grist Magazine, Al Gore was inconveniently quoted saying that *"Nobody is interested in solutions if they don't think there's a problem. Given that starting point, I believe it is appropriate to have an over-representation of factual presentations on how dangerous it is, as a predicate for opening up the audience to listen to what the solutions are, and how hopeful it is that we are going to solve this crisis."*[35]

The photo in figure 57 below shows just one of many ancient tree stumps found in the Arctic, far north of the current treeline. This stump was radiocarbon dated to 5,000 years ago — *yet it is located 100 km north of the current treeline* in Northwest Canada in what is currently permanently frozen tundra.[36] Likewise, from 9000 to 7000 years ago, Russia's boreal forest reached up to or near the current arctic coastline[37] — it doesn't even come close today. In other words, from at least 9,000 to 5,000 years ago, forests grew significantly further north across the entire Northern Hemisphere on soils that were not permanently frozen then, but which are permanently frozen today.

---

[35] "An inconvenient quote." (2007, Mar 14). *The Economist*. https://www.economist.com/democracy-in-america/2007/03/14/an-inconvenient-quote

[36] Watts, A. (2019, Apr 12). Inconvenient stumps. *Watts Up With That*. https://wattsupwiththat.com/2019/04/12/inconvenient-stumps/

[37] MacDonald et al. (2000). Holocene Treeline History and Climate Change Across Northern Eurasia. *Quaternary Research*. 53(3) 302-311. https://doi.org/10.1006/qres.1999.2123

PLATE IV Tree stump (*Picea glauca*) in the north Canadian tundra.
The stump, radiocarbon dated about 4940 years (± 140) B.P., is seen still standing on a steep bank on the Tuktoyaktuk Peninsula (69°7′N 133°16′W) which borders the Arctic Ocean (Beaufort Sea) east of the delta of the Mackenzie River in extreme northwest Canada. This tree in what is now tundra shows wider growth rings than the nearest present-day spruce forest 80–100 km farther south, near Inuvik in the lowest part of the Mackenzie River valley.
(*Photograph kindly supplied by Professor J. C. Ritchie of Scarborough College, Toronto University, and reproduced with his kind permission.*)

Figure 57. A 5,000-year-old tre stump near Inuvik, 100 km north of the current treeline in what is now permanently frozen tundra.

Source: WattsUpWithThat.com
https://wattsupwiththat.com/2019/04/12/inconvenient-stumps/

And it's not just the arctic treeline that has shifted south. The same phenomenon has been well documented in mountainous areas as the alpine treeline has gradually moved to lower elevations over the last 3,000 years, in keeping with a gradually cooling climate (see figure 58).[38]

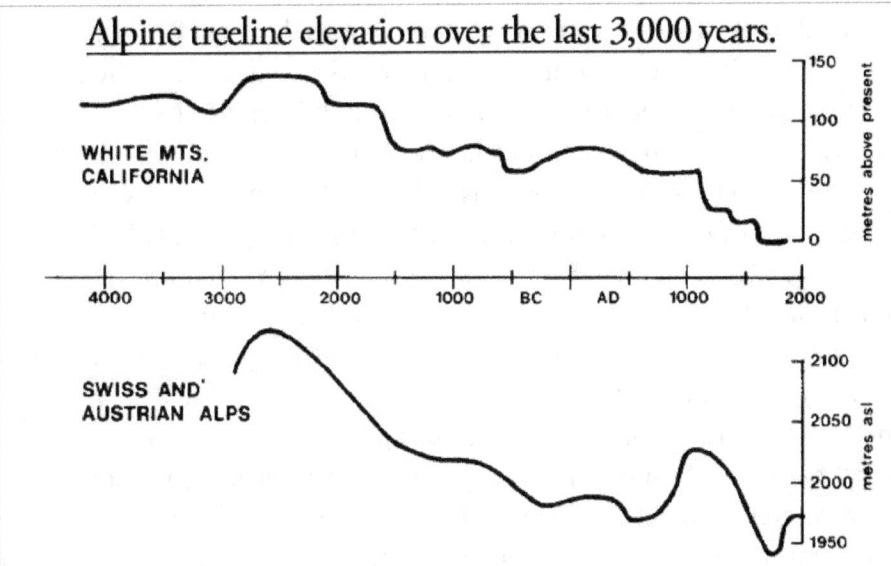

Changes in the height of the upper tree line in two areas in the White Mountains, California and in the Alps in Switzerland and Austria. (From work by V.C.La Marche and V.Markgraf.)

Figure 58. The elevation of the treeline in alpine areas has gradually declined over the last 3,000 years, indicative of a gradually cooling climate.

Source: H.H. Lamb (1995, p. 128). *Climate, History and the Modern World*, second edition. London: Routledge Publishing

Now let's have a closer look at the supposed correlation between rising temperatures and atmospheric carbon dioxide. Once again, let's turn to the Greenland Ice Core data shown in back in figure 55, only this time let's focus on the lower line, which shows how atmospheric carbon dioxide levels changed over the last 10,000 years. The chart shows carbon dioxide gradually declining since the end of the Ice Age 10,000 year ago until around 7,000 years ago. Then, even as temperatures began to cool starting

---

[38] Lamb, H. H. (1995). *Climate, History and the Modern World*. Routledge Publishing. https://ens9004-infd.mendoza.edu.ar/sitio/climatologia-meteorologia/upload/08-%20LAMB,%20H.H.%20-%20LIBRO%20-%20Climate,%20History%20and%20the.pdf

7,000 years ago, carbon dioxide began a slow but steady rise. So much for the idea that carbon dioxide drives temperature — the ice core data clearly shows them moving in opposite directions. The hottest temperatures revealed by the Greenland ice core data happened almost 8,000 years ago when carbon dioxide was only 260 ppm (at a time when the Arctic enjoyed ice-free summers and trees grew much far further north than they do today). In other words, it was clearly considerably warmer when carbon dioxide was only 260 ppm than it is today at 416 ppm. Oops. And, finally, the chart clearly shows dramatic temperature swings over this 10,000-year period but these dramatic temperature swings are not mirrored by the much smoother curve of changing carbon dioxide. Another oops.

The Greenland ice core data clearly demonstrates that carbon dioxide does not function as the control knob on the climate. Whatever insignificant role that carbon dioxide plays in regulating Earth's temperatures at a miniscule 416 parts per million is being completely overpowered by other far more powerful factors that drive both short-term and long-term changes in global temperatures, which I'll explain momentarily. But now you understand the reason why they had to cool the past and warm the present — the only way to create a simple narrative to link temperature to the relatively smooth rise in atmospheric carbon dioxide levels is to cool the past, warm the present, <u>and erase all the wild temperature swings of the last 10,000 years</u> (a.k.a. *"We need to get rid of the Medieval Warm Period"*). This is what it looks like when scientists are so committed to proving a theory that they blind themselves to all other explanations and then manipulate the data to match the theory. This is more than just sloppy science — this is a contemptible disregard for scientific integrity.

In 1988, Senator Al Gore (who went on to become Vice-President before starting a lucrative investment firm selling "carbon credits" on his way to amassing a $200 million net worth[39]) organized a historic Senate hearing, with testimonies from prominent climate scientists like James Hansen, to convince policymakers and the media of the correlation between carbon dioxide and global temperatures. Two years earlier, the Associated Press reported from another Senate hearing that Hansen

---

[39] "How Al Gore amassed a $200-million fortune after presidential defeat." (2013, May 6). *Financial Post*. https://financialpost.com/news/how-al-gore-amassed-a-200-million-fortune-after-presidential-defeat

"predicted global temperatures should be nearly 2 degrees higher in 20 years [by 2006]" and predicted *"an increase of an additional 3 or 4 degrees sometime between 2010 and 2020"*.[40] His alarmist predictions turned out to be an almost ten-fold overstatement of the actual warming recorded by the University of Alabama's satellite measurements, discussed earlier, which have only recorded a meagre 0.69 °C of warming from 1979 to 2023.

Al Gore's infamous 1988 hearing was wildly successful — climate change has been front page news ever since. In his 2010 book, *The Climate Fix,* political scientist Roger Pielke Jr. describes the tactics used at the hearing to convince policymakers: *"The hearing that day was carefully stage-managed to present a bit of political theater, as was later explained by Senator Tim Wirth (D-CO), who served alongside Gore in the Senate and, like Gore, was also interested in the topic of global warming. 'We called the Weather Bureau and found out what historically was the hottest day of the summer. Well, it was June 6th or June 9th or whatever it was. So we scheduled the hearing that day, and bingo, it was the hottest day on record in Washington, or close to it. What we did is that we went in the night before and opened all the windows, I will admit, right, so that the air conditioning wasn't working inside the room.'"*[41]

After leaving the Senate, Tim Worth went on to serve as the president of the United Nations Foundation. In 1990, he candidly stated that *"We've got to ride the global warming issue. Even if the theory of global warming is wrong, we'll be doing the right thing, in terms of economic policy and environmental policy."*[42] His words echo those of physicist Sir John Houghton, who served as the Chairman and Co-Chairman of the Intergovernmental Panel on Climate Change from 1988 to 2002 (he also served as the lead editor of the first three IPCC reports); Houghton stated that *"Unless we announce disasters no one will listen"* and that *"The impacts of global warming are such that I have no hesitation in describing*

---

[40] Bailey, R. (2016, Jun 17). Climate Change Prediction Fail. *Reason Magazine.* https://reason.com/2016/06/17/climate-change-prediction-fail/

[41] Pielke, R. Jr. (2010). *The Climate Fix.* Basic Books Publishing.

[42] Williams, W. (2010, Mar 17). Walter Williams: Global warmers apparently don't intend to go quietly. *The State Journal-Register.* https://www.sj-r.com/story/opinion/columns/2010/03/17/walter-williams-global-warmers-apparently/42946506007/

it as a '*weapon of mass destruction*'".[43] Likewise, Stanford professor and environmental activist Stephen Schneider stated during a 1989 interview with Discover Magazine that, *"We have to offer up scary scenarios, make simplified dramatic statements, and make little mention of any doubts we may have. Each of us has to decide what the right balance is between being effective and being honest."*[44] Schneider served as a consultant to the Whitehouse under Richard Nixon, Jimmy Carter, Ronald Reagan, George H.W. Bush, Bill Clinton, George W. Bush, and Barack Obama.

The next chart in figure 59 shows annual carbon dioxide emissions since 1880. Can you see how this chart completely destroys the narrative that "carbon dioxide is the control knob on the climate?" Hint: remember the temperature spike of the 1930s and the subsequent cooling of the 1970s?

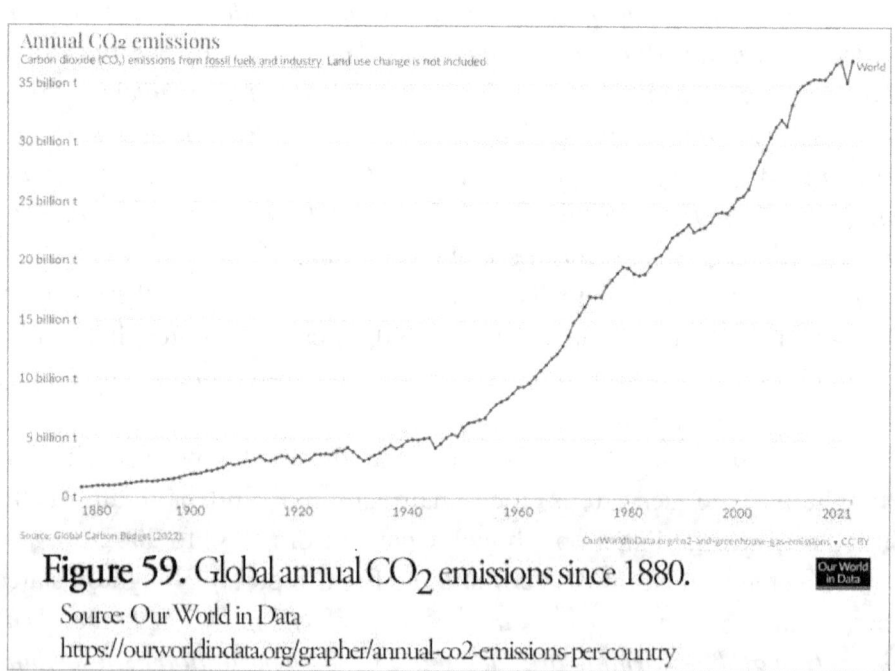

Figure 59. Global annual CO_2 emissions since 1880.
Source: Our World in Data
https://ourworldindata.org/grapher/annual-co2-emissions-per-country

---

[43] Bratby, P. (2008, May 15). Memorandum by Dr. Phillip Bratby. *UK government (www.parliament.uk).* https://publications.parliament.uk/pa/ld200708/ldselect/ldeconaf/195/195we07.htm

[44] Williams, W. (2010, Mar 17). Walter Williams: Global warmers apparently don't intend to go quietly. *The State Journal-Register.* https://www.sj-r.com/story/opinion/columns/2010/03/17/walter-williams-global-warmers-apparently/42946506007/

The lack of correlation between temperature and carbon dioxide levels becomes crystal clear when I overlay CO_2 emissions over the unaltered NASA temperature chart from 1999, as I've done in figure 60 below with time scales matched up.

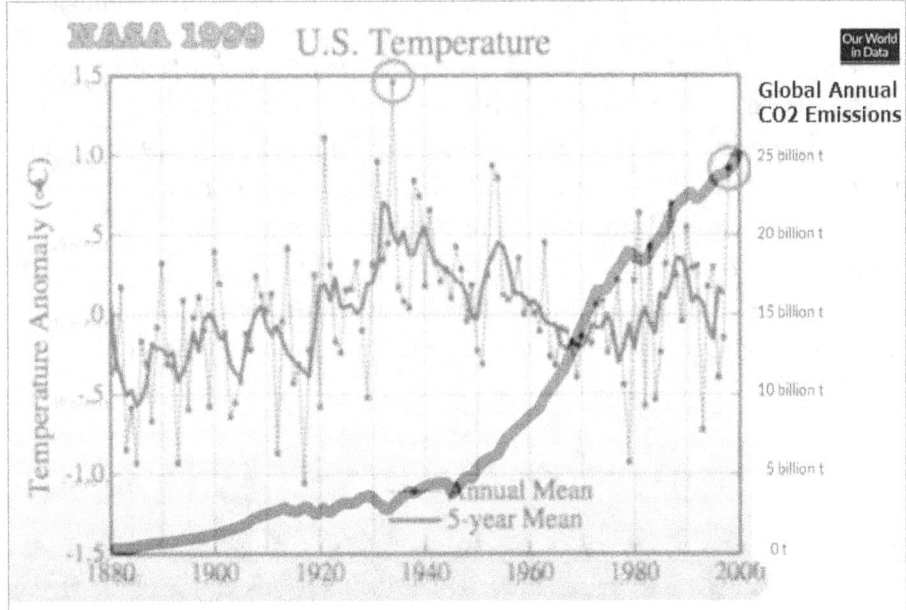

Figure 60. Global annual CO_2 emissions (from figure 68) overlaid on NASA's unaltered temperature chart from 1999 (from figure 56). Clearly, there is no correlation between CO_2 emissions and temperature.

As the chart in figure 60 shows, global CO_2 emissions don't really take off in earnest until the 1950s yet the massive temperature spike of the 1930s and 1940s happened *before* the dramatic increase in CO_2 emissions that began in the 1950s. In other words, the warmest temperature on the chart (1934) happened *before* the sharp ramping up in CO_2 emissions in the 1950s — oops. Then, just as CO_2 emissions start ramping up in the 1950s, temperatures declined sharply for thirty years (the global cooling scare of the 1970s) — another oops. Furthermore, annual CO_2 emissions increased by more than 500% between 1934 (the hottest year in the series) and 1999, yet 1999 is still colder than the 1934 peak — another big oops. The correlation doesn't work out — clearly CO_2 is not the control knob on the climate that it is made out to be. The only way to make the narrative work is if you hide the decline in temperatures from

1940 to 1970 by cooling the past and warming the present — in other words, "adjust the data" until the temperature record is bent to fit the rise in CO_2.

In his famous dystopian novel, *1984*, George Orwell described a totalitarian society based on mass surveillance and thought control, *"Every record has been destroyed or falsified, every book rewritten, every picture has been repainted, every statue and street building has been renamed, every date has been altered. And the process is continuing day by day and minute by minute. History has stopped. Nothing exists except an endless present in which the Party is always right."* In contrast to Orwell's authoritarian control measures, in Aldous Huxley's dystopian novel, *Brave New World*, society is not enslaved by repressive tyranny but by psychological manipulation and superficial illusions of pleasure: *"One believes things because one has been conditioned to believe them."*

Chapter 9

Icarus Writes an Algorithm

"To err is human, but to really foul things up you need a computer."

— Paul R. Ehrlich

Super computers changed the way the world works because they gave programmers the processing power to use complex algorithms to analyse vast volumes of historical data, like global temperature records, the spread of a new strain of a virus, or the behaviour of the stock market. By creating complex algorithms that attempt to account for a near infinite number of variables and then using those algorithms to back-test historical data, programmers can test whether the assumptions and mathematical formulas that they used to build their algorithm match the historical record. But it's a dangerous game — garbage in, garbage out.

Algorithms can serve as a useful tool for developing theories about how the world works, but they have a notoriously abysmal track record for predicting the future. It's easy to form fit the past using incorrect variables that assign too much weight to some variables, not enough weight to others, omit things that we didn't know about or didn't think mattered, and falsely assign weight to things that we thought mattered but don't, yet still come up with some combination that matches the historic data. The end result is often a perfectly tweaked form-fitted reconstruction of the past, which gives us the illusion that we understand cause and effect, only to have the illusion of understanding fall apart when our crystal ball is asked to make predictions about things that haven't happened yet (like Al Gore's ice-free Arctic Ocean). As baseball legend Yogi Berra once said, *"It's tough to make predictions, especially about the future."*

In essence, the algorithm simply becomes a more high-tech version of the fortune teller's crystal ball — a kind of black box wrapped in a cloak of technobabble to legitimize a prediction of the future that is, in essence, little more than the sum of all the assumptions and biases of the fortune teller — gut instinct dressed up as "science". At least the fortune teller knows she's merely a performer; what's truly dangerous are scientists who believe in the infallibility of their little black box and have the hubris to think they know every single variable and the exact magnitude of each variable. Compound that with the political power to impose policies on others based on these "scientifically legitimized" forecasts churned out by their little black box, and you soon discover that your life is held hostage by an algorithmic monstrosity that is dumb, blind, and completely wrong, yet the black box gives the algorithm so much credibility that those wielding it (and the politicians they advise) feel completely justified to ruthlessly squash dissenting voices. Trust the Scienz™.

Anyone who has tried trading the stock market with an algorithm, only to lose their shirt when the market failed to behave according to the algorithm's predictions, knows how easy it is to fool ourselves into believing that we have a formula to explain the behaviour of a complex dynamic system, only to be confronted by our own hubris as we watch it all fall apart in the real world. As physicist Professor Richard Feynman once said, *"The first principle is that you must not fool yourself, and you are the easiest person to fool."*

Mao's land reforms during the Great Leap Forward were informed by the best scientists and ecologists that the Soviet Union could offer. The plow that turned the Great Plains into the Dust Bowl had the blessing of America's most prominent scientific voices. One of the most prominent players in America's early 20th century weather modification programs was Irving Langmuir, a chemist and physicist who had won the Nobel Prize in Chemistry in 1932 before whole-heartedly dedicating his life to the science of weather modification — yet the multi-million-dollar boondoggle of weather modification only proved once again that theories that seem to work in the lab often completely fail to deliver in the real world because the real world is far more complex than we think.

In 2022, Ian Plimer, professor emeritus of geology at the University of Melbourne, summed up the problem with climate models and pointed out the fundamental flaw embedded within them that views CO_2 as the control knob on the climate. Here are a few key excerpts from his tongue-in-cheek speech:[1]

> "We hear about climate scientists, whatever that is. Now, in geology, we have a 250-year track record of arguing about climate.

---

[1] Plimer, I. (2022). Lecture at CPAC 2022. *Twitter (@wideawake_media)*. https://x.com/wideawake_media/status/1747593340143415743?s=20

Textbooks are full of it. We've been labouring about climate for a long while and then we see a sudden new invention of "climate science." [...] So, there's one group of people that use models, another group of people ... and this is really sinful... we use evidence. And the two are not in accord. And if they're not in accord, you've got to throw out the models, which we've seen time and time again are incorrect.

So, we can look back in the past and we can see that we've had six great ice ages. During that ice age, we'll have the ice expand, that's a glaciation, or it will contract, that's an interglacial. We are currently in an interglacial of an ice age that started on a Thursday, 34 million years ago. And the ice has come and gone. In our last interglacial, sea level was about 7 meters higher. Temperature was 5 degrees warmer. So, if someone says this is the hottest day on record, you have to ask, since when? If it's the hottest day on record in the last 120,000 years, then that is a record, but, um, since when? So, if we go to the peak of our interglacial, which was about 4,000 years ago, it was about 5 degrees warmer [than it is today]. [...] If we go to the time of Jesus when it was warm, it's about 4 degrees cooler [now] than [it was] then. If we go to the Dark Ages, go to the Viking Age, we've actually warmed up since then. If we go to the Medieval Warming, we've cooled down since then. And if we go to the Little Ice Age, we've warmed up since then. So, since when?

And, I know this is going to surprise you, but we've just come out of a Little Ice Age. What do you think temperature is going to do? Fall, or rise? It's been rising since the Maunder Minimum [a prolonged low in solar activity] more than three-hundred years ago. So, it is no surprise that if you have cut-off times for temperature, or for sea level, or for hurricanes, or for whatever, you can spin whatever yarn you want to spin. These six great ice ages started when we had more carbon dioxide in the atmosphere than we have now. We have 0.04% of that gas in the atmosphere and we hear words like "emissions", well, that means nothing to me because the atmosphere has changed in its carbon dioxide content from over 20% to now [at 0.04%], which is really low in geological time. If we halved it, all plant life would die, and animals would die."

In so many ways, climate scientists and their models are a lot like the people who ran weather modification programs during the 1940s, 50s, and 60s — they have perfected the art of sensationalism in order to commandeer political attention and keep their funding flowing. Many meteorologists, geologists, astronomers, and other serious scientists watch in horror as these showmen convince governments and taxpayers to throw good money after bad to fund crackpot ideas that are completely at odds with demonstrable historical reality. Sadly, many serious scientists have also fallen prey to the peer pressure and groupthink of the "narrative" — its pretty hard to build a career or secure research funding nowadays without paying homage to the climate crusade. Even Galileo was ultimately prodded by the Inquisition to publicly renounce *"with sincere heart and unfeigned faith"* the idea that the Sun rather than the Earth was at the center of the Universe in order to spare himself from the fate that befell his fellow heretical astronomer, Giordano Bruno, who was burned at the stake for challenging the consensus and refusing to recant.

The discrepancy between algorithmic predictions and reality reveals just how poorly climate scientists understand the underlying forces that shape our world. But with reputations, funding, and even careers on the line, the scientific soothsayer has a lot of incentives to convince himself (and those around him) why it's the world rather than his algorithms that are flawed and why it's the heretics and their heretical evidence-based challenges that are the true threat to society. We're still waiting for Al Gore's infamous recurring predictions of an ice-free Arctic Ocean to come true, but he remains unphased, declaring at the World Economic Forum's 2023 meeting in Davos that our accumulated carbon dioxide emissions are *"trapping as much extra heat as would be released by 600,000 Hiroshima-class atomic bombs exploding every single day on the Earth; that's what's boiling the oceans, creating these atmospheric rivers, and the rain bombs, and sucking the moisture out of the land, and creating the droughts, and melting the ice, and raising the sea level, and causing these waves of climate refugees..."*[2]

I've already shown you the Heat Wave Magnitude Index published by the U.S. government's 2018 Fourth National Climate Assessment, which

---

[2] World Economic Forum Video. (2023, Jan 18). *Al Gore – The Equivalent of 600 Thousand Hiroshima Bombs.* [Video]. YouTube.
https://www.youtube.com/watch?v=rfAYLSQIxTI&ab_channel=WorldEconomicForumVideo

showed that, contrary to the doomsday predictions, heat waves in the U.S. are not getting worse and are significantly less intense today than they were during the 1920s, 1930s, and 1940s. Additional graphs published in the same report, reproduced in figure 61 below, show that while winters (upper graph) have gotten warmer over the last century, the warmest summer temperatures (lower graph) have essentially remained unchanged since the 1970s and are far cooler today than they were a century ago.

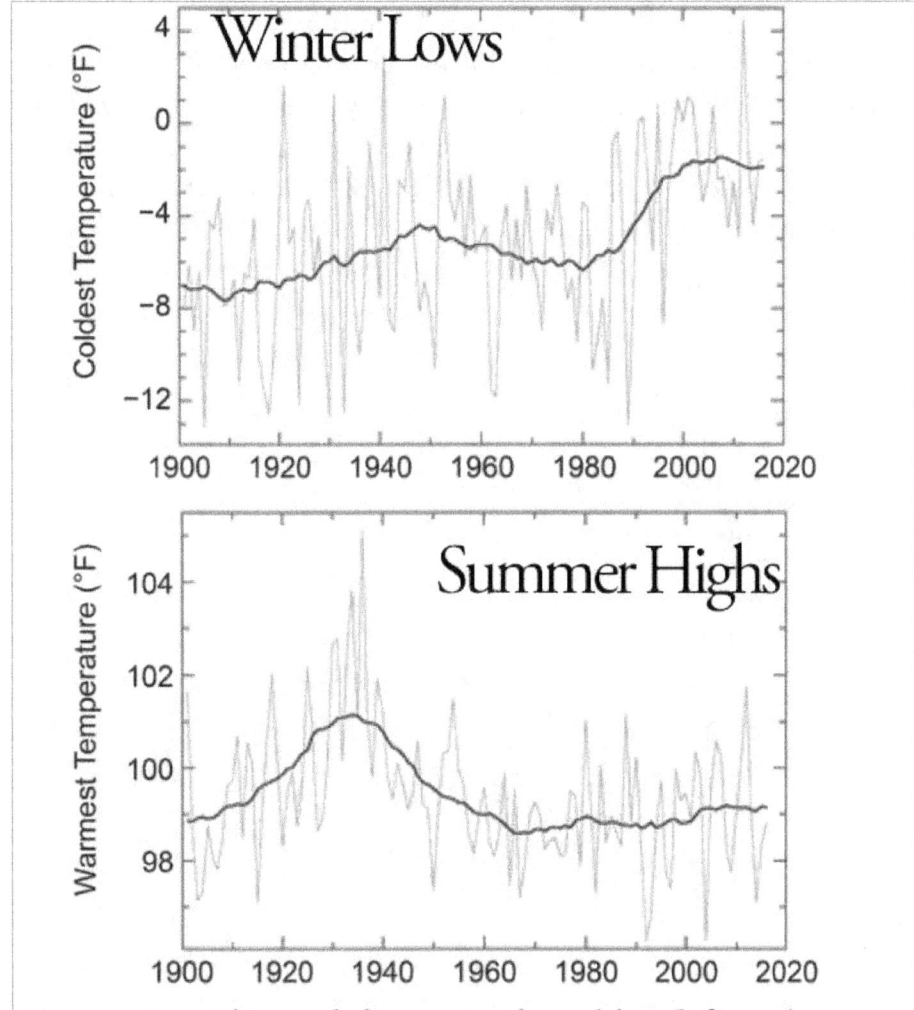

Figure 61. Observed changes in the coldest (left) and warmest (right) daily temperatures in the contiguous United States, published in 2018 by the Fourth National Climate Assessment. https://science2017.globalchange.gov/chapter/6/

Another method to track changes in climate, which provides a more useful measure than a single hot temperature record (a single hot day is not significant if the rest of the month is cold), is to look at the number of days each year that temperatures exceed a temperature threshold (i.e., the number of days above 95 °F). As shown in figure 62, climate historian and geologist Tony Heller has combined official raw temperature data (published by NOAA's National Centers for Environmental Information) for all historical U.S. weather stations that have operated continuously since 1895.[3] The story told by the raw temperature record is crystal clear — the percent of days each year with temperatures above 95 °F are getting less and less frequent (the thin grey line shows the percent of days above 95 °F, the thick black line shows the 10-year moving average). Summers are getting cooler irregardless of what a few isolated local heat records may suggest.

---

[3] Heller, T. (2023, Nov 27). *Denying Scientific Principles*. [Video]. YouTube. https://www.youtube.com/watch?v=3dY2OxXR5us&ab_channel=TonyHeller

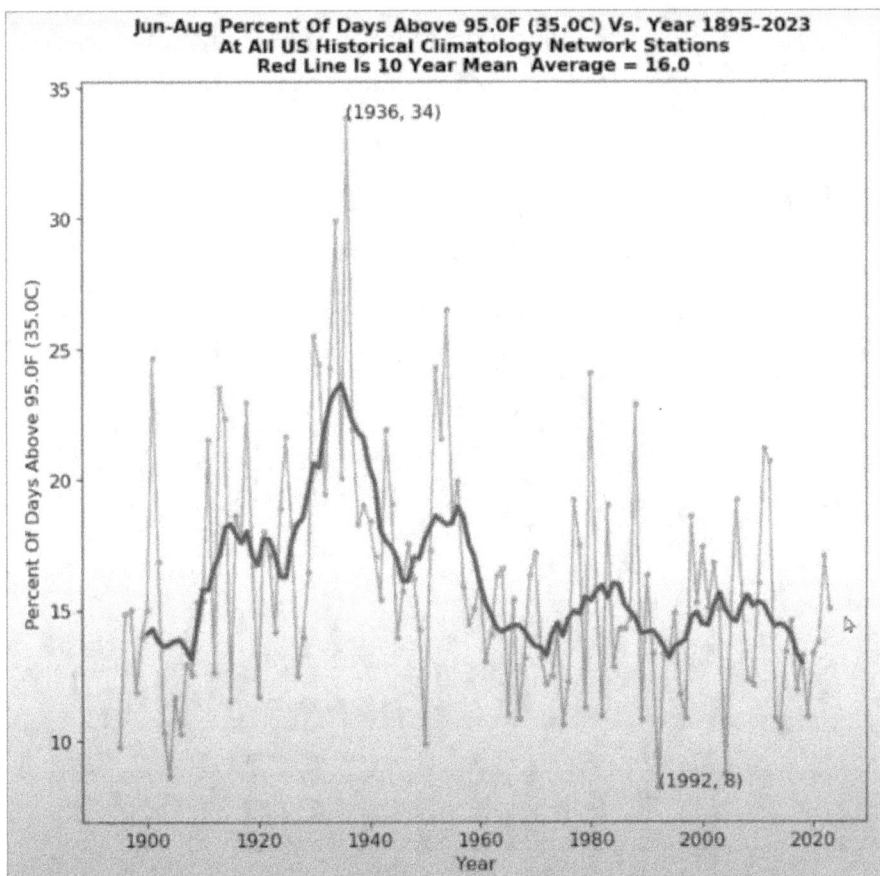

Figure 62. Days above 95 °F from 1895 to 2023 at all U.S. historical weather stations.
Courtesy of Tony Heller
https://www.youtube.com/watch?v=3dY2OxXR5us&ab_channel=TonyHeller

The graphic in figure 63 was published by the National Oceanic and Atmospheric Administration (NOAA) on their state climate summaries website.[4] The black line shows the number of days recording temperatures above 90 °F, while the bars show the 5-year average. Once again, like in the previous chart compiled by Tony Heller, summers today are not experiencing the long, hot heat waves that were common during the 20s, 30s, and 40s.

---

[4] "State Climate Summaries 2022." *NOAA National Centers for Environmental Information.* National Oceanic and Atmospheric Administration. Accessed 2024, Mar 9. https://statesummaries.ncics.org/chapter/ct/

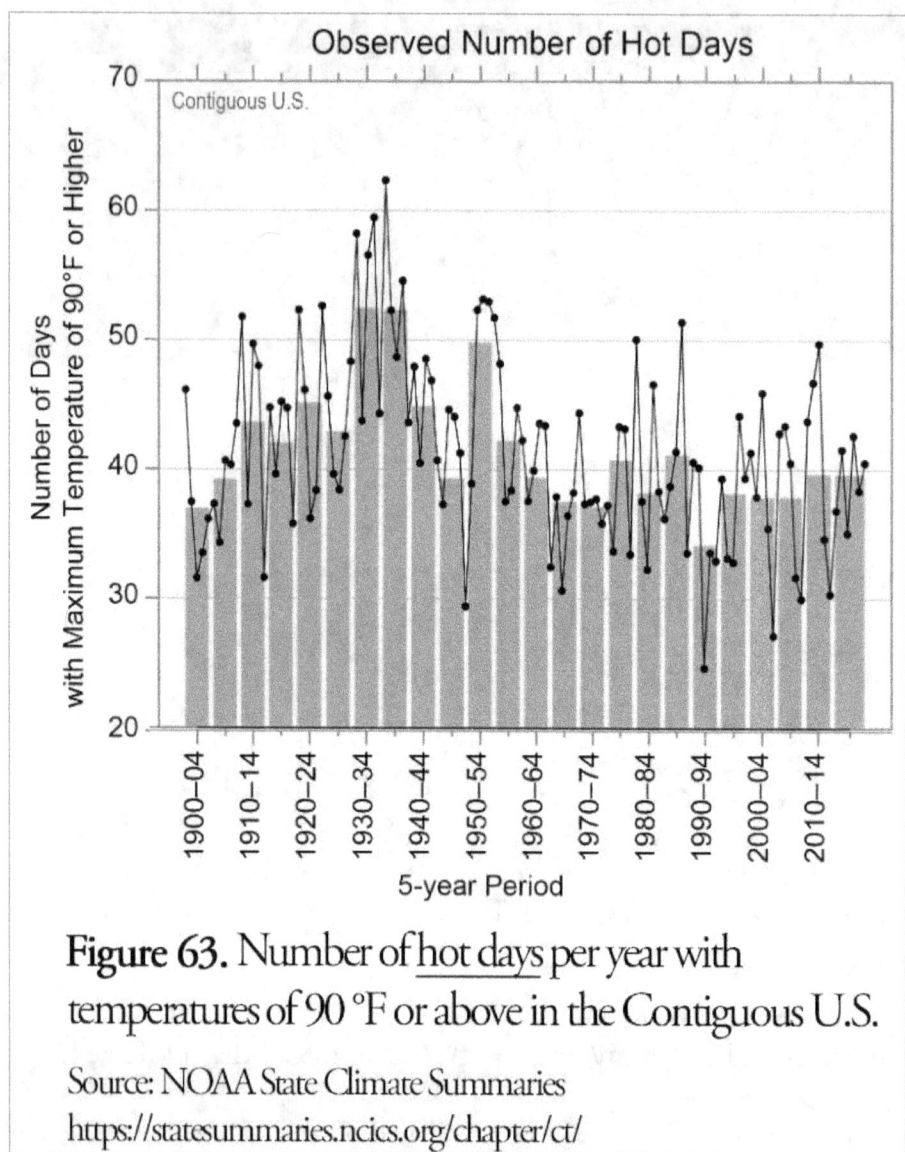

Figure 63. Number of hot days per year with temperatures of 90 °F or above in the Contiguous U.S.

Source: NOAA State Climate Summaries
https://statesummaries.ncics.org/chapter/ct/

NOAA's state summaries website also shows us what is happening at night (figure 64 below) — once again we see that while days are not getting as hot, nighttime temperatures have warmed since the 1970s. The chart shows the number of days each year during which the lowest temperature of the 24-hour period remains at or above 70 °F. Nights are getting warmer even as days cool. But even so, the number of warm nights today are equivalent to but not more common than they were during the 1930s.

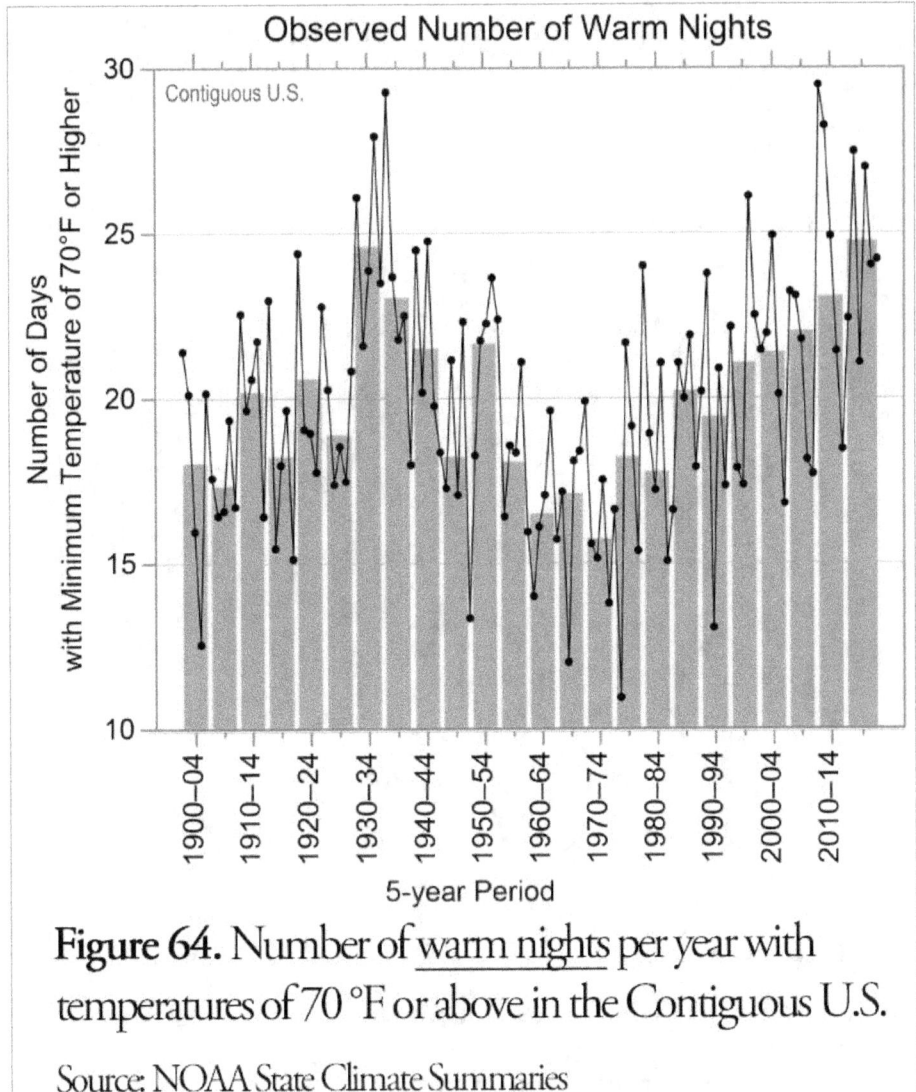

Figure 64. Number of warm nights per year with temperatures of 70 °F or above in the Contiguous U.S.

Source: NOAA State Climate Summaries
https://statesummaries.ncics.org/chapter/ct/

And the chart in figure 65, also from NOAA, shows the number of very cold winter nights (below 0 °F) in the contiguous United States. Cold winters are becoming less common. In other words, winters and nights are getting warmer, but hot summer days are getting cooler. Is the world really getting hotter, or is it getting more comfortable? Isn't this the perfect Goldilocks scenario — not too hot and not too cold. Once you see that the warming is happening at night and during the winter, not during hot summer days, the scary headlines don't have quite the same ring anymore, do they?

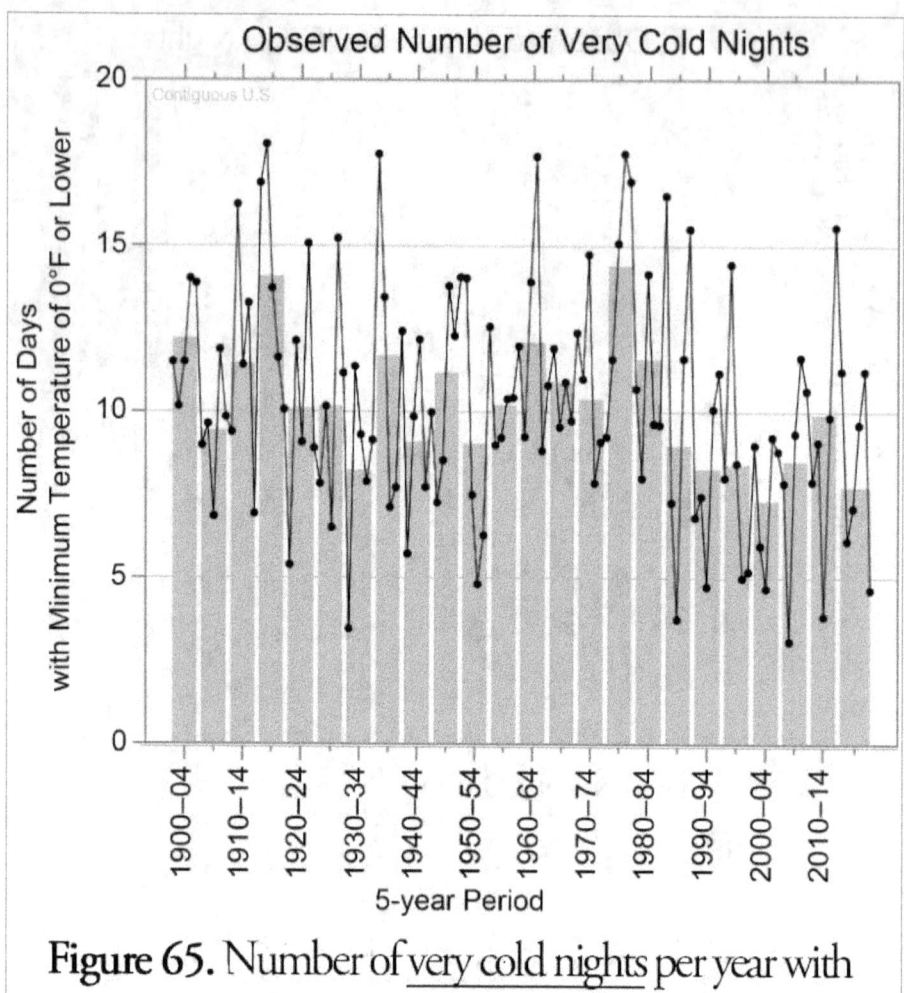

Figure 65. Number of very cold nights per year with temperatures of 0 °F or below in the Contiguous U.S.

Source: NOAA State Climate Summaries
https://statesummaries.ncics.org/chapter/ct/

We see a similar story when it comes to precipitation. To see how precipitation has changed in the contiguous U.S. over the past century, we have to go to a chart published by the Environmental Protection Agency, reproduced in figure 66 below.[5] The chart clearly shows that rainfall has become more abundant over time, with far more precipitation today than

---

[5] "Climate Change Indicators: U.S. and Global Precipitation." (2022, Jul). *United States Environmental Protection Agency*. https://www.epa.gov/climate-indicators/climate-change-indicators-us-and-global-precipitation

during the dry years of the 1930s and 1950s. Of course, there are local variations, but the big picture simply doesn't support the narrative of decreasing rainfall — on the contrary, rainfall is increasing ever so slightly and years with below-average rainfall are becoming less common.

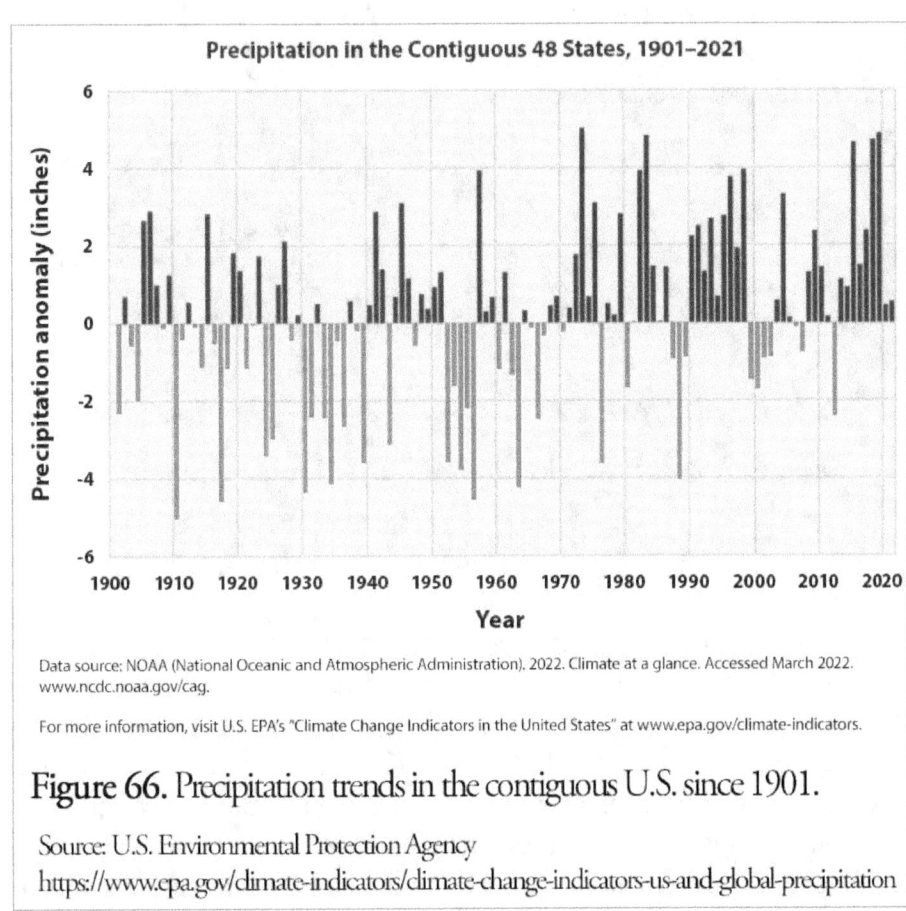

Figure 66. Precipitation trends in the contiguous U.S. since 1901.

Source: U.S. Environmental Protection Agency
https://www.epa.gov/climate-indicators/climate-change-indicators-us-and-global-precipitation

What about sea level rise? A study published in 2016 called *Coastal planning should be based on proven sea level data*[6] showed that the appearance of rapidly rising sea levels are an illusion created by theoretically calculated after-the-fact data adjustments, which in the words of the authors is *"a rather abstract computation, far from being reliable, and is preferred by activists and politicians for no scientific reason."* In the study, the authors demonstrated that theoretically calculated sea levels

---

[6] Parker, A. and Ollier, C.D. (2016). Coastal planning should be based on proven sea level data. *Ocean & Coastal Management*. 124. 1-9.
https://doi.org/10.1016/j.ocecoaman.2016.02.005

don't match proven local tide gauge data. Sound familiar? The chart in figure 67 below is the crowning graph of the study (my annotations added for clarity).

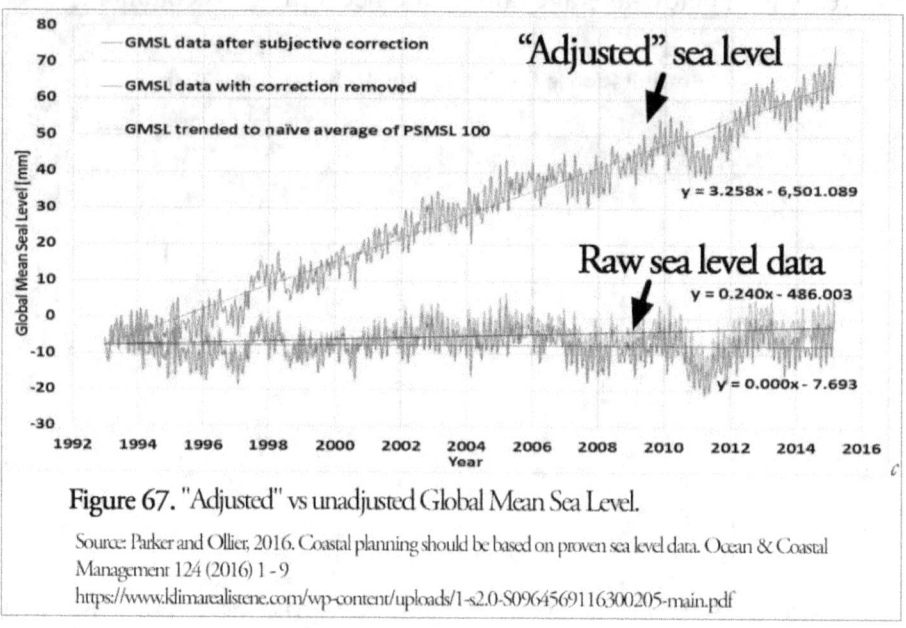

Figure 67. "Adjusted" vs unadjusted Global Mean Sea Level.

Source: Parker and Ollier, 2016. Coastal planning should be based on proven sea level data. Ocean & Coastal Management 124 (2016) 1 - 9
https://www.klimarealistene.com/wp-content/uploads/1-s2.0-S0964569116300205-main.pdf

The upper slopped trendline shows the Global Mean Sea Level (measured by satellites) *after the satellite data has been "corrected"* — the rising sea level looks scary! But the lower trendline shows the raw *uncorrected* satellite-based altimeter data that was used to calculate this "theoretical" Global Mean Sea Level. Once again, the "corrected" data has been tortured to tell an entirely different story — the uncorrected trend in sea level measurements is almost flat and, contrary to claims made by activists and politicians, it is not accelerating.

In a diplomatically worded statement reflective of the wrath than is unleashed against heretics who dare to poke holes in the climate change narrative, the authors of the study sum up their findings as follows: *"We conclude that if the sea levels are only oscillating about constant trends everywhere as suggested by the tide gauges, then the effects of climate change are negligible, and the local patterns may be used for local coastal planning without any need of purely speculative global trends based on emission scenarios. Ocean and coastal management should acknowledge all these facts. As the relative rates of rises are stable worldwide, coastal protection should be introduced only where the rate of rise of sea levels as determined from historical data show a tangible short term threat. As the first signs the sea levels will rise catastrophically within few years are nowhere to be seen, people should start really thinking about the warnings not to demolish everything for a case nobody knows will indeed happen."*

Remember when the Maldives island chain was predicted to be overwhelmed by global warming-induced sea level rise? In 1988, Agence France-Presse reported that government officials had declared that a *"gradual rise in average sea level is threatening to completely cover this Indian Ocean nation of 1196 small islands within the next 30 years"* and that drinking water would run out in the Maldives by 1992.[7] In 1989, the Canberra Times reported that an Expert Group from the Commonwealth of Nations (an international association of 56 member states, the vast majority of which are former territories of the British Empire) was set up to look into climate change; this Expert Group predicted that sea levels would rise between 1 and 4 meters (3 to 13 feet) by the year 2030 (see figure 68 below).[8] The article also reported that the Secretary-General of the Commonwealth had declared that *"Governments must yield national sovereignty [!] to multilevel authorities able to enforce laws 'across environmentally invisible frontiers' if the greenhouse effect, which threatens the future of whole nations, is to be overcome [my emphasis]."*

---

[7] "Threat to islands" (1988, Sept 26). *Agence France-Press (AP News)*.
https://trove.nla.gov.au/newspaper/article/102074798

[8] Ardill, J. (1989, Jan 26). Call for anti-greenhouse action. *The Canberra Times*.
https://trove.nla.gov.au/newspaper/article/120906718

> # Call for anti-greenhouse action
>
> **From JOHN ARDILL, in London**
>
> GOVERNMENTS must yield national sovereignty to multilateral authorities able to enforce laws "across environmentally invisible frontiers" if the greenhouse effect, which threatens the future of whole nations, is to be overcome, the Commonwealth Secretary-General, Sir Shridath Ramphal, said on Tuesday.
>
> A Commonwealth Expert Group set up to look at climate change estimated there was a 90 per cent certainty that the planet would become warmer by at least 1-2 degrees, perhaps much more, and that sea levels would rise by between one and four metres, by the year 2030. Global warming and sea level rises would continue for decades, perhaps centuries.
>
> There was a prospect of widespread, perhaps catastrophic flooding across large areas of Egypt, India, China, the United States, Britain and Holland, and atolls in the Indian and Pacific oceans.
>
> "Surveys of some of these areas conducted for the Commonwealth Group suggest brutal options. One is the large scale abandonment of land; conceivably whole countries," Sir Shridath said.
>
> Who would house the displaced populations of low lying areas like the Maldives, a chain of 1200 islands barely above sea level? Current attitudes to refugees and immigrants in most countries did not suggest that large population movements were feasible. Acceptance of an enhanced risk of large scale drowning was clearly not an option. Building defences was simply beyond the means of most poor countries.
>
> "The cost of doing nothing to prevent climate change is simply unacceptable," he said. "But the problems of progressing from collective study to collective action are immense.
>
> "The need to curb emissions of carbon dioxide, the main greenhouse gas, has prompted many environmentalists to advocate a world of slower economic growth. So long as large-scale poverty and rapid population growth remained, this was no solution."
>
> He added, "A large and growing number of environmental issues are cross-border problems which simply cannot be solved nationally. Unless there is a regional or global framework for handling such issues we will see them escalating dangerously, in some cases to conflict."
>
> Sir Shridath was giving the first of a series of Cambridge lectures on the theme of the Brundtland Report on environment and development. He was a member of the UN commission headed by the Norwegian Premier, Gro Harlem Brundtland, whose 1987 report argued that environmental problems could be tackled successfully only in parallel with economic growth.
>
> "Underlining the report's message of a common future," he said, "Is the unspoken premise ... that we must be ready to nurture tomorrow's concepts of global governance."
>
> This required a change of habit by some of the major powers, which were undermining embryonic forms of multilateral control. An effective law of the sea had been frustrated by the refusal of the US to conform; Russia and Japan had shown a cavalier disregard for the need to observe fishing agreements, and African states were denied a voice in the future of Antarctica.
>
> "Regulation of all the world's commons face similar problems of inequity and unrepresentative control," he added.
>
> — The Guardian

Figure 68. Canberra Times, January 26th, 1989.
https://trove.nla.gov.au/newspaper/article/120906718

While the erosion of national sovereignty and individual autonomy continues unabated, the islands are doing just fine. Far from being underwater, the 1196 islands that comprise the Maldives are being flooded by investors building beachfront homes and resorts.[9] And the Maldives government is investing over $1 billion into their airport — not to relocate it to higher ground but rather to expand it to be able to handle the growing tourist volumes[10] — the airport is quite literally at sea level (see photo in figure 69 below). That hasn't stopped the doomsday prophets from plying their craft, but the ever-moving doomsday deadline has been delayed by another 20 years. In May of 2023, the media once again reported that

---

[9] Moosa, H. and Anand, G. (2017, Mar 26). Inhabitants of Maldives Atoll Fear a Flood of Saudi Money. *The New York Times*.
https://www.nytimes.com/2017/03/26/world/asia/maldives-atoll-saudi-money.html

[10] "Velena International Airport Expansion." (2023, Oct 19). *Airport Technology*.
https://www.airport-technology.com/projects/velena-international-airport-expansion/

"scientists say 80% of the Maldives could be uninhabitable by 2050."[11] Beware of false prophets; actions speak louder than words.

Figure 69. Valena International Airport is owned by the Maldives government and is currently undergoing a $1 billion expansion. Image credit: Ondřej Havelka/Wikimedia Commons, CC BY-SA 4.0 DEED

In 2016, President Obama tweeted that *"We're seeing the fastest rise in sea-levels in nearly 3,000 years"*.[12] But in 2015 Obama bought a beachfront property in Hawaii for $8.7 million (right at sea level) and then proceeded to build a multi-million-dollar house right off the edge of the beach.[13] And in 2019, he put even more of his hard-earned money at risk from rising sea levels when he bought an oceanfront house in Martha's Vineyard for around $11.75 million.[14] Figure 70 shows Obama's oceanfront mansion in Martha's Vineyard smack dab in the crosshairs of "rising sea levels". Once again, actions speak louder than words.

---

[11] Karayel, F. (2023, May 5). Paradise at risk: Maldives in danger of disappearing due to climate change. *Daily Sabah*. https://www.dailysabah.com/life/environment/paradise-at-risk-maldives-in-danger-of-disappearing-due-to-climate-change

[12] Obama, B. (2016, Mar 1). *We're seeing the fastest rise in sea-levels in nearly 3,000 years…"* Twitter. https://x.com/BarackObama/status/704770760259166208

[13] "The Obamas Awesome Hawaiian Mansion Nearing Completion…" (2022, Feb 22). *TMZ*. https://www.tmz.com/2022/02/22/barack-obama-mansion-hawaii-sea-wall-oahu/

[14] Hendrickson, V.L. (2019, Dec 6). Barack and Michelle Obama Reportedly Close Deal for $11.75 Millon Martha's Vineyard Estate. *Barron's PENTA*. https://www.barrons.com/articles/barack-and-michelle-obama-reportedly-close-deal-for-11-75-million-marthas-vineyard-estate-01575648643

Figure 70. In 2019, President Obama purchased a second beachfront house. Would a man worried about "the fastest rise in sea-levels in nearly 3,000 years" put his money here?

Source: Barron's, Dec 6th, 2019

https://www.barrons.com/articles/barack-and-michelle-obama-reportedly-close-deal-for-11-75-million-marthas-vineyard-estate-01575648643

This pattern of endless doomsday predictions that fail to come true once you dig below the headlines repeats itself, over and over again. Hurricanes. Tornados. Flooding. Drought. Sea level rise. The ice-free Arctic (this one seems to be a particular favorite). The end of polar bears. The vanishing corals. The end of snow.[15] There's a highway that runs along the Hudson River, called Westside Highway, which was supposed to below sea level by 2020 — of course, it's doing just fine.[16] New York

---

[15] "Failed Prediction Timeline". (2023 Apr 15). *Watts Up With That.*
https://wattsupwiththat.com/failed-prediction-timeline/

[16] Hansen, S. (2001, Oct 23). Stormy weather. *Salon Magazine.*
https://web.archive.org/web/20110202162233/https://www.salon.com/books/int/2001/10/23/weather/

was supposed to be underwater as early as 2015.[17] In 1970, *Life* magazine reported that, *"In a decade, urban dwellers will have to wear gas masks to survive air pollution. [...] At the present rate of nitrogen buildup, it's only a matter of time before light will be filtered out of the atmosphere and none of our land will be usable."*[18] And in 2004 the Pentagon informed President Bush that Britain would be plunged into a "Siberian" climate by 2020 (figure 71).[19]

> **The Guardian**
>
> ## Now the Pentagon tells Bush: climate change will destroy us
>
> · Secret report warns of rioting and nuclear war
> · Britain will be 'Siberian' in less than 20 years
> · Threat to the world is greater than terrorism
>
> **Mark Townsend** *and* **Paul Harris** *in New York*
> Sun 22 Feb 2004 01.33 GMT
>
> Climate change over the next 20 years could result in a global catastrophe costing millions of lives in wars and natural disasters..
>
> A secret report, suppressed by US defence chiefs and obtained by The Observer, warns that major European cities will be sunk beneath rising seas as Britain is plunged into a 'Siberian' climate by 2020. Nuclear conflict, mega-droughts, famine and widespread rioting will erupt across the world.
>
> The document predicts that abrupt climate change could bring the planet to the edge of anarchy as countries develop a nuclear threat to defend and secure dwindling food, water and energy supplies. The threat to global stability vastly eclipses that of terrorism, say the few experts privy to its

Figure 71. In 2004, the Pentagon warned President Bush that climate change would plunge Britain into a "Siberian" climate by 2020.

Source: The Guardian, February 21st, 2004
https://www.theguardian.com/environment/2004/feb/22/usnews.theobserver

---

[17] Whitlock, S. (2015, Jun 12). FLASHBACK: ABC's '08 Prediction: NYC Under Water from Climate Change By June 2015. *mrcNewsBusters (America's Media Watchdog)*. https://www.newsbusters.org/blogs/scott-whitlock/2015/06/12/flashback-abcs-08-prediction-nyc-under-water-climate-change-june

[18] "Failed Prediction Timeline". (2023 Apr 15). *Watts Up With That*. https://wattsupwiththat.com/failed-prediction-timeline/

[19] Townsend, M. and Harris, P. (2004, Feb 22). Now the Pentagon tells Bush: climate change will destroy us. *The Guardian*. https://www.theguardian.com/environment/2004/feb/22/usnews.theobserver

As a side note, hurricane activity is closely correlated with solar activity. Retired petrophysicist Andy May recently shared the graph reproduced in figure 72 below on his blog[20] — it shows the clear inverse correlation between the number of major hurricanes each year (round dots) versus the annual sunspot count (thin line). As ABC's WQAD8 news channel explained in 2022,[21] researchers believe that as increased solar activity warms the upper atmosphere, this reduces the temperature difference between the upper and lower atmosphere, which robs hurricanes of the vertical temperature differences from which they derive their energy.[22] In other words, <u>increased solar activity reduces hurricanes</u>, while decreased solar activity increases hurricanes — precisely the opposite of what the "climate narrative" would seem to imply about the alleged link between temperature and hurricanes. Oh, and if you look at the number of hurricanes in 2023 — allegedly the "hottest year in over 120,000 years" and the year in which "climate breakdown has begun" — it had the lowest number of hurricanes in over 40 years. Oops.

---

[20] May, A. (2024, Jan 24) Hurricane Frequency and Sunspots. *Andy May Petrophysicist.* https://andymaypetrophysicist.com/2024/01/24/hurricane-frequency-and-sunspots/

[21] Stutzke, A. (2022, Sept 5). Why sunspot activity could actually influence hurricane season. *WQAD8.* https://www.wqad.com/article/weather/ask-andrew/sunspot-hurricane-seasons/526-886c00c1-bf25-4f4b-96d8-8522d628614d

[22] Hodges, R., Jagger, T.H., and Elsner, J. (2014). The sun-hurricane connection: Diagnosing the solar impacts on hurricane frequency over the North Atlantic basin using a space–time model. *Natural Hazards* 73, 1063–1084. https://doi.org/10.1007/s11069-014-1120-9

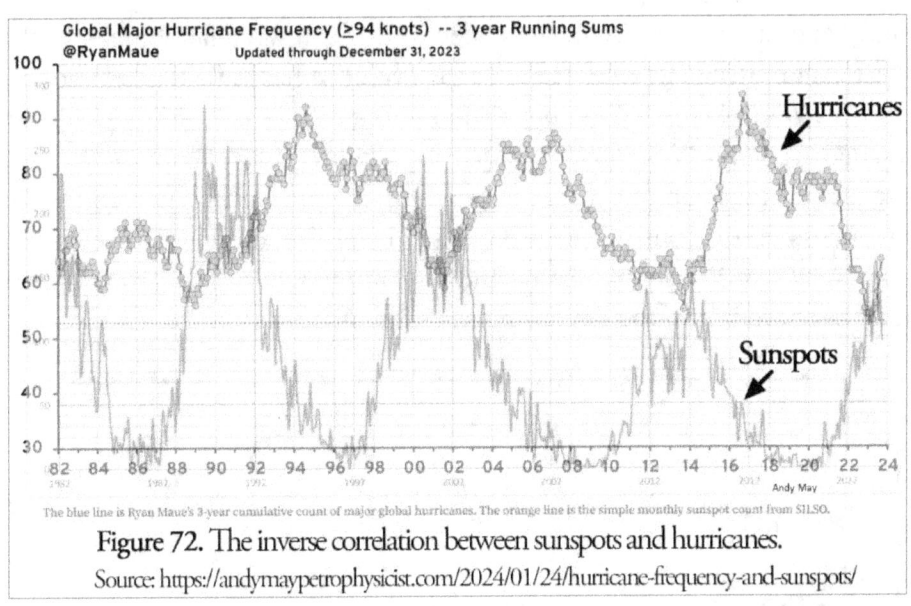

Figure 72. The inverse correlation between sunspots and hurricanes.
Source: https://andymaypetrophysicist.com/2024/01/24/hurricane-frequency-and-sunspots/

As this chapter reaches its close, I'll give one last example to demonstrate the disconnect between the doomsday proclamations made by our media, activist-scientists, politicians, and intergovernmental agencies (including the supposedly impartial and apolitical United Nations) versus the reality playing out in the raw data. In 2017, a paper published in the Journal of Glaciology demonstrated that both NASA's ICESat data and Europe's Remote-Sensing Satellite data (ERS) showed that in Antarctica, *"mass gains from snow accumulation exceeded discharge losses."*[23] In other words, snowfall is exceeding melting and glacier calving in Antarctica. Similarly, in 2015 NASA published an analysis of its satellite data of the Antarctic icesheet, which found that *"an increase in Antarctic snow accumulation that began 10,000 years ago is currently adding enough ice to the continent to outweigh the increased losses from its thinning glaciers. [...] The research challenges the conclusions of other studies, including the Intergovernmental Panel on Climate Change's (IPCC) 2013 report, which says that Antarctica is overall losing land ice."*[24] The chart in figure 73 below is from that NASA

---

[23] Zwally et al. (2015). Mass gains of the Antarctic ice sheet exceed losses. *Journal of Glaciology*. 61(230). 1019-1036. doi:10.3189/2015JoG15J071

[24] NASA Science Editorial Team. (2015, Nov 5). Study: Mass gains of Antarctic ice sheet greater than losses. *NASA*. https://science.nasa.gov/science-research/earth-science/water-energy-cycle/cryosphere/study-mass-gains-of-antarctic-ice-sheet-greater-than-losses/

study and shows that, with the exception of a few isolated areas in the west, Antarctica is <u>gaining</u> ice and that overall it has gained a net average of *+112 billion tons of ice <u>per year</u> between 1992 and 2001* and a net average of *+82 billion tons of ice <u>per year</u> between 2003 and 2008*! If the biggest ice mass on our planet is growing, that doesn't bode well for the breathless predictions of catastrophic sea level rise, does it? However, it's great news for President Obama and his family, whose beachfront homes may be spared from the devouring sea after all.

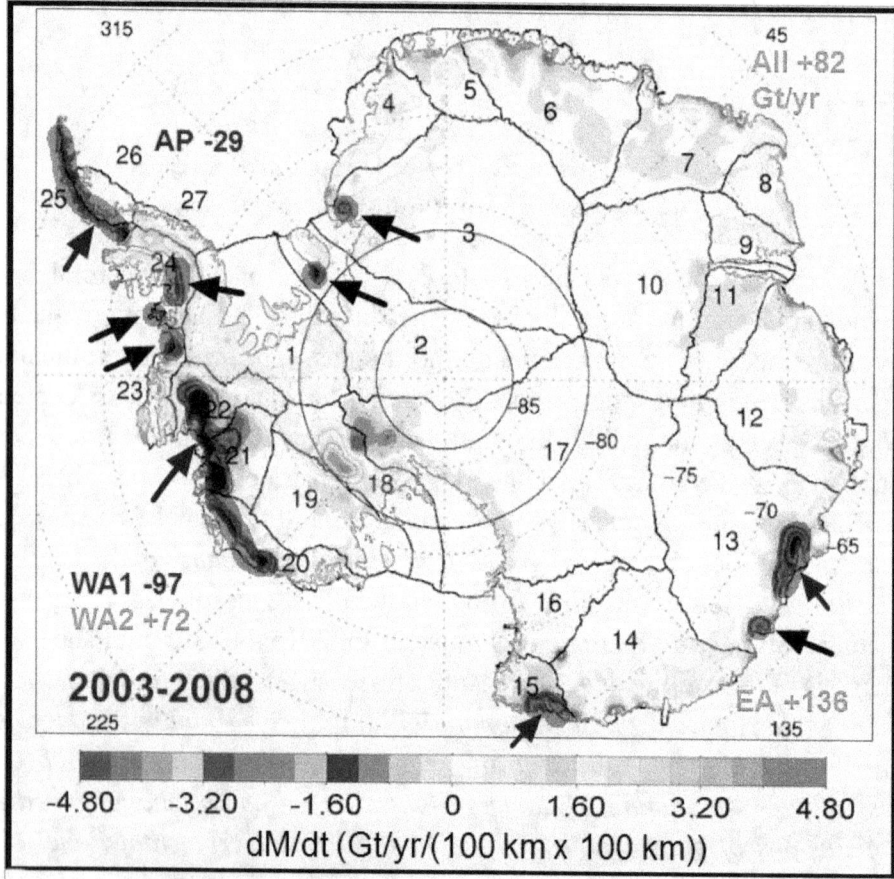

Figure 73. Map showing changes in Antarctic ice mass from 2003 to 2008 based on satellite measurements (I have highlighted the few areas that lost ice and marked them with arrows; everywhere else has gained ice.) Between 2003 and 2008, Antarctica gained a <u>net</u> average of 82 billion tons of ice <u>per year</u>.
Source: NASA Study: Mass Gains of Antarctic Ice Sheet Greater than Losses
https://science.nasa.gov/science-research/earth-science/water-energy-cycle/cryosphere/study-mass-gains-of-antarctic-ice-sheet-greater-than-losses/

Furthermore, in 2021 Antarctica recorded the second coldest winter on record[25] (second only to 2004) and the coldest 6-month period on record.[26] That's hardly what you would expect from a planet experiencing "global boiling." The predictions of doom and gloom may be falling woefully short and the headlines may be completely disconnected from reality, but they are serving to legitimize increasingly daring raids on the public purse and providing cover for an increasingly authoritarian political agenda. The public will accept anything as long as it can be convinced that there is a crisis that cannot be solved any other way. The ends justify the means.

The noble lie is completely reshaping our relationship with our governments. Time and again we see the limits that our ancestors placed on government power crumble away as classical liberal principles collide with the dire predictions of impending global catastrophe promoted by scientists and their algorithms. Time and again, we see influential people willing to compromise their scientific integrity and their intellectual curiosity in exchange for a government contract. Eisenhower's warning rings loud across the ages. Icarus' newest toy – the computer model – has given him unprecedented power to convince everyone else to go along for the ride. What could possibly go wrong?

---

[25] Howes, N. (2021, Oct 11). Coldest place on Earth just saw its second chilliest winter on record. *The Weather Network.* https://www.theweathernetwork.com/en/news/weather/forecasts/coldest-place-on-earth-just-saw-its-second-chilliest-winter-on-record-antarctica

[26] Chinchar, A. (2021, Oct 9). Antarctica's last 6 months were the coldest on record. *CNN.* https://www.cnn.com/2021/10/09/weather/weather-record-cold-antarctica-climate-change/index.html

Chapter 10

CO_2: A Tragically Misunderstood Villain

"By denying scientific principles, one may maintain any paradox."

— Galileo Galilei

As I showed you earlier, the claim that "rising carbon dioxide levels cause global temperatures to rise" breaks down when you look a little closer at temperatures over the last 10,000 years — the only way to demonstrate a correlation between rising CO_2 and temperature is to create a fraudulent temperature record. And yet, when we zoom out to look at much longer timespans on the order of hundreds of thousands of years, there is an undeniable correlation between temperature and atmospheric CO_2 that makes it easy to confuse cause and effect. It's hardly the first time that scientists have fooled themselves into seeing what they firmly believe and then reinterpret the evidence to fit those beliefs. As economist Ronald H. Coase once famously said, *"If you torture the data long enough, if will confess to anything."*

The chart in figure 74 shows ice core data from Vostok, in Antarctica. In many ways this is the single most important chart to understand the origins of the deluded idea that carbon dioxide levels are the control knob on global temperatures.[1]

---

[1] "Ice age." *Wikipedia.* 7 Jun 2024. https://en.wikipedia.org/wiki/Ice_age

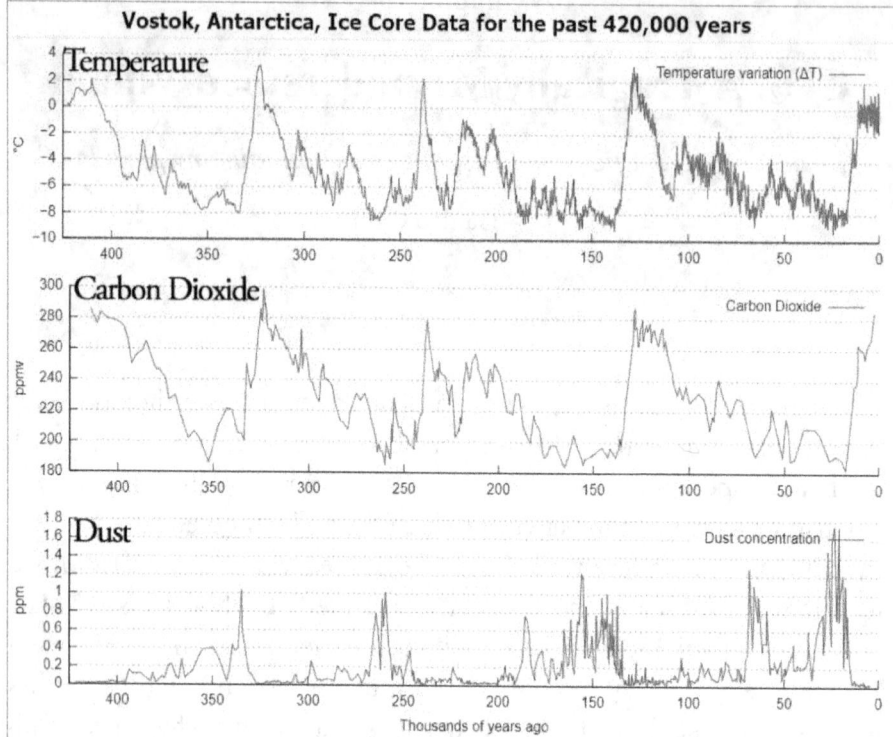

Figure 74. Ice core data from Vostok, in Antarctica, showing the correlation between temperature (top), atmospheric carbon dioxide (middle), and dust (bottom).

Source: Wikipedia https://en.m.wikipedia.org/wiki/File:Vostok_Petit_data.svg

The upper line in the chart shows how reconstructed temperatures over the last 400,000 years have cycled back and forth between long cold glacial ice ages and short warm interglacial periods — a temperature difference of around 12 °C. The middle line shows how reconstructed atmospheric carbon dioxide levels have matched these irregular ice age cycles — low CO_2 when it's cold, higher CO_2 when it's warm — as CO_2 bounces up and down within a narrow range of around 100 ppm (from 180 to 280 ppm). For now, you can disregard the lowest line (dust trapped in the ice cores); we'll come back to the dust and the surprising story it tells us about global aridity in later chapters).

Although the temperature and CO_2 lines are not an absolutely perfect match, there is clearly a strong correlation between them. But which is the cart, and which is the horse? And why did this correlation not show up in previous chapters when we zoomed in on much shorter timescales over the last 10,000 years? And why hasn't there been another 12 °C increase (or more) from the start of the Industrial Revolution until today as CO_2 levels soared by another 140 ppm beyond the peak of pre-Industrial levels (from 280ppm to 420 ppm)?

A closer analysis of the 400,000-year-long Antarctic ice core data shows that atmospheric carbon dioxide levels *lag* temperature by around 800 years (studies range from 600 to 1,300 years). Carbon dioxide is the cart, not the horse. However, many prominent climate scientists claim that, despite the lag, carbon dioxide nevertheless serves as a feedback mechanism to magnify all the other factors that affect global temperature changes (I'll discuss these other factors shortly). This leads to an awkward question for the climate change narrative: if carbon dioxide merely acts as an alleged feedback mechanism for other factors beyond our control, what percentage of their predicted warming would happen anyway even if carbon dioxide levels remained stable?

This simple question puts the climate hysterics in a catch-22. If they claim that the feedback from CO_2 is small, then why are they proposing economy-destroying, poverty-inducing, authoritarian policies when those other factors will keep changing the climate anyway? But if they claim that the CO_2 feedback is so powerful that it acts as the primary control knob on the climate, then they are asking us to believe that, even as the planet dipped in and out of ice ages (at least 8 times over the 850,000 years), global temperatures changed direction a full 800 years *before* a change in atmospheric CO_2 [i.e. the cart is controlling the direction of the horse]. By their logic, why did temperatures stop rising and begin to drop at the peak of prior warm interglacial periods and precipitate the beginning of the next ice age a full 800 years *before* carbon dioxide began to drop? Likewise, why would temperatures begin to rise sharply at the bottom of prior ice ages a full 800 years *before* atmospheric carbon dioxide began to rise?

If, as they claim, carbon dioxide serves as the control knob on the climate, higher CO_2 levels would sustain warm temperatures indefinitely while low CO_2 levels (like those seen at the bottom of the last ice age) would sustain cold temperatures indefinitely unless some other more powerful influence on the climate were to overpower the influence of CO_2, which brings us back to the beginning of this circular dilemma — what is this other more powerful factor that is more powerful than the alleged control knob on the climate and how much will this other factor impact temperatures irregardless of what happens to the amount of CO_2 in the atmosphere? As Galileo once pointed out, the only way to maintain such a paradox is to deny scientific principles. Clearly, cause and effect have been reversed.

It's an indisputable scientific fact taught in high school chemistry class[2] that the solubility of gases in water decreases as water temperature increases. The chart in figure 75 (published by the parent organization of the National Center for Atmospheric Research) demonstrates this inverse relationship between CO_2 solubility and water temperature.[3] It shows how the solubility of carbon dioxide *decreases* as water temperature *increases*. In other words, *cold water is capable of storing far more dissolved carbon dioxide than warm water*. As oceans cool, they absorb CO_2 from the atmosphere, causing atmospheric carbon dioxide to fall. And as oceans gradually warm up again after an ice age, they begin to release (degas) carbon dioxide back into the atmosphere. This relationship between water temperature and CO_2 solubility explains the correlation between global temperatures and atmospheric carbon dioxide in the ice core data (it also explains the lag – more on that in a moment). Temperature is the horse; carbon dioxide is the cart. Basic chemistry won't have it any other way.

---

[2] "Gas Solubility in a Liquid | Overview, Factors & Examples." *Study.com*. Accessed 15 May 2024. https://study.com/academy/lesson/the-solubility-of-gases-in-a-liquid.html

[3] "What do soda and the oceans have in common?" *UCAR Center for Science Education*. Accessed 8 Jun 2024. https://scied.ucar.edu/activity/what-do-soda-and-oceans-have-common

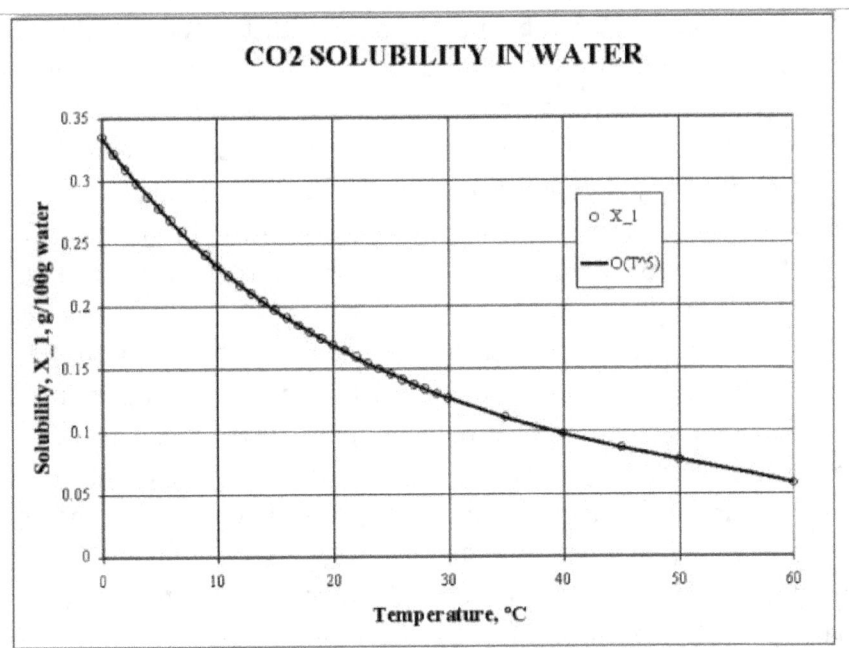

Figure 75. CO₂ solubility decreases as water temperature increases, which means that as water warms up carbon dioxide degasses into the atmosphere, and as water cools down it absorbs carbon dioxide from the air back into the water.

Source: UCAR Center for Science Education
https://scied.ucar.edu/activity/what-do-soda-and-oceans-have-common

You can verify how water temperature controls CO_2 solubility with a can of soda pop or carbonated water. As you pour a glass of soda, you can watch bubbles of carbon dioxide slowly emerge from the soda as it warms up and outgasses the dissolved CO_2.

Another test is to remove an unopened bottle of carbonated water from the freezer and measure the pressure inside the bottle as it warms up; the pressure increases as the warming water loses its ability to store dissolved carbon dioxide — in extreme cases, if a bottle of soda or carbonated water is left in a hot car on a hot summer's day, it will eventually explode.

Yet another test is to take two identical bottles of soda, keep one chilled while allowing the other to warm up to room temperature before opening, and then open them both at the same time. The warm one will froth like crazy and then go flat. However, the cold one will only fizz lightly when it is opened but will keep bubbling gently until the soda reaches room temperature, at which point the fizzing will also stop. That's the basic principle that produces the effect of lagging CO_2 levels as the planet has gone in and out of ice ages — the "fizzing" continues until ocean temperatures re-adjust to a warmer climate, and that re-adjustment takes many centuries. Furthermore, even after the soda (or the ocean) has reached room temperature and gone flat, if you warm it up some more, the fizzing (degassing) will restart.

This well-established relationship between water temperature and CO_2 solubility perfectly explains the correlation between temperature and atmospheric carbon dioxide shown in the Antarctic ice core data. As global temperatures rose after the end of ice ages, oceans warmed up and released dissolved carbon dioxide back into the atmosphere. Then, as the cooler temperatures of the next ice age began, carbon dioxide from the atmosphere was slowly re-absorbed (dissolved) back into the oceans as oceans gradually cooled.

This CO_2 degassing from the oceans is not an insignificant phenomenon considering that the ocean stores 50 times more carbon dioxide than the atmosphere and 20 times more carbon dioxide than land plants and soil combined.[4] The CO_2 solubility chart in figure 75 shows that *as water temperature increases from 0 °C to 10 °C, carbon dioxide solubility decreases by roughly 33%!* Considering the size of our oceans, that's a colossal amount of CO_2 that necessarily has to degas every time ocean temperatures increase after the end of an ice age.

By fraudulently reversing cause and effect, scientists are able to terrify the public by suggesting that we must reduce atmospheric CO_2 to prevent the planet from warming in order to keep the oceans from degassing — check out the link to the shameless article from PBS on March 25th, 2022, entitled: *"When it comes to sucking up carbon emissions, 'the ocean has been forgiving.' That might not last"*.[5] The article is a remarkable example

---

[4] Isaacs-Thomas, B. (2022, Mar 25). When it comes to sucking up carbon emissions, 'the ocean has been forgiving.' That might not last. *PBS News* https://www.pbs.org/newshour/science/the-ocean-helps-absorb-our-carbon-emissions-we-may-be-pushing-it-too-far

[5] Isaacs-Thomas, B. (2022, Mar 25). When it comes to sucking up carbon emissions, 'the ocean has been forgiving.' That might not last. *PBS News* https://www.pbs.org/newshour/science/the-ocean-helps-absorb-our-carbon-emissions-we-may-be-pushing-it-too-far

of circular reasoning that takes gaslighting to a whole new level — a half-truth designed to tell a whopper of a lie that only works as long as the cause-and-effect relationship between temperature and atmospheric CO_2 is reversed. And, surprise, surprise, the article then poses the question: *"Can humans intentionally alter the ocean to be even more efficient at soaking up carbon dioxide in order to stave off ecological catastrophe?"* I wonder how long until some billionaire funds some research project to pour some high-tech concoction of chemicals into the ocean to try to increase CO_2 uptake in another scheme to sell carbon credits? The academic brainstorming of how to tamper with our ocean chemistry has already begun, with ideas ranging from fertilizing phytoplankton with iron, changing the pH of the ocean by adding mineral powders to the seawater, and using electrolysis to transform dissolved CO_2 into solid carbonates (at a cost of $56/tonne). Nor should we be surprised by the academic research money that is being poured down this rabbit hole after the International Energy Agency and the US National Academy of Sciences declared that *"it will be necessary to capture and store 7 gigatons of CO_2 per year by 2050 [10 GT per year according to the National Academy of Sciences] in order to achieve carbon neutrality."*[6] All that's missing now is some heavy-hitting investors and some legislation to reward them with carbon credits in exchange for our tax dollars.

This correlation between water temperature and CO_2 solubility also perfectly explains the 800-year lag between temperature and atmospheric carbon dioxide levels revealed by the ice core data because, although ocean surface temperatures change rather quickly from season to season, our oceans are so deep and there's so little circulation between surface waters and the deepest parts of the ocean that it takes hundreds of years for the deep ocean to warm or cool in response to ice ages.

A shallow pond warms up entirely during the summer, but a deep lake does not — the surface of a deep lake will warm up during the summer, but the cold bottom barely changes at all. You can experience this if you go swimming in a deep mountain lake — by end of summer the surface temperatures are quite pleasant for swimming but if you let your toes hang down as you tread water you can feel that the deeper water near your toes

---

[6] Bopp, L. and Hatton, A.T. (2023, Oct 13). How to optimise CO2 capture by the ocean. *Polytechnique insights.* https://www.polytechnique-insights.com/en/columns/planet/how-to-optimise-co2-capture-by-the-ocean/

is much colder. And if you dive deep under the surface, you can feel the lake get colder and colder as you descend. If you grew up swimming in cold mountain lakes, you will be very familiar with this temperature gradient in large bodies of water. Winter is still lurking down in the deep. Those deep waters still need to warm up before they begin to degas. And the ocean is vastly deeper than any lake.

It takes thousands of years for global temperature changes to affect temperatures in the deep ocean, which is why the Vostok Antarctic ice core data shows the 800-year lag (+/-) created by dissolved carbon dioxide slowly being released from deep waters as the oceans gradually warm up after an ice age (just like the slow fizz of a carbonated drink after it comes out of the fridge). Likewise, as new ice ages begin, there is a corresponding 800-year lag as atmospheric carbon dioxide is gradually reabsorbed back into the oceans as they slowly cool.

~

Ah, you might say, but carbon dioxide is a "greenhouse gas" — isn't it a scientific fact that water vapour, carbon dioxide, methane, nitrogen dioxide, and other greenhouse gases act as a blanket of insolation by trapping heat from our atmosphere? Aren't they the reason our planet is habitable instead of a frozen globe? Haven't we been told that without the moderating effect of greenhouse gases in our atmosphere, the average global temperature would be around -18 °C instead of the comfortable +16 °C we enjoy today? And look at how hot our sister planet Venus is with its atmosphere composed of 96% carbon dioxide — who can deny the greenhouse effect after that? So why wouldn't temperatures increase as CO_2 increases?

Figure 76. Earth's thin atmosphere as seen from space.

Again, this is where the simplistic portrayal of how the greenhouse effect works collides with a much more complex reality. The most powerful lies contain a kernel of truth. Yes, these gases are greenhouse gases and part of the reason why our planet is habitable in the first place. But our atmosphere is also extremely thin; there isn't a lot of gas up there to act as a blanket. The atmosphere on Venus is 93x denser than Earth's atmosphere (the air pressure on Venus is 92 bars (1,350 psi, 9322 kPa); on Earth it is 1 bar (14.7 psi, 101 kPa). There are 66x more gas molecules in every square meter of air on Venus as there are here on Earth.[7] In effect, Venus is being smothered by many layers of heavy wool blankets. Our atmosphere is little more than a wispy negligee by comparison.

But just how wispy is it? Just how weak is this greenhouse effect in our thin atmosphere? We only need to dip just a little deeper into the scientific literature to understand what is going on...

---

[7] Soper, D. E. (n.d.) Atmosphere of Venus. *University of Oregon, Institute of Theoretical Science.* Accessed 2 Apr 2024.
https://pages.uoregon.edu/soper/Venus/atmosphere.html

Over 160 scientific research papers have calculated this greenhouse effect and have come to the conclusion that the actual "feedback effect" on our climate is negligible.[8] A few sample quotes sum up the findings:

"The line-by-line method gives the change of the global temperature (0.4 ± 0.1) K as a result of doubling the carbon dioxide concentration. The contribution to the global temperature change due to anthropogenic injection of carbon dioxide in the atmosphere, i.e. resulted from combustion of fossil fuels, is approximately 0.02 K now."

— Smirnov, 2018.[9]

What Smirnov was saying, to translate from scientific lingo into English, is that if atmospheric carbon dioxide were to double from current levels, we would only see temperatures rise by a modest 0.4 °C, and that all of our CO_2 emissions from 1950 to 2018 levels have only caused temperatures to rise by a paltry 0.02 °C.

"This result strongly suggests that increasing levels of CO_2 will not lead to significant changes in earth temperature and that increases in CH_4 [methane] and N_2O [released by fertilizers] will have very little discernable impact."

— Coe et al, 2021.[10]

"If CO_2 causes global warming, then CO_2 should always precede warming when the Earth's climate warms up after an ice age. However, in all cases, CO_2 lags warming by ~800 years. Shorter time spans show the same thing—warming always precedes an

---

[8] "160 Papers Find Extremely Low CO_2 Climate Sensitivity." *No Tricks Zone*. Accessed 9 Jun 2024. https://notrickszone.com/50-papers-low-sensitivity/

[9] Smirnov, B.M. (2018) Collision and radiative processes in emission of atmospheric carbon dioxide. *Journal of Physics D: Applied Physics.* 51(21). DOI 10.1088/1361-6463/aabac6

[10] Coe, D., Fabinski, W., and Wiegleb, G. (2021). The Impact of CO2, H2O and Other "Greenhouse Gases" on Equilibrium Earth Temperatures. *International Journal of Atmospheric and Oceanic Sciences,* 5(2), 29-40. https://doi.org/10.11648/j.ijaos.20210502.12

increase in CO_2 and therefore it cannot be the cause of the warming."

— Easterbrook, 2016.[11]

*"Most of the greenhouse effect takes place early [at low CO_2 levels] (Fig. 9.7). After that, the effect **decreases exponentially, so the rise in atmospheric CO_2 from 0.030% to 0.038% from 1950 to 2016 could have caused warming of only about 0.01 °C**. The total change in CO_2 of the atmosphere [from 1950 to 2016] amounted to an addition of only one molecule of CO_2 per 10,000 molecules of air." [my emphasis]*

— Easterbrook, 2016[12]

Nor are these observations restricted to recent discoveries. Many scientists were already pointing this out back in the 1970s, long before the hysteria and censorship reached current levels:

"The conclusion is that at low latitudes the influence of doubling CO_2 on surface temperatures is less than 0.25 K, smaller by a factor of 8 than the findings generally accepted. Our finding is comparable to that by Zdunkowski et al. (1975)."

— Newell and Dopplick, 1979[13]

In 1981, a study by Gates, Cook, and Schlesinger found that *"In January the globally averaged tropospheric temperature is increased with respect to the control mean by 0.30 °C (0.48 °C) for doubled (quadrupled) CO_2, which may be compared with an interannual January temperature*

---

[11] Easterbrook, D. (2016). *Chapter 9 – Greenhouse Gases.* Evidence-Based Climate Science (Second Edition). Easterbrook, D. (Ed.) Elsevier Publishing. https://doi.org/10.1016/B978-0-12-804588-6.00009-4

[12] Easterbrook, D. (2016).

[13] Newell, R. E., Dopplick, T. G. (1979). Questions Concerning the Possible Influence of Anthropogenic CO2 on Atmospheric Temperature. *Journal of Applied Meteorology and Climatology.* 18(6). https://journals.ametsoc.org/view/journals/apme/18/6/1520-0450_1979_018_0822_qctpio_2_0_co_2.xml

variability of 0.15 °C in the control..."[14] Translated into English, what the study is saying is that doubling CO_2 from current levels would produce only 0.30 °C of warming while quadrupling it would produce only 0.48 °C of warming, which they point out is in line with the normal temperature variations that we experience from year to year. That's quite a lot less dramatic than the "climate forcing" doomsday predictions of 2.6 to 8.5 degrees that are being advertised by the IPCC (I'll discuss how the IPCC came up with these doomsday numbers shortly — it has everything to do with using computer models that overestimate the role of carbon dioxide by ignoring the impact of cloud cover, as physicist John Clauser pointed out in a previous chapter).

The referenced studies are pointing out that the greenhouse effect does not simply keep increasing with higher CO_2. On the contrary, the first tiny addition of carbon dioxide in the atmosphere has a strong insulating effect — it can absorb a lot of the incoming infrared radiation emitted by our sun. But additional increases of CO_2 to our atmosphere don't just keep absorbing more and more solar radiation — additional CO_2 increases have progressively weaker and weaker insulating power. In other words, the initial CO_2 in Earth's atmosphere contributed a small amount towards creating our habitable planet, but this greenhouse effect is extremely limited in scope — once CO_2 levels rise above a certain level, further increases are incapable of having any meaningful impact on global temperatures.

The following excerpt from a research paper from 2020 explains how carbon dioxide's ability to absorb infrared radiation reaches its saturation point around 300 ppm (that was at pre-industrial levels, we are at 416 ppm today). Above 300 ppm, CO_2 isn't able to absorb any meaningful additional amounts of infrared radiation.

> *"The absorption [of infrared radiation] reaches values close to 100% for a realistic CO_2 content of 0.03% [300 ppm], it is concluded that any further increase of (anthropogenic) CO_2 cannot*

---

[14] Gates, W.L., Cook, K. H., and Schlesinger, M. E. (1981). Preliminary analysis of experiments on the climatic effects of increased CO2 with an atmospheric general circulation model and a climatological ocean. *Journal of Geophysical Research.* 87(C7). 6385-6393. https://doi.org/10.1029/JC086iC07p06385

lead to an appreciably stronger absorption of radiation, and consequently cannot affect the earth's climate."

— Schildknecht, 2020.[15]

The graphic in figure 77 is from Easterbrook, 2016.[16] Each bar represents the increasing saturation of carbon dioxide in the atmosphere as it increases from left to right in 20 ppm increments and shows how much infrared radiation (warming) is absorbed by CO_2 as CO_2 increases.

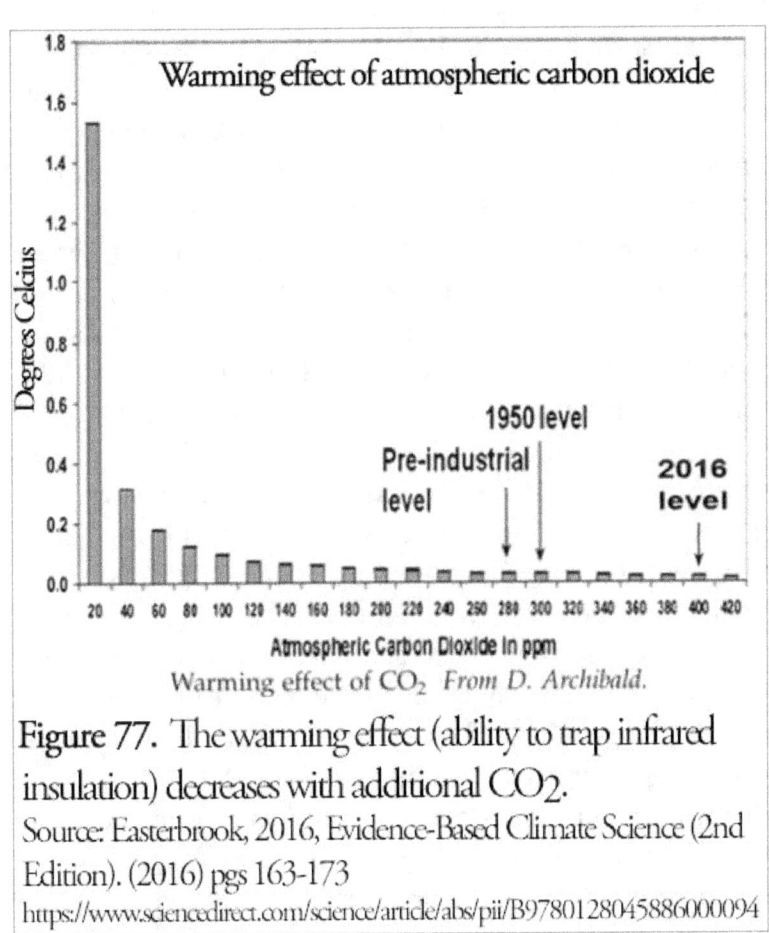

Warming effect of CO_2 From D. Archibald.

Figure 77. The warming effect (ability to trap infrared insulation) decreases with additional CO_2.
Source: Easterbrook, 2016, Evidence-Based Climate Science (2nd Edition). (2016) pgs 163-173
https://www.sciencedirect.com/science/article/abs/pii/B9780128045886000094

---

[15] Schildknecht, D. (2020). Saturation of the Infrared Absorption by Carbon Dioxide in the Atmosphere. *International Journal of Modern Physics B.* 34(30) https://doi.org/10.1142/S0217979220502938

[16] Easterbrook, D. (2016). *Chapter 9 – Greenhouse Gases.* Evidence-Based Climate Science (Second Edition). Easterbrook, D. (Ed.) Elsevier Publishing. https://doi.org/10.1016/B978-0-12-804588-6.00009-4

The chart clearly demonstrates that the first addition of CO_2 to the atmosphere had a noticeable impact — those first few ppm helped make our planet warm enough to be habitable. But almost all of the greenhouse warming comes from the first 20 ppm of CO_2 — there is a rapidly diminishing greenhouse effect after those initial 20 ppm. By the time CO_2 reaches 300 ppm (i.e., pre-Industrial levels), additional increases in CO_2 provide virtually no meaningful further insulation. It is a classic example of the *law of diminishing returns*. As Schildknecht pointed out in his study, the greenhouse effect effectively tops out around 300 ppm CO_2. In essence, the first addition of CO_2 to our atmosphere was like putting on a goose-down jacket, but each additional increase had less insulating value so that by the time you reach 300 ppm CO_2, adding additional CO_2 is like adding a fishnet stocking under your down jacket — insignificant. At 416 ppm CO_2 (the level in 2023), any additional CO_2 emissions are utterly irrelevant. At current levels of CO_2, changes in global temperatures are wholly controlled by other factors — it is thus physically impossible for carbon dioxide to act as the control knob on the climate. Full stop.

Despite the claims made by prominent climate scientists and the media today about the alleged scary impact of CO_2 on global temperatures, scientists have known for a very, very long time that the greenhouse effect from CO_2 is extremely limited because CO_2 only absorbs energy from a very narrow band of infrared radiation — a band that is far too narrow to have a large effect. Furthermore, this narrow wavelength overlaps with the wavelength absorbed by water vapour, which far exceeds CO_2 concentrations in the atmosphere and thus cancels out any meaningful effect from CO_2. In other words, before our planet had an ocean (and thus, before large concentrations of water vapour began filling our atmosphere from water evaporating out of those early oceans), CO_2 had some impact on global temperatures. But once our atmosphere began to fill with water vapour from water evaporating out of Earth's early oceans, any greenhouse effect from CO_2 was rendered completely meaningless because, with or without CO_2 present, water would absorb all of the same infrared radiation. The already miniscule greenhouse effect from CO_2 only matters in an atmosphere that is completely devoid of water vapour, which hasn't been the case since the first oceans began to cover our planet more than 3.8 *billion* years ago. In other words, thanks to water vapour in our atmosphere, CO_2's ability to absorb infrared radiation has been utterly meaningless for at least 3.8 billion years.

The following quote is from page 94 of a book called *Climate and Man: Yearbook of Agriculture, 1941* published by the United States Department of Agriculture in 1941:[17] "*Much has been written about varying amounts of carbon dioxide in the atmosphere as a possible cause of glacial periods.* **The theory received a fata blow when it was realized that carbon dioxide is very selective as to the wave lengths of radiant energy it will absorb, filtering out only such waves as even very minute quantities of water vapour dispose of anyway.** *No probable increase in atmospheric carbon dioxide could materially affect either the amount of insolation reaching the surface or the amount of terrestrial radiation lost to space [my emphasis].*" Hat tip once again to climate historian and geologist Tony Heller for finding this quote and sharing it on his Twitter feed — if you're not following his work already, you really should be following his Twitter feed (@TonyClimate) and his YouTube videos (@TonyHeller).

But why is the greenhouse effect so much stronger on Venus (with an average planetary temperature of 464 °C) if the greenhouse effect is so feeble above 300 ppm here on Earth? The answer lies in how much denser Venus' atmosphere is than our atmosphere on Earth. The ability to trap infrared radiation increases with atmospheric pressure (i.e. as the atmospheric gases become more dense). As air molecules are pressed closer together, their ability to absorb infrared radiation increases. Atmospheric pressure, not CO_2 concentration, is the key driver of the greenhouse effect. The concept is summarized in the following excerpt from a research paper entitled *Atmospheric pressure as a natural climate regulator for a terrestrial planet with a biosphere* (Li et al, 2009),[18] which states: "*... atmospheric pressure also plays a critical role in the greenhouse effect through broadening of the infrared absorption lines of these gases by collisional interaction with other molecules (mainly N_2 and O_2 in the present atmosphere). In other words, if the total atmospheric pressure were lower, the climate forcing of greenhouse gases would be smaller, the magnitude of the greenhouse effect would be less, and the global mean temperature would drop.*"

---

[17] U.S. Department of Agriculture. (1941). *Climate and Man: Yearbook of Agriculture, 1941*. U.S. Government publication. https://archive.org/details/yoa1941/mode/2up

[18] Li, K-F., Pahlevan, K., Kirshvink, J. L. and Yung, Y. L. (2009). Atmospheric pressure as a natural climate regulator for a terrestrial planet with a biosphere. *PNAS*. 106(24). https://doi.org/10.1073/pnas.0809436106

In other words, without Venus' high atmospheric pressure, its high concentration of greenhouse gases (96.5% carbon dioxide) would not have produced a runaway greenhouse effect. For context, like Venus, the Martian atmosphere is also mostly made of CO_2 (95.3%), yet the Martian atmosphere is so thin (less than 1% of Earth's atmospheric pressure) that the average planetary temperature on Mars is -53 °C (-64 °F). Pressure matters, CO_2 concentration does not.

A runaway greenhouse effects requires much higher atmospheric pressures, not more greenhouse gases. Throughout most of Earth's history, concentrations of greenhouse gases in Earth's atmosphere were far higher than what they are today despite multiple severe ice ages, some of which lasted hundreds of millions of years (more on those later). 4 billion years ago Earth's atmosphere had 1000x more methane[19] and 750x more CO_2 than it has today. In those days CO_2 made up approximately 30% (300,000 ppm)[20] of the Earth's atmosphere, in contrast to today where CO_2 makes up a paltry 0.04% of our atmosphere (just over 400 ppm). Yet despite these extraordinarily high concentrations of greenhouse gases in the Earth's early atmosphere, there was no runaway greenhouse effect, and the planet was even tipped into several ice ages lasting millions of years at a time because the atmospheric pressure here on Earth was simply too low to trap a large amount of infrared heat. At Earth's atmospheric pressure of 1 bar, air molecules are spaced too far apart to serve as an efficient insulation blanket.

Because atmospheric pressure decreases the higher you go, air temperature falls with increasing altitude, which is why air temperatures outside your airplane window (at 30,000 ft) hover around -44 °C (-48 °F) and why it's cooler in the mountains than down in the valleys — the thinner layer of atmospheric gases provides less insulation from the bitter cold of outer space despite the fact that atmospheric CO_2 concentrations are identical at all elevations. It's also a big part of the reason why the bottom of the 6,000-ft deep Grand Canyon is around 11 °C (20 °F) hotter than the rim — atmospheric pressure increases and thus the atmosphere gets denser as you descend into the canyon. You can test this yourself if

---

[19] "Atmospheric Methane." *NASA Earth Observatory*. Accessed 2 Jun 2024. https://earthobservatory.nasa.gov/images/5270/atmospheric-methane

[20] "Why Earth wasn't one big ball of ice 4 billion years ago when Sun's radiations was weaker." (2010, Apr 1). *Science Daily*. https://www.sciencedaily.com/releases/2010/03/100331141415.htm

you go for a drive in the mountains — watch your car's thermometer begin to fall as your car starts climbing up a mountain road. Temperature falls at an average rate of 0.65 °C for every 100 meters of elevation gain (up to 1 °C if the air is very dry).[21] Dress for the correct altitude, not for the amount of CO_2 in the air.

Scientists call this phenomenon the "adiabatic lapse rate", which describes the temperature change per kilometer of elevation change. On Venus the adiabatic lapse rate is 10.468 °C/km of elevation. The adiabatic lapse rate on Earth is remarkably similar at 9.76 °C/km of elevation despite the fact that the atmosphere on Venus made of a whopping 96.5% CO_2 while Earth's atmosphere only has paltry 0.04% CO_2. The rest of Earth's atmosphere is made of 78% nitrogen gas, 21% oxygen, and 1% argon, none of which are greenhouse gases. The slight difference between the lapse rate on Venus vs Earth is explained by the fact that Earth is slightly further from the Sun, so our planet receives slightly less solar radiation.

The fact that the adiabatic lapse rate is almost identical between two similar sized planets despite Venus having an atmosphere that is almost entirely composed of greenhouse gases while Earth has an atmosphere almost entirely devoid of greenhouse gases, demonstrates that the greenhouse effect of CO_2 is *exceedingly* insignificant. The temperature difference between our two planets is created almost entirely by the effect of atmospheric pressure. Venus simply has a much thicker blanket of atmospheric gases — gravity compresses that thick layer of gas to create a much denser layer of gas near Venus' surface, thus allowing those gases to trap much more heat from the Sun. At sea level on Earth, the weight of all the gases in our atmosphere pressing down on us is 1 atmospheric pressure (approximately 14 psi) — we don't even feel it because it's "normal". If you were to stand on surface of Venus, the crushing weight of all of Venus' atmospheric gases would be 93x higher (93 atm / 1366 psi) — standing on Venus' surface is like experiencing the crushing pressure of being surrounded by water if you dive 940 m (3080 ft) below our oceans. But, if you ascend 50 km above Venus' surface to where the atmosphere thins to the same 1 atm that we experience here on Earth's

---

[21] Federal Office of Meteorology and Climatology MeteoSwiss. (n.d.). Decreases in temperature with altitude. *Government of Switzerland*. Accessed 17 Jun 2024. https://www.meteoswiss.admin.ch/weather/weather-and-climate-from-a-to-z/temperature/decreases-in-temperature-with-altitude.html

surface at sea level, at that pressure you will also find comfortable Earth-like temperatures in the 0 to 50 °C range despite the fact that Venus' atmosphere at that elevation is nonetheless composed of 96.5% CO_2 instead of the 0.04% we have here on Earth.

Thus, all those scary scenarios about a runaway greenhouse effect published by the Intergovernmental Panel on Climate Change are physically impossible. Quite simply put, the United Nations is lying through its teeth when they say that humanity has "opened the gates to hell" and that a "climate breakdown has begun." They are manufacturing a purely fictitious crusade to fight phantoms of the imagination.

~

The various climate-changing forces that actually drive global temperature changes on our planet are not a mystery. And these forces haven't stopped impacting the climate just because we're living in the modern era — they are all continually pushing our climate one way or another on nearly every time scale. Once we put aside the hysteria of the carbon dioxide crusade, we return to the question asked by National Geographic back in 1976 — which way do we go next, warmer or cooler?

For example, there are cyclical fluctuations in ocean currents that bring warm tropical waters north, like the short 2- to 7-year El Nino cycle, the 40- to 50-year cycle in ocean currents produced by the previously discussed Pacific Decadal Oscillation, and a corresponding 60-year cycle in the Atlantic Ocean called the Atlantic Multidecadal Oscillation. Scientists have also identified a 1,000-year "global conveyor belt" of deep sea and sea surface currents that circle the globe on a 1,000 year time span, which (ever so slowly) bring cold waters from the deep up to the surface and send warm surface waters down into the ocean deeps.[22] The oceans are our planet's primary heat storage mechanism, so anything that affects how that stored heat is spread around the planet will ultimately affect both the local and the global climate.

Just as there are ocean currents here on Earth, there are currents acting inside the sun that continually change how much heat is radiated out towards the Earth. These changes in solar activity are known as solar cycles. The most well-known are the 11-year and 22-year sunspot cycles. There's also a 30-year solar cycle,[23] a 200-year cycle known as the de Vries solar cycle,[24] and a 400-year grand solar cycle (the lows of the Little Ice Age line up perfectly with a grand solar minimum called the Maunder Minimum, suggesting that this 400-year cycle deserves a large part of the blame for the frozen crops and cold weather that drove the Vikings out of Greenland and caused the canals to freeze in the Netherlands).[25] There's also the 976-year Eddy solar cycle — it's warm phases over the last two thousand years line up with the warm weather experienced during the Roman Warm Period, the Medieval Warm Period, and the Modern Warm Period.[26] And there's also the 2,500-year Bray/Hallstatt solar cycle,[27] whose repeating cycle has been identified in the ice core data going back for more than 30,000 years.

As if this all wasn't complicated enough, there are also wobbles in the Earth's orbit, known as Milankovitch Cycles, which change the intensity of sunlight reaching our planet. These cyclical changes in the Earth's orbit

---

[22] "The Global Conveyor Belt." *National Ocean Service – National Oceanic and Atmospheric Administration.* Accessed 4 Jun 2024. https://oceanservice.noaa.gov/education/tutorial_currents/05conveyor2.html

[23] Pérez-Peraza, J., Velasco, V., Libin, Igor Y., and Yudakhin, K. F. (2012) Thirty-Year Periodicity of Cosmic Rays. *Advances in Astronomy* https://doi.org/10.1155/2012/691408

[24] Raspopov et al. (2008). The influence of the de Vries (~ 200-year) solar cycle on climate variations: Results from the Central Asian Mountains and their global link. *Paleogeography, Palaeoclimatology, Palaeoecology.* 259(1). https://doi.org/10.1016/j.palaeo.2006.12.017

[25] Zharkova et al. (2023). Periodicities of solar activity and solar radiation derived from observations and their links with the terrestrial environment. *Natural Science.* 15(03). 111-147. http://dx.doi.org/10.4236/ns.2023.153010

[26] Curry, J. (2017, Dec 2). Nature Unbound VI – Centennial to millennial solar cycles. *Climate Etc.* https://judithcurry.com/2017/12/02/nature-unbound-vi-centennial-to-millennial-solar-cycles/

[27] Mearns, E. (2016, May 11). Periodicities in solar variability and climate change: A simple model (guest post). *Energy Matters.* https://euanmearns.com/periodicities-in-solar-variability-and-climate-change-a-simple-model/

are the driving forces that created eight glacial advances over the last 800,000 years, as shown earlier in the National Geographic charts.

Changes in the tilt of the Earth's axis during its orbit around the sun, known as **obliquity**, run on a 41,000-year cycle, from a maximum of 24.5 degrees to a minimum of 22.1 degrees. A steeper tilt leads to more solar radiation hitting the Earth during the summer but colder winters. The maximum tilt happened 10,000 years ago — it maximized the amount of sunlight hitting the poles during the summer, which brought an end to the last Ice Age, melted the ice sheets, and created the warm temperatures that led to trees growing in the High Arctic 8,000 years ago, far north of where trees are able to grow today. Today, the Earth is tilted at 23.4 degrees and that tilt is slowly decreasing. As the tilt decreases, winters will keep getting milder, but summers will simultaneously get cooler, which will eventually trigger icesheets to start building up again on land in polar regions and in the high mountain passes as summer melting fails to keep up with winter snows accumulating on the ice caps. The trend of increasingly warm winters and cooler summers of recent years, which I demonstrated in the previous chapter, may to some small degree be influenced by this gradually decreasing tilt in the Earth's axis. Prior to 1 million years ago, this 41,000-year cycle was the dominant cycle that triggered the beginning and end of ice ages.

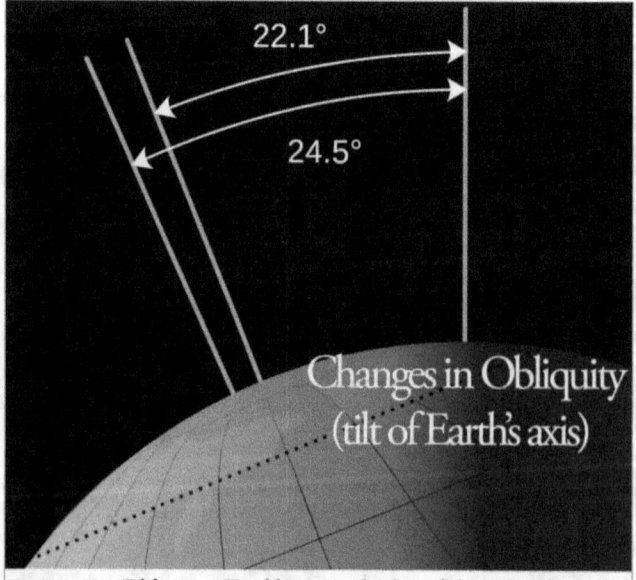

Figure 78. Obliquity: Earth's axis is tilted, and that tilt varies on a 41,000-year cycle from 22.1 to 24.5 degrees.

Another wobble in the Earth's orbit is known as **eccentricity** — it runs on a 100,000-year cycle. Our orbit around the sun isn't perfectly round — sometimes it's closer to round, sometimes it's more stretched out (elliptical). This stretching and shrinking in our orbit is caused by the gravitational pull of Jupiter and Saturn tugging on our planet. As Earth's orbit gets stretched into a more elliptical shape, the seasonal difference between how much sunlight reaches the Earth during summer versus winter gets more extreme (making summers hotter and winters cooler). At it's most elliptical, 23% more solar radiation reaches the Earth from the Sun during the summer compared to when the orbit is more circular[28] — that too plays a big role in ending ice ages. But we're currently going the opposite direction. Our orbit around the sun is currently becoming less elliptical. In other words, as our orbit becomes more round again, it is once again reducing the heat coming from the sun during the summer (even as it makes our winters warmer), again pushing the balance between summer ice melt versus winter ice build-up ever so slowly towards the next ice age on a timescale that our great-great-great-great-great-great-great-grandchildren probably still won't have to worry about. Since about 1 million years ago, ice age cycles lengthened to match this 100,000-year cycle, creating an alternating cycle of 10,000- to 15,000-year warm interglacial periods when the ice sheets retreated, followed by 85,000- to 90,000-year cold glacial periods when the ice sheets advanced. During our current interglacial period, we've already had 10,000 years of warm weather. Based on the timing of prior glacial cycles, around 3,000 years from now things are going to start getting very chilly again.[29] And once it does, it will stay cold for an unimaginably long time.

---

[28] NASA Science Editorial Team. (2020, Feb 27). Milankovitch (Orbital) Cycles and Their Role in Earth's Climate. *NASA*. https://science.nasa.gov/science-research/earth-science/milankovitch-orbital-cycles-and-their-role-in-earths-climate/

[29] Pleistocene Epoch. (2014, May 5). Pleistocene Epoch. *Geology Page*. https://www.geologypage.com/2014/05/pleistocene-epoch.html

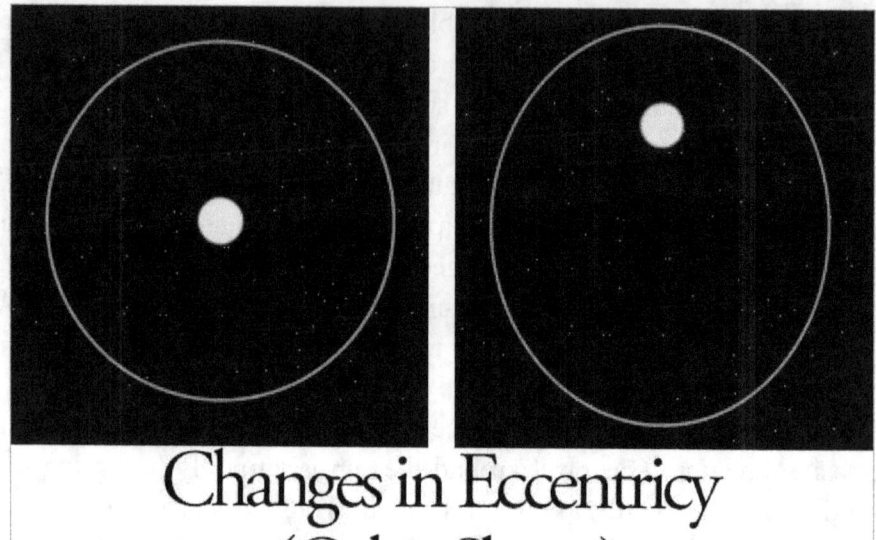

Changes in Eccentricy (Orbit Shape)

Figure 79. Eccentricity: Earth's orbit around the Sun is sometimes closer to round, and sometimes it gets stretched out because of the gravitational pull from Jupiter and Saturn. These changes in orbital shape vary on a 100,000-year cycle.

And finally, there is one other wobble in the Earth's orbit, called **precession**, which changes whether the tilt of the Earth's axis faces towards or away from the sun at the closest point during our planet's annual orbit around the sun. This wobble is also known as the "spinning top" effect; it runs on a 25,771-year cycle. This wobble makes seasons more extreme in one hemisphere versus the other depending on whether the axis is pointing towards or away from the sun at the closest point during the Earth's orbit around the sun. Currently, precession is making summers cooler in the Northern hemisphere while making summers hotter in the Southern Hemisphere, but in around 13,000 years this will flip.[30]

---

[30] NASA Science Editorial Team. (2020, Feb 27). Milankovitch (Orbital) Cycles and Their Role in Earth's Climate. *NASA*. https://science.nasa.gov/science-research/earth-science/milankovitch-orbital-cycles-and-their-role-in-earths-climate/

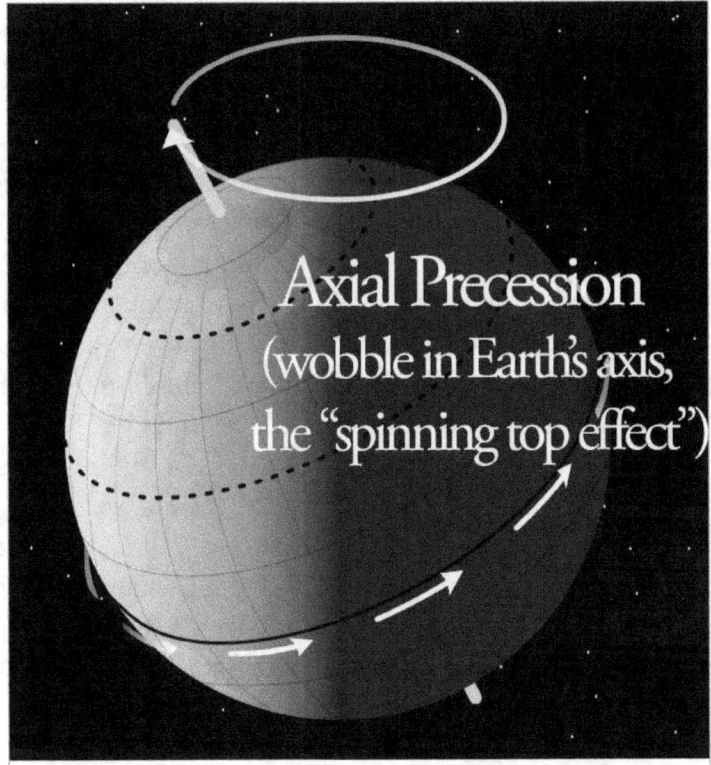

Figure 80. Precession (the "spinning top" effect) determines whether the Earth's axis points towards or away from the Sun at the closest point in our annual orbit around the Sun. The "wobble" takes 25,771 years to complete a single cycle.

The sum of all these different factors (and there are even more that I have not included in this partial list) all add up to constantly push or pull Earth's climate in different directions. Sometimes these various cycles reinforce one another; sometimes they cancel each other out. But they never stop. The lesser cycles shape the short-term wiggles and wobbles in our global temperatures to produce short-term climate variations like the heat of the 1930s or the cold centuries of the Little Ice Age, while the powerful Milankovitch cycles dominate the big sweeping temperature changes between glacial and interglacial periods. Our 19th century ancestors didn't understand any of these cycles, so when the rains increased out on the Great Plains in the 1870s and 1880s, they pointed at the most obvious change that they could see as the cause — the arrival of the plow. For better or for worse, in the absence of a better explanation we always find a way to make ourselves responsible for all that happens in our universe.

However, those who are familiar with Earth's history also know that there are long periods — hundreds of millions of years at a time — when Earth's temperature remained relatively stable and was not plunged into repeating cycles of ice ages despite the never-ending orbital wobbles produced by Milankovitch Cycles. For example, during the entire 165-million-year span when dinosaurs ruled the Earth, temperatures never got anywhere near as cold as they are today. Our current back-and-forth cycles of long cold ice ages interrupted by short warm interglacials (collectively known as the Quaternary Glaciation) only began around 2.58 million years ago (Antarctica has been in an ice age for the last 34 million years but the Northern Hemisphere only began to freeze 2.58 million years ago). Before that you have to go back another 260 million years (before the age of the dinosaurs) before you find geological evidence for another long period of ice ages, known as the Karoo Ice Age, which lasted for 100 million years (from 360 to 260 million years ago)! But then there's another 70-million-year pause of warm weather uninterrupted by ice ages before there's geological evidence for the bitterly cold Ordovician Glaciation, which lasted from 460 to 430 million years ago. And on and on it goes as you reach ever further back in time. Why the long pauses? Why the long ice ages that last millions of years? And why isn't there a clear cyclical pattern on these much longer time scales?

Remember back in the introduction when I mentioned Alfred Wegener's discovery of plate tectonics? The continents are constantly moving around the Earth, ever so slowly, at an average rate of around 1.5 centimeters (0.6 inches) per year. Sometimes these continents drift closer to the poles. Sometimes they drift closer to the equator. Sometimes they collide and get bunched up together. Sometimes they fragment and get scattered all around the globe with large oceans between them. It's difficult to build continental ice sheets unless you have large continental landmasses near the poles. That's why ice ages tend to only happen whenever there are large land masses near the poles, as is the case today (Eurasia, Greenland, North America, and Antarctica).

Whenever continental drift shifts large land masses towards the poles, the summer melt near the poles will struggle to keep up with winter snowfall during periods when the Milankovitch Cycles favor cooling, leading to ice sheets that gradually spread southwards to cover entire continents. But as continental drift brings those large landmasses closer to the equator, ocean currents are able to equalize temperatures between tropical and northern regions by circulating cold water from the north back down to temperate or tropical regions — in that case although global temperatures might dip slightly, you won't get the large ice sheets that turn the planet into a deep freeze. With so many of our continents currently stretching far up towards the poles in the Northern Hemisphere and at the slow rate that the continents are moving, it's a safe bet that over the next few million years our planet will continue to experience many more chilly waves of ice sheets that cover entire continents. Geological time works on a scale that is utterly incomprehensible to the short perspective offered by a human lifespan.

Chapter 11

Fixing the Algorithm with Hypothetical Feedback Loops

"Today's scientists have substituted mathematics for experiments, and they wander off through equation after equation, and eventually build a structure which has no relation to reality."

— Nikola Tesla

Slowly but surely, I am dismantling the beliefs that are an obstacle to what I want to show you over the course of this detective story. Next, I want to turn to the story of how greenhouse gases (and "feedback loops") became the scapegoats in our mad crusade to try to keep our climate in the Goldilocks Zone in defiance of everything I just showed you about the how greenhouse gases actually work. Again, we have to return to the gloomy mood of the 1970s and to the scientific discoveries that had (*and had not!*) been made by the 1970s. All the things we have learned about the complex global-scale physical processes that shape our planet, our atmosphere, our oceans, and our solar system are extremely recent discoveries. During the 1970s, much of that knowledge was in its infancy.

The 11-year solar cycle and its short-term impact on climate was first identified in 1843. But the 22-year cycle was only discovered in 1925. The 976-year Eddy solar cycle was not identified until the 1970s. It took even longer to identify the cyclical ocean currents that impact our climate — the 60-year Atlantic Multidecadal Oscillation was not identified until 1994 while the discovery of the 40- to 50-year Pacific Decadal Oscillation only happened in 1997 while I was at university studying geology.

Even the most powerful forces that create ice ages — Milankovitch Cycles — are only a very recent discovery. Geophysicist and astronomer Milutin Milanković recognized them in the 1920s. By the 1930s and 40s most European geologists were won over by evidence that these Milankovitch were playing a role in creating ice ages. But then, in an amusing twist of irony that perfectly highlights how messy science actually is, during the 1950s and 60s new radiocarbon evidence emerged that seemed at odds with Milanković's theory, which led the majority of scientists to firmly reject the role of Milankovitch Cycles in creating ice ages. But then the pendulum began to swing the other way again as the contradictions were resolved so that by 1976 the majority of scientists were once again taking Milankovitch Cycles seriously.[1]

The first evidence for ice ages and other big shifts in Earth's climate was only discovered by geologists in 1836, though it took until the 1870s before the idea was broadly accepted that Earth's seemingly benign climate had been periodically plunged into a series of long ice ages over the course of its geologic history. However, even during the 1950s and 60s, many scientists still assumed that things like the Little Ice Age from 1300 to around 1850 and the Dust Bowl during the 1930s were abnormal stand-alone "climate events" under the assumption that the slow planet-wide forces that created the repeated glacial and interglacial periods over the last 2.58 million years would be so slow as to be virtually undetectable within the span of a single human lifespan.

In other words, without understanding solar cycles, ocean currents, and all the other recently discovered short-term drivers that are constantly pulling climate in one direction or another, there was a general assumption that the climate in our era should be a relatively stable unchanging thing. And yet, by the cool 1970s it was also clear that the changing climate in our era was anything but stable, though the reason why was a hotly debated mystery. And so, it should surprise no-one that the explanation which emerged was very much in keeping with the pessimistic spirit of the 60s and 70s: it was probably all our fault because of the drastic impact of industrialization, pollution, and overpopulation — we were allegedly destabilizing the planetary ecosystem in the same way that we were undeniably destabilizing many local ecosystems. As Mark Twain once remarked, *"every complex problem has a solution which is simple, obvious, and wrong."*

By linking climate change to our industrial activities, it naturally follows to project any short-term climate trend indefinitely into the future in keeping with our increasing industrial footprint and growing

---

[1] Richet, P. (2021). The temperature-CO2 climate connection: an epistemological reappraisal of ice core messages. *Hist. Geo Space. Sci.*, 12, 97-110. https://doi.org/10.5194/hgss-12-97-2021 (see also https://wattsupwiththat.com/2021/05/27/the-temperature-co2-climate-connection-an-epistemological-reappraisal-of-ice-core-messages/)

population. By the 1970s, those predictions had already crystalized into visions of an apocalyptic future if we failed to take drastic steps to deindustrialize and depopulate before the climate spirals out of control, although they were still having trouble agreeing as to whether that man-made apocalypse would be hot or cold. But eventually the language evolved from "global cooling" to "global warming" until finally settling on "climate change", which helped cover all bases in case temperatures change direction once again. Either way, by ignoring all the short-term drivers of climate change, any deviation from absolutely stable was assumed to be our fault. And once the issue caught the eyes of politicians and began capturing funding from the public treasury, any chance of settling it via evidence-based scientific debate was scuttled.

In addition to all the profiteering and all the self-serving political agendas that have grafted themselves onto this narrative, it should also not surprise us that many well-meaning people have been caught up by this narrative and are willing to support censorship, participate in data fraud, justify blatant propaganda, collaborate in the subversion of democracy, and strip their fellow citizens of their individual autonomy. Anyone who sees themselves on a crusade to save the world from the apocalypse will not be troubled by their moral conscience — their virtuous crusade makes them blind to what they have become. The end justifies the means. Many roads to Hell are paved with good intensions.

> *"Of all tyrannies, a tyranny sincerely exercised for the good of its victims may be the most oppressive. It may be better to live under robber barons than under omnipotent moral busybodies. The robber baron's cruelty may sometimes sleep, his cupidity may at some point be satiated; but those who torment us for our own good will torment us without end for they do so with the approval of their own conscience. They may be more likely to go to Heaven yet at the same time likelier to make a Hell of earth. This very kindness stings with intolerable insult. To be "cured" against one's will and cured of states which we may not regard as disease is to be put on a level of those who have not yet reached the age of reason or those who never will; to be classed with infants, imbeciles, and domestic animals."*
>
> — C. S. Lewis

Correlation does not automatically prove causation, but politicizing an issue and throwing vast quantities of money at researchers to investigate a potential link has a way of taking science where evidence cannot. As Eisenhower said in his farewell speech, *"a government contract becomes virtually a substitute for intellectual curiosity."* Money flows to the most publicized problems — just think of what would have happened to all the funding being poured into the climate sciences if researchers had come back with a conclusion that "everything is just fine, there's nothing to worry about, and the planetary-scale changes in our climate are all natural, non-threatening, and completely outside of our control." I recently ran across a wonderful cartoon mocking the IPCC, which sums up the situation perfectly —in the cartoon the speaker at the climate change conference asks the scientists in the audience, *"Hands up, who here thinks greenhouse gases have no effect, and therefore we all need new jobs? Anyone?"*

From its earliest beginnings in the 1960s and 70s, climate science has always been inseparable from politics. From day one, it has been an academic discipline riven with shortcomings in scientific objectivity because of its intimate entanglement with the masterful administrative state and the politicized nature of its funding.

But why hasn't the whole house of cards fallen in upon itself with so much overwhelming evidence about all the natural drivers of climate cycles and everything we've learned about the feeble greenhouse effect? How can any scientist keep a straight face while pointing at the "feedback loop" of greenhouse gases as the alleged control knob on our climate? The answer lies in how climate modelers came up with a value for the mysterious and much-hyped "feedback loop" that has been programmed into the algorithms of their climate models.

Ah, the feedback loop — that mysterious X-factor, which arises from the idea that while individual climate drivers are quantifiable (orbital wobbles, solar cycles, ocean currents, etc.), when multiple forces act simultaneously on a complex dynamic system, they begin to reinforce one another in a theoretical feedback loop. It's the perfect example of the apocalyptic thinking that fuels hysteria. For example, according to the "feedback" theory, our CO_2 emissions may only cause a mild increase to how much infrared radiation is absorbed by our atmosphere, but the mild warming that it causes will thaw some permafrost in the high arctic, which will release additional methane as the thawed permafrost begins to decompose, and that extra methane has an additional warming effect on our planet, which in turn will cause still more methane-rich permafrost to melt, which will lead to still more warming. And soon Chicken Little is racing around the coop, screaming his bloody head off that the sky is falling because all those theoretical feedback loops will cause the entire system to spiral out of control in a spectacular climate breakdown.

How did the climate modelers come up with a number for this feedback loop multiplier in order to plug it into their algorithms? Hold on to your chair… because when they calculated values for all the solar cycles, ocean currents, Milankovitch Cycles, etc. and plugged those numbers into their climate algorithms, the algorithmically-reconstructed climate model that came out of the other end of the little black box didn't match the historic ice core data… so they added a theoretical X-factor to the algorithm and then tweaked that X-factor until the model matched the historic record.

It's the height of hubris to claim that their algorithm is so perfect and that their understanding of all the variables that affect our global climate is so complete that when the algorithm fails to accurately reconstruct the historic record of our complex dynamic global climate, the only explanation for the difference is a hypothetical X-factor ("climate sensitivity" created by "feedback loops") rather than some flaw or omission in the assumptions that they programmed into their algorithm. Bear in mind that the complex dynamic processes going on in our atmosphere are still so poorly understood that we can't even reliably predict next Thursday's local weather, much less whether next year's winter will be mild or bitter cold or whether next summer will be wet, dry, hot, or cool. The ultimate test of a scientific theory is its ability to predict the future, not to create a form-fitted reconstruction of the past.

It's also the height of gaslighting to claim that some theoretical feedback loop created by rising carbon dioxide was able to warm the planet's temperature at the bottom of the last ice age a full 800 years before carbon dioxide levels began to rise. A feedback loop linked to CO_2 faces the exact same catch-22 of cause-and-effect discussed in the previous chapter: a bowling ball cannot knock over pins at the end of the bowling alley before it begins to roll. Tying feedback loops to greenhouse gases creates an impossible paradox because it resurrects the problem of cause-and-effect between temperature and atmospheric CO_2.

Furthermore, it's a transparent lie to claim that we are facing a dangerous feedback loop today, at atmospheric carbon dioxide levels of around 416 ppm, when most of Earth's history had orders of magnitude higher levels of carbon dioxide and other greenhouse gases without any detrimental impact on the climate — that alone disproves their theoretical feedback loop and demonstrates that their assumptions about what drives our climate are fundamentally flawed.

As I will show you momentarily once we dive in the untold story of CO_2 in part two of this book, there are long periods in the geologic record (millions of years at a time) when the climate stayed stable or even cooled even as carbon dioxide levels were rising sharply due to volcanic emissions (such as during the mid-Cambrian period when multi-cellular life began, during the late Ordovician period when the planet froze over, or during the mid-Jurassic period while dinosaurs ruled the Earth, just to name a few). These examples would be impossible if CO_2 drives temperature. Likewise, there are also long periods of time in the geologic record (again millions of years at a time) when atmospheric carbon dioxide levels were falling sharply even as the climate warmed (like during the early Silurian period when plants colonized the continents). And there are periods when carbon dioxide levels were plummeting against the backdrop of a stable climate (like during the late Devonian and early Carboniferous periods — the period when most of the world's coalbeds were formed). The entire premise of linking greenhouse gases to feedback loops in climate models is fundamentally at odds with the story told by the geologic record. To sustain the models, you only have to systematically ignore 4.54 billion years of geological evidence.

Obviously, there are plenty of dissenting scientists who haven't been bamboozled by the climate model algorithms. As the aforementioned Nobel-prize-winning physicist John Clauser pointed out, an obvious area where the climate models fall short is that they have failed to account for changing cloud cover, which has a dramatic impact on how much infrared radiation reaches the surface of the planet and handily accounts for the theoretical feedback loops that are being programed into the climate algorithms. It's a rather enormous oversight considering that cloud cover can vary from 5% to 95% of the entire planet at any given time, thus dramatically impacting how much solar energy reaches the Earth's surface and how much heat is prevented from radiating back out into space.

Ignoring cloud cover effects is not a trivial omission considering the pivotal role that clouds play in the weather of our planet. But as the scientists running weather modification programs found out during the

40s, 50s, and 60s, cloud modelling is exceedingly complex — no one knows exactly what they will do tomorrow, much less a decade or a year from today. So, climate modelers simply left them out of the equation because they assumed that on balance, total cloud cover from year to year and from decade to decade will be relatively consistent despite the chaos happening above our heads on shorter time scales. That assumption is fundamentally, disastrously, atrociously wrong.

It's not just the effect of clouds from day-to-day weather changes that is missing from the algorithms. Research has also clearly demonstrated that long-term changes in cloud cover are linked to solar cycles, ocean current cycles, and other extremely long-term factors that affect our climate. As I mentioned in Chapter 4, long-term cycles in ocean currents that bring warm water into northern regions are strongly correlated with cloud formation at all time scales (i.e. the 2- to 7-year El Nino cycle,[2] the 40- to 50-year Pacific Decadal Oscillation,[3] and the 60-year Atlantic Multidecadal Oscillation[4]). Whenever warmer water penetrates further into cold regions, there is a corresponding increase in cloud cover.

Furthermore, in recent years scientists (including the IPCC,[5] though you won't know it from the press summaries they give to the media and to politicians) have documented strong evidence that the amount of cloud cover on Earth is strongly correlated with solar cycles[6] — it is thought that high-energy particles coming from the sun and from space play a role in ionizing gas molecules in our atmosphere. These ionized gas molecules form aerosols that serve as condensation nuclei for water vapour, leading

---

[2] Yirka, B. (2016, Jan 5). Clouds may have more of an impact on El Nino than thought. *Phys.org*. https://phys.org/news/2016-01-clouds-impact-el-nino-thought.html

[3] Chen et al. (2019). Distinct Patterns of Cloud Changes Associated with Decadal Variability and Their Contribution to Observed Cloud Cover Trends. *American Meteorological Society – Journal of Climate*. 32(21). 7281-7301. https://doi.org/10.1175/JCLI-D-18-0443.1

[4] Chen et al. (2019).

[5] IPCC Working Group I: The Scientific Basis. (n.d.) *6.11.2.2 Cosmic rays and clouds*. Intergovernmental Panel on Climate Change. https://archive.ipcc.ch/ipccreports/tar/wg1/246.htm

[6] Kumar et al. (2023). The influence of solar-modulated regional circulations and galactic cosmic rays on global cloud distribution. *Scientific Reports* 13, 3707. https://doi.org/10.1038/s41598-023-30447-9

to cloud formation.[7] In other words, the models fall short because the global energy balance calculations only calculated changes in the direct heat coming from the sun but failed to account for how much solar radiation indirectly impacts the ever-changing number of clouds covering the globe, which sometimes reflect heat, sometimes trap heat, and sometimes do both at the same time depending on temperature and how extensive the cloud cover is.

Surprisingly, even orbital changes (Milankovitch cycles) are thought to affect cloud cover,[8] also presumably because changes in the Earth's tilt and wobbles in the Earth's orbit impact how many high-energy particles from the sun are able to reach the Earth's atmosphere where they ionize gas molecules to create aerosols that serve as condensation nuclei for clouds. There's even evidence that our Earth's continually changing geomagnetic field plays an important role in influencing global cloud cover. A strong geomagnetic field helps shield our planet from getting fried by high-energy particles coming from the sun and from cosmic radiation coming from space. So, whenever the geomagnetic field weakens (as is happening today[9]), more high energy particles from the sun and from space are able pass through the magnetic field and penetrate the Earth's atmosphere, which enhances the formation of low-lying clouds and increases global cloud cover, leading to cooling.[10]

There is also evidence that as our orbit around the Sun wobbles (Milankovitch Cycles), this leads to slight variations in the gravitational tug on our planet from the Sun and from other planets (especially Jupiter and Saturn), which causes strains on the Earth's crust and on the mantle deep below our feet, which in turn triggers cyclical changes in volcanic

---

[7] Svensmark, H., Enghoff, M.B., Shaviv, N.J. et al. (2017) Increased ionization supports growth of aerosols into cloud condensation nuclei. *Nature Communications.* 8, 2199. https://doi.org/10.1038/s41467-017-02082-2

[8] Berger, A. (1988). Milankovitch Theory and Climate. *Review of Geophysics.* 26(4). 654-657.
https://ebme.marine.rutgers.edu/HistoryEarthSystems/HistEarthSystems_Fall2008/Week12a/Berger_Reviews_Geophysics_1988.pdf

[9] Irion, R. (2003, Dec 12). Earth's Waning Magnet. *Science.*
https://www.science.org/content/article/earths-waning-magnet

[10] Campuzano et al. (2018). New perspectives in the study of the Earth's magnetic field and climate connection: The use of transfer entropy. *PLoS One.* 13(11) https://doi.org/10.1371%2Fjournal.pone.0207270

activity on the planet[11] — volcanic dust doesn't only directly filter how much sunlight can penetrate down to the Earth's surface, these dust particles launched high into our Earth's atmosphere also provide condensation nuclei for cloud formation.

There's no getting around the fact that clouds are not just weather, they are deeply integral to the complex dynamic processes that create our Earth's ever-changing climate. As long as climate model algorithms cannot fully account for cloud formation and cloud behaviour, from day-to-day weather pattern all the way out to multi-centuries-long cyclical changes in how many clouds are covering the globe, they can't claim to understand the weather, much less accurately model the global climate.

By ignoring cloud effects and focusing attention on CO_2 instead, activist-scientists are able to stoke public fears about an impending apocalypse in order to strong-arm society into accepting a complete rewrite of the long-established limits on government power and usher in a new global world order. And, once activist-scientists started down this path, there was no easy way to pull back from their increasingly absurd claims without damaging their credibility, blowing up their careers, vaporizing their access to public funding, and utterly destroying the credibility of their scientific institutions. Once they started down that dishonest path, the fear of being held accountable meant there was no way out except to manipulate the historic record in order to sustain the fiction. As the old German proverb says, *"Once the Devil catches you by a thread, he'll soon be leading you by a rope."*

---

[11] Kutterolf et al. (2019). Milankovitch frequencies in tephra records at volcanic arcs: The relation of kyr-scale cyclic variations in volcanism to global climate changes. *Quaternary Science Review.* 204. 1-16.
https://doi.org/10.1016/j.quascirev.2018.11.004

PART TWO

Reconstruction

Chapter 12

The Elusive Fingerprint of Fossil Fuels

"Geology is intimately related to almost all the physical sciences, as history is to the moral. An historian should, if possible, be at once profoundly acquainted with ethics, politics, jurisprudence, the military art, theology; in a word, with all branches of knowledge by which any insight into human affairs, or into the moral and intellectual nature of man, can be obtained. It would be no less desirable that a geologist should be well versed in chemistry, natural philosophy, mineralogy, zoology, comparative anatomy, botany; in short, in every science relating to organic and inorganic nature."

— Charles Lyell, Principles of Geology

What has emerged so far is a complicated story about hubris, conflicts of interests, Malthusian fears about overpopulation, and political agendas that cloak themselves in an aura of scientific legitimacy in order to shepherd society into accepting exploitative and authoritarian policies that would otherwise be firmly rejected as unacceptable assaults on our individual liberty.

With the conspiracies and intrigues unmasked, it is tempting to wash our hands of the whole sordid mess, congratulate ourselves on having pulled back the curtain on the mad crusade against CO_2, and comfort ourselves in the knowledge that by a stroke of good fortune, our industrial activities are pumping a life-giving gas (carbon dioxide) back into the atmosphere to the benefit of all life on Earth. Despite all the toxic chemicals and local environmental destruction that have haunted us since the beginning of the Industrial Era, at least there's some kind of silver lining, right? After all, photosynthesis cannot work without CO_2 — it is our planet's single most important plant nutrient, and at 420 ppm (0.04% of the atmosphere) it has been running dangerously low when looked at on geological time frames. It's tempting to simply stop the story here and move on to the urgent business of wrestling democracy back out of the hands of the manipulators, liars, propagandists, ideologues, authoritarians, and doom-mongers who are leading society down the net-zero rabbit hole.

But in the introduction, I promised you that there is more to this story than just a thorough debunking of a nonsensical climate theory — it is also a detective story with far-reaching ecological and political implications for civilization. And so, our meandering trail of cookie crumbs brings us to the untold story of carbon dioxide — to the story of CO_2-starved plants and degassing soils, to the complex carbon cycle that circulates CO_2 in and out of our atmosphere, and to the devastating ecological story unfolding beneath our feet.

I will begin this story by challenging one of the core assumptions that almost everyone on both sides of the debate takes as the whole and gospel truth — that our sharply rising atmospheric CO_2 levels are the result of our sharply rising global CO_2 emissions. Once again, correlation does not automatically mean that it's a simple case of cause and effect.

Over the past 800,000 years, atmospheric CO_2 levels have fluctuated up and down within a fairly narrow 100-ppm range. At the bottom of ice ages (at least 8 over the last 800,000 years), levels reached as low as 180 ppm. During the warm interglacials, levels consistently topped out somewhere between 260 and 280 ppm, as shown in the chart from NASA in figure 81 below. But starting in the mid-1800s, atmospheric CO_2 levels climbed above the upper boundary of the last 800,000 years, and then just kept climbing. The climb accelerated in the 1950s and has relentlessly continued ever since. By 2023, levels had reached 416 ppm, 140 ppm above the top of the narrow range of the last 800,000 years. And they're not showing any signs of slowing down.

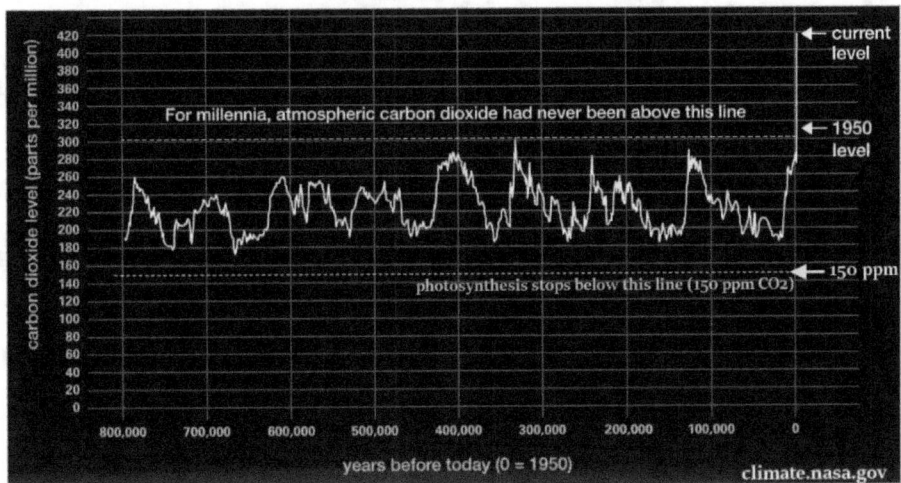

Figure 81. Over the past 800,000 years, atmospheric carbon dioxide levels have stayed confined within a clear range as temperatures alternated between cold ice ages and warm interglacials. But in our modern era, CO2 levels have broken far outside of this range.

Adapted from climate.nasa.gov https://science.nasa.gov/resource/graphic-the-relentless-rise-of-carbon-dioxide/

According to the consensus climate change narrative, 78% of this increase in atmospheric CO_2 has been caused by fossil fuel emissions (they assume that the remaining 22% comes from agriculture, deforestation, and other land-use changes).[1] The tight correlation between rising atmospheric CO_2 levels and our rising CO_2 emissions is unmistakable, as shown in the next two charts — the upper chart in figure 82 shows annual CO_2 emissions over the past 270 years, while the lower chart shows atmospheric CO_2 levels over this same time period. But is it really such a straightforward case of cause and effect?

---

[1] "Global Greenhouse Gas Overview." *United States Environmental Protection Agency.* Accessed 5 Jun 2024. https://www.epa.gov/ghgemissions/global-greenhouse-gas-overview

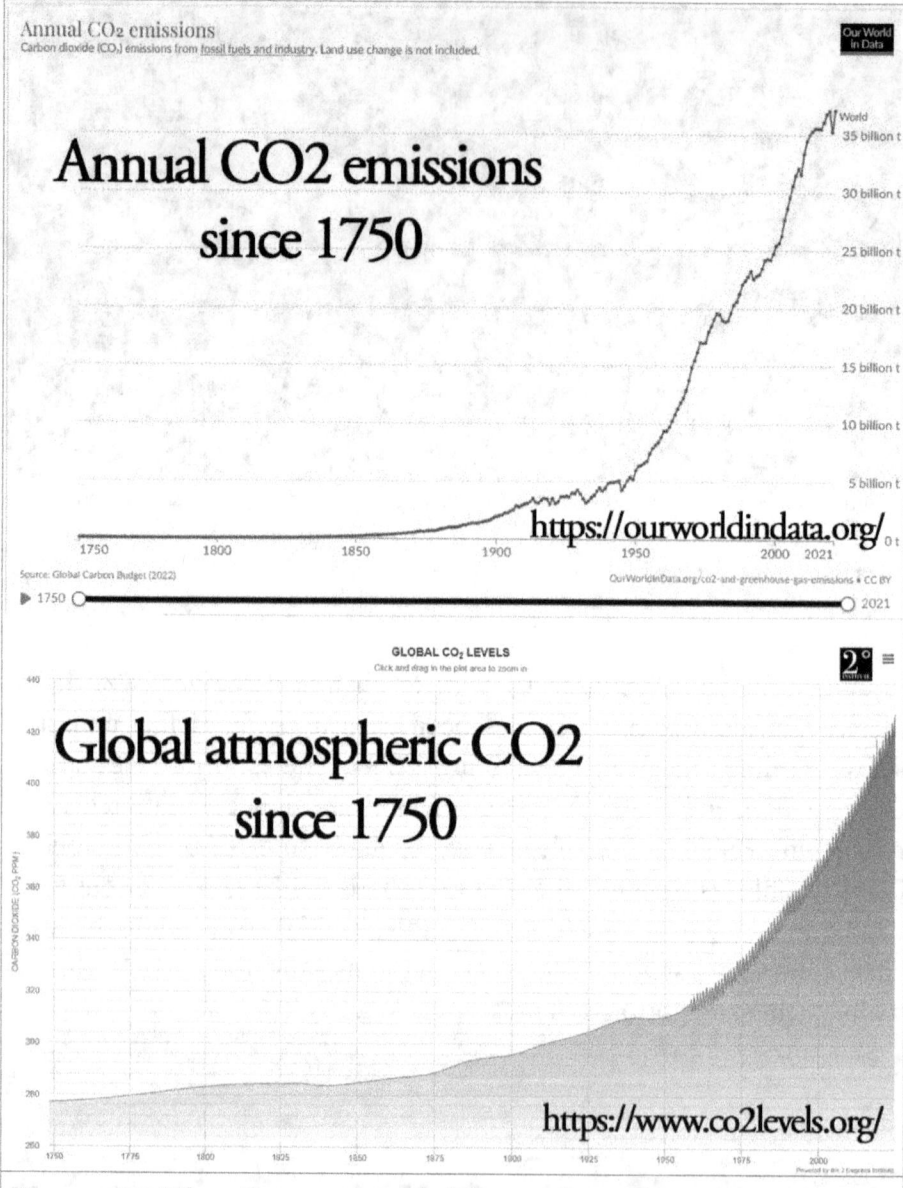

Figure 82. There is an unmistakable correlation between emissions and CO_2. But is it a straightforward case of cause and effect?

If we overlay the two previous charts, as shown in figure 83 below, a few niggling inconsistencies pop out — loose threads in a scientific theory that inject a whisper of doubt.

The Elusive Fingerprint of Fossil Fuels

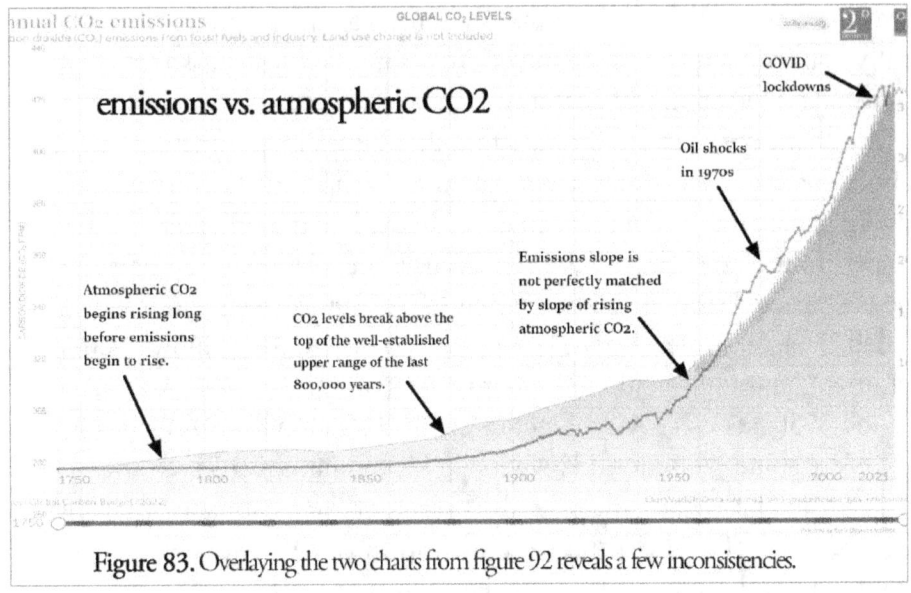

Figure 83. Overlaying the two charts from figure 92 reveals a few inconsistencies.

Some heretics pointed out that there must be more to the story because CO_2 levels did not begin to rise in 1750 but had already been rising steadily since around 7,000 years ago — as demonstrated by the Greenland ice core data shown back in figure 55 (chapter 8) — and that from their interglacial lows 7,000 years ago they had already risen by at least 25 ppm (!) *before* the Industrial Era began. That clashes with the commonly accepted narrative about fossil fuels being to blame since this large increase began long before we started burning fossil fuels. Nor can we blame carbon dioxide degassing from the oceans because, as demonstrated by the ice core data (and by archaeological evidence, 5,000-year-old arctic tree stumps, and so on), this 25 ppm increase in CO_2 *happened in spite of the fact that the climate was gradually cooling throughout this 7,000-year period* (this cooling should have increased CO_2 solubility in the oceans and thus it should have pulled CO_2 *out* of the air rather than putting it in).

In other words, the fact that CO_2 levels unexpectedly decoupled from global temperatures and instead began to rise 7,000 years ago, and merely accelerated during the Industrial Era, indicates that there is another powerful yet hitherto unidentified contributing factor (or factors) that began exerting a massive influence on atmospheric carbon dioxide levels as early as 7,000 years ago. Since this earlier rise cannot be explained away either by the burning of fossil fuels nor by the relationship between temperature and CO_2 solubility in our oceans, we are forced to conclude that there is another major untold element to the story that needs to be fully understood (and quantified) before we can assume that we understand the complete picture.

Furthermore, despite the fact that there is clearly something different about our current interglacial period compared to all other prior interglacials, fossil fuel emissions fail to explain the full story because the upper limit of the last 800,000 years was breached *before* our emissions began to accelerate at the beginning of the 1900s (before our world was filled with cars, diesel generators, and tractors pulling plows across the prairies), and long before the sharp acceleration in emissions that began in the 1950s as electricity generation, steel smelting, manufacturing, and mass consumer automobile ownership kicked into overdrive.

It would be easy to dismiss this breach of the established range as the consequence of coal burning… until you take a closer look at the slope of both lines and see that fossil fuel emissions in the late 1800s are *lagging* atmospheric CO_2 (during the 1800s the slope of rising CO_2 is steeper than the slope of rising emissions), which is puzzling if fossil fuel emissions are meant to be the driver of rising atmospheric CO_2 and even more puzzling when you consider that after the 1950s the correlation inverts so that suddenly emissions rise much more steeply than atmospheric carbon dioxide levels — for a cause-and-effect relationship to be true, it must follow consistent, not arbitrary rules. While these niggling inconsistencies are hardly conclusive, they are red flags alerting us to the possibility that there might be a lot more to the story than the tidy consensus explanation of fossil fuels as the principal driver of rising carbon dioxide levels.

Furthermore, starting in the 1950s, fossil fuel emissions begin to follow a much more variable track full of surges interrupted by temporary plateaus and pullbacks, in line with how we built out and expanded our industrial capacity. Yet atmospheric carbon dioxide levels don't reflect the jagged path taken by rising emissions — instead they just keep on rising year after year, as steady as a ticking clock, unbothered and unfazed by anything going on around them, even during times when humanity periodically took its foot off the gas pedal. If fossil fuels are allegedly responsible for a full 78% of the rise of atmospheric CO_2, then rising atmospheric CO_2 should reflect the bumpiness of the path taken by rising emissions. But they don't.

Whenever an explanation for a scientific phenomenon has a loose thread, you have to test your assumptions by pulling on that loose thread to see what falls out. In 2020, the world pulled decisively on that thread. The emergence of the Covid virus left world leaders scrambling in a crude medieval experiment to see who could impose the most draconian lockdowns. The world economy sputtered and stalled. It was the perfect unwitting experiment to decisively test the link between fossil fuel emissions and rising atmospheric CO_2.

Our global annual CO_2 emissions from fossil fuels typically increase by approximately 1% each year. 2020 was the exception to that rule. In 2021, NASA published its observations about the effect of the pandemic on the atmosphere: *"The most surprising result is that **while carbon dioxide (CO_2) emissions fell by 5.4% in 2020, the amount of CO_2 in the atmosphere continued to grow at about the same rate as in preceding years**. During previous socioeconomic disruptions, like the 1973 oil shortage, you could immediately see a change in the growth rate of CO_2 [...] We all expected to see it this time, too."*[2]

That isn't what should have happened in a straightforward case of cause and effect. Figure 84 zooms in on the last decade to show rising atmospheric CO_2 levels, as measured at the Mauna Loa Observatory in Hawaii. An effect from Covid-era lockdowns is nowhere to be seen.

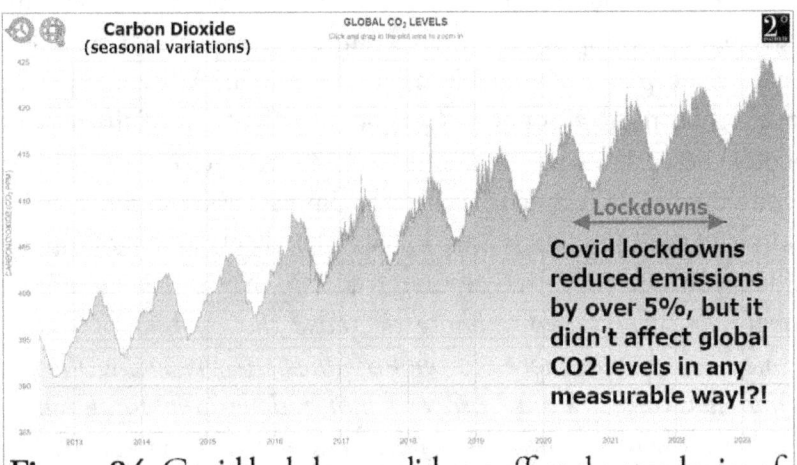

Figure 84. Covid lockdowns did not affect the steady rise of atmospheric CO_2 levels. Adapted from www.co2levels.org/

---

[2] "Emission Reductions From Pandemic Had Unexpected Effects on Atmosphere" (2021, Nov 9) *NASA Jet Propulsion Laboratory.*
https://www.jpl.nasa.gov/news/emission-reductions-from-pandemic-had-unexpected-effects-on-atmosphere

The sawtooth pattern in figure 84 reveals the carbon cycle in action — the sawtooth pattern is created because during the growing season in the Northern Hemisphere (where the bulk of our continental land masses are located) plants draw more CO_2 out of the atmosphere than what is returned to the atmosphere by rotting plant debris, thus causing global CO_2 levels to decline during the summer months. But then, as plant growth stalls during the Northern Hemisphere's winter, the amount of CO_2 consumed by photosynthesis drops below the amount of CO_2 being recycled back into the atmosphere by rotting plant debris, which causes atmospheric CO_2 to go back up. Far from being a permanent steady exchange, this rapid seasonal change in growing biomass makes global CO_2 levels bounce up and down by a remarkable 6 to 7 ppm as the seasons flip back and forth between summer and winter. Clearly CO_2 levels are extremely sensitive to subtle changes in the amount of CO_2 entering and leaving the atmosphere — Covid lockdowns should therefore also have had a measurable impact if fossil fuels were the primary driver of rising CO_2 levels.

This natural seasonal fluctuation of CO_2 entering and leaving the atmosphere is also a rather striking reminder that fossil fuel emissions, warming oceans, and volcanoes are not the only forces that can have big impacts on the composition of our atmosphere. Even small seasonal changes to the planet's total biomass can dramatically alter the natural balance between the amount of CO_2 being pulled out of the atmosphere by photosynthesis versus the amount of CO_2 re-entering the atmosphere as a consequence of rotting plant debris, animals and microbes breathing, and other natural processes that exhale CO_2 back into the atmosphere. Whether you increase the amount of CO_2 being "exhaled" by our biosphere *or whether you simply decrease the amount of CO_2 being consumed by our biosphere via photosynthesis (by reducing the planet's total biomass through any process that reduces the total amount of vegetation growing on the planet),* either way can fundamentally change the long-term balance of carbon dioxide cycling in and out of our atmosphere... with big consequences to the amount of CO_2 building up in our atmosphere. As we'll soon see in upcoming chapters, both sides of this equation are playing pivotal roles in fundamentally altering our atmospheric composition.

The lack of impact that Covid lockdowns had on rising CO_2 levels foreshadowed the bombshell that was dropped in 2022 by a group of scientists who set out to measure what percentage of atmospheric carbon dioxide originates from the burning of fossil fuels. In 2022, a controversial study (Skrable et al, 2022)[3] was published, which, if true, throws a giant wrench into the entire CO_2 emissions story. Instead of merely assuming that rising fossil fuel emissions are to blame for rising CO_2 levels, its authors decided to use radioactive carbon isotopes to directly measure how much of the increase in atmospheric CO_2 can be directly traced back to fossil fuels versus non-fossil-fuel sources. No-one had measured this before (kind of a major oversight!) — previously scientists had merely calculated how much CO_2 has increased in the atmosphere since the beginning of the industrial revolution and compared that to their calculations for how much CO_2 has been produced by burning fossil fuels and then assigned the balance to agriculture, deforestation, and other land-use changes.

The authors of the controversial study decided not to assume but to measure instead ("trust but verify"). But how can you differentiate between CO_2 that originally comes from burning fossil fuels versus CO_2 released into the atmosphere from non-fossil-fuel sources? Isn't all CO_2 the same? As it turns out, no. I promise not to get too technical, but since this study is so important I think it is worth dedicating a few paragraphs to explain the core physics underpinning it rather than just quoting the study's conclusion.

The study comes down to the fact that there are several different versions of carbon atoms (known as isotopes in chemistry), most importantly carbon-12 versus carbon-14. The number indicates the sum of protons and neutrons contained inside the atom: carbon-12 has 6 protons and 6 neutrons, while carbon-14 has 6 protons and 8 neutrons. Most of the world's carbon atoms are carbon-12 isotopes — it is the most stable and abundant form of carbon. However, as radiation from space (a.k.a. cosmic rays) enter our atmosphere, they sometimes collide with nitrogen-14 atoms (nitrogen-14 has 7 protons and 7 neutrons) — the collision knocks off one proton and adds one extra neutron to the original nucleus of the

---

[3] Skrable, K., Chabot, G., French, C. (2022). World Atmospheric CO2, Its 14C Specific Activity, Non-fossil Component, Anthropogenic Fossil Component, and Emissions (1750–2018). *Health Physics*. 122(2). 291-305.
https://web.archive.org/web/20220202154330/https:/journals.lww.com/health-physics/Fulltext/2022/02000/World_Atmospheric_CO2,_Its_14C_Specific_Activity,.2.aspx

nitrogen-14 atom, which temporarily converts the nitrogen-14 atom into a carbon-14 atom.

But carbon-14 is not stable; it is a radioactive isotope that gradually decays to return to its original form (nitrogen-14). During this decay process, a small amount of radiation is released as one of the neutrons is converted back into a proton to restore the original balance of 7 protons and 7 neutrons, thus transforming carbon-14 back into nitrogen-14. Carbon-14 decays with a predictable half-life of 5,730 years. After 60,000 years (approximately 10 half-life cycles), there's essentially no carbon-14 left — it has all reverted back to nitrogen-14.

However, since cosmic radiation never stops colliding with nitrogen-14 in our atmosphere, carbon dioxide gas in our atmosphere always contains a constant predictable ratio of carbon-12 to carbon-14 atoms. And as plants draw carbon dioxide out of the atmosphere during photosynthesis, this atmospheric ratio of carbon-12 to carbon-14 is integrated into the carbon contained inside plant tissues (and into the tissues of animals that feed on those plants.).

Once an animal or plant dies, no more new carbon gets incorporated into its tissues. And if those dead plant or animal tissues get buried, the existing carbon inside those tissues is locked away, forever. But since carbon-14 continues to slowly decay to revert back to its original nitrogen-14 form, after 60,000 years all the carbon-14 atoms originally locked inside these plant and animal tissues will have decayed away, leaving only carbon-12 behind. Like an atomic clock, this predictable radioactive decay of carbon-14 inside plant and animal tissues (like in fossils, plant residues, and the bones found in archaeological remains) allows scientists to calculate how old carbon-based materials are by measuring the ratio of carbon-14 to carbon-12 trapped inside the carbon-rich tissues — it's called radiocarbon dating. This carbon-14 dating technique works for anything younger than 60,000 years old. To establish the age of older things, like the rock layers studied by geologists, other radioactive elements with longer half-lives are used (for example, uranium-235 decays to become lead-207 with a half-life of around 700 million years while uranium-238 decays to become lead-206 with a half-life of 4.5 billion years).

The authors of the controversial study realized that since the carbon trapped inside fossil fuels was produced by photosynthesis that happened millions of years ago and that, because these fossil fuels have been buried underground for many millions of years, these fossil fuels should have absolutely no carbon-14 left within them — all the carbon-14 reverted back to nitrogen-14 long ago. *So, when you burn fossil fuels, this produces CO_2 gas made exclusively of carbon-12 atoms.* And as these emissions are added to the atmospheric mix, they dilute the ratio of carbon-14 to carbon-12 in the atmosphere. Thus, by measuring the current ratio of carbon-14 to carbon-12 in our atmosphere and by comparing that ratio to what it was before the beginning of the Industrial Era (i.e. around the year 1750), the authors of the controversial study were able calculate how much of the CO_2 in our atmosphere comes from burning fossil fuels. [There are a few extra complications that the authors had to overcome while performing their calculations, such as accounting for the speed at which CO_2 cycles back and forth between plants, soils, the atmosphere, and the oceans, and accounting for extra carbon-14 molecules created by the burst of radiation released during nuclear test blasts during the mid-20th century, but I think I've provided enough details to explain the essence of their study].

Their results were astounding: *"Our results show that the percentage of the total CO_2 due to the use of fossil fuels from 1750 to 2018 increased from 0% in 1750 to 12% in 2018, much too low to be the cause of global warming."*[4]

So, despite the fact that atmospheric CO_2 levels soared from around 280 ppm in 1750 to over 420 ppm today, the study found that only 12% of that increase comes from fossil fuel emissions rather than the 78% assumed by the consensus narrative! Oops. Nor was this study the only one to find that fossil fuel emissions are only responsible for a very small proportion of rising CO_2 levels — several additional heretical studies using different methodologies have come to very similar conclusions. Harde and Salby (2021)[5] found that fossil fuel emissions are only responsible for a maximum of 5% of the CO_2 increase since the beginning of the Industrial Era. Berry (2023)[6] found that fossil fuel emissions can

---

[4] Skrable, K., Chabot, G., French, C. (2022). World Atmospheric CO2, Its 14C Specific Activity, Non-fossil Component, Anthropogenic Fossil Component, and Emissions (1750–2018). *Health Physics.* 122(2). 291-305. https://web.archive.org/web/20220202154330/https:/journals.lww.com/health-physics/Fulltext/2022/02000/World_Atmospheric_CO2,_Its_14C_Specific_Activity,.2.aspx

[5] Harde, H. and Salby, M. L. (2021). What Controls the Atmospheric CO_2 Level? *Science of Climate Change.* https://doi.org/10.53234/scc202106/22

[6] Berry, E. X. (2023). Nature Controls the CO_2 Increase. *Science of Climate Change* 3(1) 68-91. https://doi.org/10.53234/scc202301/21

only account for a maximum of 14% of the increase in atmospheric CO_2 since the beginning of the Industrial Era (18 ppm of the total 133 ppm rise). And Harde (2023)[7] calculated a maximum of 15%.

Predictably, the criticisms launched against these controversial consensus-challenging studies have been ferocious. By challenging the consensus climate change narrative, these researchers kicked a veritable hornet's nest. However, the authors and their supporters continue to stand by their results while their critics continue to point at the "consensus". Who's right?

Instead of arguing about whether the formulas used by the authors of these controversial studies are correct, let's turn our attention to some of the other things happening on our planet that correlate with rising atmospheric CO_2.

Sometimes the best way to test a scientific result is to see how those unexpected results fit into the big picture to see if they make any sense in relation to what we observe happening around the world — if they make sense, we gain important new scientific insights; and if they do not, then we can conclude that there is a flaw in the calculations even without necessarily having to hunt down that flaw in the mathematical formulas. As famous soil scientist Dr. William Albrecht once said, *"read books and observe nature. If the two disagree, throw away the book."* For example, you don't need to be an expert in calculating hockey stick graphs in order to see that the shape of the hockey stick (and the claims about CO_2 controlling the climate) are utterly incompatible with archaeological data (such as what the northernmost limit of vineyards in medieval Europe tells us about the history of our climate), or that the hockey stick is clearly contradicted by glacial advances and retreats since the Little Ice Age, or that it is fundamentally at odds with the story told by ice core data spanning the 10,000-year interval since the last ice age, or that it defies the logic of finding prehistoric tree stumps in what is now permanently frozen arctic permafrost, and that it is refuted by the lack of correlation between historic CO_2 data and the raw unaltered temperature data. And so, as our story evolves over the coming chapters, we will likewise be putting the controversial results found by *Skrable et al* through a similar

---

[7] Harde, H. (2023). Understanding Increasing Atmospheric CO_2. *Science of Climate Change*. 3(1) 46-67. https://doi.org/10.53234/scc202301/23

gauntlet to see how they stack up with what we observe happening in the real world.

To date, these is no evidence of any natural atmospheric changes (like increased volcanic eruptions) that can explain why CO_2 levels during our current warm interglacial cycle should behave so differently than during previous ice age cycles. Besides, the timing of the rise strongly suggests that it has our fingerprints all over it. Human "progress" would seem to be responsible, but if fossil fuels aren't the primary culprit, then what is?

If non-fossil fuel sources are indeed responsible for the vast majority of the increase in atmospheric CO_2, this demands a fundamental rethink of our understanding of our atmosphere and of the processes affecting the gas mix in our air. If the authors of these studies are right, then we need to have a much closer look at all the other things we are doing (besides burning fossil fuels) that could disrupt the balance of CO_2 entering *and leaving* the atmosphere. Although I don't want to spoil the story, I will foreshadow by pointing out that not only do soil erosion and soil degassing add vast quantities of CO_2 to the atmosphere that were previously stored as carbon in the soil — the numbers will surprise you — but reducing the total biomass of the planet (i.e. through deforestation and desertification) also has the power to radically alter the chemistry of our atmosphere by changing the balance of how much CO_2 is *removed* from the atmosphere as CO_2 is consumed by photosynthesis. The total biomass of our planet matters in so many surprising ways — as this story unfolds, brace for the unexpected!

Even though rising CO_2 has no meaningful effect on global temperatures and even though plants undoubtedly benefit from more CO_2 in the air, it is nonetheless vitally important to understand what is causing the atmospheric mix to go through such a fundamentally transformative change. If it's not caused by fossil fuels but it is instead caused by something else that we are doing to our planet (and not merely the consequence of gas being burped out by a volcano or smoke pouring out of a car exhaust), then maybe that rising CO_2 isn't so benign after all — not because that carbon dioxide itself is dangerous to the climate but because it is alerting us that we are damaging our fragile ecosystems in ways that are so destructive that consequences of that destruction are quite literally the altering the chemistry of our atmosphere! Perhaps rising CO_2 is merely the warning signal alerting us to something much more important happening right beneath our feet.

We would do well to remember the painful lessons of the 1930s when the subtle, unrecognized, and unintended consequences of plowing up the sod that once protected the Great Plains only became visible when the entire ecosystem suddenly collapsed during the Dust Bowl as the fragile prairie was left utterly incapable of dealing with the dry part of the climate cycle. It's also worth remembering that our biomass not only gives our planet the thin protective green skin that protects our soils from erosion, flood, and drought, but that reducing the planet's biomass also has significant "downstream" consequences because it fundamentally changes local weather patterns (through evapotranspiration, through soil moisture absorption, through cloud and rainfall behaviour, and so on). In short, by reducing biomass, you can manufacture severe <u>local</u> droughts *in spite of an otherwise stable and benign global climate*. Thus, the answer to the question of what is causing CO_2 to rise is fundamentally important to the future of civilization — as you will soon see, it truly is the story of the century.

Slowly but surely, the stories I have told throughout the first half of this book are coming full circle: the story of the plow that broke the Great Plains, the erosion and nutrient loss of our deteriorating forest soils, the wildfire cycle in the American West, the disappearing agricultural soils on our farms, the story of Mao's ill-fated obsession with deep plowing and cutting down trees to fuel China's industrialization, the impact of pollution and other industrial processes on our ecosystems, the rapid ecological collapse in Zimbabwe following Mugabe's embrace of socialism, the Malthusian Trap that is created when population growth and environmental degradation destroy a country's ability to feed itself (and the effort to counteract those Malthusian forces through colonial expansion and emigration), the environmental movement's crusade against cows and fertilizers, and the ecological consequences of stripping individuals and local decision makers of their local autonomy as the global post-WWII world order has transferred more and more decision-making into the hands of distant central planners, institutional bureaucracies, and intergovernmental organizations. This is the point in our detective story where things get truly interesting as, layer by layer, the pieces of this massive all-encompassing puzzle slowly begin to fall into place. Everything is connected.

Chapter 13

The Blind Men and the Geological Elephant

"I look at the natural geological record as a history of the world imperfectly kept, and written in a changing dialect; of this history we possess the last volume alone, relating only to two or three countries. Of this volume, only here and there a short chapter has been preserved; and of each page, only here and there a few lines."

— Charles Darwin, On the Origin of Species, 1859

As academia splits into ever more specialized fields of expertise, each person's narrow focus becomes the lens through which they view the world even as the bigger picture that ties everything together has gradually been lost. We have a lot of knowledge about a lot of things, but we have lost sight of how all those things are interconnected. In a recent interview,[1] Richard Teague, retired professor of grazing ecosystems ecology at Texas A&M University, described the phenomenon as such: *"If you do a small plot experiment on a couple of meters by a couple of meters, that'll tell you something about that [plot] that's going to be accurate. But what does that mean to the whole farming unit? That's what's missing in the world today. We're producing papers and papers and papers about smaller and smaller things."*

In essence, science is suffering from the same blindness as what is described in the well-known parable of The Blind Men and the Elephant. 19th century poet John Godfrey Saxe converted the ancient parable into a poem, which goes something like this:

> *It was six men of Indostan*
> *To learning much inclined,*
> *Who went to see the Elephant*
> *(Though all of them were blind),*

---

[1] Carbon Cowboys. (2023, Sept 11). *Richard Teague (PhD), retired professor of Grazing Ecosystems Ecology at Texas A&M University speaking about science.* [Video & quote in description]. TikTok. https://www.tiktok.com/@carbon_cowboys/video/7277664345805393194

That each by observation
Might satisfy his mind.

The First approached the Elephant,
And happening to fall
Against his broad and sturdy side,
At once began to bawl:
"God bless me! but the Elephant
Is very like a wall!"

The Second, feeling of the tusk,
Cried, "Ho! What have we here?
So very round and smooth and sharp?
To me 'tis mighty clear
This wonder of an Elephant
Is very like a spear!"

The Third approached the animal,
And happening to take
The squirming trunk within his hands,
Thus boldly up and spake:
"I see," quoth he, "the Elephant
Is very like a snake!"

The Fourth reached out an eager hand,
And felt about the knee.
"What most this wondrous beast is like
Is mighty plain," quoth he;
"'Tis clear enough the Elephant
Is very like a tree!"

The Fifth who chanced to touch the ear,
Said: "E'en the blindest man
Can tell what this resembles most;
Deny the fact who can,
This marvel of an Elephant
Is very like a fan!"

The Sixth no sooner had begun
About the beast to grope,
Then, seizing on the swinging tail
That fell within his scope,
"I see," quoth he, "the Elephant
Is very like a rope!"

And so, these men of Indostan
Disputed loud and long,
Each in his own opinion
Exceeding stiff and strong,
Though each was partly in the right
And all were in the wrong!

So oft in theologic wars
The disputants, I ween,
Rail on in utter ignorance
Of what each other mean,
And prate about an Elephant
Not one of them has seen!

The Blind Men and the Elephant
1907 American illustration

Nature is not a laboratory where things happen in carefully controlled and isolated conditions. Rather, it is a messy interplay between many overlapping pieces where the whole that emerges is far greater than the sum of its parts. We have lost sight of the elephant, and the noisiest blind man controls the beliefs that rule our world.

In a detective story, we're always looking for a smoking gun to link a perpetrator to a crime. In the neat, controlled conditions of a laboratory, finding a smoking gun seems deceptively easy as scientists reduce each question to a simple set of variables to find explanations for how the world works. But as the weather modification programs of the mid-20th century so clearly demonstrated, all that breaks down once you step out of the controlled conditions of the laboratory into the messy real world where a dizzyingly complex array of dynamic overlapping forces all come together without respecting the neat and tidy boundaries of our academic fields. Sometimes those forces amplify each another, sometimes they cancel each other out, frequently they overlap, and the balance between them is constantly changing because they all work on entirely different time scales.

The narrow focus and the politicization of climate science has essentially recreated a carbon-copy of what led mid-20th century weather modification programs to failure as high hopes were dashed by Mother Nature's complexity. Greenhouse gases seemed like such a tidy explanation for how our climate works — but only if you ignore everything else. And, when climate scientists began measuring the composition of our atmosphere, fossil fuel emissions emerged as a simple and tidy explanation for rising atmospheric CO_2 levels — take one planet-sized atmospheric balloon, pump it full of CO_2 emissions, and watch atmospheric CO_2 levels climb, just as the chart from NASA at the beginning of the previous chapter seems to show.

But geologists, chemists, meteorologists, astronomers and physicists have a much more complicated story to tell about the complex biological, chemical, and geological processes that turned our once uninhabitable world into a hospitable planet — and kept it habitable. That story not only confirms that there is no conceivable way that CO_2 is acting as a control knob on the climate, but it also completely overturns the idea that adding large volumes of CO_2 to the atmosphere will have any kind of sustained impact on atmospheric CO_2 levels. Photosynthesis and other biological and geological processes would quickly remove the extra CO_2 from the air, as they have throughout Earth's long geological history. Rising CO_2 levels in our modern era are the consequence of a far more complicated and dynamic tug-of-war between competing counter-acting forces that are continually adding and subtracting CO_2 from the atmospheric balloon.

In recent years, activist-scientists, politicians, and the mainstream media (and Wikipedia) have engaged in a shameless campaign to rewrite geological history in the public imagination in order to defend the "consensus climate narrative". In this distorted version of our planet's history — written as a kind of simplistic, almost cartoonish version of the natural geological history of our planet — CO_2 is portrayed as the central character allegedly driving our planet's ever-changing climate. High CO_2 levels are blamed for the hot climate of the early Earth, low CO_2 levels are blamed for the multi-million-year cryogenic freeze that turned our planet into a giant snowball 720 million years ago, rising CO_2 levels from volcanic eruptions are credited for the warm climate that enabled dinosaurs to roam the Earth for 165 million years, and falling CO_2 levels caused by the weathering of rocks are blamed as the trigger for the glaciations that started in Antarctica 34 million years ago and expanded to the Northern Hemisphere 2.58 million years ago. Every twist and turn in temperature is somehow blamed on CO_2.

This revisionist version of geologic history, legitimized by the form-fitted algorithms of climate models that interpret everything through the narrow lens of CO_2, is blinding everyone to a much more complex (and interesting) story. Climate models say one thing. But the twisted layers of rock beneath our feet read like the pages of a book, and they have an altogether different story to tell.

Geologists have forever been a thorn in the side of activist climate scientists who try to pull the wool over the world's eyes with half-truths and lies. As I set the record straight by retelling the strange and fascinating history of the geological, chemical, and biological forces at work during our planet's long evolution, everything about our own era will become clear — from the forces controlling our current climate, to the profound changes happening in our soils, to the complex reasons behind the rise in atmospheric CO_2 that began 7,000 years ago and began to accelerate in the mid-1800s.

When the history of civilization is portrayed as a series of sequential events, this leaves us scrambling to find the links to tie these events together and frequently seduces us into fabricating links that simply aren't there. Historians pour though the biographies of historical figures in search of clues about why history played out the way it did, thinking that their unique personalities somehow shaped the course of history. Yet these historical figures are mere bit players, unique only to some very limited degree, but ultimately they are merely acting according to a script created by the hard realities and societal beliefs of each era. If you want to understand history, you largely have to ignore the players and focus instead on the forces that shape the evolution of ideas. At best, a biography might help explain why certain leaders reflected the hopes, fears, and dreams of their nations better than others and thus rose to prominence because of their ability to channel the ideas boiling beneath the surface of society. But if you want to understand the course of nations, you have to study geography, the political map of the era, the advantages and vulnerabilities of nations and their neighbors, the religious and philosophical beliefs underlying their cultures, and the hopes and fears of a nation's people. The bit players are interchangeable, the ideas that shape a culture are not.

The same thing is true when studying the natural history of our planet, of which our current climate is merely an extension. The history of our planet is the evolution of various *processes* — the events (like ice ages, extinctions, meteorite impacts, climate changes, volcanic eruptions, evolutionary adaptations, etc.) are all merely the inevitable consequence of those processes playing out over time. Focusing on the events instead of focusing on the processes seduces people with false correlations, leading to absurd beliefs like "the rain follows the plow" or that "carbon dioxide is the control knob on the climate." Once you understand the processes and how they play out over time within the context of the big picture, these sorts of glib explanations lose their power. Once you see the whole elephant, all the pieces naturally fall into place.

The best place to start this story is at our planet's earliest beginnings 4.5 billion years ago (the geological time scales involved in our planet's history are truly mind bending!). During the entire first half of its history, our planet was a wholly alien world dominated by chemical rather than biological forces. It was a fragile world, hostile to Life, and extremely vulnerable to subtle changes in the atmosphere, the oceans, and the soils.

Two things become apparent as this story unfolds. On the one hand, if any number of small things had been different about our young Earth, Life would have been extinguished long ago — we are truly fortunate to be here. As we follow the remarkable history told by the rocks and fossils beneath our feet and as we see the endless chain of catastrophic extinctions that have stalked Life at every turn, it's easy to understand why people like Thomas Malthus and his fellow doomsday prophets came to view our planet (and our climate) as a fragile, finite world with finite resources, inhabited by a thin biological crust that is forever doomed to cling to a brief and precarious existence, and forever at the mercy of tiny changes that spiral into existential planet-altering events.

And yet, a closer look at the details of this story turns this pessimistic view on its head. Far from being a fragile passenger being blown about by the winds of time, Life evolved from the primordial goop to gradually transform itself from a fragile and largely inconsequential sidenote on an otherwise barren planet into the dominant force shaping virtually every aspect of our planetary ecosystem. Life's relentless ability to turn limiting factors into new opportunities and exploit new ecological niches gradually converted our once fragile young planet into the complex and dynamic world we live in today. Each limiting factor was transformed into a new source of abundance as Life found a way to turn problems and setbacks into new opportunities. Our hospitable climate, our breathable air, the mix of gases in our atmosphere, our fertile soils, the fertilizer elements that feed our plants, our drinkable water, and even the color of our skies and oceans — we owe them all to biological processes that began billions of years ago. Life itself is the force responsible for turning a hostile young planet into a thriving biosphere. Understanding those processes gives us the keys to understanding (and mastering) our world.

As late as 2.7 billion years ago — after almost half of our planet's history had already passed —our planet still had none of the core things that make our planet recognizable and habitable today. The sun was 30% weaker. The sky was virtually cloudless. The oceans were a balmy 40 °C. And there was little to no land peaking above the surface of the oceans — it was a hot and extremely humid, watery world. In those early days, our planet was also spinning faster —in the earliest days our planet had a 17-hour day, which gradually lengthened because, bit by bit, the gravitational drag from our moon worked to slow our planet's rotation. In those early days the moon was also 16x closer to our planet, which created absolutely huge tides in our oceans (tides of 12 to 25 meters in height versus the modest average of 0.6 to 1 meter today). Imagine the size of the tidal zone between the land and the sea in those early days — it is not an accident that much of Life's early history played out in tidepools!

Most important to our story was the air on our young planet — it was completely unbreathable. As I mentioned in a previous chapter, Earth's early atmosphere contained toxic levels of CO_2 (at least 750x higher than CO_2 levels today) thanks to all the CO_2 continually being released by volcanoes as our planet's still-molten crust constantly allowed molten gas-filled lava to leak out onto the surface. And there was absolutely no oxygen whatsoever in Earth's early atmosphere (compared to 21% today). Without oxygen in our atmosphere, there also was no oxygen in our oceans.

Oxygen is an extremely reactive gas. Without oxygen in the air and without oxygen dissolved in seawater, iron did not rust. Thanks to this absence of oxygen, during the first half of Earth's history the oceans were completely saturated with dissolved (soluble) forms of iron, which turned the ocean red. Likewise, in this world without oxygen, carbon also did not oxidize to become carbon dioxide. There was no carbon cycle. Any carbon-rich tissues produced by Earth's earliest microscopic life forms stayed as carbon even after they died —carbon did not return to the atmosphere because there was no oxygen to react with it to convert (oxidize) it back into CO_2. In those days, if carbon-rich material washed up on a beach or collected in a tidepool and sat exposed to the atmosphere, it simply sat there, unaltered. By contrast, today when carbon-based materials (i.e. plant or animal tissues) are exposed to the air, oxygen slowly oxidizes the carbon, which converts it back into CO_2, thus enabling it to degas back into the atmosphere. That's why wooden fences, dry grass, and dead trees gradually get lighter over time even if they aren't eaten by bugs and bacteria — oxygen is very slowly converting them back into CO_2. Likewise, today whenever bare soil is exposed to the oxygen-rich atmosphere, *oxygen will react with carbon-based organic matter in the soil (i.e. the humus layer) to turn it back into CO_2* — as you will soon see, the plow that broke the Great Plains did far more harm than just release moisture from the soil and expose the soil to wind, it also oxidized colossal volumes of humus turn it back into CO_2. But I'm getting ahead of myself — we have a few billion years of Earth's history to cover before we return to the plowman's folly that defines our own brief chapter of history.

The early Earth's oxygen-free atmosphere was a *chemical* concoction created by volcanoes, hydrothermal vents, and other geological and chemical forces at work in our planet's early history. By contrast, today our oxygen-rich atmosphere is a *biological* construction created by the balance between photosynthesis and oxygen-breathing organisms, which continually recycle both carbon dioxide and oxygen in and out of the atmosphere. How we got from there to here is a vital part of our detective story.

Based on microscopic fossils and chemical traces in ancient sedimentary rocks, geologists think the first life forms emerged 3.7 billion years ago as tiny single-celled organisms clinging to the edges of hydrothermal vents where underwater volcanoes released hot gas-filled water into the ocean. These hydrothermal fluids from deep below the Earth's surface were hyper-saturated with carbon dioxide, sulfides, and sulfur gases. The oldest geological evidence of Life on Earth is from opportunistic microbes that took advantage of these hydrothermal fluids by chemically converting carbon dioxide, iron sulfides, and hydrogen sulfide (that's the toxic volcanic gas that smells like rotten eggs) into sugars, with elemental sulfur and water as byproducts. The energy driving these opportunistic chemical reactions did not come from the sun but rather from the heat released by these chemical reactions. This process of making food using chemical reactions as the energy source is called chemosynthesis, in contrast to photosynthesis that uses sunlight as the energy source to turn CO_2 into food ("photo" means "light" in Greek). And, unlike photosynthesis, the primary byproduct of these chemical reactions was sulfur, not oxygen. This truly was an alien world.

But never-ending genetic mutations ensure that Life never stays unchanged for long. Not long after these initial microbes learned to opportunistically harness the chemical energy released by hydrothermal vents, new mutations allowed some microbes to survive away from the vents as they evolved to take advantage of other opportunities out in the wider ocean. Three of these genetic mutations in particular deserve our attention.

The first of these three mutations was the emergence of a group of microbes known as methanogens (a.k.a. methane producers) — they evolved a genetic mutation to extract energy from a chemical reaction between hydrogen gas and carbon dioxide, which produces methane gas and water as byproducts (hence the name methanogen = "methane producer"). These methanogens still thrive today in oxygen-free environments where they extract energy from complex carbon molecules (like decomposing plant material). For example, they are found in the decomposing sludge that accumulates in waterlogged swamps, at the bottom of lakes, and at the bottom of the ocean. And when rampant soil erosion washes vast quantities of carbon-rich mud into our lakes and oceans, this creates a bonanza of carbon-rich food for methanogens to feed on (this is one of the many keys to solving the puzzle of sharply rising CO_2 levels in the modern era — more on that story later).

Methanogens are also found inside the stomachs of all ruminant animals where they play an essential role in the fermentation process by breaking down tough grass and plant fibers into simple sugars and simple carbohydrates. The methane gas produced by burping cows as a byproduct of this fermentation process is created by methanogens living inside their stomachs.

This fermentation process inside a cow's stomach, courtesy of these methanogens, is the reason why soils in grasslands and semi-arid deserts are able to build up humus in the soil. The soil is too dry in these arid climates to sustain large populations of soil bacteria capable of breaking down grass fibers — without this fermentation process in the ruminant gut to pre-digest these tough fibers, dead grass residues would simply dry out and get oxidized by the air, and the carbon from this dead plant debris would instead slowly degas back into the atmosphere as carbon dioxide. Consequently, whether a cow eats the grass or whether the gras remains uneaten and instead continues to forever blow in the wind, either way, all the carbon will eventually be converted back into CO_2 — the cow just does it faster than oxygen in the air. Uneaten grass in a dry low-rainfall climate slowly turns grey, becomes increasingly brittle, and within 50 years it all disappears as it reacts with oxygen, which turns it back into CO_2. Like a fire (fire is fast oxidation), this slow process of oxidation through exposure to the air is continually converting carbon into carbon dioxide, just without either the dramatic rapid combustion of a fire or the peaceful chomp-chomping of cows and other ruminants chewing their cuds.

However, in the moist environment found inside a ruminant stomach, methanogens ferment tough grass fibers to slowly break them down so that by the time these tough grass fibers re-emerge at the back end of a cow (or some other grazing animal), these fibers are already semi-digested so that feeble soil microbes in the dry soil can nevertheless easily break down the manure to convert that carbon into humus. By this process, not all the carbon returns to the atmosphere as some is instead transformed into fertile life-giving soil thanks to the collaborative work of cud-chewing ruminants, the methanogens living inside their rumens, and microorganisms in the soil.

In dry climates, tough grass fibers would be too difficult for soil microbes to break down without the help of grazing animals (and without the methanogens living in their stomachs) so much of the carbon from these plant residues would be oxidized and degas back into the atmosphere before soil microbes could turn them into humus. Without the help of methanogens living in the stomachs of grazing animals, dry climates would have very little humus in their soils. And without that humus, soils in these dry climates would contain far less nutrients and would be able to absorb far less rainwater — grazing animals and the methanogens living in their stomachs are essential for creating and sustaining fertile soils in low-rainfall climates. To put it bluntly, without grazing animals and the methanogens in their rumens, dry climates would soon turn into deserts, just as we are seeing in so many of our dry ecosystems around the world today as the large migrating grazing herds that once kept those ecosystems healthy are replaced by scattered continuously grazing animals or have their animal populations removed altogether.

When environmentalists say that livestock are a leading cause of desertification, it is yet another half truth obscuring a whole lie — in the absence of grazing strategies designed to mimic the *behaviour* of wild grazing herds, livestock are indeed a leading cause of soil degradation and soil erosion. And yet, grazing animals are nevertheless essential for creating humus in dry climates — *their power to create humus depends entirely on whether their grazing behaviour is designed to mimic the behaviour of wild grazing herds*. Livestock are neither good nor bad — it's their grazing behaviour that either turns them into a destructive or a constructive ecological force. You simply cannot reverse desertification in low rainfall climates without reintroducing migratory grazing herds (or rotational grazing in a farm context) — it is high time we ended the crusade against livestock and instead learned the grazing techniques that harness their impact on the land as a powerful beneficial tool for building humus. Although I have introduced this concept of grazing behaviour here, we will explore this topic in much greater detail in later chapters as it is integral to understanding the "ecological story of the century" that is playing out in the soils beneath our feet.

~

It might be worthwhile to take a moment to define what humus actually is — we talk a lot about it, and our civilization and our ecosystems are completely dependent upon it, yet this mysterious substance is one of the most poorly understood and most overlooked building blocks of Life on Earth. In essence, humus is the stable organic component of soil; all humus is organic material but not all organic material is humus — compost and other organic debris that is still rotting/decomposing has not yet become humus because it has not yet been broken down into its most elemental stable form. In its most basic form, think of humus as a tiny clay particle, except made of carbon. In order to create humus, you have to break apart the long carbon chains in plant and animal tissues and then you have to bury that carbon in the soil to prevent atmospheric oxygen from reacting with it to turn it back into CO_2. Thanks to the reactive properties of oxygen as well as the ease with which humus is eroded away by water and wind, humus is one of the most delicate and vulnerable substances on our planet's surface, yet it is also the key ingredient of soil fertility and soil moisture retention.

Without humus, soil loses much of its ability to store soil nutrients, thus leaving plants increasingly dependent on synthetic man-made fertilizers (even more than clay, humus has a phenomenal nutrient storage capability that we will be exploring in greater detail in later chapters as we see the impact of soil degradation on human health). And without humus, soil loses much of its capacity to absorb and retain soil moisture (the elemental carbon particles in humus have a phenomenal moisture-absorbing capability), which leaves the soil increasingly vulnerable to drought, increasingly dependent on irrigation, and increasingly incapable of capturing rainwater, which then runs downslope to create floods whenever it rains even as the soil quickly returns to drought conditions not long after the rain ends. Humus is what gives soil its ability to sustain abundant Life in spite of a continually changing cyclical climate.

The deep humus-rich soils of the dry North American prairie, the Ukrainian and Russian steppe, and the East African plains all owe their existence to methanogens working inside the stomachs of ruminant grazing animals. It's worth noting that soils in prairie ecosystems around the world contain much more carbon than any other ecosystem on the planet today (even more than forests!)[2] — these vast carbon sinks all owe their deep carbon reserves to the symbiotic relationship between grass, grazing animals, and the methanogens living inside ruminant digestive systems. Even forests do not store as much carbon as grasslands because in a forest most of the carbon is tied up in the fibers of the trees. But in a grassland that is repeated grazed, the carbon slowly builds up in the soil, layer upon layer, thanks to the positive impact that repeated grazing has on creating a healthy grass sod, thanks to the manure left in the herd's wake, and thanks to the cyclical die-back of roots after grazing, thus creating vast carbon reserves and extraordinarily deep, rich, fertile soils.

But when those fertile grasslands begin to degrade because they are plowed, because of the mismanagement of grazing animals, or because the animals that once kept those grasslands healthy have been removed altogether, without the protective skin of a healthy sod the humus soon begins to degas away as it is exposed to the atmosphere even as still more of the humus is eroded away by the rains to end up at the muddy bottom of a swamp, lake, or ocean, where methanogens break down that carbon-rich mud into methane. That methane is subsequently converted back into CO_2 as soon as it comes in contact with atmospheric oxygen (methane is not a stable gas in our atmosphere — oxygen reacts with and destroys methane by converting it into CO_2 and water — more on that in a moment). Everything is connected.

And so, there is a bitter irony to the war that has been declared by climate activists and central planners like John Kerry against livestock (and against burping cows in particular) — ruminant livestock are by far the most efficient and cost-effective tool for rebuilding humus and for storing carbon and fertilizer nutrients in the soil, especially in dry climates. Yet, instead of being celebrated for what comes out from under their tails, for the positive impact they have on the land when their grazing impact mimics the behaviour of wild grazing herds, and for the fermentation that happens inside their stomachs, all of which help turn sterile dirt into fertile humus-rich soil, livestock are being demonized for their burps. We live in a world run by scientifically illiterate clowns that can't see beyond the narrow reductionist thinking of the laboratory.

---

[2] "Carbon Sequestration in Grasslands." *Minnesota Board of Water and Soil Resources (BWSR)*. Accessed 3 May 2024. https://bwsr.state.mn.us/carbon-sequestration-grasslands

Back in the early Earth, as methanogen populations expanded across the oxygen-free oceans, they began producing vast quantities of methane. Since there was no oxygen yet in the atmosphere to react with this methane to oxidize it back into carbon dioxide, methane began to build up in the atmosphere, adding to the methane released into the atmosphere by volcanoes. And so, for the first time in Earth's history, a biological process began altering the mix of gases in the atmosphere.

The second important branch to emerge in the microbial family tree emerged around 3.4 billion years ago when another series of genetic mutations produced the first microbes capable of photosynthesis. Instead of relying on chemical energy, these microbes tapped into sunlight to fuel their chemical reactions ("photo" means "light" in Latin). Since sunlight was not able to penetrate very far below the surface of the oceans, these photosynthetic microbes colonized vast tide pools where they had access to both nutrient-rich ocean waters and ample supplies of sunlight, even as the water provided them with a measure of protection against harmful UV radiation.

Although this new genetic adaptation allowed microbes to venture away from hydrothermal vents into the wider ocean, it is not a form of photosynthesis that we would recognize today. When we think of photosynthesis, we think of *aerobic* photosynthesis (aerobic means involving oxygen), which uses the energy from sunlight to convert carbon dioxide and water into sugar, with oxygen as a byproduct. The initial form of photosynthesis that emerged 3.4 billion years ago, called *anaerobic* or *anoxygenic* photosynthesis (anoxygenic means without oxygen), worked a lot more like the chemistry of the microbes living alongside hydrothermal vents, which converted carbon dioxide *and sulfides* (i.e., iron-sulfides) into sugar, water, and elemental sulfur. In time some of these strange microbes capable of anoxygenic photosynthesis would acquire further genetic mutations to go from anoxygenic to aerobic photosynthetic. Some of their descendants will ultimately evolve into the ancestors of all photosynthetic organisms, from simple single-celled phytoplankton to complex plants and everything in between. But 3.4 billion years ago, the genetic mutations for that next evolutionary step were yet to come.

In our modern world, we recognize the oceans by their bluish-green color. That color is a consequence of our oxygen-rich oceans — the blue-green colors are produced by photosynthetic microorganisms living in our ocean waters that turn carbon dioxide into chlorophyl. But the oxygen-free young Earth didn't look anything like that. Before the emergence of Life, the iron-saturated ocean waters would initially have been red but then, as these first simple microbes capable of anoxygenic photosynthesis emerged to feed on iron sulfides, these primitive, purple-colored bacteria slowly turned the oceans purple.

The third group of microbes to evolve from those earliest beginnings emerged around 3.1 billion years ago. This time a different genetic mutation among those initial microbes feeding on hydrothermal fluids gave rise to a family of organisms adapted to using the highly reactive nature of oxygen to extract energy from food — these microbes are the ancestors to all oxygen-breathing organisms that use oxygen to metabolize (oxidize) food and then exhale carbon dioxide as a byproduct. In time, the complementary relationship between photosynthesis (which breaks CO_2 apart into carbon and oxygen) and oxygen-breathers (biologists refer to all oxygen-breathing organisms as "respirators" because they recombine carbon and oxygen to re-create CO_2) will become the carbon cycle that is constantly cycling carbon dioxide in and out of the atmosphere.

These oxygen-breathing organisms extract energy from carbon-based tissues by using the reactive properties of oxygen to break apart complex carbon molecules via oxidation, releasing energy (and CO_2) as the byproduct. Our metabolisms quite literally run on the corrosive power of oxygen. Unfortunately, there is a price to be paid for this highly efficient metabolic strategy — the very same oxygen that makes it easy to extract energy from food also slowly breaks down the cells in our bodies. The stress that oxygen puts on our cells makes aging and death inevitable. Nevertheless, it was an ingenious evolutionary adaptation that took advantage of oxygen's highly reactive chemical properties to make it possible to evolve much more active high-energy lifeforms — it set the stage for the evolution of animals. The only problem was that it was an adaptation ahead of its time — in those early days, oxygen was extremely scarce. Volcanic gases sometimes contained a bit, but that oxygen didn't hang around for very long before it was scrubbed from ocean waters by chemically reacting with iron in the water or by reacting with methane in the atmosphere. The oxygen-breathers' time would come, but the first 2 billion years of their existence were a precarious time as they were forced to scrape by in a world almost completely devoid of oxygen.

For now, we're going to set aside the oxygen-breathers to have a closer look at the sky (and the climate) above the young Earth's early oceans. The early Earth's atmosphere was packed full of methane released by volcanic eruptions. Then, as methanogens evolved, they too contributed vast quantities of methane to the atmosphere. Without oxygen in the atmosphere, methane didn't oxidize (just like iron didn't rust and carbon didn't oxidize), so the methane in the early atmosphere just stayed there, building to levels that were 1,000x higher than they are today, which enveloped the Earth in a thick global haze of methane. Instead of the blue sky we see today, this methane would have colored the early Earth's sky a smoggy orange,[3] much like the sky recorded by the European Space Agency's Huygens probe when it landed on Saturn's methane-rich moon Titan in 2005 and discovered Titan's methane-saturated atmosphere.

Earth's early methane-rich atmosphere also provides a *partial* answer to another nagging question — if the sun was 30% weaker during Earth's early history, why didn't the planet's oceans freeze into an ice ball? On the contrary, the planet was considerably warmer than it is today, with ocean temperatures hovering around 40 °C! Unsurprisingly, activist-scientists have a ready solution to the paradox: according to them, the Earth allegedly didn't freeze thanks to the greenhouse effect produced by extremely high atmospheric carbon dioxide levels — as high as 70% CO_2 according to "climate models".[4] I've already told you about carbon dioxide's diminishing returns as a greenhouse gas as CO_2 concentrations rise above 300 ppm. Further disproving these outlandish climate models is geological evidence in rocks from that time period — chemical weathering on rocks from 2.2 billion years ago provide hard evidence that CO_2 levels had already fallen from around 30%[5] in the Earth's early

---

[3] "Why is the sky blue?" *National Geographic Kids*.
https://kids.nationalgeographic.com/books/article/sky

[4] Carlson, E. K. (2020, Jan 22). Ancient Earth's Atmosphere May Have Been Over 70 Percent Carbon Dioxide. *Discover Magazine*.
https://www.discovermagazine.com/the-sciences/ancient-earths-atmosphere-may-have-been-over-70-percent-carbon-dioxide

[5] "Why Earth wasn't one big ball of ice 4 billion years ago when Sun's radiation was weaker." (2010, Apr 1). *Science Daily*.
https://www.sciencedaily.com/releases/2010/03/100331141415.htm

atmosphere to around 4% by 2.2 billion years ago.[6] This geological evidence comes from sedimentary minerals from those earlier eras because different types of carbonate minerals form at different concentrations of atmospheric CO_2 as sediments are exposed to the atmosphere — if CO_2 levels had been as high as the models claim, entirely different minerals would have formed in those sediments. The geological evidence is irrefutable and puts clear upper bounds on the maximum concentration of CO_2 in the atmosphere during those earlier eras, yet it is systematically ignored by the modelers — many may not even be aware of this geological evidence as they work inside their echo chambers. Nor is a speck of CO_2 needed as a mechanism to explain the warm climate once you see all the other processes at work on our young Earth, which handily explain Earth's warm early history despite the cooler sun of that early era.

One of those other factors is the greenhouse effect from methane, which has a much more potent greenhouse effect (28x stronger than CO_2). Today, methane concentrations are so low that they have virtually no impact on planetary temperatures. But the young Earth didn't have any oxygen to break down (oxidize) methane so our early atmosphere was a completely different beast — it was absolutely brimming with methane released by volcanoes and by methanogens living in the oceans. That brown smoggy haze didn't just paint the sky orange, it was also trapped some heat. However, even those high concentrations of methane gas were merely playing a minor contributing role in heating the Earth's early atmosphere — methane was merely one among many forces and far from the most important one.

As I mentioned earlier, the early Earth was a watery world without large land masses. Compared to 'land', water has a 4x higher capacity to act as a heat sink by absorbing and storing solar energy (in this context, "land" is defined as the combination of rocks, soil, and vegetation).[7] So,

---

[6] Rye R., Kuo P.H., Holland H.D. (1995) Atmospheric carbon dioxide concentrations before 2.2 billion years ago. Nature. 378(6557):603-5. https://doi.org/10.1038/378603a0

[7] Kenkeit-Meezan, K.A., Foothill College (n.d.). 4.3: Ocean and Continent Effects. *LibreTexts Geosciences.* https://geo.libretexts.org/Bookshelves/Geography_(Physical)/Physical_Geography_(Lenkeit-Meezan)/04%3A_Global_Circulation/4.03%3A_Ocean_and_Continent_Effects

while the Sun in those days was weaker, the planet could store a lot more incoming solar energy because the entire surface of the planet was covered by oceans.

Without large continental land masses, it was also much more difficult to accumulate snow and ice —it's much harder to freeze salty seawater than to accumulate freshwater snow on land (salt water only freezes below -12 °C and it is always being melted from below, so it is much more difficult to trigger a self-reinforcing chain reaction of ice sheets. Moreover, land is highly reflective, especially land without vegetation, so land acts like a mirror to reflect a lot of solar energy back out to space. But since the young Earth didn't have any continents, the entire watery surface of the young Earth was capable of capturing and absorbing far more of the Sun's heat instead of reflecting it back out into space.

And then we come to clouds — ah, those pesky clouds again. One of the ways that clouds form is that dust particles are picked up by winds on land that then get blown up into the higher levels of the atmosphere where they serve as nuclei to condense water vapor into clouds. But there's not much dust in a world made of water. Thus, a 2010 study from the University of Copenhagen was able to calculate that our young Earth had significantly less cloud cover to reflect sunlight back into space. The lead author of the 2010 study, Professor Minik Rosing, summed up his findings as follows: *"What prevented an ice age back then was not high CO_2 concentration in the atmosphere, but the fact that the cloud layer was much thinner than it is today. In addition to this, Earth's surface was covered by water. This meant that the Sun's rays could warm the oceans unobstructed, which in turn could layer the heat, thereby preventing Earth's watery surface from freezing into ice."*[8] His second point about "layering the heat" is an important one — without large continents to disrupt ocean currents, warm water would have been able to spread and mix easily across the entire Earth's surface, unlike today where warm waters pool near the tropics even as oceans cool (and begin to freeze) near the poles.

And then there's the greenhouse effect from water vapour — water vapour is a much more powerful greenhouse gas than either methane or

---

[8] "Why Earth wasn't one big ball of ice 4 billion years ago when Sun's radiation was weaker." (2010, Apr 1). *Science Daily*.
https://www.sciencedaily.com/releases/2010/03/100331141415.htm

carbon dioxide. With less clouds covering the surface of the young Earth, there would also have been less rain to wash humidity back out of the air. Furthermore, before continents formed on the Earth's crust, the young Earth lacked the mountains ranges that trigger the formation of clouds and rainfall as moisture-laden air masses are pushed up into the cold air found at higher elevations where the humid air will begin to condense into raindrops. Likewise, without large continents to disrupt ocean circulation, the planet's oceans would all have been roughly the same temperature, from equator to poles, so there wouldn't have been large temperature differences to trigger cloud formation and rainfall as moisture-saturated air was blown across ocean waters. Instead, the humidity evaporating out of the oceans would simply have stayed in the air, effectively wrapping the entire planet in a giant supersaturated blanket of warm moist air. All in all, the young Earth must have been an *extremely* warm and humid place with very little cooling at night as water vapour trapped heat from escaping back out into space at night. Oxygen isotope data from ancient sedimentary rocks suggest that surface temperatures on the young Earth ranged between 45 and 85 °C![9] In a world this humid, even the greenhouse warming produced by methane would have been all but irrelevant compared to the greenhouse effect of an entire planet sweltering under a blanket of stifling humidity.

Humidity is the most important insulation blanket that protects our planet from the bitter cold of outer space whenever the sun isn't shinning. Consider that Florida and the Sahara Desert both have the exact same amount of CO_2 in the air, and both lie about the same distance from the equator. But in the dry Sahara, temperatures plummet as soon as the sun goes down (the average nighttime temperature is -4 °C (25 °F) despite average summer daytime temperatures well above +40 °C (104 °F)). By contrast, the thermometer barely budges on those hot humid Florida nights. Now imagine Florida's hot humid climate stretching from pole to pole. Water vapor is the only greenhouse gas that matters to our global climate.

~

---

[9] Kastling, J.F. and Howard, M. T. (2006). Atmospheric composition and climate on the early Earth. *Philos Trans R Soc Lond B Biol Sci.* 361(1474). https://doi.org/10.1098%2Frstb.2006.1902

2.7 billion years ago, this alien world came to an abrupt end. Water is a lot more abundant than hydrogen sulfide so any photosynthetic bacteria that acquired the necessary genetic mutations to replace hydrogen sulfide with water in the chemical reaction that turns carbon dioxide into sugar would gain a seemingly unlimited resource to exploit — and would gain a major competitive advantage over their hydrogen-sulfide-dependent peers. 2.7 billion years ago a new class of photosynthetic bacteria emerged, called cyanobacteria (a type of blue-green algae that contains chlorophyl), which did just that. They used sunlight and water to turn carbon dioxide into sugars and, instead of churning out sulfur as a byproduct, this new class of bacteria churned out vast quantities of oxygen as a byproduct of that chemical reaction. And so, the aerobic photosynthesis that powers our world today was born. These early cyanobacteria quickly colonized the tidal flats where they had access to ample supplies of water, CO_2, and sunlight. Bit by bit, oxygen levels began to rise in the oceans and in the atmosphere. Geologists call this the Great Oxygenation Event. It completely changed our planet. Forever.

In the absence of oxygen, ferrous iron is soluble in water. But as soon as this ferrous iron encounters oxygen, it reacts with the oxygen to form various iron oxides (rust), which are not soluble in water. And so, the oxygen produced by the cyanobacteria began to oxidize the vast quantities of dissolved iron that had accumulated in the oceans over billions of years, turning them into rust. By 2.4 billion years ago, these cyanobacteria were producing so much oxygen that vast quantities of iron-oxides began raining down out of the oceans to settle as thick layers of rust all over the seafloor. Year after year, century after century, millennia after millennia, these iron-rich sediments just kept raining down onto the bottom of the ocean.

Geologists call these purplish layers of iron-rich sediments "banded iron formations." In many places they are hundreds of meters thick. And geologists have found them on every single continent, highlighting that the oxygenation of the oceans that began 2.4 billion years ago was truly a global phenomenon. Today the overwhelming majority of the world's iron and steel is mined from these banded iron formations; they are easily recognizable by their distinct purplish-red sedimentary layers. As iron settled out of the oceans and as cyanobacteria colonized the seas, the reddish-purple oceans gradually changed color to become a recognizable bluish green (cyanobacteria contain the same blue-green photosynthetic pigment that plants use, called chlorophyll).

The oxygenation of the oceans wasn't the only monumental change that took place on our young Earth at that time, which would change our planet forever. Ever since our planet formed 4.5 billion years ago, the Earth's boiling hot crust was gradually cooling. By 2.8 billion years ago, the cooling crust began to thicken and crack into giant tectonic plates, called "oceanic plates", which began to float around on top of the semi-molten layers of rock below. As the crust thickened, it grew strong enough to support the weight of lava from volcanic eruptions without that lava simply sinking back down into the crust under its own weight. And so, the first continental land masses began to form in a process similar to how Hawaii is formed today as plumes of molten rock flooded out onto the ocean floor in giant flood volcanoes. As these lava layers built up, the first large land masses began to rise up out of the oceans.

Furthermore, the oldest and coldest parts of these oceanic plates began to sink back down into the molten rock below (cold lava is more dense and thus heavier than hot lava), setting up the massive convection cycles that continually bring new hot molten rock to the surface at the points where these crustal plates spread apart (like at the Mid-Atlantic Ridge that runs the length of the Atlantic Ocean) even as older, cooler sections of these crustal plates begin to sink beneath the hotter and more buoyant younger plates. Geologists call this process "subduction". The start of this subduction process ensured that hundreds of millions of years of accumulated sediments that had built up on the seafloor would now get scraped off the sinking plates as the colder plates slid beneath the hotter more buoyant plates in these subduction zones, causing these sedimentary rock layers to begin to pile up, not unlike the way froth floats on top of a pot of soup during a rolling boil. As those floating chunks of lighter materials began to collide and glue themselves together, the world's first true continental landmasses began to form.

Because these emerging "continental plates" were too light to sink down into the boiling mass of dense semi-liquid magma below, for the first time in Earth's history permanent landmasses began drifting around the Earth's surface, carried along by the motion of the heavier "oceanic plates" upon which they were floating, setting in motion the conveyer-belt of tectonic plates and the slow drift of continents that define our planet's surface today. And thus, Alfred Wegener's "continental drift" was born.

And so, for the first time in our planet's history, this watery world was interrupted by large barren continents peeking above the waters. For the first time, winds could sweep dust up off the land masses, thus seeding the atmosphere with large quantities of dust particles to serve as nuclei to condense water vapour into clouds. For the first time, Earth had large landmasses to disrupt ocean currents that had previously circulated freely to mix ocean waters from all around the globe. For the first time, the planet contained large landmasses where freshwater snow and ice could accumulate. For the first time, large areas of the planet contained highly reflective rock that reflected solar radiation back into space instead of absorbing all of that heat into the oceans (oceans are a heat sink, rock is a heat reflector). And, for the first time, the Earth developed continental mountain ranges that forced moisture-laden clouds up into cooler higher elevations where that moisture was forced to condense into raindrops.

With clouds being pushed up and over these emerging mountains, an efficient mechanism was set in motion for triggering rainstorms to regularly suck humidity out of the atmosphere on the windward side of these new mountain ranges while creating dry rain shadows on the leeward sides, thereby bringing an end to the era when the entire planet was permanently encased in a thick insulating blanket of water vapor. All the ingredients were falling into place for a much colder climate.

The development of continental landmasses also set another chain of events in motion. As rainfall began weathering rocks up on land, large quantities of minerals (including fertilizer elements like phosphorus) began dissolving out of the rocks and were washed into the oceans to feed cyanobacteria growing in the deltas and tidepools, ultimately contributing to the world's first global algal bloom, with devastating results.

Carbon dioxide dissolves in rainwater to form carbonic acid, which makes all rain slightly acidic. As rain falls on land, this slightly acidic rain gradually dissolves minerals in exposed rocks and washes these minerals into the oceans where they accumulate in tidal flats and deltas along the continents' edges. One of the minerals that began washing off the young Earth's new continents was phosphorus[10] — one of the three vital fertilizer elements that makes all plants grow. And we've all heard what happens

---

[10] Kopp et al. (2005). The Paleoproterozoic snowball Earth: A climate disaster triggered by the evolution of oxygenic photosynthesis. *PNAS*. 102(32) 11131-11136. https://doi.org/10.1073/pnas.0504878102

when you combine algae and phosphorus (phosphorus is a key ingredient in washing machine detergent) — you get massive algal blooms. The flush of fertilizers washing off the continents kicked photosynthesis into overdrive.

As photosynthesis from these algal blooms began raising oxygen in the world's oceans, the high levels of dissolved iron in seawater initially served as a buffer to absorb all that extra oxygen, leading to enormous quantities of the aforementioned rusty sediments (banded iron formations) raining down onto the ocean floor. However, rising oxygen in seawater proved toxic to many of the Earth's early microbes, including many methanogens. This accumulating oxygen in the ocean didn't only oxidize iron, it oxidized everything in the water, including other metals like nickel, effectively removing those from the oceans too. But nickel plays an essential role in the chemical reactions used by methanogens to produce methane, so as rising oxygen depleted nickel in the oceans, methane-producing methanogens were starved of nutrients and began dying en masse. This one-two punch of oxygen toxicity and iron-depletion (and nickel-depletion) triggered the world's first global mass extinction as the majority of the ocean's anaerobic microbes died off. This great die-off made room for still more oxygen-producing blue-green algae to colonize tidepools formerly occupied by other non-oxygen-producing microbes. And, as populations of methane-producing microbes crashed, they stopped pumping methane into the atmosphere, thus turning off the tap on the methane-fueled greenhouse effect that was helping to keep the cooling planet from turning into an ice cube. But it gets worse.

The runaway algal bloom of cyanobacteria pumped so much oxygen into the oceans that once all the soluble iron was oxidized and scrubbed out of the oceans, oxygen was free to start bubbling out of the water and into the atmosphere. Atmospheric oxygen levels soared. If you add up the total amount of oxygen in our present-day atmosphere, it's estimated that the Great Oxygenation Event flooded the surface of the planet with 12 to 22 times that amount![11]

On a side note that will become extremely relevant later in our story, as the oxygen accumulating in the atmosphere began reacting with UV

---

[11] Bekker, A. and Holland, H.D. (2012). Oxygen overshoot and recovery during the early Paleoproterozoic. *Earth and Planetary Science Letters*. 317-318. Pg 295-304. https://doi.org/10.1016/j.epsl.2011.12.012

radiation in sunlight, the world's first ozone layer began to form. For the first time, the planet was shielded from life-extinguishing levels of harmful UV radiation, which had until then made life outside of the oceans impossible. In time, this ozone layer would make it possible for microbes and then plants and animals to colonize the continents without being fried by solar radiation. Bit by bit, biological forces were transforming our planetary chemistry to make our world more habitable. Life was making the planet more suitable for more Life.

As oxygen flooded into the atmosphere, it began reacting with methane. When methane is oxidized, it turns into carbon dioxide and water and so rising oxygen levels began scrubbing billions of years of accumulated methane out of the air. Methane levels plummeted, and the protective greenhouse effect from that methane plummeted in lockstep. Without the greenhouse effect from methane and without the high levels of water vapour that had once wrapped the planet in an insulating blanket, the dim Sun couldn't provide enough heat to keep the planet warm. Combined with the effects triggered by the emergence of continents (including the shade cast by proliferating clouds and the disruption of ocean currents caused by continental landmasses), all these factors tipped the Earth into a global ice age that lasted an unimaginable 300 million years! Geologists call this the Huronian Glaciation — it is by far the longest glaciation in our planet's history. It encased almost the entire globe under a thick crust of ice, including the oceans, except for a small, exposed band of slushy seawater that remained open in the tropics.

Methane and oxygen cannot coexist in the atmosphere because (by virtue of oxygen's reactive nature) as soon as these two gases come in contact with one another, oxygen reacts with methane, destroying it. The methane molecule is ripped apart by oxygen, creating carbon dioxide and water, as shown in the simple chemical formula below:

$$CH_4 \text{ (methane)} + O_2 \text{ (oxygen)} >>>> CO_2 + H_2O \text{ (water)}$$

In the early days while methane was still abundant, and oxygen was still rare, high atmospheric methane levels effectively scrubbed all oxygen from the atmosphere. Likewise, as long as the seas were still full of iron and oxygen was still rare, high iron levels in the oceans effectively scrubbed all oxygen from the oceans. Earth was condemned to remain an oxygen-free world as long as methane and iron were plentiful and as long as oxygen was rare. But as soon as the global algal bloom began producing such vast quantities of oxygen, oxygen gained the upper hand as it overwhelmed the iron in the oceans and the methane in the atmosphere. Henceforth, any speck of new methane that enters the atmosphere and any speck of new iron that is added to the oceans is quickly scrubbed out by the vast quantities of oxygen in our atmosphere and oceans. Photosynthesis gave oxygen to the world while simultaneously destroying the ability for methane to exist in the atmosphere and for iron to exist in the oceans.

Any methane released into the atmosphere today is temporary. Thanks to its reaction with oxygen, it is gone (fully converted to CO_2) *within only 12 years*. Thus, the vision of methane building up in our atmosphere today from burping cows and burning fossil fuels is yet another nonsensical claim made by activist-scientists, journalists, and politicians. It's a complete non-issue — an absurd fabrication that is completely at odds with basic chemistry. There's so much oxygen in our atmosphere that methane has no chance of building up to levels that would have any meaningful greenhouse effect — oxygen systematically scrubs it out of the atmosphere as fast as we can produce it. Oxygen levels on our planet would have to be completely depleted before methane can ever begin to accumulate in our atmosphere.

That's why, despite millions of years of methane bubbling out of bogs, despite millions of years of methane seeping out of carbon-rich sludge accumulating on the ocean floor, and despite millions of years of methane continually percolating out of volcanic vents and leaking out of melting permafrost, our atmosphere today still only contains microscopic traces of methane — a mere 0.00017% (1.7 ppm), which reflects the fast rate at which it is oxidized and scrubbed from the air by oxygen. Whether it comes from burping cows, rotting forests, degassing wetlands, carbon-rich mud being digested by methanogens living at the bottom of lakes, river deltas, and oceans, or from fossil fuel emissions, our oxygen-rich atmosphere prevents any of this methane from sticking around for long — and at 0.00017% of the atmosphere, it's impact as a greenhouse gas is utterly insignificant. The fact that methane has increased from 0.00007% (0.7 ppm) before the industrial revolution to around 0.00017% (1.7 ppm) today merely reflects the fact that thanks our activities that erode vast quantities of carbon-rich muds and deposit them at the bottom of bogs, lakes, river deltas, and oceans, methanogens are being given a bit more "food" to eat — the methane measured in our atmosphere isn't "building up", it's merely recording the fleeting existence of a slightly larger quantity of methane during its brief 12-year lifespan in our atmosphere.

The emergence of continental land masses was likely the most significant factor driving the cooling of the planet and triggering the onset of the Huronian Glaciation because the emergence of the continents triggered a drop in global humidity levels, disrupted the circulation of the oceans, reflected solar energy back into space instead of absorbing it, and provided a place for ice to start to accumulate. However, the collapse of the methane-fueled greenhouse effect during the Great Oxygenation Event was one of the many straws that broke the proverbial camel's back, helping to tip the balance towards glaciation in an already significantly cooler world. Over the next 300 million years, the ice advanced and retreated across the continents and out across the seas at least three times. The sunlight-blocking crust of ice and the bitter cold brought photosynthesis to a screeching halt. As photosynthesis was cut off by cold and ice, oxygen levels fell again, allowing methane levels to rebuild via volcanic eruptions and via the dwindling number of methanogens still surviving in the oceans. As methane returned to the atmosphere, the greenhouse effect melted the ice, only for the warm weather and the fertilizer-rich glacial dust scraped off the continents by ice sheets to fuel the next wave of the algal bloom, which flooded the atmosphere with a fresh burst of oxygen that caused methane levels to collapse once again, which triggered the next surge of the ice sheets. Rinse and repeat at least three times over 300 million years.

~

The runaway global algal bloom had another transformative effect on the atmosphere. Even before the Great Oxygenation Event, photosynthesis had been steadily draining carbon dioxide from the atmosphere, even though the CO_2 consumed by photosynthesis was continually offset by CO_2 released by volcanoes and by lava leaking onto the seafloor at tectonic spreading zones. But as aerobic photosynthesis kicked into overdrive and as repeated surges of runaway global algal blooms spread around the globe, the vast amounts of CO_2 consumed by all this photosynthesis caused atmospheric carbon dioxide levels to plunge. As algal populations bloomed and died, a steady stream of carbon-rich organic residue settled onto the ocean floor to create gigantic layers of carbon-rich organic shales at the bottom of the ocean, thus permanently removing large quantities of carbon from the atmosphere, never to return — think "carbon sinks" on steroids!

Photosynthesis isn't the only force that began scrubbing vast quantities of CO_2 from the atmosphere 2.7 billion years ago. Remember the acidic rain I described earlier whereby carbon dioxide is dissolved in rainwater to produce carbonic acid (which began washing phosphorus into the oceans)? One of the most common minerals in the Earth's crust is calcium. Calcium is extremely easily dissolved by carbonic acid and so, as the continents formed, rainfall began reacting with calcium-rich rocks on the continents to form calcium carbonate as that calcium was bonded with carbon dioxide, which was subsequently carried by rivers and streams to the oceans. Rinse and repeat for millions of years and the oceans soon became supersaturated with calcium carbonate minerals. This calcium carbonate continually precipitates out of seawater to settle as thick layers of calcium carbonate sediments on the bottom of the ocean. As these sediments build up over time, the pressure and heat from being buried compresses these calcium carbonate sediments into limestone. And so, the carbon that once filled the atmosphere was slowly being locked up in calcium carbonate sediments raining down on the bottom of the oceans, lost forever from the carbon cycle.

Incidentally, anyone whose house is supplied by groundwater from a well is extremely familiar with calcium carbonate — it's the white crust that clogs faucets, builds up in toilet bowls, and appears as white stains in the kitchen sink when water spots dry. Calcium carbonate is also the material that all seashells are made of — supersaturated oceans provide the perfect building material for multicellular plants and animals to create seashells and coral reefs. These seashells and corals also rain down onto the ocean floor as they die, and then get buried to create still more limestone. Although multi-cellular organisms haven't evolved yet at this point of the story, they're coming, and when they do, it will only accelerate the amount of carbon accumulating in limestone beds at the bottom of the oceans. The days of abundant CO_2 in our young Earth's atmosphere were over, forever.

This calcium carbonate cycle slowly but relentlessly scrubs carbon dioxide from the atmosphere. Limestone is made of 44% carbon dioxide, *all of which was originally sourced from the atmosphere*. This calcium carbonate cycle essentially acts as a non-stop conveyor belt that continually scrubs carbon from the atmosphere and deposits it in sediments at the bottom of the ocean.

Evidence of chemical weathering preserved in sedimentary layers shows that by the time the Great Oxygenation Event and the Huronian Glaciation ended 2 billion years ago, the combination of continental weathering and photosynthesis had reduced atmospheric CO_2 levels from over 30% (300,000 ppm) to around 4% (40,000 ppm),[12] still 100x higher than current levels (420 ppm) but a dramatic 87% reduction from the 300,000 ppm that existed at the very beginning of our story. The entire remaining history of our planet is one long story of falling atmospheric CO_2 levels interrupted by short intervals when atmospheric CO_2 briefly recovered before the relentless scrubbing of CO_2 from our atmosphere resumed anew.

The Great Oxygenation Event marks the inflection point in our Earth's history when the dual processes of photosynthesis and continental weathering became the key drivers in the story of atmospheric CO_2 — as you will soon see, this gradual but relentless removal of carbon dioxide from the atmosphere will, in time, become an existential challenge for all Life on Earth as carbon dioxide levels threaten to fall to levels below which all photosynthesis stops.

~

The end of the Huronian Glaciation 2.2 billion years ago marked the end of Earth's first inglorious experiment with oxygen and aerobic photosynthesis. Why things settled back down and why the cycle of Huronian ice ages ended is still a hotly debated question, though we can make an educated guess based on the conditions that came afterwards, which at least provide a partial explanation. Geologists have described the long period that followed the end of the Huronian Glaciation as "the dullest time in Earth's history". They nicknamed it "the Boring Billion" — a mindboggling long period from around 1.8 billion to around 800 million years ago marked by weak levels of photosynthesis, slow biological evolution, a relatively stable warm climate without any more glaciations, relatively stable carbon dioxide levels, and extremely low but

---

[12] Rye, R., Kuo, P.H., and Holland, H.D. (1995). Atmospheric carbon dioxide concentrations before 2.2 billion years ago. *Nature*. 7(338) https://doi.org/10.1038/378603a0

gradually climbing oxygen levels. The continents also didn't move much yet during this era — plate tectonics had started but were still in their infancy so continental drift was still hesitant and intermittent. The first single-celled microbes likely colonized land during this era.[13]

The simplest explanation for why the excitement of the Great Oxygenation Event and the wild temperature swings of the Huronian Glaciation were followed by the long and unexciting "Boring Billion" is that by draining all the iron from the oceans, by adding oxygen to seawater, and by flooding oceans with calcium carbonate weathered from continental rocks, the ocean chemistry was altered so rapidly and so profoundly that algae were no longer able to bloom as aggressively in the chemically-altered world they had helped to create. In essence, they were a victim of their own initial success. Perhaps the chemical changes they triggered in the oceans were so profound that Life had to spend the next billion years playing evolutionary catchup to adapt to this chemically transformed oxygen-infused world.

Aerobic photosynthesis slowed to a crawl after the end of the Huronian Glaciation and so, with less blue-green algae churning out oxygen, oxygen levels in the oceans and in the atmosphere fell back down to extremely low levels, making it possible for methanogen populations to recover some of their lost territory and begin pumping methane back into the atmosphere, thus helping to stabilize and re-warm the climate. Geological evidence confirms that anoxygenic photosynthesis returned to many parts of the ocean during the Boring Billions (and turned parts of the ocean purple again). However, the thick slurry of sulfur- and iron-rich seawater that had been the norm since our planet's earliest beginnings was gone forever, depleted by the oxidizing effect of the Great Oxygenation Event. Life now had to adapt to a much less chemically rich ocean and a world where aerobic photosynthesis would forever keep pumping oxygen into the air. The primordial soup was transforming into a much more ordinary ocean full of water.

Our modern world runs on the familiar carbon cycle that continually recycles carbon dioxide in and out of the atmosphere. Photosynthesis is on one side of the equation; oxygen breathers are on the other. Like Yin

---

[13] Yoon et al. (2004). A Molecular Timeline for the Origin of Photosynthetic Eukaryotes. *Molecular Biology and Evolution*. 21(5) 809–818. https://doi.org/10.1093/molbev/msh075

and Yang, these two counter-balancing and inseparable lifeforces sustain Life on our planet today. One side of this equation uses the energy from the sun to draw carbon dioxide out of the air to create sugars and complex carbohydrates, while creating oxygen as a byproduct. The other side uses oxygen to release energy from sugars and complex carbohydrates, while creating carbon dioxide as a byproduct. The Great Oxygenation Event marks the point in history when the photosynthesis portion of the carbon cycle sputtered into action. But to get to its counter-balancing lifeforce — oxygen breathers — we have to cross to the other side of the Boring Billion to the series of events that brought the "dullest period in history" to a dramatic end around 800 million years ago.

Chapter 14

The Boring Billion

"Civilization exists by geological consent, subject to change without notice."

— Will Durant

From our perspective here in the 21st century, 800 million years ago seems terribly long ago, but it's worthwhile to take a moment to put all this geologic time into perspective. By this point in the story, we've already covered more than 83% of Earth's long history. The Earth is starting to look old by this point — yet we haven't even evolved the first multicellular life at this point in the story! If Earth's long history was stretched out along a ruler running the full length of Canada's 7,000-kilometer TransCanada Highway, from the lighthouse in St. John's Newfoundland on the Atlantic Coast to the Horseshoe Bay Ferry Terminal in Vancouver on the Pacific Coast, then our journey through geological time has already taken us all the way from the Atlantic coast to the foothills of the Rocky Mountains. The Great Oxygenation Event happened somewhere near the middle of that long journey. The Boring Billion comes to an end just as we approach the foothills of the Rocky Mountains. And yet, all we've seen along the way are chemical changes and single-celled microbes — we still haven't even evolved any plants or animals! Life has been taking its merry old time getting going. But things are about to get extremely busy along the last 800-million-year stretch of Earth's history.

All life on Earth depends on nitrogen — it is one of the most important macronutrients on the planet. It is essential for making proteins and nucleic acids, and it plays a vital role in carbon metabolism — all oxygen-breathing organisms depend on nitrogen for their metabolic processes while all photosynthetic organisms require nitrogen in order for the chlorophyll molecule to capture sunlight energy. Nitrogen controls virtually all aspects of plant growth and development, including seed germination, rooting, branching, and flowering, as well as plant metabolism, gene expression, and hormone signaling.[1] When nitrogen is in short supply, plant growth stalls and yields fall. In short, Life is not possible without nitrogen.

The problem for Life on Earth is that although our atmosphere is absolutely full of nitrogen gas, it's not a form of nitrogen that plants and animals can use — they can't simply absorb or inhale nitrogen gas to make it useful. Before plants can make use of nitrogen for photosynthesis, nitrogen needs to be bonded with hydrogen atoms (to create ammonium) or with oxygen atoms (to create nitrates and nitrites). Likewise, animals depend on biologically useable forms of nitrogen, which they can only get from their diet by consuming nitrogen-rich plant and animal tissues. [In another stroke of grim irony, the hysterical climate change brigade has widened its crusade on greenhouse gases by declaring war against nitrogen fertilizers — specifically ammonium, nitrates, and nitrites.]

During the early history of our planet, there were only very few sources of these biologically useable nitrogen compounds. Volcanoes emit some ammonia and some nitrates during eruptions. Some more were added over time as nitrogen-rich meteorites collided with the Earth during its early history (ammonia-rich meteorites are thought to be the source of the first ammonia for life on Earth[2]). But the majority of biologically useful nitrogen compounds on the young Earth were created by lightning during thunderstorms. The high temperature of a lightning bolt breaks the bonds of atmospheric nitrogen gas, which then immediately react with oxygen to form nitrates before being washed from the sky by rain. And to get from nitrates to nitrites, both sunlight[3] and bacteria[4] can each independently convert nitrates into nitrates. Farmers love the free nitrogen fertilization

---

[1] Vega, A., O'Brien, J.A., and Gutiérrez, R.A. (2019). Nitrate and hormonal signaling crosstalk for plant growth and development. *Current Opinion in Plant Biology.* 52: 155-163. https://doi.org/10.1016/j.pbi.2019.10.001

[2] Minard, A. (2011, Feb 28). Meteorites may have delivered first ammonia for life on earth: new study. *PHYS.org.* https://phys.org/news/2011-02-meteorites-ammonia-life-earth.html

[3] Moore, B. (1917). The Formation of Nitrites from Nitrates in Aqueous Solution by the Action of Sunlight, and the Assimilation of the Nitrites in Green Leaves in Sunlight. *Proceedings of the Royal Society B. Biological Sciences.* 90: 158-167. https://doi.org/10.1098/rspb.1918.0007

[4] Kevil, C.G. and Lefer, D.J. (2010). Nitrite Therapy for Ischemic Syndromes. *Nitric Oxide (Second Edition).* Elsevier Inc. https://www.sciencedirect.com/topics/agricultural-and-biological-sciences/nitrite

that comes with a good thunderstorm, which gives their crops a boost of growth far beyond what rain alone can produce. Once in the soil or in the water, plants and microbes absorb these nitrogen compounds to fertilize their growth.

Before nitrogen-fixing bacteria evolved to directly convert atmospheric nitrogen into biologically useful nitrogen compounds, the scarcity of these nitrogen compounds was a severe bottleneck for Life on Earth. The algal bloom that launched the Great Oxygenation Event depended on nitrates and ammonium in seawater to fuel their growth but as these sources of nitrogen were depleted, a nitrogen crisis emerged — this is yet another factor why the Great Oxygenation Event and Huronian Glaciation may have come to an end — the nitrogen depletion of the oceans starved blue-green algae and caused their populations to collapse. Oxygen production collapsed in lockstep with collapsing algae populations, which allowed methane levels to bounce back a bit. A lack of biologically-available nitrogen goes a long way in helping to explain the low levels of photosynthesis during the Boring Billion — a nitrogen shortage helped turn the global cyanobacterial algal bloom into a global algal bust.

As the nitrogen shortage after the end of the Huronian Glaciation put a massive damper on photosynthesis, evolutionary pressure began to favor any microbes capable of capturing nitrogen gas directly from the air (after all, our atmosphere is 78% nitrogen gas!) in order to *biologically* convert atmospheric nitrogen into ammonium and nitrates. And so, as the long millennia of the Boring Billion went by, increasingly efficient nitrogen-fixing bacteria evolved to do just that.[5] Initially they were restricted to coastal tide pools, but by the end of the Boring Billion nitrogen-fixing bacteria had spread around the planet, both on land and in the oceans. As these nitrogen-fixing bacteria died and left behind their nitrogen to accumulate in the soil and in the oceans, photosynthesis became increasingly efficient and algal growth slowly picked back up. And, most important of all, as nitrogen became more readily available, it became easier for microbes to make proteins, which would soon help open the door for the evolution of the complex cells required for multicellular life.

---

[5] Navarro-González, R., McKay, C. & Mvondo, D. (2001) A possible nitrogen crisis for Archaean life due to reduced nitrogen fixation by lightning. *Nature* 412, 61–64. https://doi.org/10.1038/35083537

Another group of microbes that benefited from rising levels of biologically available nitrogen at the end of the Boring Billion were the oxygen-breathing microbes, discussed briefly in the last chapter, which first emerged 3.1 billion years ago but had not yet played any significant role in recycling carbon dioxide back into the atmosphere. But as biologically available nitrogen became more common, the chemical reaction used by oxygen breathers to convert carbon-rich material into energy also became increasingly efficient. And so, as photosynthesis ramped up towards the end of the Boring Billion and began once again to pump vast volumes of oxygen into the oceans and atmosphere, this time readily available quantities of nitrogen helped prime oxygen-breathers to be able to take advantage of all that oxygen. And so, oxygen breathers finally also become a dominant lifeforce on the planet. Yin and Yang were finding each other — the complete carbon cycle was finally poised to take over the world.

All the ingredients were falling into place for a tremendous new chapter in the history of Life on our planet. At this point in the story (600 million years ago), we are standing on the cusp of the Cambrian Explosion of Life when multicellular lifeforms emerged to colonize the world. By this time, we still have lots of available carbon dioxide in the atmosphere as the primary food for photosynthesis. We have already evolved an efficient form of photosynthesis capable of using solar energy to turn CO_2 into sugars and carbohydrates, with oxygen as a byproduct. We have oxygen breathers capable of using oxygen to break the molecular bonds in sugars and carbohydrates to release energy, thus returning CO_2 to the atmosphere to complete the carbon cycle. And we have readily available quantities of biologically useful nitrogen compounds for building proteins. The last step towards the emergence of multicellular life was for some of those single-celled microbes to begin "collaborating" with one another to give themselves a competitive advantage. In other words, multicellular life began when some single-celled microbes figured out how to attach themselves (or even engulf) other useful microbes in order to parasitize them. These early symbiotic relationships would then evolve to become increasingly complex until plants and animals appear on the landscape.

~

Our story is about to explode in a burst of life and drama. But before it does, I want to take a brief moment to point out a few things about the carbon cycle — the symbiotic Yin and Yang relationship between photosynthesis and oxygen breathers. The early atmosphere was essentially a byproduct of chemical processes and volcanic gases, with one-way contributions from biological sources (like methane from methanogens or oxygen from photosynthetic organisms). But as the Great Oxygenation Event demonstrates, in those early days the atmospheric mix could easily be permanently altered by some "event" that added or removed gases from the air. During the Great Oxygenation Event, rising oxygen levels catastrophically changed the chemistry of both the air and the oceans and ultimately made the oceans toxic for many existing microorganisms. As such, it was an exceedingly fragile world — the evolutionary success of early photosynthetic microbes also proved to be their downfall. The Great Oxygenation Event and the Boring Billion that followed demonstrate how easy it was to destabilize this world.

But once oxygen breathing populations expanded, their symbiotic relationship with photosynthetic organisms helped stabilize gas ratios in the atmosphere and oceans. Today, every drop of carbon dioxide that is removed from the air by photosynthesis will eventually be turned back into CO_2 by oxygen-breathing bacteria, microbes, and animals, with the exception of any carbon that escapes the carbon cycle as it gets trapped in permanent carbon sinks like sedimentary layers, wetland bogs, and humus in the soil. In other words, the idea that burping cows and other oxygen breathers could possibly contribute to rising atmospheric CO_2 levels is patently ridiculous. They are merely returning CO_2 back to the atmosphere that was previously removed from that same atmosphere by plants. Even if burping cows and other oxygen breathers were capable of converting every single molecule of carbon created by plants back into CO_2 (thus ignoring all the carbon that is constantly lost to permanent carbon sinks), at most oxygen breathers could merely stabilize atmospheric CO_2 levels to prevent them from falling lower — it is chemically impossible for oxygen-breathers to raise atmospheric CO_2 levels because they cannot add more CO_2 to the atmosphere than plants removed from the atmosphere via photosynthesis in the first place. The methane they produce is also soon converted back into CO_2. And if you don't send plant material through a cow, then microbes and/or direct oxidization by atmospheric oxygen will still relentlessly convert every last molecule of carbon from those plant tissues back into CO_2. Either way, you end up right back at the same place. Except with livestock in the equation, we get a beneficial source of protein, soils get lots of pre-digested organic material to turn into fertile soil, and the skin of grass sod that keeps the soil protected from the elements is kept healthy by the impact of grazing livestock. And by using up oxygen in the atmosphere every time they inhale, it likewise helps stabilize oxygen levels in the atmosphere — as you'll soon see in upcoming chapters, a world with too much oxygen becomes very, very flammable indeed.

The only way to add new sources of carbon dioxide to the atmosphere from outside of this carbon cycle is for volcanoes to release fresh CO_2 from inside the Earth or if carbon previously trapped in a carbon sink is converted back into CO_2 (i.e. by burning fossil fuels, degassing limestones, or exposing humus to the atmosphere). Cows do none of these things. Scapegoating livestock as a source of greenhouse gas emissions is a cheap fairy tale that ignores basic chemistry while serving some altogether different agenda.

Another part of the story that is left out of the "narrative" are all the ways in which plants benefit from rising CO_2 as they take advantage of every extra drop of fresh CO_2 that is added to the cycle. Whenever carbon dioxide increases, plant growth surges because at higher CO_2 concentrations it becomes easier for plants to absorb CO_2 through the walls of the photosynthetic plant cell. We call this effect "CO_2 fertilization". Without the symbiotic stabilizing impact of oxygen breathers (a.k.a. "respirators", like cows) and the CO_2 they recycle back into the atmosphere, Earth would have run out of atmospheric CO_2 long ago even as oxygen levels would have surged to toxic levels, much like what happened during the Great Oxygenation Event.

In chapter 18 we will see what happens when plants evolved two new type of plant fiber (cellulose and lignin) that were undigestible to oxygen breathers, which once again gave photosynthesis the upper hand over oxygen breathers for a period lasting many millions of years, causing global atmospheric CO_2 levels to plunge to near fatally low levels while flooding our planet with never-before-seen levels of oxygen until, finally, oxygen breathers were able to play evolutionary catchup by evolving new strategies to digest these undigestible plant fibers to turn all that plant matter back into CO_2.

Still later in the story we will see how Life was forced to adapt once more as falling atmospheric CO_2 became increasingly challenging to plants living in an increasingly cold and dry carbon-depleted world. And in the final chapters we will see what happens to atmospheric CO_2 levels when a number of forces (both man-made and natural) come together to reduce our planet's total biomass to such low levels that photosynthesis can no longer remove as much CO_2 from the atmosphere as is released by volcanoes and other fresh sources of CO_2 that extra carbon to the carbon cycle.

As CO_2 increases in our atmosphere today, we can see the positive effects of CO_2 fertilization all around us. Thanks to the increase in CO_2 in our atmosphere over the last few centuries, our planet is currently experiencing a massive greening, leading to much higher yields and increased leaf growth. Plants and other photosynthetic organisms are loving all the extra CO_2 in the air. It is part of the reason why corn and other crop yields keep rising year after year after year. It is even part of the reason why food is so abundant today despite our rapidly growing population and despite the rampant soil erosion and soil degradation that has destroyed so much agricultural land in so many countries around the world. If CO_2 levels still hovered down at 280 ppm instead of rising to today's 420 ppm, it is questionable whether we would be capable of producing enough food to feed everyone on our planet and we would have undoubtedly had to clear vast additional tracts of forests to expand our agricultural land base to compensate for lower yields.

NASA's satellites have documented this greening of our planet by measuring the increase in leaf area from the 1980s until today — almost the entire planet has greened (most areas by as much as 25 to 50%!), as shown in figure 85 below. The dark areas on the map are all areas that have experienced increased leaf growth thanks to CO_2 fertilization. For the color version of this graph, check out NASA's website or Google "carbon dioxide fertilization greening Earth".[6]

---

[6] Hille, K. B. (2016, Apr 26). Carbon Dioxide Fertilization Greening Earth, Study Finds. *NASA*. https://www.nasa.gov/technology/carbon-dioxide-fertilization-greening-earth-study-finds/

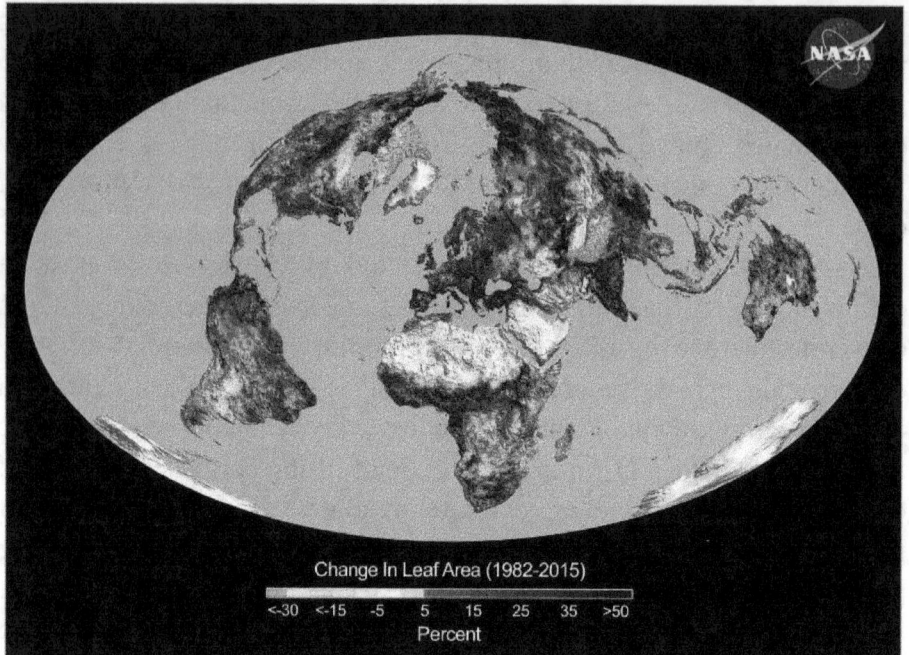

Figure 85. Carbon dioxide fertilization is greening the Earth.
https://www.nasa.gov/technology/carbon-dioxide-fertilization-greening-earth-study-finds/

Plant yields aren't the only thing to benefit from higher CO_2 levels. Because it gets easier for plant cells to absorb CO_2 through the cell wall as CO_2 levels rise, plants don't have to open their pores as widely to get their fill, *which helps them conserve water and makes them more tolerant of drought as CO_2 levels rise.* In other words, rising CO_2 is simultaneously making plants grow bigger and faster while also helping them conserve water and making them less vulnerable to drought. If CO_2 levels still hovered around 280 ppm, our world would be *significantly* more vulnerable to drought than it is today.

~

Geologist Andrew Knoll has spent his life studying the evolution of Life on our young Earth by looking at the rocks and minerals that reveal how these chemical and biological processes evolved. In his 2004 book, *Life on a Young Planet: The First Three Billion Years of Evolution on Earth*, he makes an eye-opening statement that forces us to reevaluate how we look at our atmosphere. "*Volcanoes supply CO_2 to the oceans and atmosphere, but photosynthesis removes it at a far faster clip. So much* *faster, in fact, that **photosynthetic organisms could strip the present-day atmosphere of its CO_2 in little more than a decade.** They don't of course, principally because of respiration..."*[7]

During the Great Oxygenation Event, photosynthesis pulled CO_2 out of the atmosphere for 300 million years without any meaningful populations of oxygen breathing organisms to put any back in. Although the atmosphere never completely ran out of CO_2, without an abundance of oxygen breathers [a.k.a. respirators] to recycle CO_2 back into the air, photosynthesis depleted the atmosphere of 87% of its CO_2 (CO_2 dropped from 30% to 4% over the course of those 300 million years). *Yet today, biological life has become so much more abundant that without oxygen breathers to replace CO_2 in the atmosphere, photosynthesis would pull every last molecule of CO_2 out of our atmosphere in only 10 years!* Biomass isn't just a feeble crust on the surface of the planet anymore, it has become the dominant force controlling the mix of gases in our atmosphere. As you will soon see as our journey through geological time continues, anything that messes with the total biomass on the planet triggers rapid and dramatic changes to the composition of our planet's atmosphere.

The amount of CO_2 added to the atmosphere by our industrial emissions is not that important — a healthy Yin/Yang relationship between photosynthesis and oxygen breathers would quickly convert that extra CO_2 into plants, seashells, plankton, coal, humus, and other forms of carbon, just as it has done over and over again during the last 800 million years whenever volcanoes released extra CO_2 into the biosphere. Some of these past volcanic eruptions were so huge and lasted so many millions of years that our paltry 150-year history of rising fossil fuel emissions look truly insignificant by comparison. And yet, despite these periodic volcanic outbursts that temporarily flooded our atmosphere with vast quantities of extra CO_2, overall the story of the last 800 million years is a story in which we see carbon dioxide levels slowly but steadily decline as CO_2 slowly but continually bleeds out of the system as carbon gets trapped in carbon sinks like shales and limestones forming at the bottom of the ocean, as it is deposited as coal forming at the bottom of swamps, and as it is stored as humus in the soil. At current levels of around 400 ppm (0.04% of the

---

[7] Knoll, A.H. (2004). *Life on a Young Planet: The First Three Billion Years of Evolution on Earth*. Princeton University Press.

atmosphere), atmospheric CO_2 is so depleted (by historical geologic standards) that plants are quite literally starving for CO_2 and are being forced to adapt to a CO_2-depleted world. Plants evolved at much higher levels of CO_2 — commercial greenhouse growers actually pump CO_2 into their greenhouses to raise CO_2 above 1,000 ppm in order to increase yields so that plants can grow in optimal conditions — in later chapters we will see the extreme lengths that plants have gone to in order to adapt to our current CO_2-depleted world. But I'm getting ahead of the story...

~

As the Boring Billion came to an end 800 million years ago, increasing photosynthesis (helped by more readily available nitrogen) led to a Second Great Oxygenation Event that lasted from 850 to 540 million years ago. Once again, iron-rich sediments rained out of the ocean as oxygen once again flooded the seas, though this second wave of banded iron formations was not nearly as thick as those formed during the first Great Oxygenation Event, as shown in figure 86 below. But this Second Great Oxygenation Event was accompanied by a rise in oxygen-breathing microbes to recycle carbon dioxide back into the atmosphere. After the first Great Oxygenation Event, atmospheric CO_2 stood around 40,000 ppm (4%), approximately 100x higher than they stand today. By the time the Second Great Oxygenation Event ended, atmospheric carbon dioxide had plunged by another 83%, from 40,000 ppm (4%) to around 7,000 ppm (0.7%),[8] which is still approximately 17x higher than they stand today. This dramatic plunge in CO_2 was matched by a corresponding permanent rise in atmospheric oxygen levels to around 10% (approximately half of where they are today). However, instead of another biological crash, by the end of the Second Great Oxygenation Event the emerging balance between photosynthesis and oxygen breathers finally brought a measure of stability to the balance of oxygen and carbon dioxide in the atmosphere. This time Life was ready to take advantage of an oxygen-rich world — this time Life was not going to go back to sleep like it had at the start of the Boring Billion. This is becoming a much more familiar world, even if the fossils from that time are still wholly unrecognizable to us today.

---

[8] Foscolos, A.E. (2017). Climate Changes: Anthropogenic Influence or Naturally Induced Phenomenon. *Bulletin of the Geological Society of Greece.* 43(1):8. http://dx.doi.org/10.12681/bgsg.11157

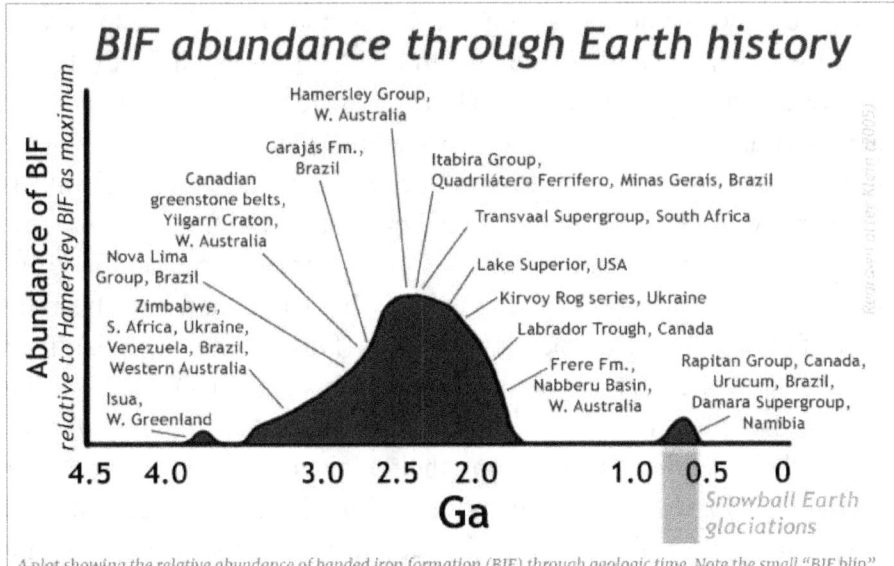

Figure 86. Most banded iron formations (BIFs) were formed during the Great Oxygenation Event 2.7 billion years ago, followed by a smaller episode during the Second Great Oxygenation Event 800 million years ago.

Source: https://opengeology.org/historicalgeology/case-studies/snowball-earth/

During first 100 million years of the Second Great Oxygenation Event, Earth continued to enjoy a very warm climate. But then, starting around 720 million years ago, Earth was plunged into a second global ice age (the most severe in Earth's history), known as the Cryogenic Glaciation, which lasted over 85 million years with at least three separate advances and retreats of the ice. Geologists nicknamed this time period "Snowball Earth". The average global temperature dropped to a bone-chilling -40 °C. The entire ocean, *including the tropics*, was covered by more than 200 meters (660 ft) of ice.[9]

This time methane levels cannot be assigned any of the blame — the atmospheric methane that had once helped warm the planet had long since been scrubbed from the atmosphere by rising oxygen levels during the tail end of the Boring Billion. Instead, activist-scientists predictably point to

---

[9] Earle, S. (n.d.) *Environmental Geology, 3.2 Plate Tectonics and Climate Change*. Thompson Rivers University. Accessed 17 Mar 2024. https://environmental-geology-dev.pressbooks.tru.ca/chapter/plate-tectonics-and-climate-change/

their models to blame the Cryogenic Glaciation on falling CO_2 levels. Using their climate models, they claim that CO_2 levels allegedly fell all the way down to present-day levels (0.04%) to trigger the start of the glaciation) and then go on to claim that the glaciation ended because volcanoes erupting through the ice allegedly caused CO_2 levels to spike to 14%, approximately 350x higher than they are today. After everything we know about the diminishing returns of CO_2 as a greenhouse gas above 300 ppm, it's completely unrealistic to believe that CO_2 could have been the driver of these extreme temperature changes. Furthermore, there isn't any direct geological evidence for such dramatic swings in CO_2 levels — the minerals that formed in sedimentary rocks from that era simply don't support their claims. Once again, their claims are purely theoretical speculations driven by climate models. Garbage in, garbage out — a theoretical model built to give legitimacy to the modelers' biases while ignoring everything else.

As with all the other ice ages yet to come, there are many different contributing factors that came together to turn the planet into Snowball Earth. One of those factors was the position of the continents. Geologists have been able to reconstruct the position of the continents in the lead-up to the Cryogenic Glaciation — at that time, all the continents were clustered together near the equator, which interfered with the circulation of equatorial ocean currents and made it more difficult to transport warm equatorial waters to the poles. Furthermore, solar radiation from the Sun is strongest near the equator, so with the land masses all concentrated near the equator there was less ocean water to absorb solar heat near the equator. And, in an era before plants, the lack of vegetation on the large continents had a similar effect as the reflective surface of polar ice caps, which reflect solar radiation back into space instead of absorbing that heat. Furthermore, the collision of continental plates as they collided with one another near the equator created extremely high plateaus and massive mountain ranges upon which ice sheets began to grow.

However, the location of the continents was merely a contributing factor to the Cryogenic Glaciation, not the sole cause. Here our story takes a strange and unexpected turn as changes in cloud cover come back into the picture, driven by events happening far outside of our solar system.

Chapter 15

Clouds and Cosmic Radiation

"I may conclude this chapter by quoting a saying of Professor Agassiz, that whenever a new and startling fact is brought to light in science, people first say, 'it is not true,' then that 'it is contrary to religion,' and lastly, 'that everybody knew it before.'"

— Charles Lyell, The Antiquity of Man

I previously discussed how clouds (and rainfall) cannot form unless water vapour is able to condense around some nucleus. Evaporation lifts moisture up into the atmosphere but to turn that moisture into clouds and rainfall requires condensation. Dust, pollen, volcanic ash, sea spray, smoke particles, soot and air pollution, and even ice crystals are all common aerosols that serve as nuclei for cloud formation. I also previously mentioned that solar cycles trigger changes in cloud cover because high-energy particles coming from the sun collide with various gases in our atmosphere, ionizing them to form aerosols that serve as nuclei for cloud formation. However, thanks to the work of some extremely perceptive physicists and astronomers, our cloud formation story is about to get a lot weirder...

First, let's start with a little background info about how this ionization process works — I promise not to get too technical so bear with me, it's worth it! Ionization happens when high-energy particles coming from the Sun and when cosmic radiation coming from space (such as protons, neutrons, electrons, gamma rays, and the nuclei of former atoms that have been stripped of their electrons during supernova explosions) penetrate our lower atmosphere where they collide with gas molecules like nitrogen (N_2), oxygen (O_2), water vapour (H_2O), carbon dioxide (CO_2), argon (Ar), sulfur dioxide (SO_2), and so on. These high-speed collisions can either knock electrons off these gas molecules or, in some cases, the collision results in a spare electron temporarily being captured by one of these various atmospheric gas molecules — either way, the resulting ion gains either a positive or negative electrostatic charge not unlike the static electricity that builds up on your hair when you rub a balloon on your head.

You can sometimes see the consequences of ionization happening in real time in our upper atmosphere — the aurora dancing across the skies in our planet's polar regions (and which sometimes penetrate further south during intense solar storms, like the once-in-a-generation solar storm that happened on May 10th, 2024) are created as high energy particles from the sun collide with oxygen or nitrogen molecules in the upper atmosphere; orange or green light is released by oxygen ions while blue or purple/red light is emitted by nitrogen ions as they throw off energy/light during their return to their normal neutral atomic state.

Our planet is continually being bombarded by radiation coming from the sun and from outside of our solar system. The ionization that creates aurora in the upper atmosphere is also happening in the lower atmosphere at the level where clouds form (though it doesn't produce aurora at this lower level). The balloon analogy is particularly appropriate to explain the behaviour of these ions because as these gas molecules become electrostatically charged (+ or -) as a result of their collision with high energy particles from space, they begin to stick to one another just like a balloon will stick to your hair or to the wall after you rub it on your hair. As they stick to one another, they form much larger temporary molecules, also called aerosols. As these clusters of electrostatically charged aerosols grow larger, they get big enough to allow water vapour to begin to condense around them, thereby forming the nuclei for the water droplets that create clouds and raindrops. *That's how radiation coming from our sun and cosmic radiation coming from beyond our solar system both contribute to cloud formation here on Earth.*

But here's where things get weirder. We instinctively tend to think of solar radiation as the primary source of the high energy particles hitting our planet — after all, that's where the heat that warms our planet comes from. But space is absolutely flooded with high-energy particles — this cosmic radiation (a.k.a. cosmic rays) is left over from dying stars that exploded as supernova. Every year, astronomers observe a few hundred supernovas outside our galaxy and at least four dozen happening within our own Milky Way Galaxy. The main reason why we haven't been fried by all that cosmic radiation — the main reason why Life endures on our planet — is because the Sun's *magnetic field* shields us from the majority of that incoming cosmic radiation, deflecting it away from our planet. But some of that cosmic radiation nevertheless makes it through — it is constantly all around us. When those cosmic rays collide with gas molecules in our atmosphere, the resulting aerosols become the nuclei for clouds. And, as the Sun's magnetic field rises and falls across its ever-changing solar cycles (like the 11-year sunspot cycle, the 80-year Gleissberg cycle, the 400-year grand solar cycle, or the 976-year Eddy cycle), the amount of cosmic radiation that is deflected by the Sun's magnetic field rises and falls in sync with our Sun's solar activity. By consequence, cloud formation increases and decreases depending on how many of these cosmic rays are deflected by the Sun's magnetic field. This introduces two unexpected new players into the forces that create our planet's short-term weather and long-term climate — the cyclical changes of the Sun's magnetic field and the waves of cosmic radiation coming from space as our solar system travels through different areas of our galaxy.

You can witness this cloud formation process, driven by cosmic rays, with your own eyes with a simple DIY experiment that you can create in your own home — the European Organization for Nuclear Research (known as CERN) has posted instructions online for how to build your own DIY *cloud chamber*.[1] In simple terms, you put dry ice at the bottom of an empty fish tank, attach a rag to the inside of the lid, and soak the rag in isopropyl alcohol before sealing the tank. The isopropyl alcohol evaporates into the tank even as the dry ice begins to cool the air sealed inside the tank. The evaporated isopropyl alcohol trapped inside the sealed fishtank begins to cool but it has no nuclei to condense around so the alcohol cools far below its condensation point without being able to turn back into a liquid. And then the magic happens. Suddenly a brief streak of white appears inside the tank as a tiny cloud briefly forms inside the chamber. Then another and another. These cloud streaks are created as incoming cosmic radiation collides with the gas molecules inside the fish tank to create temporary aerosols that serve as condensation nuclei for the supercooled gaseous isopropyl alcohol. Check out the YouTube video on the webpage where CERN has posted its DIY cloud chamber instructions — it's quite a remarkable thing to watch radiation from deep space creating clouds inside the cloud chamber. The cloud chamber at the multi-billion-dollar CERN facility (maintained by the European Union) is just a fancier version of this DIY cloud chamber.

---

[1] Hetherton, S. (2015, Jan 21). How to make your own cloud chamber. *CERN*.
https://home.cern/news/news/experiments/how-make-your-own-cloud-chamber

When Nobel Prize winner John Clauser said that climate models fail to account for clouds, this is part of what he is referring to. The upcoming charts in this chapter tell a remarkable story as we see how the rise and fall of the Sun's magnetic field perfectly correlates with cloud cover and temperature here on Earth. Although there are still many questions about the exact ionization pathway to explain how cosmic rays create the aerosols involved in cloud nucleation (as with early 20th-century cloud seeding experiments, nature's complex dynamics are extremely difficult to reproduce in laboratory conditions), the strong correlation between temperature, cloud cover, and the strength of the Sun's magnetic field makes it abundantly clear that solar activity and cosmic rays are both playing a central role in driving climate changes on our planet.

Figure 87 shows the correlation between cloud cover and cosmic radiation from 1980 to 2003.[2] Cloud cover tracks cosmic rays almost perfectly; it's an almost perfect correlation with the 11-year and 22-year solar cycles. It has long been known that summer temperatures change by between 0.5 to 1.5 °C between the top and the bottom of the 11-year solar cycle — cloud cover explains why.[3] If you want to forecast global temperature trends for the coming decade, have a look where we are in the 11- and 22- year solar cycle.

---

[2] Shaviv, N.J. (n.d.) Cosmic Rays and Climate. *Science Bits*. http://www.sciencebits.com/CosmicRaysClimate

[3] Shaviv, N.J. (2003). The spiral structure of the Milky Way, cosmic rays, and ice age epochs on Earth. *New Astronomy*. 8(1). 39-77. https://doi.org/10.1016/S1384-1076(02)00193-8

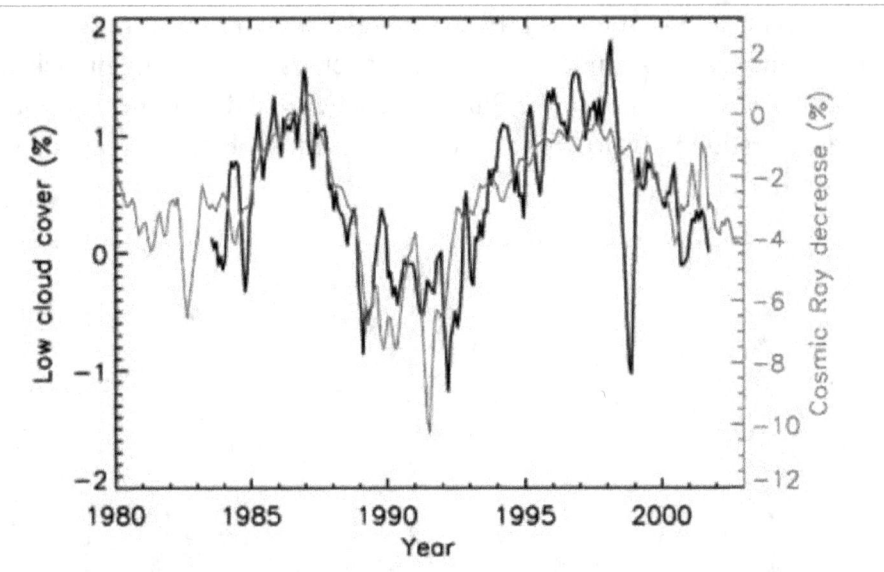

The correlation between cosmic ray flux (orange) as measured in Neutron count monitors in low magnetic latitudes, and the low altitude cloud cover (blue) using ISCCP satellite data set, following Marsh & Svensmark, 2003.

Figure 87. Correlation between cloud cover (black line) and cosmic rays (grey line).

Source: Astronomer Nir Shaviv's website
http://www.sciencebits.com/CosmicRaysClimate#MSJGR
with data sourced from Marsh and Svensmark, 2003. Galactic cosmic ray and El Niño-Southern Oscillation..., J. Geophys. Res., 108, 4195.
https://agupubs.onlinelibrary.wiley.com/doi/10.1029/2001JD001264

But let's zoom out a bit. Since we have no direct record of cloud cover reaching back thousands of years, another group of researchers chose instead to track the correlation between temperature and solar activity from 9,600 until 6,000 years ago — their chart is reproduced in figure 88 below.[4] The area in dark grey (labeled $\delta^{18}O$) shows how temperatures have changed over that 2,600-year period. The area in white (labeled $\Delta^{14}C$) shows how cosmic radiation has fluctuated over the same time period.

---

[4] Shaviv, N.J. (n.d.) Cosmic Rays and Climate. *Science Bits.*
http://www.sciencebits.com/CosmicRaysClimate

Once again, temperature clearly tracks solar activity even at this much longer time scale — temperature and solar activity are moving in lockstep in part because changes in the Sun's magnetic field determine how much ionizing cosmic radiation reaches our lower atmosphere to create clouds here on Earth.

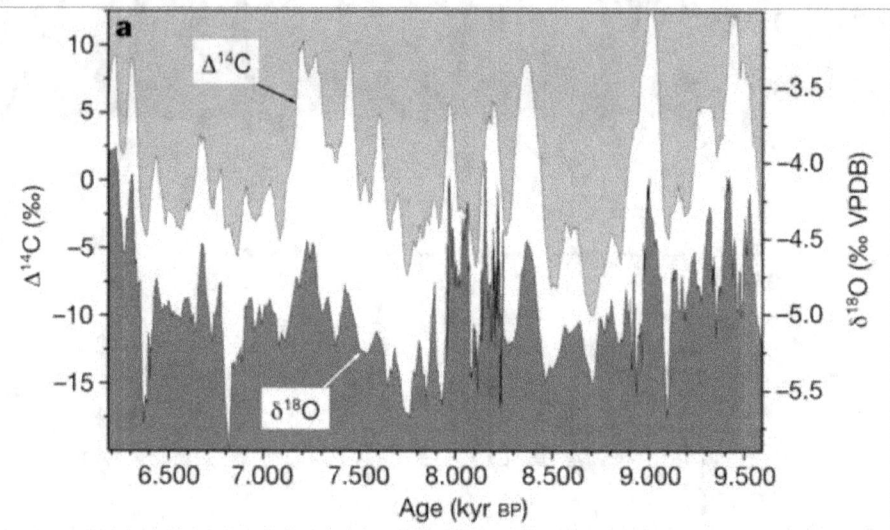

The correlation between solar activity—as mirrored in the ^{14}C flux, and a climate sensitivity variable, the $^{18}O/^{16}O$ isotope ratio from stalagmites in a cave in Oman, on a centennial to millennial time scale. The ^{14}C is reconstructed from tree rings. It is a proxy of solar activity since a more active sun has a stronger solar wind which reduces the flux of cosmic rays reaching Earth from outside the solar system. A reduced cosmic ray flux, will in turn reduce the spallation of nitrogen and oxygen and with it the formation of ^{14}C. On the other hand, $^{18}O/^{16}O$ reflects the temperature of the Indian ocean—the source of the water that formed the stalagmites. (Graph from Neff et al., 2001

Figure 88. Solar activity (in white) correlates extremely closely with ocean temperatures (dark grey).
Source: Astronomer Nir Shaviv's Science website:
http://www.sciencebits.com/CosmicRaysClimate#MSJGR
with data from Neff et al., 2001. Nature., 411.
https://www.geo.umass.edu/climate/papers2/neffetalnature2001.pdf

The story told by the previous two charts makes it clear that the Sun is the primary driver of climate changes on our planet, at least on a short time scale of a few thousand years — the Sun's changing magnetic field controls how much solar and cosmic radiation reaches our planet and, by consequence, strongly influences cloud cover on our planet. As Henrik Svensmark, a physicist at the Danish National Space Institute in Copenhagen, points out in a Danish television documentary about his research, *"Instead of thinking of clouds as something being a result of the climate, it's actually sort of upside down. The climate is a result of changes in the clouds."*[5]

Once again, direct geological evidence preserved in the rocks beneath our feet confirm the link between our climate and solar activity. In the Dead Sea region of Israel there are huge deposits of sediments, called the Lisan Formation (figure 89), which formed at the bottom of a prehistoric lake called Lake Lisan as it slowly dried up starting 70,000 years ago until it disappeared 12,000 years ago. The alternating sedimentary layers were created by evaporation as freshwater drained into the basin during the rainy season and then dried out during the hot summer. Lake Lisan is gone now — all that remains are the hypersaline Dead Sea at the southern tip and the Sea of Galilee at the northern tip. Layer upon layer, these evaporite sediments created a permanent record of the changing climate over almost 60,000 years, as shown in the photo in figure 89. Dark layers formed during the cooler winter when evaporation was slow, while light layers formed during the hot summer when evaporation was fast. Thin light-colored layers are evidence of cool summers while thick light-colored layers are evidence of hot summers. Layer by layer, the thickness of the evaporite bands show how the climate warmed and cooled over 60,000 years. And once again, those patterns correspond perfectly with well-documented short- and long-term cyclical changes in solar activity.[6]

---

[5] Swiss Policy Research. (2016, May 2). *The Cloud Mystery (2008 Documentary).* [Video]. Odysee. https://odysee.com/@swprs:3/the-cloud-mystery-svensmark-2008:7

[6] Prasad et al. (2004). Evidence from Lake Lisan of solar influence on decadal- to centennial-scale climate variability during marine oxygen isotope stage 2. *Geology* 32(7): 581–584. https://doi.org/10.1130/G20553.1

Figure 89. Lisan Formation evaporite layers, Dead Sea, Israel. Dark sediment bands were deposited in winter, light bands were deposited in summer — together these alternating layers provide a climate record spanning over 60,000 years.
Source: Wikimedia Commons, CC BY 3.0 DEED,
https://upload.wikimedia.org/wikipedia/commons/e/eb/Lisan_Formation.jpg

But Professor Svensmark and his team went still further by asking a simple question:[7] If cosmic radiation has such a profound effect on cloud cover, is the Sun's changing magnetic field the only thing that affects how many of those cosmic rays reach our planet, or is there anything outside of our solar system that could cause cyclical changes in how much cosmic radiation reaches our solar system, which could explain even longer-scale climatic changes here on Earth? In other words, as our solar system travels through space, does it encounter regions of space where there are higher concentrations of cosmic radiation? It turns out that there are. This is where things get really weird!

As our solar system orbits our Milky Way Galaxy, it passes through the spiral arms of our galaxy. These spiral arms are dense congested zones with much higher concentrations of stars than elsewhere in the galaxy. They are also flooded with massive volumes of leftover radiation from dying stars that exploded as supernova. Most of the more than four dozen supernova that happen within our Milky Way Galaxy every year happen

---

[7] "Hendrik Svensmark." *Wikipedia.* 13 Feb 2024.
  https://en.wikipedia.org/wiki/Henrik_Svensmark

somewhere inside these spiral arms. So, every time our solar system's orbit takes us through one of these spiral arms, our solar system is bombarded by a massive increase in cosmic radiation. And that means a lot more clouds forming in our lower atmosphere. As Professor Svensmark points out, while our Sun's changing magnetic field can increase or decrease the amount cosmic radiation reaching our planet by around 10%, when our solar system passes through one of these spiral arms the cosmic radiation bombarding our planet can increase by a whopping 100%.[8] That means a lot more clouds!

The picture shown in figure 90 from NASA shows our disc-shaped Milky Way Galaxy, along with an arrow pointing to the current location of our solar system within our galaxy. The four main spiral arms are clearly visible in the image, as are some of the minor spurs that branch off those spiral arms. The whole galaxy is spinning around its center, but our solar system is orbiting the center faster than those spiral arms.

---

[8] European Institute for Climate and Energy EIKE. (2023, Jun 30). *Henrik Svensmark - The Impact of Solar Activities and Cosmic Rays on the World Climate.* [Video]. YouTube.
https://www.youtube.com/watch?v=e_8w6JPLEl8&t=94s&ab_channel=European InstituteforClimateandEnergyEIKE

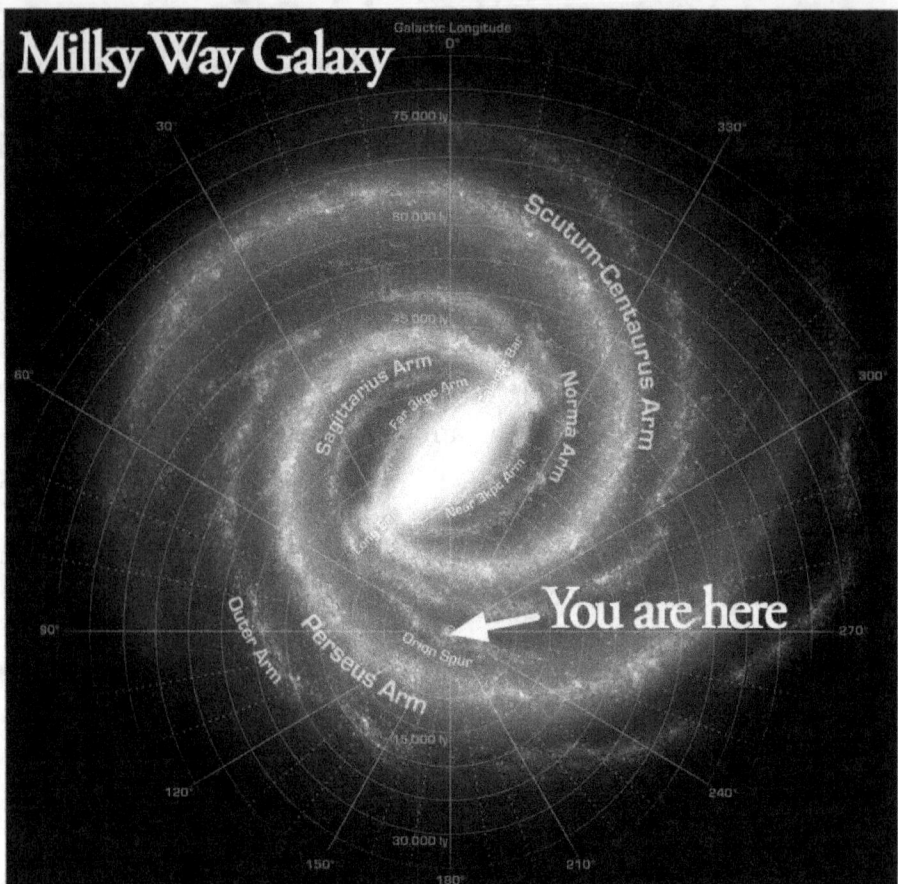

Figure 90. Our Milky Way Galaxy, showing the four main spiral arms of our galaxy along with the approximate location of our solar system on the Orion Spur of the Sagittarius Arm.

Adapted from NASA/JPL-Caltech/R.Hurt, CC BY 4.0 DEED

We are currently passing through the Orion-Cygnus Spur of the Sagittarius Arm. On average, we pass through one of these spiral arms every 140 million years or so. Professor Svensmark and his team discovered that each of our traverses through those spiral arms correlates perfectly with the long cold periods that plunged our planet into Icehouse conditions every 140 million years or so. Even the current glacial-interglacial cycles that we have been experiencing over the last 2.58 million years, collectively known as the Quaternary Glaciation, began after our solar system began traversing the Orion-Cygnus Spur.

The chart in figure 91 below from Professor Jan Veizer (Professor of Geology at the University of Ottawa) and Astronomer Nir Shaviv (Professor of Physics at the Racah Institute of Physics in Jerusalem) demonstrates the tight correlation between global temperatures and cosmic rays over the last 500 million years.[9] Professor Shaviv states that as our solar system orbits around our galaxy, the increased cloud cover caused by the higher levels of cosmic radiation inside these spiral arms has the effect of lowering global temperatures by at least 5 °C.[10]

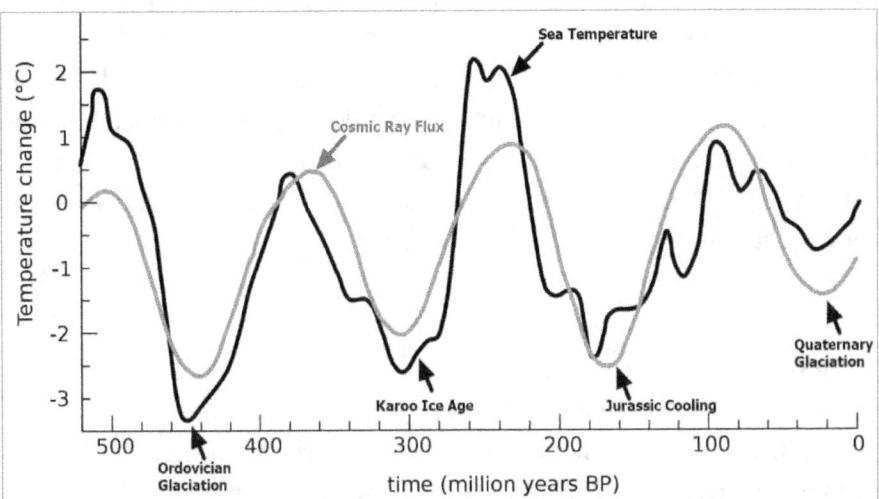

Figure 91. Correlation between variations in cosmic ray flux (grey) and change in sea temperature (black).

Adapted from Wikipedia (https://en.wikipedia.org/wiki/Henrik_Svensmark#), with data sourced from Shaviv and Veizer, 2003. Celestial driver of Phanerozoic climate? July 2003, GSA Today.
https://rock.geosociety.org/net/gsatoday/archive/13/7/pdf/i1052-5173-13-7-4.pdf

This link between glaciations and increased cosmic rays when our solar system crosses the spiral arms of our galaxy is further confirmed by geologists who have found a much higher abundance of isotopes like Beryllium-10, Carbon-14, and Oxygen-18 in sediments corresponding to ice age periods — these isotopes are all created by collisions with cosmic rays.[11] And so, the timing of our passages through the spiral arms of our

---

[9] "Hendrik Svensmark." *Wikipedia.* 13 Feb 2024.
https://en.wikipedia.org/wiki/Henrik_Svensmark

[10] Swiss Policy Research. (2016, May 2). *The Cloud Mystery (2008 Documentary).* [Video]. Odysee. https://odysee.com/@swprs:3/the-cloud-mystery-svensmark-2008:7?t=1947

[11] Sankaran, A. V. (2008). Galactic triggering of geologic events in earth's history. *Current Science*, 95(6), 714–716. http://www.jstor.org/stable/24102576

galaxy has been recorded by isotopes trapped within the sedimentary layers in the rocks beneath our feet.

Another way to test the correlation is to look at the fraction of marine sediments that contain organic matter — there's more organic matter that will get trapped in sediments accumulating at the bottom of the ocean whenever the planet's biomass is higher (during warm periods) and less when biomass falls (during cold periods). In other words, as biomass falls during cold weather, global photosynthesis falls, which reduces the amount of carbon that gets removed from the atmosphere and subsequently gets trapped in carbon sinks at the bottom of the ocean. Once again, this record of the planet's fluctuating biomass corresponds with the multimillion-year cosmic ray flux as our solar system traversed the spiral arms of our galaxy — in this case the correlation holds up over the last 3.5 billion years of our planet's history![12] Our path through our galaxy is quite literally shaping temperatures here on Earth.

~

These spiral arm crossings are not just linked to cloud cover. They also correlate with the big mass extinction events on Earth — after all, these spiral arms are littered with all sorts of debris from exploded stars.[13] Although the many asteroids that have bombarded our planet (like the one that wiped out the dinosaurs 65 million years ago) are thought to have a more "local" origin in the shell of icy rocks that surrounds our solar system just beyond the last of our planets (this swarm of icy bodies is known as the Oort cloud), asteroid impacts during our planet's long geologic history are not randomly distributed in time; they also come in distinct waves that correspond with our solar system's traverses through spiral arms of the galaxy. Thus, as was pointed out by the author of a paper from 2008, *"it is suggested that these impacts may be the result of galactic quasi-periodic gravitational perturbations on the Oort Cloud, the source of the earth-impacting materials, during the orbit of the solar system along the galactic plane."*[14] In plain English, what the author of this study is saying

---

[12] Svensmark, H. (2022). Supernova Rates and Burial of Organic Matter. *Geophysical Research Letters.* 49(1). https://doi.org/10.1029/2021GL096376

[13] Sankaran, A. V. (2008). Galactic triggering of geologic events in earth's history. *Current Science*, 95(6), 714–716. http://www.jstor.org/stable/24102576

[14] Sankaran, A. V. (2008).

is that the increased gravitational pull from all the stars and cosmic debris contained inside the spiral arms of the galaxy could disturb icy rocks in the Oort cloud and launch some of those icy rocks towards us as our solar system passes through those spiral arms.

The graphic in figure 92 below is adapted from a diagram published in the peer-reviewed scientific journal *Current Science* (in collaboration with the Indian Academy of Sciences).[15] It illustrates our circular orbit through the Milky Way Galaxy as we cross spiral arms roughly every 140 million years or so. The graphic demonstrates the link between these spiral arms, cold periods on our planet, and mass extinction events as our orbit takes us through the debris and cosmic radiation left over from supernova explosions.

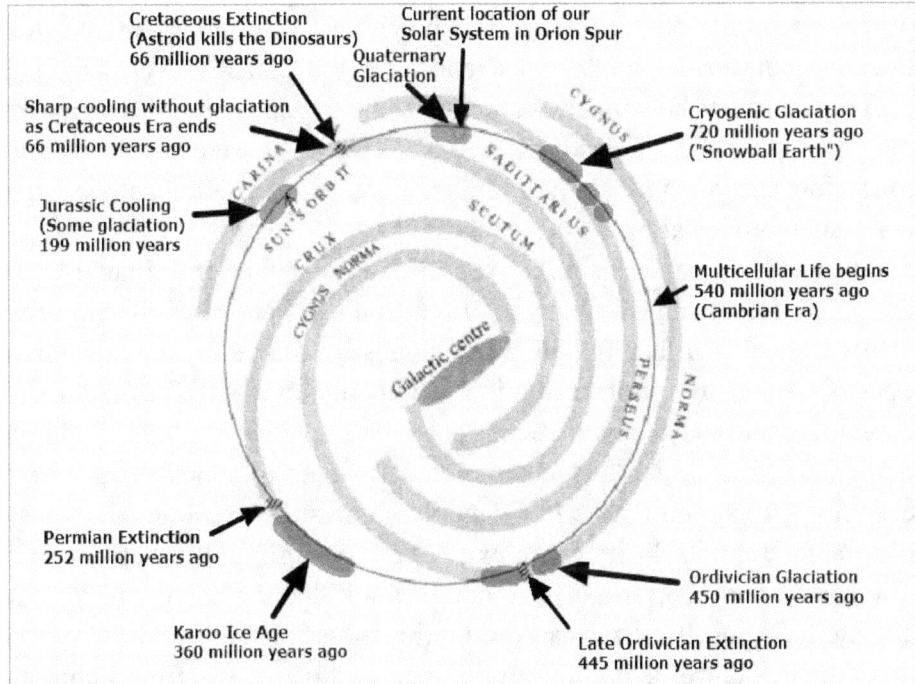

Figure 92. Annotated schematic view of the Milky Way Galaxy showing the orbit of the solar system crossing the spiral arms.
Adapted from Sankaran, 2008. Galactic triggering of geologic events in earth's history. Research News, Current Science, Vol. 95, No. 6 (25 September 2008), pp. 714-716.
https://www.currentscience.ac.in/Volumes/95/06/0714.pdf

---

[15] Sankaran, A. V. (2008).

The cooling caused by traversing these spiral arms is compounded by fluctuations in our Sun's solar activity (on both long and short timescales), which change how strongly we are protected from cosmic radiation by our Sun's ever-changing magnetic field. For example, over the last 100 years our Sun's magnetic field has doubled, which explains much of the warming of the past century as less cosmic radiation is able to penetrate our atmosphere to trigger the formation of clouds.[16] A stronger magnetic field = less cosmic rays reaching Earth's lower atmosphere = less cloud formation = a warmer climate.

Then factor in the slow drift of the continents (over millions of years), which changes how easily ice sheets can form, how easy it is for ocean currents to equalize water temperatures between the tropics and the poles, how many mountain ranges exist to trigger rainfall as trade winds push water vapour around the planet, how much heat is absorbed or reflected by large continental landmasses, and how the planet's ever-changing geography affects how much dust, pollen, and organic matter is carried up into the atmosphere by strong winds to serve as aerosols for cloud formation, and you have all the complex ingredients to periodically plunge our planet into ice ages.

Then add in wobbles in our Earth's orbit around our Sun (Milankovitch Cycles), which change how much of our Sun's radiative heat reaches our planet while also affecting how many clouds are formed by the ionization effect from incoming solar radiation. Compound that with fluctuations in the Earth's internal geomagnetic field, which also plays a role in shielding our planet from solar and cosmic radiation, and suddenly our ever-changing climate no longer looks nearly so mysterious, although it does illustrate the extraordinary complexity of our dynamic climate puzzle. It's a seductive and comforting fairy tale to think that we could control our climate by controlling our carbon dioxide emissions, but reality reveals that climate control is nothing more than an illusion. Far from being an isolated self-contained little biosphere rotating around our little Sun, we are, quite literally, at the mercy of interstellar winds. The powerful correlation between the Sun's magnetic field, cosmic rays, and cloud

---

[16] Lockwood, M., Stamper, R., and Wild, M.N. (1999). A Doubling of the Sun's Coronal Magnetic Field during the Last 100 Years. *Nature.* 399(6735). https://www.nature.com/articles/20867

formation should serve as the final nail in the coffin on the theory of "carbon dioxide as the control knob on the climate."

Of course, politics says otherwise.

As Professor Svensmark pointed out in the aforementioned documentary about his research, when he and his colleagues published their astonishing results in 1996, "*... we were of course very excited to present the results, but much to our surprise it was received very, very negatively. And the only thing we had done was to present scientific results that showed that the Sun, through the clouds, might be very, very important for climate. [...] The whole climate community really hated the idea that the Sun should have any major impact on climate. That was seen as a disaster.*"[17] He had difficulty getting published and has had extreme difficulty in securing funding for research ever since. The head of the IPCC even went as far as to declare that it was "irresponsible" to suggest that the Sun's effect on cloud cover, rather than CO_2, could be main driver of climate on Earth.[18]

As Professor Eugene Parker of the Department of Physics, Astronomy, and Astrophysics at the University of Chicago pointed out, "*Now that global warming is a political issue, things become politically incorrect ... we have cases of good solid research on global warming being refused for publication because somebody has made up his mind that that isn't the way that it is, and you can't publish.*"[19]

Back in 1990, Senator Al Gore infamously declared that any scientist who did not consider global warming an immediate "crisis" was being "unethical" (see figure 93 below)[20]. William Gray, professor of atmospheric science at Colorado State University and a pioneer in the science of hurricane forecasting stated that when Al Gore became Vice President in 1993, Al Gore invited Professor Gray to a global warming conference in Washington, DC. Professor Gray informed Al Gore that he would be happy to attend but wanted Gore to know that he disagreed with

---

[17] Swiss Policy Research. (2016, May 2). *The Cloud Mystery (2008 Documentary)*. [Video]. Odysee. https://odysee.com/@swprs:3/the-cloud-mystery-svensmark-2008:7

[18] Swiss Policy Research. (2016, May 2).

[19] Swiss Policy Research. (2016, May 2).

[20] *The Indianapolis Star.* Sunday, August 12, 1990. Page 102.
https://www.newspapers.com/newspage/106446342/ see also
https://realclimate.science/2023/12/11/al-gore-legitimate-science-is-unethical/

his global warming theories. Despite having receiving funding from NOAA for many decades, from that point forward Professor Gray was unable to secure another penny of federal government funding.[21]

> **THEN IN** March 1990, in the *Bulletin of the American Meteorological Society*, MIT's Richard Lindzen pointed out that whereas all of the climate models predict a warming of 5 degrees to 14 degrees Celsius in the Artic during the winter, from 1940-87, the Atlantic Artic winter temperatures have fallen 2.5 degrees.
> At his largely ignored April Interparliamentary Conference on Climate Change, Sen. Albert Gore, D-Tenn., was so frustrated by this, he charged any scientist who studied the Antarctic ice cores and did not consider global warming an immediate "crisis" was being "unethical."
>
> *The Indianapolis Star, 12 Aug 1990, Sun · Page 102*

Figure 93. In 1990, Al Gore declared that any scientist who did not view global warming as an immediate "crisis" was being "unethical."

Source: The Indianapolis Star, August 12th, 1990, page 102. (https://www.newspapers.com/image/106446342/)

As climate historian and geologist Tony Heller points out in his video about his friend, Professor Gray, Al Gore created the climate change consensus by cutting off federal funding to any academics who didn't agree with Al Gore's outrageous climate claims.[22] In a BBC article from 2000, professor Gray wrote that *"water vapour and cirrus cloudiness should be thought of as a negative rather than as positive feedback to human-induced – or anthropogenic greenhouse gas increases. No significant human-induced greenhouse gas warming can occur with such a negative feedback loop."*[23] Professor Gray has also stated that many

---

[21] Heller, T. (2023, Dec 10). *Nobel Laureate Al Gore.* [Video]. YouTube.
https://www.youtube.com/watch?v=fgHqL7qkxmg&ab_channel=TonyHeller

[22] Heller, T. (2023, Dec 10).

[23] Gray, W. M. (2000, Nov 16). Viewpoint: Get off warming bandwagon. *BBC News.* http://news.bbc.co.uk/2/hi/in_depth/sci_tech/2000/climate_change/1023334.stm

scientists only publicly support the scientific consensus on climate change because they are afraid of losing grant funding.[24] Furthermore, he pointed out that NOAA has made their grants conditional so that any research they fund <u>must</u> include some kind of algorithm-driven modelling component.[25] He who controls the consensus-determined formulas that are inputted into those algorithms controls the world.

Academia began its slow and tortured death spiral on the day that 19th-century progressives turned science into a tool of the state. Today we are reaping what our ancestors sowed as politicians like Al Gore perfect the art of weaponizing the vast power and resources of the bloated administrative state as both carrot and stick to promote their biases and strangle anything that doesn't suit their agenda. Would you risk being branded a heretic by a re-invigorated Inquisition? Would you toss an expensive education and a lucrative career in the dust bin by challenging the dominant narrative?

> *"Violence is not necessary to destroy a civilization. Each civilization dies from indifference toward the unique values which created it."*
>
> — Nicolás Gómez Dávila

---

[24] "William M. Gray." *Wikipedia.* 22 May 2024.
https://en.wikipedia.org/wiki/William_M._Gray

[25] DDPmeetings. (2016, May 28). *William Gray, PhD: Climate Change Driven by the Ocean – Not Humans.* [Video]. YouTube.
https://www.youtube.com/watch?v=beU_jVf-Z14&t=1824s&ab_channel=DDPmeetings

Chapter 16

CO_2 Fertilization: The Fuel for Life

"It is a strange fact, characteristic of the incomplete state of our current knowledge, that totally opposite conclusions are drawn about prehistoric conditions on Earth, depending on whether the problem is approached from the biological or the geophysical viewpoint."

— Alfred Wegener, The Origins of Continents and Oceans

The ice that turned the planet into Snowball Earth for much of the Cryogenic Period (from 720 to 630 million years ago) helped trigger the emergence of multicellular life as the harsh ice age conditions served as an incubator for new evolutionary mutations. Even as the planet was locked in a deep freeze, entire mountain ranges were reduced to powder by the grinding ice sheets. When the ice finally retreated, these finely ground mineral fertilizers were washed into the ocean, triggering an explosion of larger and more complex single-celled photosynthetic microbes, particularly green algae, and then, after at least one failed attempt at multicellular life and another of Earth's mass extinctions (possibly brought on by another passage through one of the spiral arms of our galaxy),[1] multicellular life finally permanently exploded onto the scene 540 million years ago in what is known as the Cambrian Explosion. It is also sometimes nicknamed the Biological Big Bang.

Figure 94 below shows a summary of the final 600 million years ago of our planet's history[2] — you might want to bookmark this chart so you can follow along visually during the events that are yet to come during the remainder of this geological detective story. By now you have a solid grasp of the dynamic forces that create our ever-changing climate here on Earth and the complex interplay between photosynthesis, oxygen

---

[1] Gillman, M. and Erenler, H. (2008). The galactic cycle of extinction. *International Journal of Astrobiology.* 7(1):17-26. https://doi.org/10.1017/S1473550408004047

[2] "Climate and the Carboniferous Period." *Geocraft.com.* 21 Mar 2009. https://geocraft.com/WVFossils/Carboniferous_climate.html

breathers, and carbon sinks that constantly circulate carbon dioxide in and out of the atmosphere. As the story proceeds, we will watch how these forces continually shape and reshape our planet as new watershed evolutionary developments emerge to disrupt the balance between them. By the time we reach the end of this journey through time, the ecological story of the century (revealed by rising CO_2 levels) will no longer be a mystery. And that knowledge will give us a clear path forward to prevent our chapter in this never-ending story from repeating the ecological collapses that turned the Great Plains into a Dust Bowl during the 1930s and turned the Soviet Union and Maoist China into graveyards for tens of millions under the watchful eye of arrogant and ideological central planners.

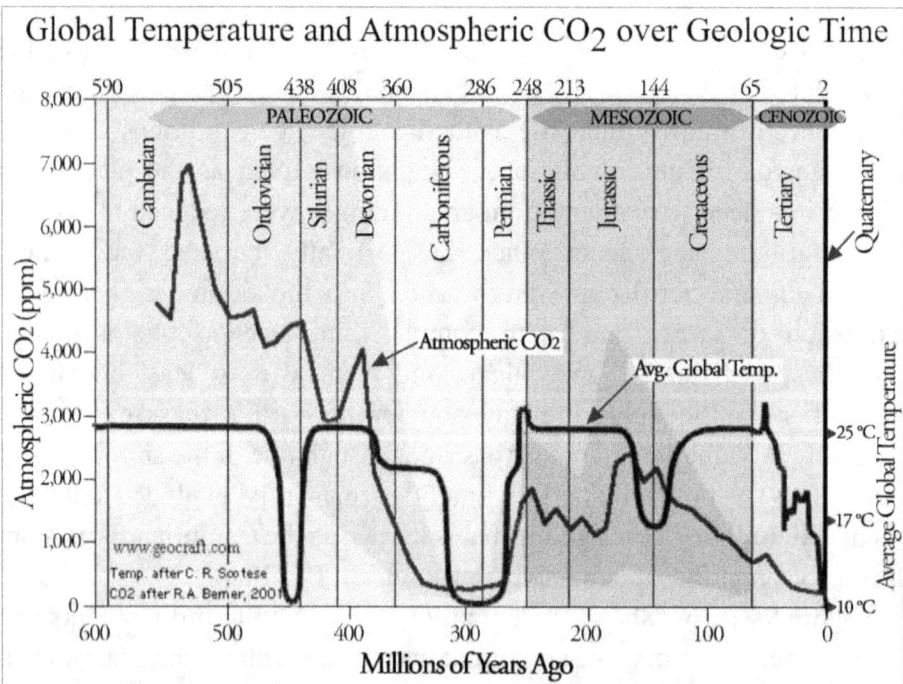

Figure 94. Geologic Time — reconstruction of the average global temperature (thick black line) and atmospheric CO2 (thin black line) over the last 600 million years.

Source: Geocraft.com (https://geocraft.com/WVFossils/Carboniferous_climate.html), with carbon dioxide after Berner (2001) and temperature after Scotese (2001; see also Boucot et al., 2004)

Before our story resumes, I just want to explain a couple of things about the chart in figure 94 to help you get oriented. There are two lines in chart that will help us follow the story: The thin black line shows how atmospheric CO_2 levels declined over these final 540 million years, from over 7,000 ppm at the time of the Cambrian Explosion to the paltry 400 ppm of today. The thick black line shows how our planet's climate has alternated back and forth between warm and cold periods, known as Hothouse Earth (with an average global temperature of around 22 °C) and Icehouse Earth (with an average global temperature of around 12 °C). Clearly, carbon dioxide levels are not controlling these alternating Hothouse and Icehouse conditions — there's absolutely no correlation between these two lines — though as we saw in the previous chapter, the timing of Icehouse conditions does correlate rather well with our solar system's traverses through the spiral arms of our galaxy.

Today, at an average annual global temperature of only 15 °C, we are still in one of those cold periods — even the warm temperatures we are experiencing today during our temporary interglacial reprieve from the bitter Icehouse conditions that have ruled our planet for most of the last 2.58 million years are still well below the tropical Hothouse conditions experienced by our planet whenever our solar system is outside of the spiral arms of our galaxy.

The labels across the top of the chart show the names that geologists use to describe different time periods during our planet's history — each period is marked by very unique planetary conditions that left their imprint in the rocks; thus, each geological period is like a different chapter in a very long book. Across the top of the chart you can see that geologic time is divided into three distinct eras — the Paleozoic Era (meaning "ancient life" in Greek in reference to the strange multicellular lifeforms that emerged to dominate our planet 540 million years ago; "paleo" means "ancient", "zoic" means "life"), the Mesozoic Era (meaning "middle life" in reference to the "Age of the Reptiles" when dinosaurs ruled the Earth), and the Cenozoic Era (which means "recent life" in reference to the mammals that emerged to rule the Earth after the dinosaurs went extinct).

Each of these eras is further divided into distinct geologic periods, such as the Cambrian Period (when multicellular life began), the Devonian Period (known as the Age of Fish and the period when plants and animals first came out of the oceans to colonize land), the Carboniferous (meaning "carbon-bearing" in reference to the long period in Earth's history when most of our planet's coal beds were formed in vast swamp forests), the Triassic Period (a period that began with a mass global extinction and ended with another mass global extinction while also ushering in the age of the dinosaurs), the Cretaceous Period (a period that ended when a large asteroid crashed into the Earth — the impact filled the atmosphere with dust and debris, which engulfed the planet in a decade-long period of cold and darkness that wiped out the dinosaurs), and the Tertiary Period (when mammals and grasslands and birds emerged, and the continents drifted to their present configuration).

As a quick side note, there's an abbreviation used by geologists, biologists, and astronomers to make it easier to discuss the massive time periods involved. "Millions of years ago" is typically abbreviated as "mya". Likewise, "bya" refers to billions of years, while "kya" refers to thousands of years ago.

The Cambrian Period (540 - 485 mya) spanned another of our planet's tropical Hothouse periods. None of the landmasses were located near the poles, sea levels were very high, and another supercontinent was breaking apart, which created vast shallow warmwater seas all around the partially submerged continental plates. All the conditions were ripe for a vast explosion of multicellular life in the oceans.

Incidentally, this is also the period when corals first evolved — they are one of the few species that have survived over the full span of these 540 million years even as the Earth's climate shifted back and forth between bitter Icehouse glaciations and tropical Hothouse periods. It's farcical to hear the climate brigade warning us that a warmer climate and higher CO_2 levels are an existential threat to corals considering that temperatures were much warmer during most of those 540 million years than they are today — corals evolved when average global temperatures were at least 10 °C warmer than they are today. Furthermore, corals first emerged at a time when CO_2 levels were at least 17x higher than they are today and thrived for over 540 million years despite the fact that CO_2 levels were significantly higher during most of those 540 million years than they are today. Yet, "everybody knows" through the force of endless repetition by media and activist-scientists that Australia's Great Barrier Reef is allegedly bleaching/dying because of climate change. And yet, you rarely hear from the dissenting voices.

As reported by Reuters, Professor Peter Ridd is the former head of the Physics department at James Cook University (JCU) in Australia as well as the former head of JCU's Marine Geophysical Laboratory. Professor Ridd points out that *"most bleached coral fully recovers"* and that *"the [Great Barrier] reef presently has record high, or near record high, coral* cover in 2020/21." According to Professor Ridd, coral is the *"least endangered of any ecosystem to future climate change."*[3] He has also written that *"the public are constantly told that reefs are being irreparably damaged by global warming, but bleaching events, about which there is so much doom-mongering, are simply corals' natural response to changes in the environment. They are an extraordinarily adaptable lifeform, and bleaching events are almost always followed by rapid recovery. [...] Corals get energy from a symbiotic relationship with various species of algae. When environmental conditions change, they can rapidly switch to a different species that is better suited to the new conditions. This shapeshifting means that most setbacks they suffer will be short-lived."*[4]

However, as reported by the Institute of Public Affairs in Australia, Professor Ridd was terminated by JCU after a 30-year career after he criticized colleagues, accused research institutions of being untrustworthy, and called for better quality assurance in Great Barrier Reef science.[5] He was initially awarded $1.2m in compensation by a federal circuit court for wrongful dismissal, arguing that he was protected by his academic freedom of speech, but JCU appealed the decision and Australia's superior court shamefully overturned his initial verdict and sided with JCU. And yet, just as Professor Ridd predicted, in 2022 the Australian Institute of Marine Science (an Australian government agency) reported that *"The northern and central Great Barrier Reef have recorded their highest amount of coral cover since the Australian Institute of Marine Science (AIMS) began monitoring 36 years ago."*[6] Oops.

Professor Ridd's case is reminiscent of the experience of another scientist, Dr. Susan Crockford, a zoologist and polar bear expert who was terminated by the University of Victoria (UVic) in Canada where she had been working as an unpaid adjunct professor. Although UVic had even

---

[3] "Peter Ridd" *DeSmog*. Accessed on 2 Mar 2024. https://www.desmog.com/peter-ridd/

[4] "Coral in a Warming World." (2023, Feb 8). *The Global Warming Policy Foundation*. https://www.thegwpf.org/publications/worlds-coral-reefs-not-declining-new-paper-reveals/

[5] Marohasy, J. (2021, June). *The Science Behind Peter Ridd's Dismissal*. Institute of Public Affairs. https://ipa.org.au/wp-content/uploads/2021/07/The-Science-Behind-Peter-Ridds-Dismissal-Jennifer-Marohasy.pdf

[6] Koopman, D. (2022, Aug 4). Highest coral cover in central, northern Reef in 36 years. [Press release]. *Australian Institute of Marine Science – Australian Government*. https://www.aims.gov.au/information-centre/news-and-stories/highest-coral-cover-central-northern-reef-36-years

previously promoted her work and had already renewed her 3-year contract multiple times, her blog (polarbearscience.com) and her public talks with school kids were ruffling feathers among the climate brigade, particularly when she pointed out that polar bears are thriving with populations now 5x higher than they were in the 1950s and that they are at no risk of extinction from climate change.[7] In 2019, UVic rejected her renewal application.[8] They did not provide a reason for refusing to renew her contract and refused to give her an opportunity to appeal their decision. Dr. Crockford believes that UVic bowed to outside pressure to terminate her position at the university. She described her experience as *"an academic hanging without a trial, conducted behind closed doors."*[9] Such is the state of freedom of speech in academia. Punish one, educate one hundred.

But let's get back to our geological story.

From the moment that the explosion of life began during the Cambrian Period, CO_2 levels have been falling, from 7,000 ppm during the Cambrian to as low as 180 ppm at the bottom of the last ice age, with only a slight recovery to 420 ppm today. But why did CO_2 levels fall across those 540 million years when, according to the carbon cycle I described earlier, oxygen breathers ought to be returning CO_2 to the atmosphere just as fast as photosynthesis removes it. For the most part, that Yin Yang balance holds true. For every molecule of CO_2 removed from the atmosphere by photosynthesis, one oxygen molecule (O_2) is released into the atmosphere, but then oxygen breathers use that oxygen molecule (O_2) to oxidize/metabolize carbon-based plant and animal tissues, thereby creating and exhaling one new CO_2 molecule back into the atmosphere. It's a perfect circle… except when some carbon leaks out of the circle to get trapped in long-term carbon sinks, like limestones and carbon-rich shales that form at the bottom of the oceans, coal beds that form in

---

[7] Miltimore, J. (2019, Sept 9). The Myth That the Polar Bear Population Is Declining. *Foundation for Economic Freedom.* https://fee.org/articles/the-myth-that-the-polar-bear-population-is-declining/

[8] Rotter, C. (2019, Oct 16). Dr. Susan Crockford: UVic bows to outside pressure and rescinds my adjunct professor status. *Watts Up With That* https://wattsupwiththat.com/2019/10/16/uvic-bows-to-outside-pressure-and-rescinds-my-dr-susan-crockford-adjunct-professor-status/

[9] Laframboise, D. (n.d.) The Academic Lynching Of Polar Bear Expert Susan Crockford. *Climate Change Dispatch.* https://climatechangedispatch.com/lynching-polar-bear-expert-crockford/

swamps, seashells that accumulate at the bottom of the ocean, rock weathering via carbonic acid in rainfall, and humus that builds up in the soil. Unfortunately for Life on Earth, there is no way to plug that leak. The long-term trend is down — at some point there won't be enough carbon left in the atmosphere to sustain life. Our planet has come close to that point several times already over these past 540 million years, most recently during each of the cold glacial intervals of our current ice age when atmospheric CO_2 levels dropped as low as 180 ppm during the coldest part of each glacial interval.

You can see evidence of this relentless drain on atmospheric carbon in the geologic time chart in figure 94, which shows carbon dioxide falling especially rapidly during the Cambrian Period from around 7,000 ppm to around 4,500 ppm. The explosion of life in the vast shallow Cambrian seas led to a rapid accumulation of buried carbon-rich sediments at the bottom of the oceans. Furthermore, without plants to cover the continents (land plants hadn't evolved yet), every inch of the exposed continents was constantly being weathered by rainfall — weathering of bare rocks is a significant source of carbon depletion from the atmosphere.

During the next geologic period, called the Ordovician (485-443 mya), CO_2 levels stabilized somewhat. That's also the period when the very first plants colonized land; these mosses, lichens, etc. slowly covered the planet's surface with vegetation, which helped slow the amount of CO_2 being lost from the atmosphere via rock weathering by coating the land with a protective shield of vegetation. Vegetation is just as important today to shield soils from oxidization — in particular the carbon-rich humus in soils, which is so easily oxidized and turned back into CO_2 when it comes in contact with the atmosphere. In upcoming chapters we'll come back to this essential role of vegetation in protecting humus from oxidation because it's so incredibly important to the greater detective story.

The vast stretches of shallow Cambrian seas disappeared during the Ordovician as the continents converged to form another vast supercontinent, which geologists call the Gondwana supercontinent. At its peak, this single supercontinent would cover one-fifth of the planet's surface! As continental drift continued, that supercontinent inched its way towards the South Pole where it came to rest by the end of the Ordovician, much like Antarctica today except that Gondwana was much, much bigger. Around the same time as Gondwana settled at the South Pole, our solar system crossed through another spiral arm of our galaxy — the Norma Arm. This one-two punch triggered another vicious ice age, called the Ordovician Glaciation. Though not quite as severe as Snowball Earth during the Cryogenic Period (the oceans didn't completely freeze over this time), the Ordovician Glaciation was far more extreme than any of our recent ice ages. Geologists estimate that 83% the planet's marine life went extinct during the Ordovician Glaciation!

In figure 94 you can see that even as temperatures fell during the Ordovician Glaciation, carbon dioxide levels in the atmosphere *rose* by around 500 ppm from around 4,000 ppm back up to 4,500 ppm. As the planet froze, photosynthesis slowed down and thus reduced the drain on carbon dioxide from the atmosphere. With less photosynthesis to remove CO_2 from the atmosphere and with ice sheets shielding most of the land from weathering, volcanic CO_2 emissions had a chance to catch up to and then outpace the amount of CO_2 lost to permanent carbon sinks and rock weathering. It's an important lesson for what is to come.

The Ordovician was a period of intense volcanic activity that lasted millions of years, triggered by the movement of the Earth's tectonic plates and by the collision of the continental plates that formed the Gondwana supercontinent. Some of the largest volcanic eruptions of the last 600 million years happened during the Ordovician. For comparison, the 1980 Mt. St. Helen's volcanic eruption ejected approximately 0.21 km^3 of rock-equivalent ash into the atmosphere. By contrast, there's a single bed of ash from a single eruption that happened 454 million years ago, which covered almost the entire eastern half of the present-day USA with a single layer of ash up to a meter thick (3.3 ft) — this enormous single blast sent over 1,140 km^3 of rock-equivalent ash into the atmosphere, more than 5,400x the amount of ash produced by Mt. St. Helen's!

A lot of things had to happen simultaneously to temporarily reverse the relentless depletion of CO_2 from the atmosphere that had been going on since the Cambrian — a surge in volcanic eruptions added more CO_2, the ice sheets protected the continents from weathering, and the cold weather and the ice reduced the drain that photosynthesis put on atmospheric CO_2. You can see from the chart in figure 94 that there was a second period of intense volcanic activity during the early part of the Devonian Period (419-359 mya), which also temporarily reversed the steady depletion of CO_2 from the atmosphere, but in either case the brief recovery didn't last long before CO_2 levels started falling again. Despite short surges, weathering and photosynthesis soon consumes any temporary CO_2 gains.

Over the next 160 million years, CO_2 levels plunged by a staggering 96% as CO_2 was systematically drained out of the atmosphere and deposited in the planet's long-term carbon sinks. By the beginning of the Silurian Period (443 mya), CO_2 had fallen to around 4,500 ppm; by the start of the Carboniferous Period (359 mya), it had dropped to around 1,300 ppm; by the time we reach the Carboniferous-Permian boundary (299 mya), atmospheric CO_2 had reached a life-threatening low of around 150 ppm[10] — even lower than the levels seen during our most recent ice age.

The image in figure 95 below shows how plant and root growth is affected by the level of CO_2 in the atmosphere — all four of the plants shown in the image are 14 days old; the only difference in their growing conditions was the amount of available CO_2 in the air while they were grown.[11] Below 150 ppm, plants are unable to absorb CO_2 from the atmosphere and photosynthesis grinds to a stop. The only reason Life survived the CO_2 plunge during the Carboniferous is because photosynthesis rates slowed in response to falling CO_2 until the amount of CO_2 consumed by photosynthesis and rock weathering reached some kind of equilibrium with the amount of new CO_2 being continually released into the atmosphere by volcanoes. But we got very, very lucky — if the volcanoes had been any less active, or if rock weathering had been faster, CO_2 would have fallen below the critical threshold for photosynthesis and Life would have been snuffed out. Life on Earth dodged a bullet.

---

[10] Beerling, D.J. (2002). Low atmospheric CO_2 levels during the Permo-Carboniferous glaciation inferred from fossil lycopsids. *PNAS* 99(20). https://doi.org/10.1073/pnas.202304999

[11] Gerhart, L. M. and Ward, J.K. (2010). Plant responses to low [CO_2] of the past. *New Phytologist*. 188(3). https://doi.org/10.1111/j.1469-8137.2010.03441.x

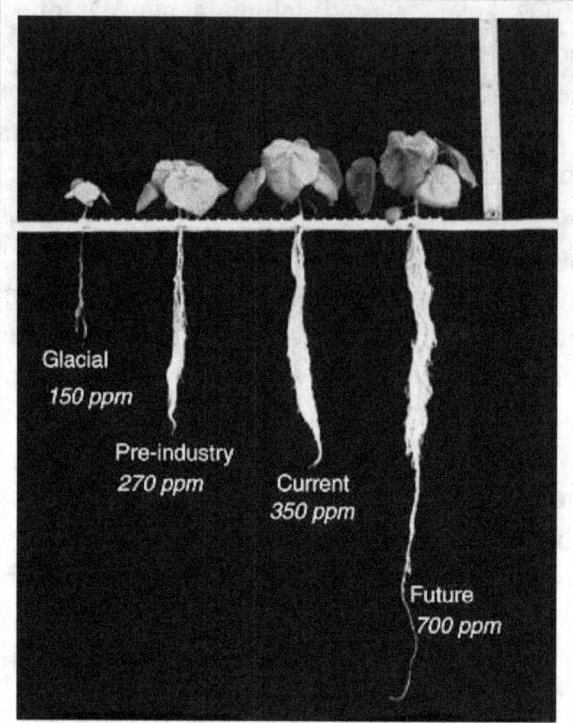

Representative plants of *Abutilon theophrasti* (C_3) grown at glacial through future [CO_2]. All plants were 14 d of age and were grown under similar water, light, and nutrient conditions. These plants were photographed during a study by Dippery et al. (1995). (Photograph is courtesy of Anne Hartley, Florida Gulf Coast University.)

Figure 95. The effect of atmospheric CO_2 concentration on plant growth. CO_2 is plant food.

Source: Gerhart and Ward, 2010. Plant responses to low [CO_2] of the past. New Phytologist, 188: 674-695.
https://nph.onlinelibrary.wiley.com/doi/10.1111/j.1469-8137.2010.03441.x

The chart in figure 96, published by the Oklahoma State University in its *Greenhouse Carbon Dioxide Supplementation Guide*, puts some numbers to the CO_2 fertilization effect.[12] A doubling of atmospheric CO_2

---

[12] Poudel, M. and Dunn, B. (2023, Sept). Greenhouse Carbon Dioxide Supplementation (HLA-6723). [Factsheet]. *Oklahoma State University Extension.*
https://extension.okstate.edu/fact-sheets/greenhouse-carbon-dioxide-supplementation.html

from current levels would effectively double plant growth rates — today, at 420 ppm, plants are still effectively being starved for CO_2. Optimal CO_2 levels for plant growth are between 1,000 to 1,300 ppm (basically the same as the level at the very beginning of the Carboniferous Period as land plants rapidly colonized the ice-free continents). CO_2 only starts to become toxic to plants above 2,000 ppm, and toxic to people above 4,000 ppm.

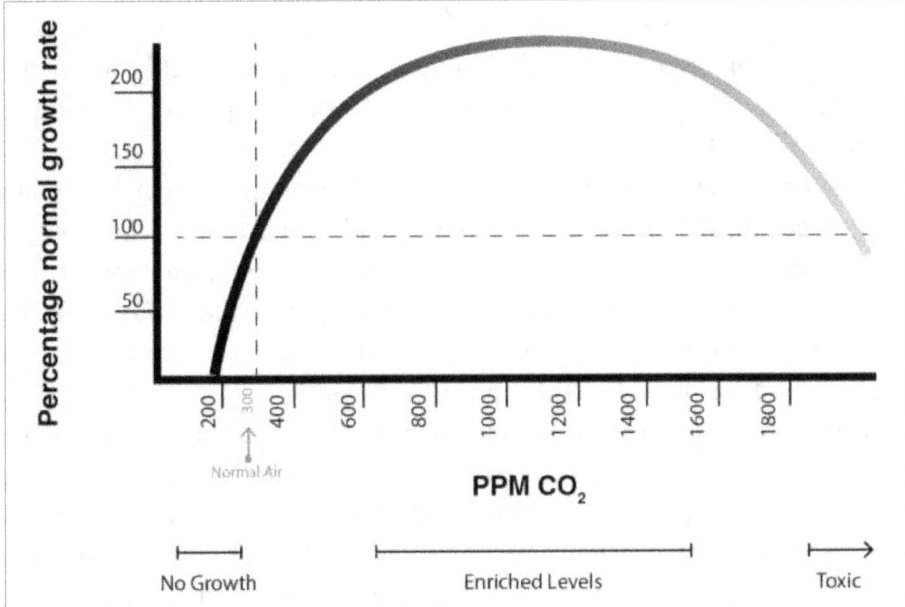

Relation between CO_2 concentration and rate of plant growth. Source: Roger H. Thayer, Eco Enterprises, hydrofarm.com.

Figure 96. The relationship between CO_2 concentration and the rate of plant growth.

Source: Oklahoma State University Agricultural Extension, Greenhouse Carbon Dioxide Supplementation
https://extension.okstate.edu/fact-sheets/greenhouse-carbon-dioxide-supplementation.html

Furthermore, at optimal levels of CO_2, plants don't have to open their pores as wide to absorb CO_2 from the atmosphere. At current CO_2 levels, most plants typically lose about 100 water molecules into the air for every single CO_2 molecule that they capture because they have to open their pores (a.k.a. stomata) so wide to capture enough CO_2.[13] As CO_2 levels rise, it becomes easier for plants to absorb all the CO_2 they need, so plants naturally respond by reducing the number of stomata. As physicist Dr. William Happer points out, if you look at elm leaves under a microscope, leaves grown in 1850 (back when CO_2 hovered near 280 ppm) had 30 to 40% more pores per square centimeter than elm leaves have today.[14] The fact that plants are able to respond to changing CO_2 levels in this way provides further evidence that atmospheric CO_2 levels have never been stable in the atmosphere — CO_2 levels have changed so often throughout history that plants have had to evolve the ability to quickly adapt the number and size of their pores in order to keep up with fluctuating atmospheric CO_2 levels.

Many commercial greenhouse growers today pump additional CO_2 into their greenhouses in order to boost yields and reduce their water consumption. Most burn propane or natural gas and then separate the pollutants from the CO_2 gas so they can pump that CO_2 into the greenhouse. CO_2 is *not* a pollutant — it is, quite literally, plant food. It is the very essence of Life on this planet.

The precipitous decline in atmospheric CO_2 that occurred from the Silurian to the Carboniferous Period was an interesting time in the evolution of our planet, with important lessons for us today. The most bizarre creatures of that era were the explosion of fish living in the oceans during the Devonian Period (the time period sandwiched between the Silurian and the Carboniferous) — including armoured fish (imagine the predatory pressures that led fish to evolve this kind of heavy armour!). Insects and other arthropods first colonized land during the early Devonian. By the late Devonian, amphibians also started leaving the water to colonize land, away from all those hungry armoured fish, but their reprieve from predatory pressures was short-lived as amphibious predators soon followed them out of the water. However, the most important evolutionary innovation during this period was the evolution of vascular plants, like ferns, trees, and flowering plants, which are able to transport nutrients and water throughout the entire plant via a complex

---

[13] Old Guard Summit. (2023, Oct 4). *"CO_2, The Gas of Life"-Dr. William Happer*. [Video]. YouTube. https://www.youtube.com/watch?v=tXJ7UZjFDHU&t=2379s&ab_channel=OldGuardSummit

[14] Old Guard Summit. (2023, Oct 4).

vascular network (that same vascular network is what mountain pine beetles target in pine trees in Western Canada today).

Plants with vascular systems were able to grow much larger than earlier more primitive plants because their vascular systems gave them the ability to transport water and nutrients to distant branches and limbs. This vascular system also enabled them to support an extensive root network to cast a much wider and deeper net in the soil from which to draw nutrients and water into the plant. By tapping into deeper soil layers, root systems enabled plants to thrive in far more challenging environments on land because they could tap into groundwater instead of having to rely on surface water to meet their needs. Thus, every square inch of the once barren continents turned green. Vascular systems also enabled plants to grow much taller than earlier plant species, which launched an evolutionary arms race to try to monopolize access to sunlight. The Carboniferous period (359–299 mya) — a golden age for mega-sized plants — was about to begin, with trees and ferns towering up to 100 feet into the sky. However, large herbivores had not yet evolved to be able to digest plant tissues and recycle CO_2 back into the atmosphere. With so many new plant varieties colonizing the continents and with so few large oxygen-breathing animals to eat them and recycle CO_2 back to the air, atmospheric CO_2 depletion accelerated.

Two additional plant innovations during this era were about to change the world forever — the evolution of lignin and cellulose. Today, cellulose is the most abundant organic compound on Earth, followed by lignin in second place. Both of these tough fibrous plant tissues are used by plants to create the durable woody fibrous material that trees, grasses, and other vascular plants use to give their stalks strength and to serve as a protective barrier to keep water in and pests and parasites out. And neither rot easily. And so, as plants died and lignin and cellulose began to accumulate in the soil, a whole new ecosystem is about to emerge underground — *humus* — that mysterious organic material at the heart of the soil ecosystem, which is ultimately at the center of a large portion of this entire detective story.

Chapter 17

The Story of Humus and the Hidden World Beneath our Feet

"How dare you treat your soil like dirt!"

— Joel Salatin, author of "Folks, This Ain't Normal: A Farmer's Advice for Happier Hens, Healthier People, and a Better World."

Lush, young, green plant growth has very little lignin and very little cellulose so when it dies (or when a farmer plows it into the ground) most of this carbon-rich tissue will be completely digested by oxygen-breathing microbes living in the soil and will be returned to the atmosphere as CO_2. Meanwhile, the nutrients that these microbes extract from the green plant tissues become fertilizer for the next generation of plants. It's common for farmers to periodically plant what are known as a "green manure" crops, which get plowed back into the ground before they reach maturity so they can decompose quickly to create a flush of nutrients for the next crop. While these green manure crops increase nitrogen, phosphorus, and other plant nutrients in the soil, they do not add much stable long-term organic. To add carbon, you need "brown matter" (as it is called in the gardening industry) — tough woody cellulose- and lignin-rich plant fibers.

The stable long-term organic component of soil — humus — is generated from tough lignin- and cellulose-rich fibers found in stalks, stems, and other woody plant tissues that form as plants mature. Humus is created as these tough fibers are broken apart into small carbon particles as they are composted/digested by soil microbes.

Humus serves as the primary ecosystem in which soil microbes, fungi, and most plant roots live. In essence, the addition of humus to the soil is what converts inorganic sediments (sand, silt, clay, etc.) into actual fertile soil — without humus, these inorganic sediments are little more than a barren mass for anchoring roots in the ground. Once humus is stripped away, the only life that still thrives on the barren soil (without the addition of manure or chemical fertilizers) are primitive cyanobacteria, mosses, and lichens (all of which we met earlier in our journey through geologic time), which survive by directly dissolving nutrients from rocks, sand, silt, and clay.

Prior to the 20th century, scientists viewed soil as little more than a medium to anchor roots so that plants wouldn't fall down. Much like a modern hydroponics system, water and manure were thought to be the chief sources of nutrients for plants. The soil itself wasn't viewed as a living mass — in their minds it was nothing more than lifeless dirt. Soil science emerged as a new scientific discipline in the late 1800s after a Russian geologist by the name of Vasily Dokuchaev published a book in 1883 called *Russian Chernozem*, named after the rich soils of the Russian steppe. Dokuchaev is widely regarded as the father of soil science, a title he shares with Charles Darwin after Darwin published his 1881 book about earthworms and their central role in the creation of soil — yes, the same Darwin famed for his theory of evolution by natural election. I'll be talking more about Darwin's earthworms shortly.

Dokuchaev recognized soil was a living ecosystem that extracted/dissolved fertilizer nutrients from inorganic sediments, stored these nutrients in a biologically available form, retained moisture, and played host to a vast symbiotic relationship between plant roots, bacteria, fungi, organic matter, insects, earthworms, and so on, all working in concert in a mutually beneficial self-reinforcing ecosystem. Humus makes this subterranean ecosystem possible — without it, the symbiotic relationship between all the living soil microorganisms would break down. The term *humus* comes from Latin — it means "living soil".

Fungi play a central role at the heart of that "living" soil ecosystem — geologists have been able to trace the start of the mutualistic relationship between plants and fungi all the way back to the time when plants were first colonizing land more than 400 million years ago. In this symbiotic relationship, fungi colonize the roots of plants and then build a massive fungal network (called mycorrhiza) *that extends the reach of plant roots by over 700x.*[1] Fungi secrete acids that dissolve nutrients from inorganic sediments and then transport these nutrients back to the plants. In return, the plants provide the fungi with sugars, carbon, and energy. These mycorrhizal networks are all interconnected, effectively linking plants

---

[1] Government of Wales. (July 2022). *Fungi and roots.* [Factsheet].
https://businesswales.gov.wales/farmingconnect/sites/farmingconnect/files/documents/CFf_Factsheet_Fungi%26Roots.pdf

together underground — a study reported on by the Guardian newspaper in 2021 demonstrated that, *"globally, the total length of fungal mycelium in the top 10cm of soil is more than 450 quadrillion km: about half the width of our galaxy."*[2]

Tillage agriculture (i.e. the plow) temporarily destroys this mycorrhizal network by chopping it up into bits. Tillage also mixes large quantities of oxygen into the soil where it can oxidize the exposed humus, along with all these delicate fungal networks. The growing shift to no-till farming, pioneered by Great Plains farmers (using herbicides to control weeds) is showing great promise not only to reverse humus degradation and soil erosion, but also to boost fertility and reduce fertilizer consumption because no-till leaves these mycorrhizal networks intact and limits the oxidation of carbon in the soil. No-till also reduces evaporation and conserves water because it doesn't bury the plant litter that helps shield the soil from the sun (effectively using plant debris as a layer of mulch). However, adapting no-till techniques to organic production systems is a challenge because, in the absence of herbicides, tillage is often the only other option to deal with weeds in large-scale monoculture crops. The invention of robot weeders, similar to robot vacuum cleaners and robot lawn mowers, may provide a technological solution to enable no-till systems to expand to organic farms.

Like tillage, clearcut logging has been shown to reduce mycorrhizal populations in the soil by up to 95% (and reduce the diversity of mycorrhizal species within the clearcut area by up to 75%) — this collapse in mycorrhizal networks does not happen when tree stands incorporate a wider range of species and when logging doesn't remove all the trees at once.[3] In practical terms, this means using select logging whenever it's feasible, using smaller block sizes when select logging is not feasible, avoiding monoculture if possible, retaining rare tree species within the mix, and trying to minimize the kind of soil disturbance that churns soil and causes erosion. In short, soil erosion doesn't just make forests more

---

[2] Kiers, T. and Sheldrake, M. ((2021, Nov 30). A powerful and underappreciated ally in the climate crisis? Fungi. *The Guardian.* https://www.theguardian.com/commentisfree/2021/nov/30/fungi-climate-crisis-ally

[3] Sterkenburg et al. (2019). The significance of retention trees for survival of ectomycorrhizal fungi in clear-cut Scots pine forests. *Journal of Applied Ecology.* 56(6). 1367-1378. https://doi.org/10.1111/1365-2664.13763

vulnerable to nutrient deficiencies and drought, it also destroys the symbiotic network of fungi that supports plant roots.

Over 80% of all carbon found in terrestrial ecosystems is found in the soil — a full 2.5 trillion tonnes of carbon![4] — the majority of this carbon is in the form of humus and the mycorrhizal fungal networks living within it. This combination of humus and fungi creates and stores the deep reservoir of nitrogen, phosphorus, potassium, trace minerals, and other nutrients that sustain complex plant life.

In addition to its role in managing soil fertility, humus also gives soil its spongey, aerated texture so that oxygen-breathing microorganisms like bacteria, fungi, and earthworms can survive. Without air in the soil, the soil becomes infertile; a compacted soil is lifeless. Without air in the soil, beneficial mycorrhizal fungi cannot grow, the aerobic soil bacteria responsible for digesting lignin and cellulose begin to die off, and plant debris will rot instead of composting as *anerobic* bacteria and molds take over, which produce methane and other compounds that are toxic to plants.

However, there is a balance to be struck. Although soil needs oxygen, when oxygen can just whistle in and out of the soil because there is no sod or plant debris to shield the humus from direct exposure to the atmosphere (or when tillage mixes vast quantities of oxygen into the soil all at once), atmospheric oxygen will quite literally oxidize away the humus, the fungi, and the soil microbes, turning them all back into CO_2. Covered humus-rich soils (covered by sod through good grazing habits or by mulch through no-till farming practices) strike the balance to protect humus from rapid oxidation while simultaneously supplying the soil with sufficient oxygen to sustain a thriving community of soil microorganisms. This should make you ponder how soil carbon levels (and atmospheric CO_2 levels) were affected starting in the late 19th and early 20th century as the steam engine and then the internal combustion engine replaced horses, which vastly expanded our ability to deep-till the soil. But I'm getting ahead of the story once again…

The spongey soil texture of a humus-rich soil also has another benefit that is equally important to the plants that grow on it and to the microbes

---

[4] Ontl, T. A. & Schulte, L. A. (2012) Soil Carbon Storage. *Nature Education Knowledge* 3(10):35 https://www.nature.com/scitable/knowledge/library/soil-carbon-storage-84223790/

living within it — humus acts as a kind of sponge to absorb and hold large quantities of water so that roots, fungi, bacteria, and other soil microorganisms can access water as they need it. Sediments without humus (especially sandy or silty sediments) retain far less water — humus (like clay) absorbs water and holds on to it, allowing plants to access water even during long periods without rain. If you destroy the humus, the soil loses its moisture-holding capacity, just like what is happening today in the forests and grasslands of western North America (and all around the world). Humus degradation is one of the ways you can manufacture a drought even without a reduction in rainfall because you destroy the soil's ability to absorb and hold on to water.

The spongey moisture-absorbing capacity of humus is what allowed historic semi-arid regions like the Great Plains to support lush grasslands despite infrequent rains. The moisture absorbed by humus also provides the necessary moisture for fungi, bacteria, and other soil microorganisms living beneath the surface so they can continue to thrive even when there's little rain — without sufficient soil moisture, these microorganisms die or become dormant and the lignin and cellulose fibers in dead plant residues are left to degas back into the atmosphere as they react with oxygen instead of being converted into humus.

Furthermore, as soil dries out the space once taken up by soil moisture is replaced by air, which accelerates how fast humus is oxidized and degases. *Moisture in the soil doesn't just feed plants; it also helps shield humus from oxidation.* In other words, a humus-rich soil protects itself from oxidation not only by sustaining a thick protective skin of plant growth and sod, but also by absorbing and holding onto moisture, which helps shield carbon particles in the soil from oxygen.

As you will see in later chapters, deforestation, soil erosion, and desertification are all major sources of CO_2 in our atmosphere as historically moist soils dry out as a consequence of poor grazing management, as trees that once shielded the soil from the hot sun are cut down, and as tillage allows moisture to escape the soil. In later chapters, I will put some numbers to these processes, but for now I am explaining the processes to bring them into the light.

This layer of moist humus is the lifelong obsession of organic and regenerative farmers — this is what they mean when they speak of "feeding the soil" — by optimizing the conditions that create and sustain humus, and by optimizing the conditions that allow fungi, bacteria, and other beneficial soil microorganisms to thrive within that humus layer, these microorganisms provide all the nutrients that plants need for growth. That humus layer also gives soil its moisture-absorbing capacity to sustain crops even as neighboring crops growing on humus-depleted soils wither away during times of drought. Humus is the black gold that brings life to soil and plants.

On the opposite end of the spectrum from the "feeding the soil" philosophy is hydroponics, which uses no soil at all and instead supplies roots with water enriched by chemical fertilizers. You can successfully grow plants in this artificial way, but you are wholly at the mercy of fertilizer companies, irrigation pipes, and the cookbook recipes used to formulate those fertilizers. Growing food this way is 100% a scientific lab experiment. And it is entirely dependent on cheap energy to continue to make fertilizers, pump water, and build the greenhouses required for this style of production.

Conventional farming is somewhere in between these two approaches. It relies on soil nutrients and humus, just as regenerative/organic farming does, but also seeks to optimize plant growth by making up for nutrient shortfalls using chemical fertilizers. Each method has its place in the quest to feed the world and each contributes to the effort made by family farmers to pass on healthy soils and profitable businesses to future generations. But as humus is gradually depleted by erosion and degassing, farmers increasingly have to compensate for the slow creeping loss of their humus by relying ever more heavily on outside fertilizer inputs and irrigation — as soils degrade, farming looks ever more like a hydroponics system, just without the benefits of growing crops inside the controlled conditions of a greenhouse. And because soil degradation is so gradual, few notice the tiny incremental increases in fertilizers, irrigation water, and other off-farm inputs that are required to compensate for the slow degradation of their soils. The slowest changes are the hardest to recognize. Instead, stressed crops are easily chalked up as the victims of "climate change", allegedly caused by burning fossil fuels.

Although climate scientists acknowledge that tillage, deforestation, erosion, and poor grazing habits all release carbon dioxide back into the atmosphere and contribute to rising CO_2 levels, as this story continues you will see how badly we are underestimating the magnitude of this degassing because the ecological process at work within the soil are so poorly understood and because research that demonstrates the scale of this non-fossil-fuel degassing is de-emphasized in order to sustain the consensus belief that most of the CO_2 increase in our atmosphere comes from burning fossil fuels.

~

If you have ever been to the Canyonlands National Park in Utah or to the Grand Canyon National Park in Arizona, you will be familiar with the signposts that line the hiking trails advising people to stay on the trail to avoid disturbing the *cryptobiotic soil crust* that holds the soil together to prevent it from blowing away. In this fragile arid ecosystem, whenever this delicate crust is disturbed, the wind and sun soon strip away the soil below. This cryptobiotic soil crust is made of primitive cyanobacteria, lichens, and mosses — organisms that emerged very early in our planet's evolutionary history, long before lignin and cellulose began creating humus in the soil. These primitive microorganisms grow on bare sand and silt sediments to form a delicate microbial crust that shields the soil from wind and oxidization even as they survive by dissolving nutrients from those inorganic sediments. Throughout most of these arid and semi-arid regions, this cryptobiotic soil crust is the soil's last line of defence against a total ecosystem collapse, like the one experienced during the 1930s Dust Bowl or during the Great Famine in Maoist China when billions of tons of soil were carried away by the dry winds. You can't go far in these parks without bumping into another signpost explaining that if you disturb this delicate cryptobiotic crust, wind and rain will soon carry away the fragile soil below. And yet, cryptobiotic soils weren't always the norm in these arid regions.

Even in this extremely dry climate, vast herds of ruminant animals (bison, antelope, deer, etc.) regularly migrated through these regions during the pre-industrial era. As these vast herds thundered across the dry landscape, their sharp hooves fractured the hard soil surface and punched innumerable hoof imprints into the desert soils — these hoofprints served as countless little rainwater collection pockets from which grass would sprout after the next rains. Their trampling feet also churned dry plant material down into the soil where microbes could turn that material into rich humus despite the dry climate. And their manure partially pre-digested tough grass fibers to make it easier for the soil microbes to turn these tough fibrous tissues into fertilizer and humus. But in the absence of these vast historic grazing herds, these dry regions are now reverting to cryptobiotic soils even as the soil degases its humus. The sea of grass that once flourished here is slowly degrading into cryptobiotic soils that is all but depleted of its humus, has no moisture retention capabilities, has no continuous sod mat to shield the soil from sun, wind, and rain, has few plants left growing on it, and produces very little plant growth to support grazing animals. But this land degradation wasn't caused by "global warming", it is the inescapable consequence of changing the way that the land is managed. It is caused by the cascading domino effect that was set in motion when *migrating* grazing animals were removed from these dry lands.

The vast herds that once migrated through these desert regions, bunched together by hungry two-legged and four-legged predators patrolling their flanks, kept these arid regions productive and healthy despite the sparse rainfall. Then, in the 1800s, the migratory herds of wild grazing animals were replaced by vast herds of migrating cattle bunched together by cowboys in the Wild West — the cattle continued where the bison left off and the sod-covered soils continued to thrive. But then, in the late 19th and early 20th century, the plow and the land developer came and pushed the migrating grazing herds away, and the remaining pastureland was divvied up with barbwire fences behind which livestock were left to continuously graze the same patch of grass for months on end.

Furthermore, in 1916 Woodrow Wilson created the National Park Service, which sought to protect the most picturesque and fragile parts of these dry regions to preserve them in their "natural state." Having themselves orchestrated the extermination of the bison, the central planners then tried to achieve their goal of preserving the land in its "natural state" by also excluding cattle, only to deprive the arid ecosystem of the migratory grazing herds that had once kept the sod healthy and had previously converted lignin- and cellulose-rich plant debris into fertile soil. Now they're building signposts to teach tourists about the devastating consequences of disturbing fragile cryptobiotic soils and to "educate" the public about the consequences of "climate change". The sea of grass of yesteryear and the complex ecological processes that created it have long been forgotten as few can imagine that these parched desert regions were once covered by an unbroken sea of grass — the slowest changes are the hardest to recognize (and the easiest to blame on false correlations).

Mob grazing (also known by other names like as short-duration grazing, rotational grazing, migratory grazing, or holistic grazing) is designed to imitate the migratory behaviour of wild grazing herds and the impact that they have on the land. The cattle are bunched together at extremely high densities (some farmers create mobs of 250 to 500+ cows per acre!), which are moved to a fresh slice of pasture every day (or even multiple times a day) so that each slice of land is briefly grazed but is then allowed to rest completely undisturbed for months (or sometimes years) until the next rains trigger a fresh flush of grass. This migratory behaviour (whether by wild migrating grazing herds or domestic livestock in a grazing mob) very rapidly transforms cellulose and lignin into humus while keeping the sod healthy, even as their churning hooves prime the soil to capture rainwater by loosening hard soils and punching countless indents into the soil surface. This behaviour is what created many of our planet's richest soils, including the rich prairie soils, the fertile plains of the Serengeti, and the fertile *chernozem* soils on the Ukrainian and Russian steppe.

By contrast, consider what happens when cattle, goats, or sheep are not grazed using these mob grazing techniques but are instead allowed to scatter aimlessly over the entire farm (or park) so they can spread out and roam around the same patches of land, day after day, week after week, and month after month, in search of the tastiest morsels while ignoring the rest — this type of *continuous* (non-migratory) grazing behaviour will utterly destroy fragile ecosystems as the best grass is systematically over-grazed even as the not-so-tasty weeds are perpetually under-grazed, all while the land is continually trampled into dusty oblivion.

Most people (and especially bureaucrats and environmentalists) who see land deteriorating when livestock are managed in this way will instinctively seek to reduce the number of animals in the belief that the land deterioration is caused by having too many animals on the land — it's precisely the wrong solution because it doesn't address the *behaviour* of the grazing animals at the heart of the problem. Even as livestock numbers are reduced, the overgrazing of the best grass will continue even as ungrazed weeds spread even faster, all while the gaps between the sod clumps grow larger. Grass sod requires regular grazing to keep the grass healthy. Grass and grazing animals depend on each other — they literally evolved together, as we will soon see in an upcoming chapter. The solution to deteriorating dryland soils is not to reduce stocking rates, but to begin bunching cattle together into *migrating* grazing mobs (a pasture rotation with daily moves to fresh slices of grass on a small farm achieves a similar effect) — farmers who do this soon discover that their forage production increases two- to four-fold or more as the sod mats return, the bare ground is re-covered by grass, and the weeds are crowded out by desirable grass species. The greater the deterioration caused by poor livestock management (or by the complete removal of livestock), the greater the recovery and boost in yields when high-density migratory grazing herds return. Many farmers in arid regions have successfully returned barren deserts back to thriving grasslands using these mob grazing techniques.

A cow in a migrating grazing herd will build soils and grasslands, while a cow left to roam across the landscape without being part of a grazing mob will gradually destroy those very same soils and grasslands and turn them into deserts. Grazing behaviour is everything. Livestock are not the enemy — on the contrary, they are our most powerful tool to manage and restore these dry regions back to health, *but only if they are managed to mimic the migratory grazing behaviour of the wild herds that once roamed these regions.*

Migration is half of the equation; being tightly bunched together is the other half of the equation — this combination is what converts grazing animals from a destructive into a constructive force. Yellowstone Park reintroduced wolves a few years ago because the scattered elk, deprived of predators, were destroying the forests and the soils as they roamed across the park in aimless scattered small groups — since the wolves were reintroduced the elk have bunched themselves back together in large tightknit herds as they seek out the protective envelope of the herd to escape the fangs of the wolf. By reintroducing the wolf, park managers dramatically changed the grazing behaviour of the elk and, by consequence, their impact on the land. And so, the willow and aspen forests have begun to recover.[5]

Without predators to keep them bunched together, a wild herd of grazing animals is equally destructive as any cattle or goat herd that is left to graze the same patch of land day after day after day. I wrote my first

---

[5] Farquhar, B. (2023, Jun 22). Wolf Reintroduction Changes Ecosystem in Yellowstone. *Yellowstone National Park Tips.* https://www.yellowstonepark.com/things-to-do/wildlife/wolf-reintroduction-changes-ecosystem/

book about this powerful relationship between grass and cattle — a farmer's guidebook called *Grass-Fed Cattle: How to Produce and Market Natural Beef* — based on my own experiences with cattle. The point I am trying to make here is that the cryptobiotic soils that have emerged in these dry desert areas are a sign that these unhealthy soils are reverting back to a more primitive state, unable to build and maintain humus and unable to sustain the grasslands that once thrived on the same soil because the grazing animals have either been removed, have been reduced to abysmally low numbers, or are being grazed in a continuous grazing system instead of being managed as a grazing mob. Cryptobiotic soils are a symptom of a slow-moving ecosystem collapse — they are the last stage of a slow-moving ecological deterioration as once-thriving grasslands gradually revert to desert.

Biologist and farmer Allan Savory, who you may know as the founder of Holistic Management, has dedicated his life to reintroducing grazing herds (both livestock and wildlife) back into dry brittle desert regions with the specific goal of using livestock as the tool to restore the soil, build humus, and reverse the process of desertification. When Zimbabwe began seizing white-owned farms in the 1980s and 90s, he pre-empted the expropriation of his own farm by donating it to the Zimbabwean government on condition that it be held in a land conservancy trust to be managed as a wildlife park and cattle ranch (not split up into small farm holdings) and that he be allowed to continue on as manager of that land conservancy. His land is thriving under his stewardship even as many of his neighboring lands are falling apart under the strain of erosion and soil degradation. Since 1992, his 3,200-hectare Dimbangombe Conservancy (located near Victoria Falls) is also home to the Africa Centre for Holistic Management, which provides both tours and training programs to the public.

Based on techniques pioneered by Allan Savory and others with similar insights about the symbiotic relationship between ruminant grazing herds, grass, and soil, there are thousands of farmers in dry regions all over the world using cattle (and wildlife) to reverse the soil degradation and desertification that has happened over the last century. The results of their efforts speak for themselves, as shown in the picture in figure 97 below. Yet many bureaucrats, environments, academics, and NGOs continue to fight them every step of the way because of their fanatic hostility towards livestock and because they reject the idea that livestock can ever be harnessed as a constructive ecological force.

Figure 97. Land restoration using cattle as a tool to improve soil. 50 years of high-density holistic grazing on the left vs. 50 years of low density continuous grazing on the right. Source: Plant Tech Associates, March 8th, 2013. https://www.planet-tech.com/blog/land-restoration-holistic-management

The grassland on the left side of the fence in figure 97 is the product of 50 years of high-density holistic grazing; the barren desert on the right side of the fence is the consequence of 50 years of low-density continuous grazing. Dry brittle ecosystems collapse when grazing animals are either removed or allowed to graze continuously; these same ecosystems are healed when tightly bunched grazing herds return. As figure 97 so clearly demonstrates, both problem and solution are intensely *local*.

The United Nations, politicians, activist-scientists, and the mainstream media all keep warning us that global climate change is causing desertification around the world and that this will lead to more than a billion climate refuges flooding north to rich northern nations to escape desertification in Africa, Asia, Central America, South America, and the Middle East. Although the desertification they are alluding to is very real and, likewise, the displacement of people caused by this desertification is not a fabricated story, in reality this desertification is being caused by war, poverty, deforestation, continuous grazing, and a tangled mess of local, centrally planned, socialist and/or crony capitalist land management policies that are destroying local ecosystems. These arid regions are slowly turning to desert because of an absence of large tightly bunched grazing herds, an abundance of roaming goats and other continuously grazed livestock that systematically trample and strip the land of its vegetation, by a desperate population driven to try to eke out a precariously living on unsustainably small subsistence farms, and a multitude of human parasites, both domestic and foreign, who squeeze local communities in service of their own agendas. It's as much an ecological land management problem as it is a political problem. And the solutions are intensely local, both ecologically and politically.

The wildebeest migration in the Serengeti is one of the last healthy grazing migrations left on the African continent — the wildebeest keep the Serengeti covered in grass, which keeps the humus intact beneath their feet and holds back the relentless advance of desertification. The story is much grimmer everywhere else as humus is systematically eroded, blown, and oxidized away. For example, in Zimbabwe, even as Mugabe blamed the CO_2 emitted by faraway capitalist nations for turning Zimbabwe's breadbasket into a basket case, it was the humus in the soil beneath his own feet, destroyed by his own socialist politics and by his destructive land management policies, that was responsible for spewing vast quantities of CO_2 into the atmosphere while driving the expanding desert to claim once-productive lands.

There's a bitter irony to the fact that the very same aid given to leaders like Mugabe out of compassion for his starving citizens is also the financial lifeline that keeps leaders like him in power and sustains those countries' broken economic models as they too learn to play the game of manipulating the heartstrings of Western gatekeepers in places like Washington, Brussels, London, Ottawa, or at the United Nations in order to secure a slice of the spoils. Zambian economist Dambisa Moyo has written a scathing condemnation of the West's aid programs and the harms they are doing to Africa — her book is called *Dead Aid: Why Aid Is Not Working and How There is Another Way for Africa*.[6] She also has a number of interviews on YouTube about the issue, which will raise the hair on your neck along with your blood pressure.

Like so many of the other greater and lesser pigs that feed at government troughs throughout the Western world, these tinpot dictators have perfected the art of weaponizing the empathy of the West to line their own pockets — most of that aid money never reaches those who need it most, but it does help keep these dictators in power even as it feeds a vast parasitic economy of western aid agencies, NGOs, charities, Western institutions, and intragovernmental organizations whose very existence depends on finding desperate people to "help" — small wonder then that few who benefit from the government trough can recognize the beneficial impact of livestock and why they are so hostile to efforts to teach local farmers how to use grazing to solve their own local ecological crises by bringing migratory grazing practices back to the land and, by consequence, rebuilding local agricultural economies that have been hollowed out by desertification. Sadly, there is a lot more than mere ego at stake in making everyone blind. The blindest man of all is the one whose mind has been blinded by his conflicts of interest.

In the earlier chapters of this book, I focused a great deal on agricultural and forest soils, but what is happening in brittle desert and semi-desert ecosystems (which cover a much larger area than either of these other ecosystems combined) is of even greater importance to understanding the surging CO_2 levels in our atmosphere. A century of removing livestock

---

[6] Moyo, D. (2009). *Dead Aid: Why Aid Is Not Working and How There Is a Better Way for Africa*. Farrar, Straus and Giroux.

and wildlife from these desert regions has turned many of these once lush grass-covered regions into the rapidly expanding almost lifeless deserts that we assume are normal today. Despite their arid climates, many of these desert regions were brimming with grass and wild grazing herds even as recently as 200 years ago — we created these barren landscapes by interrupting the complex processes that once transformed lignin- and cellulose-rich plant debris into humus-rich soils in spite of the arid conditions and that once enabled vegetation to thrive in spite of the sparse rainfall.

Our desert, prairie, and forest soils were once vast carbon sinks that continually stored an ever-growing volume of carbon in the soil. Today, the balance has tipped in the opposite direction as those same soils are now giving their carbon back to the atmosphere. Before the late 1800s, grazing animals ensured that these soils were constantly accumulating more humus, thus drawing CO_2 out of the air to permanently store a portion of that carbon in the soil as grassland soils got thicker and thicker over time. Today, these same soils are slowly degassing thousands of years of accumulated humus back into the atmosphere. As the Columbia Climate School at Columbia University pointed out in a 2018 article, *"The Earth's soils contain about 2,500 gigatons of carbon—that's more than three times the amount of carbon in the atmosphere and four times the amount stored in all living plants and animals [...] so even small changes in that pool are going to have [a] really large effects for us."*[7]

The amount of CO_2 degassing from soils back into the atmosphere is only half of the equation — the amount of CO_2 that is no longer being pulled from the atmosphere is equally important because by turning these soils from carbon sinks into carbon emitters, *the processes that create humus in the soil are thus no longer offsetting the volumes of fresh carbon that are added to the atmosphere each year by volcanic eruptions and by hot volcanic gasses leaking out of the Earth all along seafloor spreading ridges.* If soils take less carbon out of the air each year but the same amount of fresh carbon from volcanic sources continues to accumulate in the atmosphere each year, the atmospheric balance is necessarily destined to tip towards rising CO_2.

---

[7] Cho, R. (2018, Feb 21). Can Soil Help Combat Climate Change. *Columbia Climate School – State of the Planet.* https://news.climate.columbia.edu/2018/02/21/can-soil-help-combat-climate-change/

Furthermore, as our planet's total biomass shrinks because of deforestation and desertification, that also reduces how much carbon is pulled from the atmosphere by photosynthesis each year and how much of that carbon then winds up getting trapped in the soil's carbon sink. Thus, an overall reduction in the global biomass directly contributes to rising atmospheric CO_2 levels by failing to offset fresh carbon added by volcanoes, seafloor spreading ridges, and fresh carbon emissions. We will look much more closely at this second half of the equation — biomass — and the impact it has on rising CO_2 levels in later chapters.

Grazing animals were the engine that once converted fresh additions of CO_2 released by volcanoes, etc., into humus in these vast wild regions. Today that "carbon-consuming engine" is sputtering. In many places it has been turned off altogether. For someone unfamiliar with the impact that migrating cattle herds have on the soil or for someone who hasn't seen how migrating grazing herds can restore humus and grass in depleted pastures and desert soils, it's not easy to see the complex chain of events that is triggered wherever these dry landscapes are deprived of migrating herds of grazing animals. But the evidence is visible in the form of expanding deserts, with more than 4 million square kilometers of land being degraded each year, of which more than 120,000 additional square kilometers is turned into actual desert each year.[8] And it is measurable by the global loss of humus in the soil, *which is disappearing at an average rate of at least 1% per year from our agricultural soils*.[9]

According to UN estimates, we are losing around 24 *billion* tons of fertile cropland soils to erosion *every year*, and this doesn't even account for soil losses in our forests and deserts, which cover far more of our planet's surface than cropland does.[10] Some of these soil losses are washed away and accumulate as sediments at the bottom of lakes and oceans. As that erosion progresses, the carbon that was once stored in the soil as humus is either oxidized to CO_2 as it is exposed to the atmosphere (the

---

[8] "A third of the world's land surface is threatened by desertification." *The World Counts*. Accessed on 8 Jun 2024.
https://www.theworldcounts.com/challenges/planet-earth/forests-and-deserts/global-land-degradation

[9] Verso, E. (2015, Dec 9). Topsoil Erosion. *Stanford University Course Work*.
http://large.stanford.edu/courses/2015/ph240/verso2/

[10] Ritchie, H. (2021, Jan 14). Do we only have 60 harvests left? *Our World in Data*.
https://ourworldindata.org/soil-lifespans

degassing is fastest in dry climates where there's little soil moisture to shield the soil from atmospheric oxygen), or it gets washed downstream where it accumulates as carbon-rich mud at the bottom of swamps, lakes, and river deltas where methanogens quickly go to work to digest some of these carbon-rich muds to turn them into methane, which bubbles up through the water into the atmosphere where it is subsequently oxidized to break down into carbon dioxide not long after it is exposed to oxygen in the air. Thus, soil degradation and soil erosion are inextricably linked to the sharp increase of CO_2 in our atmosphere.

This soil loss has become so normalized in the agricultural industry because of centuries of tillage agriculture in both conventional and organic systems that the United States Agricultural Department considers a soil loss of *1 to 5 tons per acre per year* to be in the *acceptable* range (this is total soil loss from sediments and humus combined).[11] The average soil loss in the American Corn Belt is 3.9 tons/acre/year.[12] On Prince Edward Island — the epicenter of potato farming in Canada — the agriculture department considers 3 tons/acre/year as the upper limit of acceptable soil losses per year. During the 1980s in Iowa, the average soil loss per year was 7.4 tons/acre/year; by 2007 that had fallen to 5.1 tons/acre/year. Across the U.S. as a whole, average soil losses on agricultural soils fell from 4 tons/acre/year in 1982 to 2 tons/acre/year by 2007, thanks in part to no-till agriculture which is becoming more common in North America but is still largely unknown anywhere else in the world today.[13] Ironically, Europe (the epicenter of the climate brigade's war on agriculture) has the best control over its soil erosion, with average soil losses of only around 1.37 tons/acre/year in its agricultural areas and natural grasslands.[14] By contrast, the worst soil erosion by a wide margin is happening in overpopulated regions of the developing world. In Africa, it is over 50

---

[11] "Soil loss tolerance." *Wikipedia.* 18 May 2024.
https://en.wikipedia.org/wiki/Soil_loss_tolerance

[12] Cox, C, Hug, A., and Bruzelius, N. (n.d.), Losing Ground – Executive Summary. *Environmental Working Group.* Accessed 3 Jun 2024.
https://www.ewg.org/losingground/report/executive-summary.html

[13] Duffy, M.D. (2012, Aug.) Value of Soil Erosion to the Land Owner (File A1-75). *Iowa State University Extension and Outreach.*
https://www.extension.iastate.edu/agdm/crops/html/a1-75.html

[14] "Agri-environmental indicator – soil erosion." *Eurostat Statistics Explained.* Feb 2020. https://ec.europa.eu/eurostat/statistics-explained/index.php?title=Agri-environmental_indicator_-_soil_erosion&oldid=627451

tons/acre/year.[15] In Asia, it's a colossal 74 tons/acre/year (13 to 40x faster than the natural rate of erosion)![16]

With these kinds of soil erosion numbers, it should not surprise you to learn that 80% of the world's population is already at least partially dependent on food imports, nor should it surprise you that developing countries (where both soil degradation and population growth are the fastest) are the most vulnerable to the ongoing degradation of global agricultural soils — remarkably, nearly *half* of all calories consumed globally come from only three crops (wheat, corn, and rice), which are overwhelmingly produced in just five countries![17] The consequences of soil degradation are all around us, yet they are so normalized and they happen so slowly that few can see how brittle our world has become. Slowly the world adjusts as our globalized food system adapts to declining harvests in some regions by rerouting harvests from others, leaving the world ever more dependent on fewer and fewer countries. But no-one really notices because it happens slowly enough and because, in a global marketplace, a local crisis that would have become visible in the past can be papered over by rerouting supply chains until we slowly realize that we are now all "in this together" in a single slowly deteriorating boat.

These numbers put in perspective the climate lunacy that has infected Western countries. The groupthink of the climate brigade sees everything though the lens of fossil fuels, so their hysterical crusade is now targeting the most productive farms on the planet to force them to reduce fertilizer use by 30% (i.e. Germany, Netherlands, Canada), imposing taxes on farm fuels in order to discourage farm fuel consumption (Germany), imposing mandatory livestock culls to reduce livestock numbers (Ireland), and imposing laws that force farmers to take privately owned land out of

---

[15] Nana-Sinkam, S.C. (1995). The magnitude of the problem – Land and environmental degradation and desertification in Africa. *Food and Agriculture Organization of the United Nations.* https://www.fao.org/4/X5318E/x5318e02.htm

[16] Rahaman et al. (2015). Estimation of Annual Average Soil Loss, Based on Rusle Model in Kallar Watershed, Bhavani Basin, Tamil Nadu, India. *ISPRS Annals of the Photogrammetry, Remote Sensing and Spatial Information Sciences.* II-S/W2. https://isprs-annals.copernicus.org/articles/II-2-W2/207/2015/isprsannals-II-2-W2-207-2015.pdf

[17] Vercillo, S. and Park, A. (2022, Dec 16). Globalized food systems are making hunger worse. *Al Jazeera News.* https://www.aljazeera.com/opinions/2022/12/16/american-globalisation-is-aggravating-africas-hunger

production to meet biodiversity targets like those set by the "30 by 30" initiative agreed to at the UN's COP15 meeting, which seeks to turn 30% of all lands into protected areas. The Netherlands even tried to forcibly expropriate privately owned farms, though a change in government appears to have temporarily put that plan on pause.

The Netherlands (at 0.1 tons of soil loss per acre per year) and Germany (at 0.5 tons of soil loss per acre per year) have some of the lowest soil losses in the world (thanks to their moist climate, flat topography, and leading-edge production techniques).[18] They also having some of the highest crop yields in the world. Canada is the breadbasket of grain farming. Ireland has some of the most productive pastures in the world (and those hilly pastures are nicely protected from soil losses as long as cattle and grass are allowed to graze those pastures). Quite apart from the assault on individual property rights and individual liberties (which is outrageous enough and raises sharp questions about whether the current façade of democracy retains any of the principles that once made liberal democracy work), it is also sheer madness to roll out policies that will inevitably shift the burden to feed the world away from the most productive and most erosion-resistant soils to countries with much more brittle ecosystems and already overburdened soils that are already buckling under the strain of runaway soil erosion rates and colossal rates of humus degradation. Now their strained soils will exhaust their dwindling organic nutrient reserves even faster. And restricting access to fertilizers means you're going to need more land, not less, to feed the world, which will only increase the rate at which erosion and humus oxidation deplete our soils. As always, the fastest way to make a problem much, much worse is to strip individuals of their autonomy while empowering central planners to try to solve problems on everyone's behalf.

While the agricultural footprint of the rest of the world is still surging thanks to their growing populations, the agricultural footprints of Canada, the United States, and Europe have been falling since the 1950s. If it wasn't for the 84 million acres of corn that are grown in the USA each year to produce corn ethanol for fuel (angry face), the agricultural

---

[18] Panagos et al. (2015). The new assessment of soil loss by water erosion in Europe. *Environmental Science & Policy.* 54. 438-447. https://doi.org/10.1016/j.envsci.2015.08.012

footprint of farming in the U.S. would be falling even faster despite being the fact that the U.S. a net food exporter (the ethanol lobby is one of the strongest lobbies in Washington, so don't expect any climate policies to touch them anytime soon — incidentally growing corn to produce ethanol began in the 1970s in response to the 1970s oil crisis). This downtrend in the amount of land that is used to grow food in industrialized countries will reverse if we reduce fertilizer inputs, unless we start importing food from other less efficient countries (with higher soil erosion rates) to make up for production shortfalls.

This isn't a game. By gambling with soils and food production, the politicians and activist-scientists driving this climate crusade are gambling with ecosystems and lives. This isn't just my opinion — we've already seen a preview of how their "green" reforms play out in the real world. Under pressure from a wide range of international groups, Sri Lanka decided in 2021 (in the midst of an economic crisis) that it would become the world's first organic farming nation by imposing a complete ban on all agrochemical imports — no more chemical fertilizers and no more pesticides — effective immediately.[19] Germany's World Future Council even awarded Sri Lanka with a special award for its efforts and Sri Lanka's president released a statement in which he praised the UN's Food and Agriculture Organization along with the UN's World Food Programme for the technical assistance they provided Sri Lanka in this historic leap to *"sustainably transform its food system and ensure greater food security and better nutrition for its people."*[20] But it wasn't the success story that politics and the media initially made it out to be. Widespread protests quickly overturned the bans, but the damage was done as only a trickle of chemical fertilizers made it into the country that year. Reuters reported that rice yields plunged by 30% and that in many

---

[19] Torrella, K. (2022, Jul 15). Sri Lanka's organic farming disaster, explained. *VOX News.* https://www.vox.com/future-perfect/2022/7/15/23218969/sri-lanka-organic-fertilizer-pesticide-agriculture-farming

[20] Permanent Mission of Sri Lanka to the United Nations. (2021, Sept 23). *Statement by H.E. Gotabaya Rajapaksa, President of the Democratic Socialist Republic of Sri Lanka.* Government of Sri Lanka. https://www.un.int/srilanka/news/statement-he-gotabaya-rajapaksa-president-democratic-socialist-republic-sri-lanka-un-food

regions crop yields as a whole plunged by over 50%.[21] Some farmers saw yields decline by over 80% — "*Last year, we got 60 bags from these two acres. But this time it was just 10,*" said one farmer quoted by Reuters. A country that had once been self-sufficient in its food production suddenly witnessed huge queues to buy essential food items and the country was left scrambling to import over $450 million of rice to prevent starvation. The military even had to get involved in the emergency production of fertilizer in a desperate bid to try to make up for fertilizer shortages. By the time the government paid out compensation to the farming sector, its already declining financial reserves were decimated. After months of violent protests, the president was forced to resign (many reports suggest that the violence at these protests was primarily committed by government loyalists against peaceful protesters and by government operators infiltrating the protests in order to stoke violence to justify the government's authoritarian crackdowns against protesters — sound familiar?).

Sadly, judging by the chatter coming from the "technocratic experts", the global intelligentsia learned nothing from this failed experiment. Instead, these elites came to the conclusion that Sri Lanka's agricultural reforms only failed because the reforms should have been phased in more gradually. The idea that the administrative state should leave farming to farmers seems beyond their grasp. The anointed class always tells itself that "this time will be different" as overrules the free choices of its citizens. It never is.

This hubris, this extreme form of social engineering, this land management by technocratic diktat, it is all the result of the progressive faith in a strong centralized state — Woodrow Wilson's cherished administrative state — that has the power to force everyone else to dance to their tune. Liberal democracy didn't break because the wrong people got into power; it broke on the day society embraced a strong administrative state at the center of its system. This is the fate of all systems where decision making is centralized in the hands of a small number of gatekeepers. In time, corruption, ideology, and hubris will inevitably infect the halls of power, and the centralization of authority

---

[21] Jayasinghe, U. and Ghoshal, D. (2022, Mar 3). Fertilizer ban decimates Sri Lankan crops as government popularity ebbs. *Reuters.* https://www.reuters.com/markets/commodities/fertiliser-ban-decimates-sri-lankan-crops-government-popularity-ebbs-2022-03-03/

makes it virtually impossible to reform disastrous policies because those with their hands on the levers of power can leverage the full might of the bloated state to squash dissent. Like in Sri Lanka's example, with so much power concentrated in the hands of the administrative state, individual choices and local knowledge became irrelevant, and the course correction only happened when some idiotic policy launched the whole bus over the edge of a cliff. This is the price we pay for having been seduced by the temptation of technocratic nation building, the idea that the nanny state should be used to engineer progress, that elites should go beyond research and education and the protection of their citizens' rights, and instead be allowed to impose their social engineering schemes on society by legal decree. Where once the progressive impulse was focused on building railroads and settling the West, today's crusade is nothing less than a mission to "save the world" from fossil fuels, cows, fertilizers, and productive farmers.

Foreign Policy magazine reported in 2022 that Norway is lobbying the World Bank to stop financing all natural gas projects in Africa (and elsewhere) as soon as 2025. Meanwhile, even as that lobbying effort depriving Africa of much needed cheap energy, Norway has exempted itself from the very same standards it is pushing on Africa, arguing that Norway's own future oil and gas drilling must be allowed to continue as it is critical to Norway's "transition to renewable energy".

At COP26, world leaders pledged to stop funding all overseas fossil fuel projects and switch to funding green hydrogen technologies and smart micro-grid networks instead. Meanwhile, Germany has given itself a generous 20-year window before it wants to exit from coal; it even tore down a windfarm in 2023 to make room for a new coal mine. Likewise, despite the lofty climate targets set by the U.S. and imposed on the developing world through a combination of carrots and sticks, at the 2021 meeting of the G20 President Biden hypocritically implored the U.S. oil and gas industry to increase production to boost the U.S. economy.[22] Rules for thee, but not for me.

---

[22] The White House. (2021, Oct 29). *Background Press Call by Senior Administration Officials on the President's Upcoming G20 Visit.* [Press release]. U.S. Government. https://www.whitehouse.gov/briefing-room/press-briefings/2021/10/29/background-press-call-by-senior-administration-officials-on-the-presidents-upcoming-g20-visit/

With "friends" like this trying to help them, Africa doesn't need enemies. This is green colonialism in its most patronizing and destructive form. As Foreign Policy magazine states, *"It's the rich world telling the global south to stay poor and stop developing [...] Instead, development aid will be repackaged as climate-related transfers, keeping the global south dependent."*[23] Without rapid development, it doesn't take rocket science to predict what will happen to already overburdened soils in these poor and overcrowded countries.

The West takes pressure off its own ecosystems by relying on cheap fossil fuels at home, even as it denies others access to cheap fossil fuels, who then turn to deforestation to fuel their cooking fires and are left toiling on unsustainable subsistence farms in lieu of participating in a thriving market economy.

Nor is it only the developing world that is being crushed by self-serving green paternalism. Investigative journalist Vivian Krause has uncovered the concerted effort made by U.S. oil interests to land-lock Canadian oil and gas development as these U.S. oil interests fund climate activists and anti-oil environmental groups to put pressure on the Canadian oil industry with the stated aim of grinding oil and gas development in Canada to a halt.[24] Her shocking 2019 documentary *Over a Barrel* is still available on Facebook.[25] The narratives of the climate crusade provide many opportunities for many people, none of which have anything to do with the climate.

In the developing world, green colonialism is holding back the cheap energy required for economic development, thus ensuring that developing nations can't keep up with their population growth. It's well known by now that as soon as nations get richer, people move off the land into the cities in search of jobs and that they stop having large numbers of children as they become wealthier — by holding back economic development and

---

[23] Ramachandran, V. (2021, Nov 3). Rich Countries' Climate Policies Are Colonialism in Green. *Foreign Policy Magazine.* https://foreignpolicy.com/2021/11/03/cop26-climate-colonialism-africa-norway-world-bank-oil-gas/

[24] Corbella, L. (2019, Jan 17). Cobella: Researcher exposes money trail behind U.S.-led campaign to kill the oilsands. *Calgary Herald.* https://calgaryherald.com/news/local-news/corbella-vivian-krause-should-become-a-household-name-across-canada

[25] Over a Barrel (Krause, V.) (2020, Feb 24). *Over a Barrel Documentary 2019.* [Video]. Facebook. https://www.facebook.com/overabarreldoc/videos/over-a-barrel-documentary-2019/371240137168991/

stifling innovation to "save" the planet from "fossil fuel emissions", green colonialism is setting up a vicious Malthusian trap for these developing nations. Without rapid economic development, access to cheap energy, and modern agricultural innovations (like cheap chemical fertilizers), vast hordes of desperate people in these developing nations are pushed into subsistence farming even as their overburdened soils cannot keep up with population growth. Civilizations collapse when they overtax their soils — it will take centuries to rebuild the colossal volumes of humus that are being eroded and degassed away in developing countries thanks to the paternalistic impulse of green colonialism. Without healthy soil, you cannot feed a nation. Even the mass migration pouring into Europe is directly linked to the Malthusian Trap set in Sub-Saharan Africa by decades of paternalistic western intervention and the destructive cycle of soil degradation that emerges from it (more on that story in chapter 27).

As our geological journey through time continues, the enormity of the degradation of our planet's soils will become completely clear, as will its solutions. It would be wonderful if rising CO_2 had simply been the consequence of us liberating CO_2 trapped as fossil fuels in sedimentary rocks, so that by burning fossil fuels we ended up replenishing atmospheric CO_2 levels to reverse what has been lost to carbon sinks over millions of years — our CO_2-starved plants could certainly use all the carbon dioxide they can get. I wish that were the case, but as the numbers in later chapters will soon show, rising CO_2 is warning us about the extraordinary scale of soil degradation in our farm, forest, and desert soils. It is telling us that, although the plants growing on these soils are certainly benefiting from more CO_2 in the air, the soils supporting those plants are deeply unhealthy, leaving us ever more dependent on chemical fertilizers and irrigation canals to provide what depleted soils cannot. Humus acts as nature's built-in fertilizer system and nature's buffer against drought — we are depleting it at our peril.

Year by year, humus is being stripped away, blown away, washed away, and, most especially, oxidized away. Year by year, those soils are losing a little bit more of their moisture-storing capacities. Year by year, those soils are becoming less fertile. And, like in the lead-up to the Dust Bowl of the 1930s, year by year those soils are becoming increasingly vulnerable to the natural climate cycles that have always and will always turn some years into droughts. From a long-term geological perspective, we are very fortunate to be living in a Goldilocks climate window today, but that is not something we should take for granted. There's nothing to be feared from having more CO_2 in the atmosphere, but there's a lot to fear from having less humus in our soils. We cannot change the climate, but we can change how vulnerable we are to the next cyclical drought.

Figure 98 shows drought cycles in California over the last 1,200 years.[26] Anything above the mid-point represents a dry period while anything below the midpoint represents a wetter period. As you can see, we have been living in an unusually mild and favorable period in our climate's history — the last few centuries have been kind, but that is not the norm.

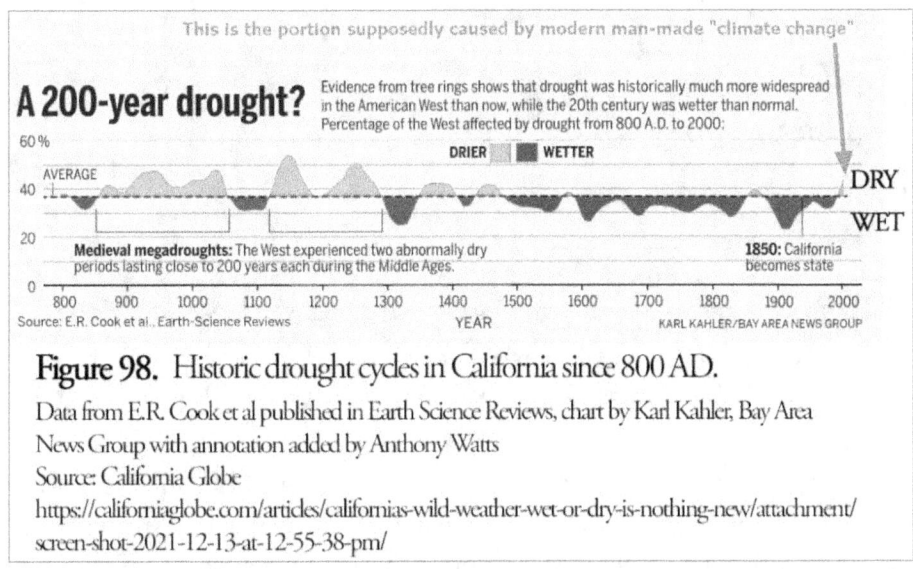

Figure 98. Historic drought cycles in California since 800 AD.
Data from E.R. Cook et al published in Earth Science Reviews, chart by Karl Kahler, Bay Area News Group with annotation added by Anthony Watts
Source: California Globe
https://californiaglobe.com/articles/californias-wild-weather-wet-or-dry-is-nothing-new/attachment/screen-shot-2021-12-13-at-12-55-38-pm/

Likewise, figure 99 shows drought cycles in New Mexico over the last 2,500 years.[27] In contrast to the previous California chart, the New Mexico chart shows dry periods below the midpoint and wet periods above the midpoint — different authors, different ways of graphing the data, but the underlying message is the same.

---

[26] Grimes, K. (2021, Dec 13). Screen Shot. *California Globe.*
https://californiaglobe.com/articles/californias-wild-weather-wet-or-dry-is-nothing-new/attachment/screen-shot-2021-12-13-at-12-55-38-pm/

[27] Oliver, J.S., Harley, G.L., and Maxwell, J.T. (2019). 2,500 Years of Hydroclimate Variability in New Mexico, USA. *Geophysical Research Letters.* 46(8). 4432-4440. https://doi.org/10.1029/2019GL082649

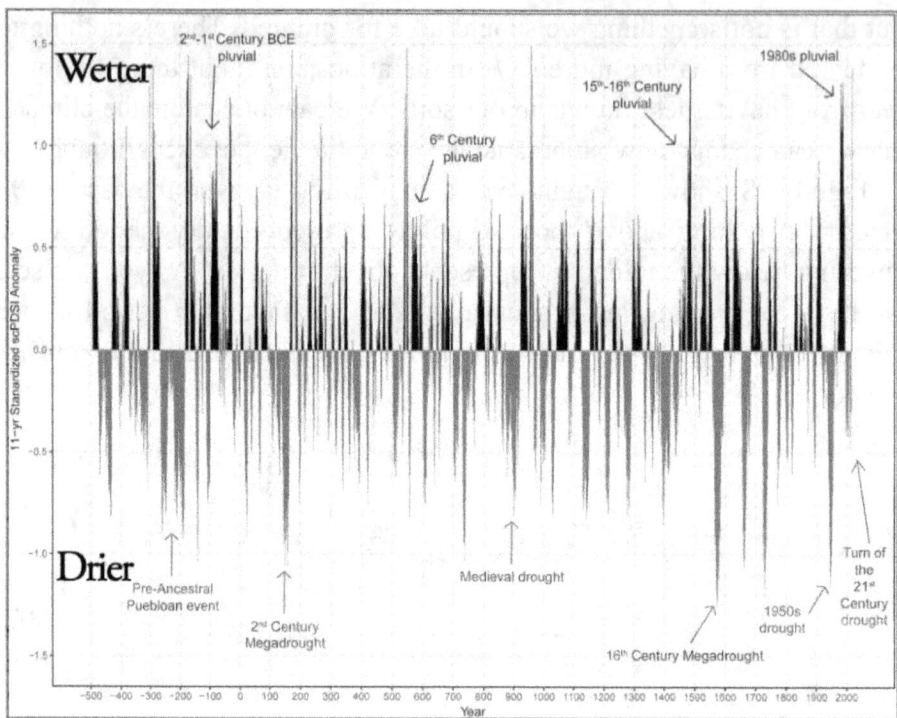

Figure 99. 2,500 years of alternating wet and drought cycles in New Mexico. Source: Oliver et al., 2019. 2,500 years of hydroclimate variability in New Mexico, USA. *Geophysical Research Letters*, 46, 4432-4440. https://doi.org/10.1029/2019GL082649

As you can see from these two charts, the recent brief interval of dry years in the American West is *not* unusual. Nor was it even especially dry by long-term historic standards. Nor is there any reason to believe that these extreme cycles won't continue to happen in the future. Once you step back to look at longer timescales, it becomes apparent that it was perfectly normal for the American West to experience vicious droughts that were not only much drier than any of the droughts we have experienced over the past century, but that some of these historic megadroughts lasted for up to 200 years at a time! And yet, trees didn't die out, the American West was not turned into a sterile Sahara-like landscape, and cryptobiotic soils didn't replace the waving sea of grass. Why not? As long as there was sod to keep the ground covered to protect against wind, sun, and rain, as long as there was plenty of humus to absorb and hold on to moisture in the soil, and as long as the region had large grazing herds to keep that sod healthy, life persevered against the backdrop of an ever-changing climate.

But remove the sod and deplete the soil's humus, and the next 200-year drought cycle will most certainly become an existential threat to the entire western ecosystem, just as the plowmen found out when they inadvertently turned the Great Plains into a Dust Bowl. Despite the fearmongering that dominates our news headlines, in reality we are living during an unusually favorable period in our climate's history — it doesn't take much imagination to recognize what will happen when our heavily eroded, degraded, and nutrient-deficient farm, forest, and desert soils are exposed to the next challenging period that makes up the normal long-term climate variability in our local ecosystems.

Some version of this unrecognized story is repeating in almost every local ecosystem all over our planet. Humus sustains civilization through good times and bad — it is the buffer that has allowed our biosphere to thrive *despite* a constantly changing climate. But as we continue to degrade our soils, we are turning our world into a terrifyingly fragile place.

Once you understand the complex processes that destroy humus and turn grasslands into deserts, you can see evidence of this destructive process happening everywhere around us. The destruction of soil humus reveals itself by the brown rivers transporting millions of tons of soil from fields, forests, and deserts to the ocean — even Africa's notoriously brown muddy rivers once ran clear, as described by the early European explorers who first ventured across the continent. That ongoing soil degradation is revealed by the bare patches of soil popping up everywhere. It's revealed by the patchiness of once thriving grasslands and the advance of weeds into the bare patches where healthy sod once grew. It's revealed by drought-stressed trees and by the pine beetles that prey on those stressed trees. And it's revealed by the subtle signs of nutrient deficiencies cropping up everywhere that a farmer's eye is trained to recognize. The subtle signs of humus destruction are all around us.

For example, Kalamalka Lake Provincial Park is a grassland park near Vernon, BC — a particularly dry and picturesque example of the grassland hillsides that separate our valley bottom from the forests up on the mountain plateaus. In 1975, the cattle ranch that owned it sold it to the provincial government and it was turned into a park. The scattered clusters of grazing cattle that had continuously grazed the area since the late 1800s were dutifully excluded in order to return the park to its "natural" state. But in the absence of regular grazing, the park degraded still further because the conditions that once made grasses thrive and sods spread in this delicate ecosystem no longer existed. The giant herds of elk and deer that once grazed these hillsides and the wolves that once stalked their flanks to keep the herds migrating as tight cohesive mobs are not coming back. With no predators to keep wildlife tightly bunched as a migrating herd, the few deer that graze there today are scattered far and wide — their impact on the land has no power to reverse the deterioration.

And so, the park is slowly being taken over by weeds, large gaps of bare soil continue to expand between the sod clumps, and there's a never-ending battle to repair the erosion after every big rainfall. Herbicides have become an irregular necessity to manage the spread of noxious weeds growing between the retreating sod clumps. As these gaps in the sod open up, oxygen is able to get in and oxidize the humus. Without regular manure to keep building humus and restore nutrients to the soil, and without a continuous sod mat to keep the soil covered, the humus layer is gradually oxidizing away. And as the soil's moisture-absorbing humus deteriorates, its vulnerability to drought increases — you can already see this drought stress in the Ponderosa Pines growing inside the park as they become increasingly stressed during dry years. It's a slow subtle process, but the entire area is ever so slowly turning into a desert. But to the untrained eye that's constantly being bombarded by climate propaganda at every turn, this looks like evidence of fossil-fuel-driven climate change.

Chapter 18

Coal and Fire in an Oxygen-Fueled World

"Despite my efforts, the [environmental] movement abandoned science and logic somewhere in the mid-1980s, just as society was adopting the more reasonable items on our environmental agenda. Ironically, this retreat from science and logic was partly a response to society's growing acceptance of environmental values. Some activists simply couldn't make the transition from confrontation to consensus; it was as if they needed a common enemy. When a majority of people decide they agree with all your reasonable ideas the only way you can remain confrontational and anti-establishment is to adopt ever more extreme positions, eventually abandoning science and logic altogether in favor of zero-tolerance policies."

— Dr. Patrick Moore, Confessions of a Greenpeace Dropout: The Making of a Sensible Environmentalist[1]

Back in the Carboniferous Period (359–299 mya), the story of humus is about to get even more interesting. The evolution of lignin and cellulose completely scrambled the balance between the amount of carbon dioxide being removed from the atmosphere by plants versus the amount returned to the atmosphere by oxygen breathers. The evolution of vascular plants with fibrous stalks kicked photosynthesis into overdrive as the world was taken over by enormous plants with woody stems, like giant ferns, the first coniferous trees (in upland regions), and a multitude of strange and bizarre tree species that are completely alien to us today (imagine clubmosses that grow up to 130 ft high!?!). Figure 100 shows what this alien world would have looked like. But animals and microbes had not yet evolved to be able to digest these tough lignin- and cellulose-rich fibers and so, as these trees died, they piled up in vast quantities on the forest and swamp floor and basically just sat there unable to rot.

---

[1] "Patrick Moore (consultant)." *Wikipedia.* 21 May 2024.
https://en.wikipedia.org/wiki/Patrick_Moore_(consultant)

Figure 100. 19th century artist's etching depicting the swamp forests that dominated the Carboniferoud Period (358.9 - 298.9 mya).

Until now, I have focused on humus in grasslands, forests, farmlands, and deserts. But when lignin- and cellulose-rich plant debris accumulates in a wetland, the lack of oxygen below the waterline prevents aerobic soil microbes from being able to compost the fibers to form humus. Instead, the acidic and oxygen-depleted conditions of the bog slowly turn plant debris into a semi-decomposed mass of organic matter called peat. And when peat gets buried, compressed, and heated for millions of years beneath the surface of the Earth, it turns into coal. The Carboniferous Period gets its name from this coal — almost all the world's coal deposits were formed during the Carboniferous. The golden age of ferns was also the golden age of coal formation — trillions of tons of coal. The carbon dioxide that once filled the Carboniferous atmosphere is locked away inside those ancient coal beds, leading to the extreme depletion of atmospheric CO_2 that I described in chapter 16 when life was almost extinguished on our planet as CO_2 levels dropped as low as 150 ppm.

During the Carboniferous, the continents drifted to form another gigantic supercontinent, only this time that supercontinent stretched from the Northern Hemisphere all the way to the South Pole (geologists have nicknamed this supercontinent "Pangaea", which means "all of Earth" in Greek). Thanks to a warm humid global climate, its lowlands were filled with lush tropical swamps — the perfect place for massive coal deposits to form. Since oxygen breathers hadn't yet evolved the ability to digest lignin and cellulose, they were unable to recycle all this carbon back into the atmosphere. Instead, the carbon was just buried at the bottom of the swamps, layer by layer, even as the oxygen produced by all this photosynthesis built up to ever higher levels in the atmosphere.

Oxygen levels surged from around 15% just before the Carboniferous kicked off to around 30-35% of the atmospheric mix at its peak during the mid Permian Period (299–251 mya), as shown in figure 101 below — the oxygen surge that happened during the Carboniferous and Permian periods stands out like a sore thumb. This dramatic rise in oxygen finally stalled in the mid-Permian as microbes capable of breaking down lignin and cellulose finally evolved. Around the same time, the ancestors of dinosaurs finally evolved the ability to digest these tough fibers with the help of stones inside their digestive system to grind the fibrous tissues down — some of the largest examples of polished grinding stones found alongside dinosaur fossils were over 10 cm (4 inches) in size. Even today, most birds have small stones in their gizzards to help them grind tough-to-digest plant fibers.

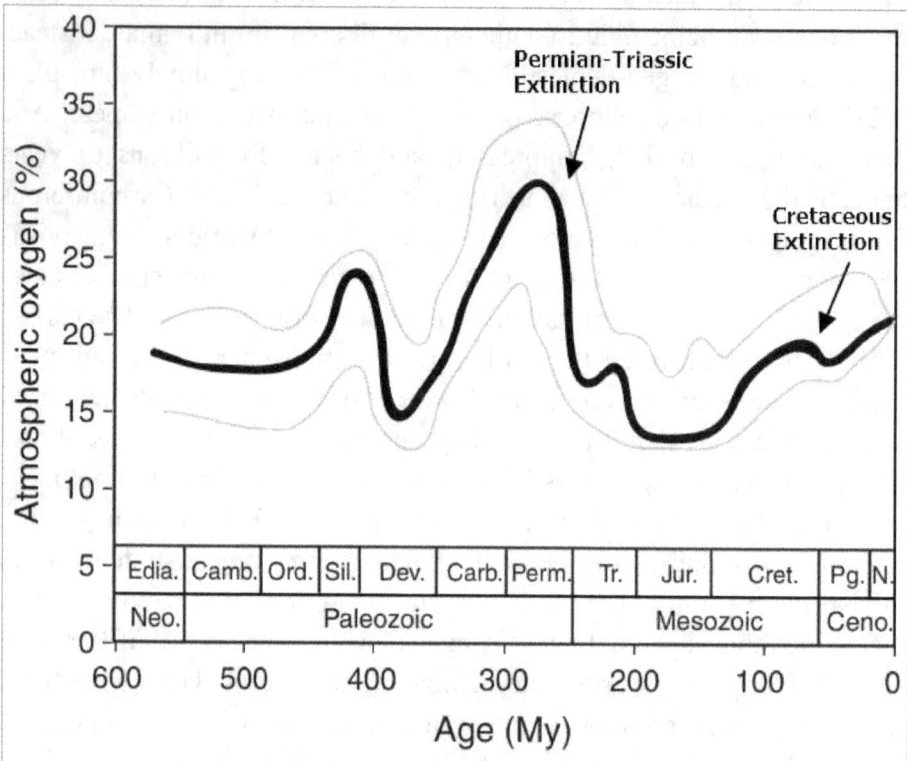

Figure 101. Reconstructed oxygen concentration over the last 600 million years (the margin of error is shown in light grey).

Adapted from: Payne et al, 2011. The evolutionary consequences of photosynthesis. Photosynthesis Research 107 (1): 37-57
https://link.springer.com/article/10.1007/s11120-010-9593-1

Fire is a form of rapid oxidization as oxygen reacts with carbon. When atmospheric oxygen levels fall below 15%, it's extremely difficult to sustain an open fire — it just won't burn. Thus, there were very few forest fires in the low-oxygen era before the Carboniferous. At today's level of 21%, it's relatively easy to sustain a fire if the fuel is dry. But by the time oxygen reached 35% during the Carboniferous and Permian periods, wildfires were running rampant through the huge stacks of dead un-rotted trees building up on the Pangaean supercontinent — sediments from that era are littered with charcoal from wildfires. These wildfires helped do what soil microbes had not yet evolved to do — break down lignin and cellulose. Dead tree trunks below the water level turned into coal but on dry land wildfires ensured that, in the absence of burping cows and other herbivores capable of digesting tough fibrous plant tissues, fire helped recycle carbon back into the atmosphere as CO_2. There is even geologic evidence from 350 million years ago that the earliest ancestors of conifer trees were already evolving fire-adaptive traits during this period as a consequence of these rampant oxygen-fueled wildfires.[2]

Even today, the fire cycle that we discussed in chapters 3, 4, and 5 is an integral part of the carbon cycle that returns CO_2 to the air so we don't lose all our atmospheric CO_2 to carbon sinks. Despite how much more efficient oxygen breathers are today at digesting plant material, fire still fills a gap in the cycle to make sure we don't run out of atmospheric CO_2.

Although we don't want to burn our precious humus soils in high-intensity wildfires, low-intensity fires that turn ungrazed plant debris back into CO_2 are a good thing so that plants don't deplete our atmosphere like they almost did during the Carboniferous. Furthermore, the charcoal left behind after a wildfire (assuming the fire didn't get hot enough to scorch the soil) is instrumental in helping build humus in the soil — adding charcoal to the soil is yet another pathway to creating humus — carbon is carbon, whether it is the product of microbes breaking apart tough plant fibers or the consequence of fires turning plant fibers into char. Forest fires and grass fires in western North America don't just help clear the underbrush of debris, low-intensity fires also help return carbon and nutrients from tough fibrous plant debris back into the soil. Research has demonstrated that the charcoal produced by wildfires stimulates beneficial soil microbial activity, increases humus production, and increases the soil's water-holding capacity; even populations of earthworms, bugs, and other microorganisms that live in the soil all surge in the year following a fire.[3]

It used to be common for farmers to burn their pastures and fields in the spring to clean up plant debris and destroy weed seeds, just as indigenous tribes once did in the pre-industrial era, but that practice has

---

[2] He, T., Belcher, C.M., Lamont, B.B., and Lim, S.L. (2015). A 350-illion-year legacy of fire adaptation among conifers. *Journal of Ecology.* 104(2) 352-363. https://doi.org/10.1111/1365-2745.12513

[3] "Fire Effects on Soil." [Online Resources] *Northern Arizona University.* Accessed 8 Jun 2024. https://www2.nau.edu/~gaud/bio300w/frsl.htm

been increasingly discouraged by bureaucrats fixated on reducing carbon emissions (and by fire departments that have to fight these fires close to residential areas whenever they burn out of control). Instead, the fuel load simply builds up wherever animals don't graze, contributing to the much hotter mid-summer wildfires that not only burn plant debris but also scorch the humus in the soil. Hot fires are detrimental to soil even as frequent low-intensity fires help build soil.

Soils in the Amazon Basin are notoriously nutrient-depleted. And yet, in the same region you can nevertheless find pockets of one of the world's most fertile agricultural soils, a rich fertile black soil called Terra Preta (which means "black soil" in Portuguese). These Terra Preta soils were created over thousands of years by indigenous farmers who enriched the region's nutrient-deficient soils with a mix of charcoal, bones, broken pottery, compost, and manure. Many small-scale organic farmers have been experimenting with ways to recreate these fertile charcoal-enriched soils and, unsurprisingly, you can now buy biochar to add to your garden at your local garden supply center. You can even find instructions on YouTube to make your own.

Another bizarre consequence of all that extra oxygen in the atmosphere during the Carboniferous and Permian periods is that it massively increased the body sizes of insects in an effect known as gigantism. Even today, modern insects raised in a lab in oxygen-enriched environments grow more than 20% larger in as little as 5 generations. During the Carboniferous, insects were absolutely massive. The largest dragonfly was roughly the size and weight of a modern-day crow (it had a 30-inch wingspan!) and geologists have found fossils of millipedes that were 2 ft wide and more than 8 ½ feet long! The Carboniferous wasn't just the Age of Coal, it is also known as the Age of Giant Insects.

High oxygen levels also had a profound effect on the evolutionary path of animals. Fish and amphibian eggs have to be laid in water so that oxygen-rich water can seep in through their jelly-like shells to keep the embryo alive. But the oxygen-rich Carboniferous atmosphere led to the evolution of the amniotic egg — eggs covered by either a leathery or hard shell that is sufficiently permeable to allow atmospheric oxygen to circulate through the egg to sustain the embryo. This allowed animals to begin laying eggs on land, well hidden or guarded by a parent to keep them out of the clutches of predators. This paved the evolutionary path from amphibians to dinosaurs (and ultimately to chickens).

Midway through the Carboniferous, the climate began to cool. Many fossilized trees from the first half of the Carboniferous do not have any growth rings, which suggests that growth did not stop during the winter.[4] But fossil trees from the second half of the Carboniferous do have clear growth rings, indicating that there was a clear winter season when growth stopped. This cooling happened at the same time as the solar system began its traverse of the Scutum-Crux spiral arm of the galaxy — it marks the beginning of another massive glaciation called the Karoo Ice Age, in honor of the Karoo region of South Africa where geologists found the first evidence for this glacial age. Vast ice sheets covered the poles for over 100 million years as the planet slipped into another Icehouse that lasted right until the end of the Permian Period, though the climate in equatorial regions remained sufficiently warm and humid to continue producing more coal.

If you look at carbon dioxide levels back in figure 94, you can see that just as the Karoo Ice Age was ending, something massive happened at the Permian-Triassic boundary (252 mya) to fundamentally disrupt the balance between photosynthesis, carbon sinks, and oxygen breathers. CO_2 levels suddenly reversed their 300-million-year decline and soared above 2,000 ppm. Likewise, you can see in figure 101 that rapidly rising oxygen levels suddenly stalled near the end of the Permian period and plunged from around 30 to 35% back down to around 15%.

The Permian-Triassic boundary marks the worst mass extinction in our planet's long history, with over 90% of all plant and animal species (including most of the forests), 83% of insects, and 96% of marine life all going extinct at the same time. Geologists have nicknamed it "The Great Dying". It has a lot to teach us about what happens to the gases in our atmosphere when something happens to dramatically reduce the planet's total biomass.

---

[4] "The Carboniferous Period." [Online Resources]. *University of California Museum of Paleontology.* Accessed 6 Jun 2024.
https://ucmp.berkeley.edu/carboniferous/carboniferous.php

Chapter 19

The Great Dying

"Some things are believed because they are demonstrably true. But many other things are believed simply because they have been asserted repeatedly—and repetition has been accepted as a substitute for evidence."

— Thomas Sowell

There are no coal beds and very few fossils from the 20-million-year period that marks the Great Dying. It took 20 million years for Life to recover from whatever caused the Great Dying. Nor was it a single cataclysmic event. During the first 5 million years of this dead zone in the fossil record, there were several waves of mass extinctions as recovering plant life was wiped out again and again by repeated pulses of whatever caused the Great Dying. Geologists have described the destructive processes at work during the Great Dying as having a "quiet and sneaking character" — a long, drawn-out ecosystem collapse that spanned millions of years.

One of the clues to unravel the mystery of the Great Dying comes from fossilized seeds and spores from the few surviving plants still growing during the mass extinction.[1] These seeds and spores show a massive increase in anatomical deformities. They also show a massive increase in the chemical compounds that normally protect plants from being harmed by UV radiation coming from the Sun. Something destroyed the global ozone layer 252 million years ago — and prevented it from re-forming — leaving the Earth to be flooded by wave after wave of harmful UV-B radiation. This destruction of the ozone layer is further reinforced by

---

[1] "Ozone Layer Collapse Contributed to End-Permian Mass Extinction." (2023, Jan 9). *Sci News*. https://www.sci.news/paleontology/ozone-layer-collapse-end-permian-mass-extinction-11550.html

geochemical evidence preserved in sediments,[2] which show that there was a massive spike in wildfires during this period, which far exceeded the wildfires that I previously described occurring during the Carboniferous Period. This lends further support to the idea that something destroyed the ozone layer during that period — without ozone, both plants and the soil would have been baked by high levels of UV radiation. Combined with high levels of atmospheric oxygen (in excess of 35%), this would have created a volatile recipe for unimaginably massive wildfires.

Even the oceans were not immune to the consequences of a world without ozone. As plankton populations and other photosynthetic lifeforms that produce oxygen in the surface layers of the ocean were killed by UV radiation, the amount of oxygen being pumped into the oceans by photosynthesis dried up. Then, as oxygen-breathing microbes fed on the mass of rotting biomass building up in the ocean, they rapidly used up all the remaining oxygen in the seawater. And so, the oxygen depletion of the oceans triggered a mass die-off of marine organisms — this marine die-off lagged behind the terrestrial dying by up to 310,000 years, which highlights the cascading chain of events set in motion by the destruction of the ozone layer.[3] "Quiet and sneaking" indeed.

The standard "consensus" explanation for the Great Dying is that an intense period of volcanic activity filled the atmosphere with volcanic gases and that these gases destroyed the ozone layer. One of these volcanic gases — hydrogen chloride — directly destroys ozone. Another volcanic gas — sulfur dioxide — reflects incoming solar radiation (UV-radiation) before it can interact with oxygen (O_2) to convert it into ozone (O_3), thus preventing ozone from forming — remember ozone is formed naturally when UV radiation collides with oxygen molecules to turn O_2 into O_3. And at first glance the geologic record seems to lend support to this consensus explanation because sediments from that era provide evidence for a truly massive series of volcanic eruptions happening right around the time that the Great Dying began. These eruptions were from a series of flood volcanoes, called the Siberian Traps, which erupted continually over

---

[2] Mays, C. and McLoughlin, S. (2022). End-Permian Burnout: The role of Permian-Triassic wildfires in extinction, carbon cycling, and environmental change in eastern Gondwana. *PALAIOS*. 00, 1-26. http://dx.doi.org/10.2110/palo.2021.051

[3] Lu et al. (2022). Diachronous end-Permian terrestrial ecosystem collapse with its origin in wildfires. *Palaeogeography, Palaeoclimatology, Palaeoecology*. 594. https://doi.org/10.1016/j.palaeo.2022.110960

1 million years. They flooded most of Siberia with basalt lava, covering over 7 million square kilometers, an area equivalent in size to the *entire* contiguous United States plus half of Alaska.

But the theory that these flood volcanoes are responsible for the Great Dying falls apart upon closer inspection. The timing is all wrong — most of the lava released by the Siberian Traps had already been released long before the mass extinction began[4], and yet wave after wave of fresh dying continued (with recoveries in between) for millions of years after the Siberian Traps stopped erupting.

Another problem is the nature of the eruptions themselves. Flood volcanoes are a very specific type of volcano that forms when a large continent-sized plume of hot magma rises up underneath a continent, burns a hole through the crust, and then begins spilling out vast quantities of runny lava that slowly spreads across the entire surface of the continent — think Hawaii but on a much, much larger scale. Although we do not have a good example of a flood volcano today, the most recent example (albeit much smaller than the massive Siberian Traps) comes from western North America where the Columbia River basalts erupted 14 to 17 million years ago — today the Columbia River basalt sheets that once flooded across the continent are still visible as the hard basalt cap that covers the surface of the mountainous plateaus throughout most of the interior of Washington, Oregon, and British Columbia, including on the highlands above my parents' farm.

Flood volcanoes are not explosive Mt.-St.-Helen's-style eruptions that spew ash high into the atmosphere. Instead, they form slow moving lava rivers and lava fountains like those seen in Iceland or Hawaii today, except on an unimaginably large scale, which slowly expand across the continent. Although they are sometimes accompanied by isolated ash eruptions that launch skyward when hot molten rock runs into a pocket of groundwater, for the most part these flood volcanoes do not release large volumes of either of these ozone-depleting gases, nor do they produce the high-intensity ash explosions that launch volcanic gases high into the upper atmosphere where they could do widespread damage to the ozone layer. Furthermore, if large volumes of sulfur dioxide had been injected into the

---

[4] Elkins-Tanton, L. (2015, Jul 31). From Russia With Magma. *Slate*. https://slate.com/technology/2015/07/siberian-flood-basalts-the-earths-largest-extinction-and-climate-change.html

atmosphere, it would have had a sharp cooling effect on the climate (like when Mt. Pinatubo erupted in the Philippines in 1991, which temporarily cooled the atmosphere by 0.2 °C to 0.7 °C[5]). But as you can see back in figure 94, temperatures during the Great Dying (at the Permian-Triassic Boundary) were *rising* sharply throughout this period (by over 10 °C) — there's no evidence of a volcanic winter in the sedimentary record to accompany these volcanic eruptions, so clearly the atmosphere did not get filled with high volumes of sulfur dioxide.

Furthermore, even if the Siberian Traps had erupted explosively and filled the upper atmosphere with ozone-threatening gases, the U.S. Geological Survey has found that hydrogen chloride from volcanic eruptions is dissolved in water droplets in the eruption cloud and quickly falls back to earth as acid rain *so that it lasts no longer than a few days to a few weeks in the atmosphere.*[6] Likewise, studies from Hawaii have shown that sulfur dioxide released by volcanic eruptions also lasts no more than a few weeks in our atmosphere.[7] And even if sulfur dioxide released by a volcanic eruption forms more complex sulfate aerosols, these complex sulfate aerosols also don't persist in the atmosphere for longer than 2 or 3 years.[8]

No matter how you look at it, it is simply impossible for the gases released by the flood volcanoes of the Siberian Traps to have destroyed the ozone layer over a span of *millions* of years, nor could they have continued to affect the ozone layer millions of years *after* the flood volcanoes stopped erupting. Even if the Siberian Traps had been explosive eruptions instead of erupting as flood volcanoes, the gases from these eruptions would have been scrubbed from the atmosphere within at most

---

[5] "30 years after Mt. Pinatubo eruption: an illustration of the relationship between volcanoes and climate." (2021, Feb 2). *Royal Belgian Institute for Space Aeronomy.* https://www.aeronomie.be/en/news/2021/30-years-after-mt-pinatubo-eruption-illustration-relationship-between-volcanoes-and

[6] McGee et al. (1997, May). Impacts of Volcanic Gases on Climate, the Environment, and People. [USGS Open-File Report 97-262]. *U.S. Geological Survey.* https://pubs.usgs.gov/of/1997/of97-262/of97-262.html

[7] Hörmann et al. (2014). Estimating the volcanic emission rate and atmospheric lifetime of SO_2 from space: a case study for Kılauea volcano, Hawai'i. *Atmos. Chem. Phys.* 14(16). https://doi.org/10.5194/acp-14-8309-2014

[8] Robock, A. (2015). Climate and Climate Change | Volcanoes: Role in Climate. *Encyclopedia of Atmospheric Sciences (Second Edition).* Rutgers University. 105-111. https://doi.org/10.1016/B978-0-12-382225-3.00448-5

a few short years after the eruptions stopped, allowing the ozone layer to repair itself. Yet the fossil record tells us that the Great Dying continued millions of years after the Siberian Traps stopped erupting, and that the extreme levels of UV radiation continued to affect seeds and spores millions of years after the volcanoes that formed the Siberian Traps had gone dormant.

Unless there's something that continues to destroy the ozone layer, fresh incoming UV radiation quickly allows the ozone layer to grow back. In Antarctica today, the absence of sunlight during the winter months causes the ozone hole to expand by around 20% each winter, but then as the sun returns the ozone hole quickly shrinks back to its former size as the flood of summer sunlight rebuilds the ozone layer. Even if the entire planet's ozone layer was completely destroyed, as long as the sun continues to shine and as long as there's oxygen in the air, the entire ozone layer would take no more than 10 years to grow back.[9] So, even if the Siberian Traps had somehow managed to destroy the ozone layer while they were erupting, it would have completely grown back within a few years — instead we have a 20-million-year dead zone in the fossil record, with plants being wiped out time and again every time they started to recover, and those wipe-outs kept happening for millions of years after the Siberian Traps had stopped erupting. You simply cannot blame volcanoes for destroying the ozone layer millions of years after the volcanoes stop erupting.

~

As a brief side note, back in chapter five I mentioned a series of studies funded by Bill Gates, George Soros, and the White House, which hope to find ways to dim the Sun in order to try to cool our planet — these studies are experimenting with injecting sulfur dioxide into the upper atmosphere to block incoming UV radiation. The chemical they are experimenting with — sulfur dioxide —is the same sulfur dioxide gas that is released by volcanoes, which "consensus science" tried to blame for the Great Dying — and yet these muppets want to release this very same gas into our upper atmosphere *at the level where ozone is formed*. As I explained previously,

---

[9] Tretkoff, E. (n.d.). Did Gamma Rays Cause Ordovician Mass Extinction? *American Physical Society.* Accessed 3 May 2024.
https://www.aps.org/archives/publications/apsnews/200407/extinction.cfm

incoming UV radiation is essential to the process that creates our ozone layer — if you block incoming UV radiation, you also inadvertently prevent ozone from forming. Ozone is an extremely unstable molecule – it reverts back to regular oxygen within 1 to 2 hours after it forms, so if you inject sulfur dioxide into the upper atmosphere to dim the sun, you also inevitably block the process that continually creates our planet's protective ozone shield. If they succeed, we will be flooded with much higher levels of UV radiation.

Our ozone layer only protects us as long as UV radiation is continually streaming in to convert oxygen (O_2) into ozone (O_3) faster than ozone reverts back to regular oxygen. So, if you block UV radiation by temporarily injecting sulfur dioxide into the atmosphere, you turn off the engine that creates the very same ozone that protects us from dangerous incoming UV radiation — probably not enough to trigger another Great Dying, but perhaps enough to increase skin cancer and other cancers here on Earth. The malformed seeds and spores found in fossils from the Great Dying provide a clear example of what happens to DNA when our atmosphere is depleted of its ozone and thus allows the full unfiltered stream of UV radiation coming from the Sun to reach the surface of our planet. Before the climate change hysteria reached its current fever pitch, protecting our ozone layer used to concern scientists so much that we banned the hydrofluorocarbons that we once used to cool our refrigerators in order to stop the expansion of the ozone hole above Antarctica. And yet, here we are only a few years later, playing around with a chemical experiment that would see our governments release ozone-destroying gases into the upper atmosphere right at the level where ozone is formed. Madness.

And it's not a case of scientists not understanding the consequences of their ozone-destroying experiment. In a CNBC article from 2022, several advocates for using sulfur dioxide for solar engineering acknowledged that spraying sulfur dioxide into our atmosphere to cool our planet poses a risk to the ozone layer (they also acknowledge that this sulfur dioxide spray could aggravate respiratory illnesses and contribute to acid rain). But the article then goes on to justify its use anyway. "'*Yes, damaging the ozone is bad, acid deposition is bad, respiratory illness is bad, absolutely. And spraying sulfur in the stratosphere would contribute in the bad direction to all of those effects,*' Parson told CNBC. '*But you also have to ask, how much and relative to what?*' [...] Meanwhile, '*the world is getting hotter, and there will be catastrophic impacts for many people in the world,'* said Pasztor. *[...] 'And even if you stop all emissions today, the global temperature will still be high and will remain high for hundreds of years. So, that's why scientists are saying maybe we need something else, in addition — not instead of — but maybe in addition to everything else that is being done,'* he said. *'The current action/nonaction of countries collectively — we are committing millions of people to death. That's what we're doing.'*[10]

At least the failed weather modification programs of the 1940s, 50s, and 60s limited their stunts to the lower atmosphere where they could only influence the behaviour of local clouds. This time Icarus has his sights set much higher in an effort to mess with the solar radiation affecting our entire planet. And we all get to go along for the ride, whether we like it or not, because the central planners who watch over us feel morally justified. Someone needs to clip Icarus' wings, and soon, before this gets any further out of hand.

~

Even the idea that *any* explosive volcanic eruption could cause a global extinction does not stand up to geological evidence, irregardless of what the climate models claim. One of the largest volcanic eruptions in our planet's long history was the Lake Toba supervolcano eruption that occurred in Indonesia 74,000 years ago. Over a period of only 9 to 14 days, Toba launched between 9,500x and 62,000x more ash into the atmosphere than what was released by the 1980 Mt. St. Helen's eruption. The entire Indian subcontinent was buried under a 6-inch blanket of ash! As geologists learned more about this massive eruption, climate modelers predictably jumped in and calculated that the sulfur dioxide released by the Toba supervolcano would have plunged the entire planet into a 6-year-long volcanic winter.

Then, in the late 1990s as efforts to map the human genome got underway, a theory emerged that 75,000 years ago some near-extinction event had caused the total human population to be reduced to between 3,000 and 10,000 surviving individuals — the Toba eruption was

---

[10] Clifford, C. (2022, Oct 13). White House is pushing ahead research to cool Earth by reflecting back sunlight. *CNBC News.* https://www.cnbc.com/2022/10/13/what-is-solar-geoengineering-sunlight-reflection-risks-and-benefits.html

identified as the likely cause of this genetic bottleneck in human evolution. The scientific community was very excited. The media made impressive documentaries about this climate crisis in our history, which had very nearly wiped out our ancestors. It was the perfect theory for the mood of the late 90s and early 2000s, which combined enthusiasm about advances in genetic sequencing with the doom and gloom of the climate hysterics. The subtext was clear — climate change is real and dangerous!

However, after the initial enthusiasm for the bottleneck theory died down, geneticists showed that the genetic bottleneck theory was a complete false alarm — the population bottleneck simply never happened.[11] Furthermore, geological research emerged from East Africa, which clearly demonstrated that the theory that the Toba supervolcano had plunged the planet into a 6-year-long global volcanic winter also did not happen — the fallout from the Toba supervolcano had most certainly devastated its local region, but it had not affected the global climate.[12] Vegetation preserved in sediments showed that East Africa was not impacted by the Toba eruption — life simply carried on in East Africa — and so the theory of a global volcanic winter following the Toba eruption also came undone.[13] Furthermore, research subsequently demonstrated that climate modelers had overestimated the amount of sulfur dioxide released by the Toba supervolcano into the atmosphere *by a factor of 10x to 100x*.[14] Oops.

As for the effect of the Toba supervolcano on the ozone layer, a 2021 climate model calculated that the eruption could only have realistically suppressed ozone formation for about a year, and only in the tropics.[15] From all this, I think we can lay to rest any idea that the flood volcanoes of the Siberian Traps at the end of the Permian Period could have

---

[11] Hawks, J. (2018, Feb 9). The so-called Toba bottleneck simply didn't happen. *John Hawks Blog*. https://johnhawks.net/weblog/the-so-called-toba-bottleneck-didnt-happen/

[12] Ambrose, S. H. (2003). Did the super-eruption of Toba cause a human population bottleneck? Reply to Gathorne-Hardy and Harcourt-Smith. *Journal of Human Evolution*. 45(3). 231-237. https://doi.org/10.1016/j.jhevol.2003.08.001

[13] Hawks, J. (2018, Feb 9).

[14] Hawks, J. (2018, Feb 9).

[15] Osipov, S. et al. (2021). The Toba supervolcano eruption caused severe tropical stratospheric ozone depletion. *Nature: Commun Earth Environ* 2, 71. https://doi.org/10.1038/s43247-021-00141-7

destroyed the ozone layer over a period of 5 to 20 million years to cause the Great Dying that wiped out almost all life on Earth 252 million years ago.

While we are on the topic of mass extinctions and before we return to solving the mystery of the Great Dying at the end of the Permian Period, I want to address the mass extinction that is allegedly happening today, commonly referred to as the "Sixth Mass Extinction." Activist-scientists, the United Nations, and the media have been telling us that we are currently in the midst of this sixth mass extinction, with species going extinct *"up to 100 times faster than is typical in the nearly 4 billion year history of life"* and that this high rate of extinction was allegedly only last seen at the time when the dinosaurs went extinct,[16] with up to 40,000 species now going extinct due to human activities each year![17] Oh my! And, like a Hollywood sequel that keep doubling down on the hyperbole to hold everyone's attention (with each sequel, the villains get badder and the explosions get bigger), by 2018 activist Greta Thunberg was using her Twitter feed to promote a petition published by the Guardian Newspaper which claimed that *"'we are in the midst of the sixth mass extinction, with about 200 species becoming extinct each day...'"*[18] (in case you are wondering, that works out to 73,000 species per year, up from the prequel at 40,000 per year). As early as 1981, the world's favorite doomsday prophet, biologist Paul Ehrlich, predicted that by 2005, 50% of the world's species would be extinct.[19]

Science writer Matt Ridley (author of one of my favorite books *The Rational Optimist*) did a little digging. Once again, it turns out that reality is far removed from the doomsday story told by those who benefit from spreading fear. Matt Ridley writes that, *"Over the past 500 years, we know of 77 mammal species (out of about 5,000) and 140 bird species (out of*

---

[16] Pelley, S. (2023, Jan 1). Scientists say planet in midst of sixth mass extinction, Earth's wildlife running out of places to live. *CBS News*. https://www.cbsnews.com/news/earth-mass-extinction-60-minutes-2023-01-01/

[17] "Holocene extinction." *Wikipedia*. 10 June 2024. https://en.wikipedia.org/wiki/Holocene_extinction

[18] Thunberg, Greta (2018, Oct. 27). *We are in the midst of the sixth mass extinction, with about 200 species becoming extinct each day..."* [Twitter post]. Twitter. https://x.com/GretaThunberg/status/1056171209874948098

[19] Ridley, M. (2019, May 9). Dismantling Free Markets Won't Solve Biodiversity Threat. *Human Progress*. https://humanprogress.org/dismantling-free-markets-wont-solve-biodiversity-threat/

about 10,000) that have gone totally extinct."[20] "Bird and mammal extinctions peaked at 1.6 a year around 1900 and have since dropped to about 0.2 a year."[21]

Likewise, we have been told that 40% of all insects are going to go extinct over the next few decades. Again, Matt Ridley provides a dose of reality, "*Of the million [insect] species described, have a guess how many are known to have gone extinct, according to the official Red List of the International Union for the Conservation of Nature. It's 63. That's not 40 per cent or 4 per cent or 0.4 per cent, or 0.04 per cent: it's more like 0.004 per cent. Most of those were on islands, which means invasive species [like rats, cats, pigs, etc.] probably caused their extinction, and of those for which a "last seen" date is given, the majority disappeared more than a century ago. Only six have a post-war "last seen" date, and the most recent is 1975.*"[22]

All this is not to minimize the threat to many wildlife species around the world — there is no doubt that many have been pushed to the brink of extinction over the last few centuries by our activities. However, on the whole, wildlife populations are beginning to recover, especially in wealthy nations that can afford to invest in the protection of their environment. Even here in BC in the forests surrounding my parent's farm, many species like elk and wolves have returned to areas where they were never seen while I was growing up here in the 70s and 80s — one of our neighbors even recently photographed a grizzly bear on his trail camera in his backyard. It is a good news story that is repeating itself everywhere as the world gets richer. A 2018 report released by the Nunavut government in Canada's High Arctic about the growing polar bear population declared that "*Inuit believe there are now so many bears*

---

[20] Ridley, M. (2015, Jun 28). Invasive species are the greatest cause of extinction. *Matt Ridley Blog, republished from his column in The Times.* https://www.mattridley.co.uk/blog/invasive-species-are-the-greatest-cause-of-extinction/

[21] Ridgley, M. (2011, Aug 27). Counting Species Out. *Matt Ridley Blog, republished from his column in The Times.* https://www.mattridley.co.uk/blog/counting-species-out/

[22] Ridley, M. (2021, Jul 29). Environmentalists have got it wrong – we're not facing an insect apocalypse. *The Telegraph.* https://www.mattridley.co.uk/blog/counting-species-out/

that public safety has become a major concern."[23] It would seem that polar bear researcher Dr. Susan Crockford stands vindicated, even if that hasn't helped restore her former job at the University of Victoria.

As Matt Ridley points out, "*On the whole what really diminishes biodiversity is a large but poor population trying to live off the land. As countries get richer and join the market economy, they generally reverse deforestation, slow species loss and reverse some species declines. Countries like Bangladesh are now rich enough to be reforesting, not deforesting, and this is happening all over the world.*"[24] Poverty, not industrialization, is the driver of species extinctions today. Once people's immediate needs for food and shelter are met, they begin taking better care their environment.

Current climate policies are on track to reverse all that. The latest buzzword of the climate change movement is "degrowth" — according to the vision championed by the World Economic Forum, "degrowth" broadly means "*shrinking rather than growing economies, to use less of the world's dwindling resources.*"[25] "Degrowth" was mentioned 28 times in the United Nations' IPCC Sixth Assessment Report published in April 2022.[26] They are, quite literally, prescribing the very thing that destroys the planet, fuels extinctions, and plunges humanity back into the Malthusian Trap.

---

[23] Cecco, L. (2018, Nov 13). Polar bear numbers in Canadian Arctic pose threat to Inuit, controversial report says. *The Guardian*. https://www.theguardian.com/world/2018/nov/13/polar-bear-numbers-canadian-arctic-inuit-controversial-report

[24] Ridley, M. (2019, May 9). Dismantling Free Markets Won't Solve Biodiversity Threat. *Human Progress*. https://humanprogress.org/dismantling-free-markets-wont-solve-biodiversity-threat/

[25] Masterson, V. (2022, Jun 15). Degrowth – what's behind the economic theory and why does it matter right now? *The World Economic Forum*. https://www.weforum.org/agenda/2022/06/what-is-degrowth-economics-climate-change/

[26] "Degrowth." *Wikipedia*. 10 Jun 2024. https://en.wikipedia.org/wiki/Degrowth

Chapter 20

Atomic Gardening

"There is nothing that living things do that cannot be understood from the point of view that they are made of atoms acting according to the laws of physics."

— Richard Feynman

There is another explanation for the "quiet and sneaking character" of the Great Dying that has nothing to do with volcanoes but explains how the ozone layer could have been destroyed over such a long time period, with multiple alternating waves of recovery and extinction. Once again, we have to turn our eyes towards our solar system's episodic traverses through the spiral arms of our galaxy.

While ozone is created by ultraviolet radiation from our Sun colliding with oxygen in the upper atmosphere, when cosmic radiation (i.e. gamma rays, x-rays, muons, and other high-energy particles from space) collides with ozone molecules, these forms of radiation tear the ozone molecules apart.[1] Research has even shown that the size of the ozone hole above Antarctica follows the 11-year solar cycle, which causes the strength of the Sun's magnetic field to rise and fall, and thus changes how many cosmic rays are able to penetrate our atmosphere to collide with ozone — the ozone hole grows larger during periods when there are more cosmic rays reaching our planet and shrinks when there are less cosmic rays reaching our planet. In short, *cosmic rays are the primary force that creates the ozone hole.*[2]

When our path through the galaxy takes our solar system through the spiral arms of the Milky Way where there are much higher levels of cosmic radiation (near dying stars, neutron stars, pulsars, supernovas,

---

[1] Lu, QB. (2023). Formulation of the cosmic ray-driven electron-induced reaction mechanism for quantitative understanding of global ozone depletion. *PNAS*. 120(27). https://doi.org/10.1073/pnas.2303048120

[2] Lu QB. (2009). Correlation between cosmic rays and ozone depletion. *Physical Review Letters*. 102(11). https://doi.org/10.1103/physrevlett.102.118501

supernova remnants, or nebulae where there is intense star formation), the high levels of cosmic radiation flooding towards us from these highly radioactive regions of our galaxy also affect how much ozone we have above us to shield us from UV radiation. And so, it turns out that our passage through the spiral arms of our galaxy affect more than just cloud cover and temperature here on Earth. The effect of cosmic radiation on the size of the ozone hole is only the beginning — buckle up, things are about to get wild...

Geologists have noticed from the fossil record that just as ice ages seem to correlate with our passage through the spiral arms of the galaxy (presumably via cloud formation driven by increased cosmic radiation), all the big extinction events also correlate with those same passages through the spiral arms. Furthermore, these passages through the spiral arms also correlate with bursts of new species appearing in the fossil record as evolution seems to speed up for a while. Part of that evolutionary acceleration, but not all, can be explained by the void left by extinctions, which open up new environmental niches that are subsequently filled by newly emerging species. But there's more to the story.

Evolution happens because of mutations in the genetic code. As DNA continually makes copies of itself, flaws happen during the copying process that lead to slight genetic mutations as the code passes from one generation to the next. And guess what scientists think causes this slow but never-ending stream of copying errors... cosmic radiation.[3] As cosmic radiation collides with DNA, these collisions introduce a slow but steady stream of small new genetic mutations to the genetic code. Mutations that provide a competitive advantage will end up having more offspring so they become more widespread in the next generation, while mutations that do not offer a competitive advantage will have fewer offspring and thus are evolutionary dead ends. [Incidentally, cosmic radiation has also been shown to degrade computer memory over time.[4]]

Genetic mutations happen at a relatively predictable rate — in humans there are approximately 64 new mutations per generation.[5] Evolution on

---

[3] Kohler, S. (2020, May 20). Cosmic Rays as the Source of Life's Handedness. *AAS NOVA*. https://aasnova.org/2020/05/20/cosmic-rays-as-the-source-of-lifes-handedness/

[4] Ziegler, J.F., Lanford, W.A. (1979). Effect of Cosmic Rays on Computer Memory. *Science*. 206(4420). https://doi.org/10.1126/science.206.4420.776

[5] "Mutation rate." *Wikipedia*. 26 Apr 2024. https://en.wikipedia.org/wiki/Mutation_rate

Earth is, quite literally, driven by the impact of cosmic radiation on our DNA — without cosmic radiation, evolution would be much, much slower, perhaps even too slow to keep up with rapid changes in our environment.

The only way to speed up the rate at which genetic mutations occur is to expose DNA to much higher levels of radiation. One way to do this is to travel into space, outside of the protective envelope of our ozone layer and outside of the Earth's protective magnetic field. A NASA study of astronauts found that all fourteen astronauts included in the study had DNA mutations in their blood-forming stem cells and concluded that astronauts should therefore be subject to periodic blood screening to monitor for a slightly elevated risk of cancer.[6] Likewise, pilots and flight attendants also have higher rates of cancer (they are 87% more likely to suffer from melanoma, 39% more likely to have thyroid cancer, and 16% more likely to contract prostrate cancer).[7] Scientists have also taken samples of bacteria into space and discovered that the frequency of genetic mutations in these bacteria increased *24-fold* compared to the control group of bacteria that remained back on Earth.[8] The frequency of genetic mutations and the degree of exposure to cosmic radiation are inextricably linked.

One of the consequences of the nuclear physics programs that emerged during the quest to develop the Atomic Bomb was that agricultural scientists like Norman Borlaug realized they could harness radiation as a means to speed up genetic mutation in seeds — they flooded plants with radiation (typically gamma rays or x-rays like those found in cosmic radiation) and then selected mutants that had new beneficial characteristics (i.e. more drought tolerance, more disease resistance, higher yields, etc.) Instead of having to wait millions of years for naturally beneficial mutations to emerge, the evolutionary process could be sped up

---

[6] Howell, E. (2022, Sept 5). Astronauts' blood shows signs of DNA mutations due to spaceflight. *SPACE.com.* https://www.space.com/astronaut-spaceflight-cancer-dna-mutations-study

[7] Hadley, G. (2023, Mar 29). Proven Higher Cancer Risk for Pilots and Ground Crew Sparks Search for Causes. *Air & Space Forces Magazine.* https://www.airandspaceforces.com/study-aviator-cancer-rates-lawmakers-dod/

[8] Fajardo-Cavazos, P. et al. (2018). Spaceflight Alters the Mutation Frequency and Mutagenic Spectrum in Bacteria. *42nd COSPAR Scientific Assembly. Held 14-22 July 2018, in Pasadena, California, USA, Abstract id. F2.4-23-18.* https://ui.adsabs.harvard.edu/abs/2018cosp...42E1031F/abstract

with a timely blast of radiation. This plant breeding strategy was nicknamed "atomic gardening."

Seventy percent of all registered agricultural crop varieties in use around the world today were produced by these classical gamma ray and x-ray irradiation techniques,[9] which works out to more than 3,000 varieties that were created in the atomic garden[10]. Atomic gardening is at the heart of the "green revolution" that occurred during the mid 20th century, which enabled humanity to feed itself with increasingly hardy, efficient, high-yield crops even as the world's population increased 5-fold from 1900 until today. Although irradiation is still used today to create some new plant varieties, its popularity for seed development decreased because of public skepticism about irradiation and because irradiation was eclipsed by new genetic technologies, starting in the 1990s with GMOs (which splice desirable individual segments of DNA from one variety into the genes of another variety), and then more recently by the gene editing technology known as CRISPR. Both GMO and CRISPR enable plant breeders to create targeted mutations with more predictable outcomes instead of having to blast seeds with radiation and then hope for a random beneficial mutation to show up.

[On a side note: Unfortunately, both GMOs and gene editing are being used for more than just developing beneficial new crop varieties — thanks to patent laws and lobbying by the big corporations involved in seed development, these techniques are being used to monopolize seed markets with varieties that make it illegal for farmers to save and replant their own seeds. Furthermore, efforts are underway to create so-called "terminator seeds", which are genetically programmed to be sterile in the second generation, thus preventing anyone from being able to replant their own grain. So far these "terminator seeds" have not been field tested or commercialized, but it is the ultimate technology by which Big Ag can hold farmers hostage. It's not altogether different from the way that farm machinery companies have learned to abuse patent laws to make it illegal for farmers to access the software inside their own tractors when trying to repair their own machinery, thus holding farmers hostage to expensive

---

[9] Ma, L. et al. (2021). From Classical Radiation to Modern Radiation: Past, Present, and Future of Radiation Mutation Breeding. *Frontiers in Public Health*. 9. https://doi.org/10.3389%2Ffpubh.2021.768071

[10] "What is 'radiation breeding'? *BBC Science Focus*. Accessed 3 Jun 2024. https://www.sciencefocus.com/science/what-is-radiation-breeding

licensed dealerships for maintenance and repair work (if you don't own the software inside your tractor, so you can't legally access it to fix your own tractor). If the work of Norman Borlaug represents the high point of the "green revolution", then terminator seed technology and other abuses of patent laws are indisputably its low point — yet another example of the new kind of corporate serfdom that is emerging on the back of Woodrow Wilson's cherished administrative state, which is manipulated by one and all in service of a never-ending stream of self-serving agendas. Like so many who idolize central planning, Wilson optimistically believed that the masterful administrate state would serve the collective good; reality has proven that, thanks to our opportunistic human nature, the masterful administrative state merely uses its expanded reach to serve itself, and that all those who learn the art of lobbying soon perfect self-serving strategies by which to extract special favors, market privileges, and government contracts from the administrative state at the expense of the collective good.

On another ironic side note: Organic certification bodies ban GMO seed varieties, but crop varieties produced via the earlier irradiation method are generally allowed. In fact, because conventional farmers now have access to so many newer varieties produced via GMO and CRISPR techniques, organic farmers rely even more heavily on seed varieties developed via early 20th-century irradiation techniques than their conventional peers. So, there's a good chance that the certified organic rice, apple, grapefruit, wheat, or barley in your pantry is a variety produced in the early 20th-century atomic garden, which used radiation to speed up the natural rate of genetic mutations. There's now a fierce unresolved debate raging within the organic community as to whether CRISPR should be allowed for organic production since gene editing does not involve gene splicing.]

Now back to our journey through Earth's long history... As geologists studied the fossil record, they noticed that in the immediate aftermath of the big extinction events that correlate with our solar system's traverses of the spiral galaxy, the fossil record suggests that evolution seemed to speed up for little while as new species emerged and existing species mutated much faster than you would expect to be possible just through the predictable background rate at which mutations normally occur.[11] The higher levels of cosmic radiation inside the spiral arms of our galaxy provide a logical explanation for how evolution could speed up via a much higher rate of genetic mutations — traversing the spiral arms is like taking a walk through the atomic garden.

But a generally higher background level of cosmic radiation is not the only thing lurking inside these spiral arms. There are other things that can temporarily crank up the dial on cosmic radiation, with disastrous consequences for life on Earth. If our solar system passes close enough to a dying star, a neutron star, a pulsar, a nebula, a supernova, or a supernova remnant, our solar system would be flooded with colossal levels of cosmic radiation, possibly for millions of years, and possibly in waves. And that radiation would systematically strip the ozone layer from our planet *and keep it off* as long as the high intensity cosmic radiation continues, just like we see in the geologic record from the Great Dying!

In the case of the Great Dying at the Permian-Triassic boundary, which stretched out over millions of years, it is thought that our solar system passed through an interstellar molecular cloud (similar to the Horsehead Nebula or the Orion Nebula, which you may be familiar with from photos from the Hubble Space Telescope).[12] These nebulae are flooded with massive volumes of electromagnetic radiation, and hiding within the nebulae are dense clusters where new stars are forming, old stars are dying, and dying stars are going supernova — in short, passing through such a nebula would take many thousands or even millions of years, all the while exposing our solar system to all the ingredients for a multi-million year extinction event punctuated by intense waves of radiation and periods of recovery, all while fueling rapid genetic mutations in the few species that manage to survive.

It's a lovely theory that tidily explains the Great Dying, but is there hard geological evidence for it? In short, yes. When supernovae explode, they

---

[11] Ebisuzaki, T. and Maruyama, S. (2015). United theory of biological evolution: Disaster-forced evolution through Supernova, radioactive ash fall-outs, genome instability, and mass extinctions. *Geoscience Frontiers.* 6(1). https://doi.org/10.1016/j.gsf.2014.04.009

[12] Detre et al. (1997). The comparison of P/Tr and K/T boundaries on the basis of cosmic spherules found in Hungary. [Conference Paper, 28th Annual Lunar and Planetary Science Conference, p. 297.] *Lunar and Planetary Science XXVIII.* https://www.lpi.usra.edu/meetings/lpsc97/pdf/1593.PDF

don't just release radiation. The shockwave also carries with it a large quantity of iron-rich dust — micrometeorites — with a very distinctive iron-rich chemistry that sets these micrometeorites apart from sediments formed here on Earth.[13] Likewise, nebulae are full of this dust left over from supernova explosions. And geologists have found these distinct micrometeorites, also known as 'cosmic spherules', all over our planet in sediments deposited during the time of the Great Dying.[14]

Nor is the Great Dying the only big mass extinction that has been linked to a supernova explosion or some other radiation-emitting cosmic phenomena happening in our vicinity of the Milky Way. The Ordovician mass extinction (445-444 mya) may have been caused by a gamma-ray burst from a nearby supernova[15] — geologists have again found elevated concentrations of cosmic spherules in sediments deposited at the time of the Ordovician extinction.[16] Likewise, the Devonian mass extinction (372-359 mya) has been linked to a nearby supernova explosion[17] — much like sediments deposited during the Great Dying, sediments from the time of the Devonian extinction contain lots of dark, misshapen spores from fernlike plants indicative of an ozone layer that was suddenly stripped away to expose unprotected plants to high levels of UV radiation[18] as well as elevated concentrations of cosmic spherules.[19] And sediments from the Triassic-Jurassic mass extinction (201 mya) once again contain evidence

---

[13] Choi, C.Q. (2016, Aug 18). Supernova Ashes Found in Fossils Hint at Extinction Event. *Space.com.* https://www.space.com/33777-supernova-ash-found-in-fossils.html

[14] Detre et al. (1999). The Permian-Triassic Supernova Impact. *Meteoritics & Planetary Science.* 34. Supplement, p.A124 https://www.lpi.usra.edu/meetings/metsoc99/pdf/5033.pdf

[15] Tretkoff, E. (n.d.). Did Gamma Rays Cause Ordovician Mass Extinction? *American Physical Society.* Accessed 3 May 2024. https://www.aps.org/archives/publications/apsnews/200407/extinction.cfm

[16] Voldman, G.G., et al. (2012). Cosmic spherules from the Ordovician of Argentina. *Geological Journal.* 48(2-3). 222-235. https://doi.org/10.1002/gj.2418

[17] Fields, B.D., et al. (2020). Supernova triggers for end-Devonian extinctions. *PNAS.* 117(35). https://doi.org/10.1073/pnas.2013774117

[18] Voosen, P. (2020, May 27). No astroids needed: ancient mass extinction tied to ozone loss, warming climate. *Science.org.* https://www.science.org/content/article/no-asteroids-or-volcanoes-needed-ancient-mass-extinction-tied-ozone-loss-warming

[19] Bai, Z. (2000). Discovery and significance of microspherules at Lower-Middle Devonian boundary of Guangxi, South China. *Sci. China Ser. D-Earth Sci.* 43. 302–307. https://doi.org/10.1007/BF02906826

of ozone depletion[20] along with a sharp increase in these cosmic spherules.[21]

Only the Cretaceous extinction that killed the dinosaurs (66 mya) doesn't fit the pattern. Although sediments from that time period again contain elevated concentrations of cosmic spherules[22], this time the killing blow was delivered by an asteroid impact, which left its mark via the 180-km-wide Chicxulub Crater near the Yucatan peninsula in Mexico — the asteroid impact also showered the planet with cosmic dust and micrometeorites, however the distinct cosmic spherules already show up in the sedimentary record *before* they are superimposed by impact-linked micrometeorites, which confirms that the solar system was once again plowing through another of the spiral arms at the time of the asteroid impact.

As I mentioned in a previous chapter, all the "Big Five" extinctions, including the Cretaceous extinction that killed the dinosaurs, happened as our solar system crossed through one of the spiral arm structures in our galaxy — the Cretaceous extinction happened as our solar system was traversing the Carina-Sagittarius spiral arm.[23] Impact craters from asteroids and meteorites are not evenly distributed throughout our Earth's history. Instead, there are episodes when bombardment surges dramatically, and those clusters of asteroid bombardment correlate with our solar system's passage through our galaxy's spiral arms. The most likely explanation for these episodic asteroid bombardments is that as we pass close to large gravitational bodies inside these spiral arms (i.e., stars, pulsars, nebulae, etc.), their gravitational pull causes gravitational disturbances that disrupt the icy rocks orbiting around the outer perimeter

---

[20] "Triassic-Jurassic extinction event." *Wikipedia*. 3 Jun 2024. https://en.wikipedia.org/wiki/Triassic%E2%80%93Jurassic_extinction_event

[21] Detre et al. (1997). The comparison of P/Tr and K/T boundaries on the basis of cosmic spherules found in Hungary. [Conference Paper, 28th Annual Lunar and Planetary Science Conference, p. 297.] *Lunar and Planetary Science XXVIII*. https://www.lpi.usra.edu/meetings/lpsc97/pdf/1593.PDF

[22] Grachev et al. (2008). „Cosmic Dust and Micrometeorites in the Tansitional Clay Layer at the Cretaceous-Paleogene Boundary in the Gams Section (Eastern Alps): Morphology and Chemical Composition". *Izvestiya Physics of the Solid Earth*. 44(7). http://dx.doi.org/10.1134/S1069351308070069

[23] Gillman, M. and Erenler, H. (2019). Reconciling the Earth's stratigraphic record with the structure of our galaxy. *Geoscience Frontiers*. 10(6) https://doi.org/10.1016/j.gsf.2019.06.001

of our solar system — the Oort cloud I spoke about previously — knocking some of them out of orbit and onto a collision course with the planets in our inner solar system.[24]

From our ozone layer to the clouds affecting our climate, to the ice ages and extinctions that completely redraw the surface of our planet every few hundred million years, and to the speed of evolution itself, one way or another our little blue marble orbiting around our Sun is part of a dynamic cosmic ecosystem that stretches right across our galaxy. Every time we cross the spiral arms of our galaxy, interesting things begin to happen as the relatively stable conditions we have on our planet are suddenly thrust into disarray[25] — ice ages, extinctions, meteorite impacts, and a sudden acceleration in genetic mutations always accompany our passage through these spiral arms of our galaxy.

There is even some limited evidence that spiral arm crossings may be contributing to periods of intense volcanic activity as these spiral arm crossings trigger the rise of especially large magma plumes (like the aforementioned Siberian Traps) that occasionally rise from the depths of our planet to erupt onto surface. In 2022 Forbes even reported on this correlation between massive flood volcano eruptions and waves of asteroid impacts, and that they correspond in time with spiral arm crossings.[26] The energy released by an asteroid impact would inject massive amounts of energy into the crust, which could trigger magma plumes melting as that energy radiates down through the crust into the deeper layers of our Earth. And the gravitational disturbances would also create friction deep within the Earth as the planet is stretched or compressed by these gravitational disturbances — that friction would also generate heat that would fuel the creation of these large hot mantle plumes. A study published in the International Journal of Geosciences in 2019 illustrated that these periods of intense volcanic activity are also mirrored

---

[24] Napier, W.M. (2006). Evidence for cometary bombardment episodes. *Monthly Notices of the Royal Astronomical Society*. 366(3). 977–982. https://doi.org/10.1111/j.1365-2966.2005.09851.x

[25] Sankaran, A. V. (2008). Galactic triggering of geologic events in earth's history. *Current Science*, 95(6), 714–716. http://www.jstor.org/stable/24102576

[26] Dorminey, B. (2022, Aug 26). Milky Way's Spiral Arms Incredibly Triggered Earth's Continent Formation. *Forbes*. https://www.forbes.com/sites/brucedorminey/2022/08/26/milky-ways-spiral-arms-incredibly-triggered-earths-continent-formation/?sh=57a74a0f35fc

by intense volcanic activity on our neighboring planets (Venus, Mars, Mercury, and even on our own Moon), which again correspond with our solar system's passage through the spiral arms of the galaxy and lend credibility to the idea that cosmic astronomical forces do indeed play an important role in fueling cyclical periods of volcanic activity not only here on Earth but even on our neighboring planets, which go above and beyond the "local" geological forces at work beneath our feet.[27]

If we want to understand our place within that cosmic ecosystem and make predictions about what the future has in store for us, we would be far better served by funding astronomers than climate scientists — it's high time that climate scientists were put out to pasture with all the other pseudo-scientific crackpots of the past so they can fade into history along with the rainmakers of the early 20th century, Lysenko's Marxist agricultural planners, and the folks who believed Mars was home to an advanced canal-building population.

---

[27] Brink, H-J. (2019). Do Near-Solar-System Supernovae Enhance Volcanic Activities on Earth and Neighbouring Planets on Their Paths through the Spiral Arms of the Milky Way, and What Might Be the Consequences for Estimations of Earth's History and Predictions for Its Future? *International Journal of Geosciences.* 10(5). https://doi.org/10.4236/ijg.2019.105032

Chapter 21

A Disturbance in the Force

"When we try to pick out anything by itself, we find it hitched to everything else in the universe."

— John Muir

The destruction of the ozone layer during the Great Dying all but turned off the planet's photosynthesis engine for up to 20 million years. There are no coal beds from this period and the fossil record is pretty sparse. Not everything died, obviously, or we wouldn't be here, but for the first time in millions of years the relentless scrubbing of CO_2 from the atmosphere by photosynthesis was dialed back to almost nothing, which made it possible for volcanoes to replenish the atmosphere with carbon dioxide — CO_2 levels rapidly surged from 150 ppm to around 2,000 ppm.

As long as volcanoes and other tectonic forces keep bringing CO_2-rich volcanic gases to the surface, there will always be some fresh CO_2 to top up the atmospheric tank that is otherwise continually being depleted as a portion of the carbon from photosynthesis gets trapped in soils, muddy sediments, peat bogs, and coal beds, and as the continental weathering of rocks (like the calcium carbonate cycle) keeps scrubbing CO_2 out of the atmosphere. As I showed you in a previous chapter during the discussion about CO_2 fertilization, plant growth decreases with falling CO_2, which helps slow the rate at which photosynthesis removes CO_2 from the atmosphere and, by consequence, slows the rate at which carbon gets trapped in carbon sinks. During the Carboniferous, the low levels of atmospheric CO_2 (as low as 150-180 ppm) nevertheless forced plant growth to slow to a crawl, which helped prevent the atmosphere from completely running out of carbon dioxide. The amount of new CO_2 being added to the carbon cycle by volcanoes was able to (just barely) keep up with how fast plants, coal beds, and rock weathering were consuming atmospheric CO_2.

Below 150 ppm atmospheric CO_2, photosynthesis grinds to a halt. If CO_2 levels had continued to fall, Life on Earth would have been extinguished. Although declining photosynthesis helped offset falling carbon dioxide levels, it's important to remember that the chemical weathering of rocks nevertheless continued to scrub CO_2 from the atmosphere. We were incredibly fortunate that the volcanic eruptions from the Siberian Traps replenished our atmosphere with fresh CO_2 when they did. And we are fortunate that the chemical weathering of rocks during that era did not happen at a faster rate (rock weathering speeds up during aggressive mountain-building phases in our tectonic history but slows down during periods when continents are breaking apart instead of colliding).

The same self-regulating process also happens at the other end of the scale because of the effect of CO_2 fertilization. As CO_2 is added to the atmosphere, it is quickly consumed by a surge of vegetation. As long as there are no other factors limiting plant growth (like soil erosion, humus degradation, deforestation, or desertification), CO_2-starved plants will quickly gobble up any extra CO_2, which invariably also increases the amount of carbon that is deposited in carbon sinks like limestones, carbon-rich muds, coal beds, and soil humus. The only way to sustain a meaningful increase in atmospheric CO_2 levels is to somehow limit the surge of photosynthesis that would normally follow an increase in atmospheric CO_2.

During the Great Dying, the destruction of the ozone layer effectively hammered the brakes on photosynthesis for millions of years, which allowed volcanic CO_2 emissions to replenish the atmosphere without immediately converting all that extra CO_2 into plant growth and coal beds. Today, deforestation, desertification, and soil degradation are likewise limiting the surge in plant biomass that would otherwise quickly consume the extra CO_2 that is being added to the atmosphere by fossil fuel emissions, volcanic emissions, and by the erosion and oxidation of humus. The scale of how much our biomass is being limited by soil degradation, deforestation, and desertification will become increasingly clear as our geological detective story reaches its conclusion.

Reducing photosynthesis doesn't only affect atmospheric CO_2. It also has a profound impact on the amount of oxygen in the air. As photosynthesis slows down, it no longer produces enough oxygen to keep up with the amount of oxygen that is continually consumed by oxygen breathers, wildfires, humus oxidation, rust, and other oxygen-consuming processes. Since oxygen reacts with just about everything (not just carbon and iron), during the Great Dying the remaining oxygen in the air continued to oxidize rocks and minerals on land, which had the effect of scrubbing oxygen out of the atmosphere faster than much-reduced levels of photosynthesis could replace it. Based on fossil evidence from the Great Dying, it is clear that the oceans became so severely depleted of oxygen that it triggered a mass extinction of marine species that depend on living in oxygenated waters. Likewise, although there is still a great degree of uncertainty as to exactly how low atmospheric oxygen levels plunged during the Great Dying, fossil and sedimentary evidence has allowed geochemists to establish that atmospheric oxygen plunged from as high as 35% before the Great Dying to a low of somewhere between 11% and 15% by the time the Great Dying reached its conclusion.

Even after plants recovered from the Great Dying, an absence of charcoal in sediments from the early Triassic Period (252–201 mya) demonstrates that oxygen levels remained below the level required to sustain wildfires during the entire first half of the Triassic. Charcoal from wildfires only begins to show up again in sediments from the second half of the Triassic. Furthermore, fossils reveal that dinosaurs evolved special air sacs in their lungs, throats, and bones during this long low-oxygen period to help them overcome the challenge of living in such a low-oxygen environment. And, once again, an evolutionary adaptation designed to survive one environmental challenge opens the door to new environmental niches as that adaptation is repurposed for something else altogether — one branch of the dinosaur family tree took advantage of its air-filled bones to colonize the skies, giving rise to birds. Their air sacs enabled them to have efficient respiration even at high altitudes where the atmosphere is thin, and their lightweight air-filled bones made them light enough to be able to fly.

Although CO_2 levels surged during the Great Dying as photosynthesis ground to a halt, CO_2 levels began to fall again once plant growth began to recover at the beginning of the Triassic, as shown in figure 94. However, the fact that CO_2 continued to rise long after the Siberian Traps stopped erupting and only leveled off and began to fall after plant growth began to recover after the end of the Great Dying illustrates once again that the planet's total biomass is *the* single most important force that must be understood in order to understand changes in the gas mix in our atmosphere — far more important than carbon dioxide emissions. When it comes to how much carbon is in our atmosphere, the rate of carbon *consumption* is far more important than the amount of CO_2 being added to the atmosphere.

This is mirrored by what we see happening to oxygen levels in figure 101, which plunged during the Great Dying (as photosynthesis shut down) and only slowly began to rise again once photosynthesis began to recover after the Great Dying.

Midway through the Jurassic Period (201-145 mya), another wave of intense volcanic activity lasting around 600,000 years raised atmospheric CO_2 levels once more, this time to around 2,500 ppm. However, as soon as these Jurassic volcanoes stopped erupting, CO_2 levels immediately began to fall once more and have been falling ever since as some carbon is continually lost to carbon sinks. In the absence of some "event" to disturb this balance, the normal state of affairs on our planet is that carbon dioxide is slowly but continually being used up.

~

It is easy to underestimate the amount of CO_2 that is continuously being added to the atmosphere by volcanic activity (and thus dismiss the impact of volcanic gases on the atmospheric mix) because we typically only think of the occasional news headline about volcanoes that erupt up on land. With so much CO_2 leaking into the atmosphere from volcanic sources, all of which is then consumed by photosynthesis and carbon sinks, it's actually quite easy to disrupt the balance of carbon in our atmosphere simply by changing the total amount of photosynthesis happening on our planet. Even a small reduction in biomass can lead to large increases in atmospheric CO_2 simply because those volcanic emissions continue to relentlessly inject fresh carbon into the atmosphere irregardless of whether forests, grasslands, and meadows can keep up. And so, we need to take a little detour beneath the surface of our planet's crust to have a closer look at where most of our planet's fresh CO_2 has come from during our planet's long 4.54-billion-year history. The mid-Jurassic volcanic eruptions provide a perfect case study to illustrate the point.

The mid-Jurassic volcanic eruptions began as the giant supercontinent of Pangaea began to break apart into the continents we recognize today. The colossal volume of lava that flooded across the surface of the fracturing supercontinent during this period is collectively known to geologists as the "Central Atlantic magmatic province" — admittedly not a very inspiring name for what was in fact the biggest volcanic episode in

Earth's history, by a wide margin. This is the period when South America "unzipped" itself from Africa, when Europe and North America began to drift apart, and when Antarctica and Australia said "adios" to the other continents and started their long drift towards the South Pole (Australia finally split away from Antarctica much later, around 30 million years ago). By the time the mid-Jurassic eruptions ended, vast lava flows covered over 11 million km^2 of land (roughly 1½ times the area of the continental United States) across what is today western Africa, southern Europe, eastern North America, and northeastern South America.

The seafloor spreading that began with the breakup of Pangaea during the mid-Jurassic continues unabated even today all along the Mid-Atlantic Ridge as the Americas continue to drift away from Europe and Africa. Even today, there are over 28,000 kilometers of uninterrupted underwater volcanic activity all along the length of this Mid-Atlantic Ridge that runs like a giant underwater zipper up the entire middle of the Atlantic Ocean, plus an additional 52,000 kilometers of spreading zones between all the other tectonic plates around the world. Furthermore, there are an additional 3.4 *million* known submarine volcanoes scattered all around the world, all driven by these tectonic forces. That's a lot of CO_2 gas that is continually leaking out from beneath the surface of the Earth all along the seams of our tectonic plates.

A similar spreading process is happening today in East Africa (albeit on a much smaller scale than the Mid-Atlantic Ridge) as East Africa slowly breaks away from the rest of the African continent all along the Rift Valley. This new spreading zone stretches from the Middle East, through East Africa, and all the way down to Mozambique. Mt. Kilimanjaro's ice-covered peak is perhaps the most iconic volcano springing up alongside this rift zone as hot molten rock pushes its way to the surface along this gradually opening continental zipper. A few million years from now the Rift Valley will become a new ocean, and at its center will be another underwater rift zone just like the Mid-Atlantic Ridge that runs through the center of the Atlantic Ocean. In 2005, over a period of only 3 weeks, a giant 60-km crack opened up in Ethiopia (up to 8 m / 26 ft wide in some areas), cutting right through freshly-planted farmers fields and across roadways — take a moment to Google some photos of the "Ethiopia split 2005"; it is truly spectacular to see these ongoing tectonic forces at work beneath our feet.

Likewise, the continent of Antarctica is also currently being split apart by a young rift zone that runs right across the entire Antarctic continent, called the West Antarctic Rift System, with many large volcanoes both above and below the ice (and *a lot* of hot rock warming and melting the ice from below). I shared a graph back in chapter 9 (figure 73) showing that most of Antarctica is gaining ice except in a few small regions in Western Antarctica — the few areas with ice loss also happen to be the same areas where the rift system is bringing heat and molten rock to the surface beneath the ice.[1]

These tectonic forces are forever shaping and reshaping the surface of our planet even as they continuously allow fresh CO_2 to bubble out from deep within our Earth. It's a process that never stops. But what drives these tectonic forces? As geology professor Ian Plimer pointed out during his keynote address on February 24th, 2023, at the 15th International Conference on Climate Change, *"There's quite a bit of evidence out there showing that seismic action in the mid-ocean ridges is related to [the gravitational pull created by] planetary alignments and lunar alignments, that this is often the trigger for the rise of molten rock and for earthquakes..."*[2] Astronomers have documented how gravitational forces caused by planetary and lunar alignments have a massive impact on some of the moons orbiting around planets in our outer solar system (Jupiter's and Saturn's moons in particular), which get massively stretched or compressed by gravitational forces as the positions of the planets and moons change relative to one another. This process is called *tidal flexing* — the same process is also at work here on Earth as the positions of the other planets change relative to the Earth. As the other planets (especially Jupiter and Satern) move along their orbital paths, the amount (and direction) of gravitational pull that they exert on our planet rises and falls. Professor Plimer describes this process like kneading bread dough — here on Earth this gravitational "kneading" causes the surface crust to fracture into tectonic plates, which breaks continents apart and drives continental drift. Whenever our solar system passes through the spiral arms of the

---

[1] "Antarctic rift valley speeds ice loss." *CBC News*. 2012, Jul 26. https://www.cbc.ca/news/science/antarctic-rift-valley-speeds-ice-loss-1.1164954

[2] The Heartland Institute. (2023, Mar 7). *Ian Plimer, a Geologist's View of Climate Change at Heartland Institute Climate Conference, 2023* [Video]. YouTube. https://www.youtube.com/watch?v=tK4LNIvlcCY&ab_channel=TheHeartlandInstitute

galaxy, this would simply add to all the gravitational disturbances driving this tidal flexing.

The friction produced by tidal flexing also causes heat to build up inside the Earth, called *tidal heating*, which generates large plumes of superheated molten rock deep below the crust that periodically rise to the surface (because hot rock is lighter than cooler rock). Unlike the volcanoes created along plate boundaries in subduction zones (like Mt. St. Helens or the Japanese island chain) or in spreading zones (like the Mid-Atlantic Ridge and along the Rift Valley in Africa), these plumes of molten rock created by tidal heating can rise up anywhere on the planet's surface, including melting their way up through the middle of both continental and oceanic tectonic plates, which is why they are known as "hotspots".

These rising plumes of molten rock create periods of especially intense volcanic activity, like those that created the Siberian Traps and the Central Atlantic magmatic province, which periodically help "top up" declining carbon dioxide levels in our atmosphere by adding even more CO_2 than is already being continually released by volcanic gases escaping the Earth's crust along rift zones and in subduction zones. Even today, our planet's crust is being perforated by thousands of smaller versions of these hotspots created by these rising mantle plumes produced by tidal flexing and tidal heating, such as well-known hotspots like the Hawaiian Islands, Iceland (the Icelandic hotspot happens to straddle the Mid-Atlantic Ridge), the Yellowstone hotspot, the entire New Zealand island chain, the island of Réunion in the Indian Ocean, and the Afar hotspot in Ethiopia, along with countless other seamounts and island chains dotting our oceans (there are over 14,500 seamounts in our oceans — some of these hotspots are still actively releasing lava and CO_2, while others have exhausted their supply of molten rock and are now extinct). If it wasn't for all this never-ending volcanic activity, continental weathering and the carbon that gets trapped in carbon sinks after photosynthesis would have depleted our atmosphere of all its CO_2 long ago — Life on our planet literally owes its existence to the CO_2 continually released by volcanoes, plate tectonics, and magma plumes.

Our smaller sister planet Mars was not so lucky. It too also once had an atmosphere — so much so that liquid water was once able to exist on the Martian surface and rain periodically fell from Martian skies. But Mars is only half the size of the Earth, so it cooled much faster than our planet. As the little red planet's interior cooled and hardened, plate tectonics ground to a halt, which put a stop to the continuous leakage of fresh volcanic gases from deep below the Martian surface that had once replenished its atmosphere to replace gases that were continually leaking away into space. And as the planet's dense semi-liquid iron core cooled, this cooling also gradually stalled the magnetic dynamo that had once wrapped Mars in a protective magnetic field. Like our magnetic field on Earth today, Mars' magnetic field once protected its delicate atmosphere from being stripped away by solar winds. But once Mars lost its magnetic field, solar winds carried away the Martian atmosphere.

Even here on Earth, solar winds are continually threatening to strip away our delicate atmosphere, but thanks to our protective magnetic field that deflects incoming charged particles streaming in from space, that hasn't happened yet. Nevertheless, we lose several hundred tons of our atmospheric gases to space *every single day*,[3] though thanks to plate tectonics that continually allow gases to leak out into our atmosphere from our planet's semi-liquid interior, these fresh volcanic gases keep our atmosphere well topped up.

However, the magnetic dynamo created by our planet's spinning iron-rich molten outer core is a bit of an unpredictable beast. From time to time our magnetic field collapses, and then reverses as it returns to full strength (i.e. the magnetic north and south pole switch places) — these magnetic reversals take from as little as 4,000 years to as long as 22,000 years and happen in unpredictable intervals, with as little as 10,000 years and up to 50 million years or more between reversals. The most recent reversal happened around 780,000 years ago.

And sometimes, in what is known as a *magnetic field excursion,* the magnetic field collapses to a mere fraction of its usual strength as though it wants to reverse, but then fails to reverse and returns to full strength at its original magnetic polarity, although the orientation of the magnetic pole can shift by up to 45° from its former location during these excursions. These excursions can likewise last from as little as a few thousand years to several tens of thousands of years. The most recent magnetic field excursion, called the Laschamps Excursion, occurred from

---

[3] NASA. (2018, Dec 17). Toward Mapping the Atmosphere's Escape from Earth. *Nasa Earth Observatory.* https://earthobservatory.nasa.gov/images/144386/toward-mapping-the-atmospheres-escape-from-earth

42,200 until 41,500 years ago, during which Earth's magnetic field plummeted to as low as 5% of its usual strength.[4]

As the magnetic field weakens during these excursions and reversals, much higher levels of cosmic and solar radiation are able to break through our weak magnetic field to penetrate our delicate atmosphere and reach our planet's surface. These reversals and excursions cause significant changes in atmospheric circulation (i.e. changes in wind patterns)[5] and they typically trigger dramatic episodes of rapid short-term climatic cooling, with major ice advances whenever they happen in the midst of an ongoing ice age.[6] Once again, the dramatic consequences of magnetic field reversals and excursions highlight the powerful link between cosmic rays and cloud cover while making it abundantly clear that even our unpredictable magnetic field has a profound impact on our planet's ever-changing climate.

Furthermore, the surge in cosmic radiation that penetrates the atmosphere during these reversals and excursions also accelerates the destruction of ozone molecules in the upper atmosphere, causing ozone levels to decrease significantly (thus exposing us to significantly higher levels of UV radiation on the planet's surface) and causing substantial changes in ozone circulation in our atmosphere.[7] The authors of a 2021 study point out that the Laschamps Excursion in particular triggered significant continental aridification and even megafauna extinctions in Australia (caused by the rapid shift to a drier climate as tropical Pacific rain belts and Southern Ocean westerly winds abruptly shifted[8]). It also caused a rapid expansion of the Laurentide Ice Sheet in North America and significant cooling across the Pacific Basin (though not in the Atlantic Basin), even as surface UV radiation levels increased by at least 10 to

---

[4] "Laschamp Event." *Wikipedia*. 23 May 2024.
https://en.wikipedia.org/wiki/Laschamp_event

[5] "Laschamp Event." *Wikipedia*.

[6] "Geomagnetic excursion." *Wikipedia*. 3 Feb 2023.
https://en.wikipedia.org/wiki/Geomagnetic_excursion

[7] Cooper et al. (2021). A global environmental crisis 42,000 years ago. *Science*. 371(6531). https://doi.org/10.1126/science.abb8677

[8] Fogwill, C., Hogg, A., Turney, C., and Thomas, Z. (2021, Feb 18). Earth's magnetic field broke down 42,000 years ago and caused massive sudden climate change. *The Conversation*. https://theconversation.com/earths-magnetic-field-broke-down-42-000-years-ago-and-caused-massive-sudden-climate-change-155580

15%, which would have compounded the aridification taking place in Australia and contributed to increased wildfire activity which is visible by increases in charcoal in the sedimentary record during that time period.[9]

The authors of the study also point out that the timing of the Laschamps Excursion is suggestive as it coincides with the *"globally widespread appearance and increase in figurative cave art, red ochre handprints, and changing use of caves"* all across Europe and on islands across Southeast Asia, leading them to speculate whether the increased risk of sunburn during this period may have driven behavioural changes in our ancestors as they sought shade in caves and increased their use of ochre as sunscreen, while transferring their art from skins and wood to the large canvases provided by cave walls — it is particularly telling that these simultaneous behavioural changes happened all across the globe in locations separated by oceans, so it is not possible for these behavioural changes to merely be the result of cultural transmission — there must be come common external trigger that everyone experienced simultaneously despite living on disconnected continents and islands.

They also point out that in addition to extinctions in Australia, the Laschamps Excursion also coincides with the timing of a cluster of megafaunal extinctions all across Eurasia, possibly including the extinction of the Neanderthals, though the exact date of the Neanderthal extinction is still highly contentious. Although the authors of the study do not mention the impact of increased cosmic radiation on the rate of genetic mutations, I am left to wonder whether the extinctions and behavioural changes were purely driven by paleoclimate shifts and ozone destruction, or whether a stroll through the atomic garden (caused by the flood of cosmic radiation and its effect on genes) may also have played some role.

Today our magnetic field is weakening once again — by approximately 9% over the last 200 years. Furthermore, the magnetic north pole is rapidly drifting north-northwest from northern Canada towards Siberia on a highly irregular path that has taken it on a 1,000-km journey since 1831. And the rate of motion has accelerated from around 9 km per year in the 1970s to around 52 km per year (2001-2007 average). All this has led to much unresolved speculation in the scientific community as to whether this signals the beginning of another magnetic field reversal (or excursion,

---

[9] Cooper et al. (2021). A global environmental crisis 42,000 years ago. *Science.* 371(6531). https://doi.org/10.1126/science.abb8677

which are even more common than full reversals) or whether this is simply part and parcel of the unpredictable and "wobbly" nature of our magnetic field.

As all these dynamic interconnected forces impacting our climate add up, I suspect you can see by now why there are so many dissenting voices among geologists, physicists, astronomers, and meteorologists mocking the simplistic greenhouse-gas-focused climate change narrative and questioning the ability of climatology algorithms to accurately model our planet's climate — I suspect many scientists in these other Earth Sciences look at the sensationalist claims made by climatologists with the same sense of horror that meteorologists must have felt as they watched mid-20th century weather modification programs capture the imaginations of politicians with their cloud seeding schemes, which only served to divert much needed funds away from serious efforts to understand the complex dynamic forces at work in our atmosphere.

~

Back in the Jurassic period, the aforementioned breakup of the Pangaean supercontinent changed the circulation of the ocean currents, thereby disrupting the currents that previously circulated warm tropical waters towards the poles. This triggered yet another period of cold weather known to geologists as the Jurassic Cooling as global temperatures dropped by around 10 °C.[10] You can see this sudden drop in temperatures in figure 94 at the Jurassic-Cretaceous boundary — the cooling began even as CO_2 levels were soaring to 2,500 ppm due to volcanic eruptions triggered by the fracturing Pangaean supercontinent (cooling against a backdrop of rising CO_2 — how's that for yet another "Inconvenient Truth," Mr. Gore?). Although this Jurassic Cooling did not trigger a full-fledged global Icehouse glaciation, there is geologic evidence that ice sheets grew in polar regions, somewhat similar to the way Greenland and Antarctica are covered by ice sheets today even as the rest of the continents largely remain ice free. That cooling likely helped suppress the total amount of photosynthesis consuming CO_2 on the planet (biomass always decreases during cold eras) so that the Jurassic volcanic eruptions

---

[10] "Scientists Shed Light on Causes behind Jurassic 'Ice Age'. (2015, Dec 14). *Sci News*. https://www.sci.news/othersciences/paleoclimatology/causes-behind-jurassic-ice-age-03511.html

were able to give atmospheric CO_2 levels a meaningful boost without that extra CO_2 immediately being consumed by aggressive plant growth.

This shift in ocean currents triggered by the breakup of the Pangaean supercontinent happened around the same time that mountain ranges began to form all along the western edge of the North and South American continents (Rocky Mountains, Andes, etc.). Like today, these mountain ranges forced clouds to dump their moisture along the western flanks of the mountains, causing the climate to their east to get drier and drier. As this drying trend continued, even the tropics became too dry to support trees. By the end of the Cretaceous Period (145–66 mya), the climate had become so dry that the giant forests that had dominated much of the Age of the Dinosaurs were in retreat. These forests were replaced by a new and extremely successful evolutionary innovation in the plant world — flowering plants. We're going to have a closer look at flowering plants and their impact on the climate in a moment. But first we need to have a look at another earlier evolutionary innovation that happened at the beginning of the Age of Dinosaurs while the climate was still warm, wet, and ideal for non-flowering plants like ferns and conifers.

Looking back at figure 94, you can see that during the Age of the Dinosaurs, CO_2 drained from the atmosphere at a much slower rate than during earlier eras. Two new forces capable of digesting lignin and cellulose emerged to dominate the planet after the Great Dying, which would forever change the atmospheric balance between CO_2-in versus CO_2-out — herbivores and fungi.

The first animals to evolve the ability to digest plant tissues were insects, though they mostly specialized in eating seeds and soft tissues. Even today, very few insects have the ability to digest cellulose or lignin. Although fungi have been around for over 1 billion years, it wasn't until the late Carboniferous Period that fungi finally cracked the code on how to break down lignin and cellulose, and even then, fungi did not play a big role in the overall ecosystem until the time of the Great Dying when vast fungal blooms emerged to take advantage of all the dying trees. One of the reasons coal formation stopped during the Great Dying and why coal formation has been very limited ever since is because fungi became so much more efficient at digesting woody plant tissues, which enabled them to become much more widespread after the Great Dying.

Before the Great Dying, dead trees either piled up and turned into coal at the bottom of swamps (thus permanently removing their carbon from the carbon cycle) or they lay about waiting for a wildfire to finally consume them. Thus, with few mechanisms to return carbon back to the atmosphere, the balance between CO_2-in versus CO_2-out was heavily tilted towards permanently scrubbing vast quantities of CO_2 from the atmosphere by depositing carbon in peat bogs and coal beds. After the Great Dying, fungi emerged as a much more powerful force to recycle dead trees by turning lignin and cellulose into humus while "exhaling" CO_2 as a byproduct.

Before fungi, soils were mostly made of clay particles and inorganic sediments. The humus created by fungi as they fed on lignin and cellulose created a brand-new soil ecosystem capable of sustaining vast volumes of microscopic plant life and biologically available plant nutrients, leading to a massive boost in the amount of plants growing on land. Fungi helped turned soil into a great storehouse for carbon. Although humus is forever vulnerable to being oxidized back into CO_2, humus is still a type of carbon sink even if it is not as permanent as a coal bed or a carbon-rich layer of mud at the bottom of a lake or ocean.

But all those tough lignin- and cellulose-rich plants growing on land were a major untapped food source. During the late Carboniferous, a few reptilian herbivores emerged that used gizzard stones to grind up tough lignin- and cellulose-rich plant fibers in order to extract energy from them, while simultaneously evolving a symbiotic relationship with various digestive enzymes and digestive microbes to help them extract energy from those tough plant fibers. However, these early herbivores were terribly inefficient, leaving most of the plant fibers undigested. But after the Great Dying (during the Triassic and Jurassic Periods), dinosaurs evolved much more efficient digestive systems capable of breaking down these tough woody lignin- and cellulose-rich plant fibers. This evolutionary breakthrough led to the 180-million-year reign of the dinosaurs as herbivorous dinosaurs cracked the code on how to live exclusively on a diet of woody plant tissues. And so, finally, herbivores emerged as yet another dominant new lifeforce, capable of efficiently converting lignin- and cellulose-rich plant tissues back into CO_2 before all that carbon could be trapped in long-term carbon sinks.

Herbivores rebalanced the off-kilter equilibrium between CO_2-consumers, CO_2-producers, and carbon sinks — by digesting and converting plant matter back into CO_2, herbivores sped up the rate at which CO_2 was recycled back into the atmosphere and, equally important, reduced how much carbon ended up trapped in coal beds and other carbon sinks. Herbivores helped recycle CO_2 back into the atmosphere almost as fast as plants were removing it, thus preventing a repeat of the rapid plunge of CO_2 that happened before the Carboniferous Period. And, by virtue of breaking apart the tough cellulose and lignin fibers in their digestive systems, herbivores helped create lots and lots of humus.

As soils improved and became more fertile, plants expanded their territory across the continent — as their territory expanded, this spreading blanket of vegetation helped shield bare rocks and sediments from weathering by carbonic acid in rainfall. Thus, the abundant vegetation supported by humus-rich soils helped slow the rate at which carbon dioxide was depleted from the atmosphere by rock weathering.

Grass (a type of flowering plant) had not yet evolved at this point in the story — that story is coming soon as grass emerges to play a dominant role in the carbon cycle after the dinosaurs went extinct. During the Age of the Dinosaurs, plant life was dominated by giant ferns, horsetails, and enormous forests full of conifers, ginkgo trees, and cycads (cycads look similar to palm trees, though they are a different species). These tough, woody, cellulose-rich plant fibers were extremely low quality, low nutrient food sources — far less nutritious than the flowering plants that emerged during the late Cretaceous. The lower the quality of the food, the more energy is wasted on digestion and the more time an animal needs to spend foraging in order to survive. Below a certain point, the amount of energy required just to live exceeds the amount of energy that can be extracted from such nutrient-deficient plant fibers.

When it comes to energy conservation, size matters. As body mass increases, it takes less energy to maintain a consistent body temperature, which means there is a significant metabolic advantage to being huge when food quality is low. Being huge takes less energy per unit of body mass to survive. The enormous size of herbivorous dinosaurs was a necessary evolutionary adaptation to a nutrient-poor diet of woody trees, leaves, and ferns. Incidentally, the long necks on many browsing dinosaurs, like Brontosaurus, also reflect the arms race between herbivorous dinosaurs competing to reach the youngest most nutritious leaves at the top of the forest canopy.

A single Brontosaurus weighed 30.6 tonnes whereas a single gecko weighs around 60 gram (2 ounces) — it's far more energy efficient to feed a single 30.6 tonne Brontosaurus than 30.6 tonnes of tiny geckos (510,000 geckos weighing 60 gram each). The same lesson applies to mammals today. Jerzy Trammer, professor emeritus at the Institute of Geology at the University of Warsaw calculated that, *"food consumption of a mammal that weighs 5,000 kg amounts to 4,257 joules/second, while 5,000 kg of mammals weighing 1 kg each requires the consumption of 54,800 J/sec – nearly thirteen times more..."*[11]

Small mammals like mice and squirrels that feed on high-density foods like nuts, seeds, and insects can afford to stay small, but these small animals cannot live exclusively on low-density foods like the woody parts of trees or the fibrous parts of grasses. Specialized herbivores that live exclusively on low density nutrition like grass or tree leaves (i.e., grazers like bison and cattle or browsers like moose, goats, or sheep) have to get huge in order to save on energy. The lower the nutritional quality of their primary food source, the larger they are as a species (that's also why the biggest moose are found in Alaska where their larger body sizes provide them with an energy advantage as browse on dormant woody plants during the long cold winter). The sheer size of the dinosaurs reflects the low-quality nutrition that they depended on to survive. Many of the early mammalian herbivores that emerged after the reign of the dinosaurs ended were similarly enormous — the biggest woolly rhinoceros reached 4 tons and was over 20 ft long, the steppe mammoth weighed as much as 10 tons, the bizarrely named Indricotherium (an ancestor of modern rhinos) reached over 20 tons, and in South America a rodent ancestor of the capybara reached weights over 1 ton.

The low quality of food during the Age of the Dinosaurs would have prevented large-bodied herbivorous dinosaurs from being warm blooded — there simply isn't enough energy in wood and in tree leaves to have provided a huge beast like Brontosaurus with enough energy to afford it the luxury of regulating its own internal body temperature. The large body size of many dinosaurs helped compensate for being cold-blooded because, by virtue of their gigantic thermal mass, cold nighttime temperatures and periods of cool weather would not have slowed them down as much as their smaller peers.

---

[11] Trammer, J. (2011). Differences in global biomass and energy use between dinosaurs and mammals. *Acta Geologica Polonica.* 61(2). https://www.researchgate.net/publication/285760348_Differences_in_global_biomass_and_energy_use_between_dinosaurs_and_mammals

This energy equation also explains the evolution of early mammals during the Age of the Dinosaurs. With so many cold-blooded predatory dinosaurs hunting during the warm daytime, it would have been a lot safer to hide during the day and come out to feed during the cold hours of the night. But that strategy doesn't work if you're cold-blooded. Thus, warm-blooded mammals evolved precisely to fill this niche — small bodied warm-blooded nocturnal omnivores feeding on energy-dense nuts, seeds, insects, and occasionally meat would have had a sufficiently energy-dense diet to generate their own internal warm body temperatures, thus enabling them to thrive at night when dinosaurs, reptiles, and snakes were lethargic. The earliest mammals were all nocturnal — mammals only came out of the shadows as dinosaur populations began to decline towards the end of the Cretaceous Period.

A warm body temperature also allowed mammals to avoid fungal infections because very few fungi can survive inside a 32 °C body temperature. By contrast, fungal infections are lethal for most reptiles. As dinosaurs were about to find out in the aftermath of Chicxulub asteroid impact, this vulnerability to fungi was a bigger deal than you might expect...

As the climate became increasingly dry during the latter half of the Cretaceous Period, the trees and especially the rainforests that dinosaurs fed on began to retreat, giving up territory to the advancing flowering plants that were better suited to take advantage of the drier climate. This dietary change helps explain why many dinosaur species were already declining sharply at least 10 million years *before* the Chicxulub asteroid impact that finally, decisively, brought the Age of Dinosaurs to a fiery end.[12]

When the Chicxulub asteroid collided with our Earth, the impact threw a huge amount of dust and ash into the atmosphere, which blocked the sun and triggered a cold period (known as an "impact winter"), which lasted until the dust cleared. Sediments from the end of the Cretaceous period contain a layer of dust and ash of the exact same age as the Chicxulub asteroid impact. That layer is extremely rich in a platinum-group mineral known as iridium. Since meteorites are rich in iridium while most rocks

---

[12] Condamine, F.L., et al. (2021). Dinosaur biodiversity declined well before the asteroid impact, influenced by ecological and environmental pressures. *Nature Communications*. 12. https://doi.org/10.1038/s41467-021-23754-0

on Earth are not, this iridium-rich sediment layer from the end of the Cretaceous period is what motivated geologists to start searching for a matching late-Cretaceous asteroid impact crater, leading to the discovery of the Chicxulub Crater in Mexico — it is the proverbial smoking gun that killed off the dinosaurs.

The cold dark impact winter that was triggered by the asteroid impact that wiped out the dinosaurs created the perfect conditions for a fungal bloom. Indeed, fossil-rich sediments from that era show that a massive global fungal bloom spread out across the planet as all the dead plants and dead animal carcasses began piling up everywhere — in one fell swoop, the darkness, cold, and dust that followed the Chicxulub asteroid impact turned the entire Earth into a massive compost heap.[13] Unfortunately for cold-blooded dinosaurs, this combination of a cold impact winter and a global fungal bloom posed a mortal risk — it is thought that fungal diseases in the aftermath of the asteroid impact played a major role in wiping out the dinosaurs.[14]

At least 75% of all species went extinct during the Cretaceous extinction. Any animal that survived was either an omnivore, an insectivore, or a carrion-eater capable of surviving on dead plant and animal tissues. Not a single species that was 100% herbivorous or 100% carnivorous survived — nutritional flexibility, not specialization, was key to survival. On the whole, warm-blooded mammals and avian dinosaurs (the ancestors of birds) both fared much better than their cold-blooded peers — avian dinosaurs had already evolved to become warm-blooded because flight requires a high body temperature and high energy outputs over long time periods, neither of which is compatible with cold-bloodedness. Very few crocodiles, turtles, or amphibians went extinct since they can survive on dead animal tissues washing up in rivers and they were probably also helped along by their ability to burrow into the mud and slow down their metabolisms to save energy during long winters or long droughts. In the oceans, anything that fed on plankton living in the upper levels of the ocean went extinct while bottom feeders that rely on eating dead plant and animal tissues raining down onto the seafloor

---

[13] Casadevall, A. (2012). Fungi and the Rise of Mammals. *PLoS Pathogens* 8(8). https://doi.org/10.1371/journal.ppat.1002808

[14] Casadevall, A. (2005). Fungal virulence, vertebrate endothermy, and dinosaur extinction: is there a connection? *Fungal Genetics and Biology.* 42(2). https://doi.org/10.1016/j.fgb.2004.11.008

survived at much higher rates. Overall, the fossil record paints a picture of the lights being violently turned out for at least two years.[15]

Once the dust cleared and fungi ran out of dead plant and animal tissues to compost, the planet began to recover. Ferns emerged to dominate the devastated planet for the first few thousand years, followed by the rise of flowering plants, which spread out to fill most environmental niches. Deciduous trees, grasses, and the ancestors of all modern agricultural crops, from corn to rice to wheat and tomatoes, are all varieties of flowering plants that emerged to fill the void left by the Cretaceous extinction. Once again, the rate of evolution was kicked into overdrive.

With a completely new group of plant species leading the charge to repopulate the devastated planet, with so many environmental niches open to repopulation after their previous inhabitants were pushed into extinction, and with so few predators left roaming the planet, the world was ripe for a completely new balance of animals to take over the planet and fill the spaces once occupied by dinosaurs. The earliest fossils of these brand-new mammal and bird species are found 185,000 to 570,000 years after the Chicxulub asteroid impact. And so, at long last, we arrive at our own geologic era, an era dominated by mammals, birds, flowering plants, grass, fungi, and fertile humus-rich soils.

---

[15] Weisberger, M. (2021). Darkness caused by dino-killing asteroid snuffed out life on Earth in 9 months. *Live Science*. https://www.livescience.com/cretaceous-extinction-darkness

Chapter 22

Sod, Soil, and Worms

"It is the little causes, long continued, which are considered as bringing about the greatest changes of the earth."

— James Hutton (1726-1797), known as the "Father of Modern Geology."

Many earlier mass extinctions in our planet's history were followed by sharp rises in CO_2 (and plunging oxygen levels) that lasted until plant populations (and photosynthesis) recovered. The Cretaceous extinction was different.

At the time of the Cretaceous extinction, most of India was being inundated by yet another episode of massive flood volcanos as yet another gigantic plume of molten rock melted its way up through the thick continental crust to spill out onto the surface. This intense period of volcanic activity at the end of the Cretaceous lasted over one million years and is known to geologists as the Deccan Traps. For a while geologists believed that the Deccan Traps were triggered by the intense energy released by the Chicxulub asteroid impact radiating through the hot interior of our planet, but a closer study of the timing revealed that although there is some overlap, the Deccan Traps began erupting approximately 300,000 years *before* the asteroid hit. And yet, despite the fact that the Deccan Traps flooded an area of over 1.5 million km^2 with molten rock (an area more than twice the size of Texas), unlike what happened to CO_2 levels during previous massive volcanic eruptions when those eruptions overlapped with a mass extinction event, this time atmospheric CO_2 failed to meaningfully reverse its steady decline. Something about this "event" was different.

As you can see back in figure 94, the one-two punch of the erupting Deccan Traps and the Cretaceous extinction only triggered a very small and very temporary hiccup in CO_2 levels (accompanied by a very brief dip in oxygen levels) before atmospheric carbon dioxide resumed its steady decline (and atmospheric oxygen resumed its steady rise). Flowering plants were simply too efficient at repopulating the planet after the asteroid hit to turn off photosynthesis for long. The extra CO_2 released by the Deccan Traps was quickly consumed by vigorous plant growth.

This leads to the obvious question that will be answered in upcoming chapters: why are CO_2 levels able to rise today despite the abundance of flowering plants and grasses that quickly respond to rising CO_2 levels by turning that extra CO_2 into vigorous plant growth?

Flowering plants emerged as a yet another planet-changing lifeforce, unique in their ability to rapidly adapt to changing environmental conditions, conquer new territory, and repopulate environmental niches left vacant by other species. Their lightweight seeds are easily carried on the winds, most are able to reach maturity and produce fertile seeds within a single growing season, and their roots are particularly adept at spreading out to claim new territory as their own. These traits enabled flowering plants to quickly take over the environmental niches devastated by the Cretaceous extinction.

One of the most important families of flowering plants to emerge after the Cretaceous extinction were grasses, which first evolved around 55 million years ago. They are, arguably, the most successful plant family in evolutionary history due to their ability to rapidly colonize new ecosystems; their ability to adapt to an extremely wide range of climates and ecosystems; their ability to withstand wind, flood, and other types of erosion; their ability to disperse seeds across vast distances; and their ability to transform the ecosystems they colonize. Scientists have nicknamed this unique combination of traits the "Viking Syndrome". Grasses also encourage frequent fires that recycle carbon back into the atmosphere and return nutrients back to the soil. Their sod mats stabilize vulnerable soils by protecting them from wind, rain, and floodwaters. They remain nutritious long after they die, which helps sustain large populations of grazing animals through the long winter months. They have extremely nutritious seeds (corn, wheat, barley, rice, and millet are all different types of grasses that have been domesticated for agriculture production). And, when combined with grazing pressure from herbivores, grasses build humus faster than any other plant species that has evolved before or since.

Since this geologic detective story is about to dive into the forces that destroy soil, let's first take a closer look at how soil is created. The primitive soils that emerged early in our planet's history were principally formed by slow chemical weathering as rocks broke down into sands, silts, and clays. Plant roots tapped into these primitive soils to access water while root acids chemically dissolved inorganic soil particles to extract nutrients like phosphorus, potassium, calcium, and other minerals.

Humus added a new dimension to soil by transforming primitive soils into a rich subterranean ecosystem for bacteria, fungi, insects, and earthworms. These microorganisms all secrete various acids that enable them to digest and extract nutrients from soil particles — this *biological weathering* of sand, silt, and clay is much faster than chemical weathering and, even more importantly, this biological activity releases nutrients from inorganic sands, silts, and clays that are then stored within complex chemical compounds that are insoluble to water moving through the soil while simultaneously being soluble to root acids. Thus, the humus ecosystem both creates and stores nutrients in a way that primitive soils could not. Consequently, humus and the microorganisms living within it *dramatically* expanded the amount of biomass living on the planet.

As humus increases, the subterranean soil ecosystem expands to support a larger volume of microorganisms below ground, which increases the amount of nutrients available to plants, which in turn supports more plants and animals up on the soil surface. As plants die, those same soil microorganisms compost the dead plant fibers and dead roots — some of the carbon gets added to the humus inventory while the nutrients extracted from the dead plant tissues are returned to the overall inventory of biologically-available plant nutrients stored within the soil. As long as erosion doesn't destroy this fragile humus-enriched crust, it is a self-improving ecosystem.

The early sodbusters plowing up virgin soils on the Great Plains didn't need chemical fertilizers because they were tapping into a deep reservoir of humus-rich soil containing thousands of years of accumulated nutrients. But as they harvested their crops, the overall inventory of soil nutrients was depleted because they were not putting back enough plant material to replenish the inventory of nutrients lost to each harvest. Furthermore, as erosion and oxidation reduced the amount of humus in the soil, *the nutrient storage capacity of the soil* declined and, just as importantly, the total volume of bacteria, fungi, earthworms, and other microorganisms living in that soil also declined, which further reduced how quickly fresh nutrients could be extracted from inorganic sand, silt, and clay particles via biological weathering. Humus, soil fertility, and the total biomass of plants and animals supported by the soil are inextricably linked.

To some degree, soil fertility can be viewed as a simple accounting balance sheet: each harvest depletes the total inventory of nutrients, but any dead plant debris that is returned to the soil (stalks, stems, or manure) helps replenish a portion of that nutrient inventory. However, since soil microorganisms and root acids are continually extracting fresh nutrients from inorganic sand, silt, and clay sediments, the nutrient balance is not necessarily exhausted by harvests as long as the amount of nutrients consumed by harvests do not exceed the rate at which fresh nutrients are released from inorganic sand, silt, and clay particles through biological and chemical weathering.

To sort out this complex symbiotic self-improving relationship, think of humus as a kind of living warehouse for nutrients. Soil nutrients are like the inventory stored inside that warehouse. Plants withdraw nutrients from the inventory while roots and soil microorganisms act like an army of microscopic minions working tirelessly to refill the warehouse by liberating biologically available fertilizer elements from inorganic sand, silt, and clay particles. The maximum storage capacity of the warehouse expands whenever humus levels increase but shrinks whenever soil is lost to erosion or whenever humus is oxidized away as CO_2 gas. In other words, although you cannot add more nutrients to the warehouse once it is full, you can expand the size of the warehouse by increasing the amount of humus in the soil. Likewise, as humus is lost to erosion or oxidation, the size of the warehouse shrinks and thus the inventory of nutrients that can be stored in the warehouse declines in lockstep.

Similarly, the size of the minion army working in the soil to extract fresh nutrients from rock particles also increases or decreases depending on how well their living conditions are optimized — the erosion of humus, a lack of water, excessive soil compaction, waterlogged soil conditions, salt build-up in the soil, or toxic chemicals can all reduce the size of this minion army that is continually working to refill the warehouse with fresh nutrients extracted from sand, silt, and clay particles. In so many ways, soil and soil fertility are a complex and dynamic ecosystem composed of a vast array of moving parts. Managing all these interconnected parts is as much an art as it is a science.

Charles Darwin is most famous for his book about natural selection as the driver of evolution, but his last (and most under-appreciated) book, published in 1881, was all about earthworms and the biological processes that create soil. It was called *The Formation of Vegetable Mould Through the Action of Worms, With Observation on their Habits.* "Vegetable mould" was the 19th century term for "humus".

For over forty years (in between voyages and writing other books), Darwin obsessively studied the habits of worms in the soils on his farm and in the surrounding English countryside. In those days, soil was seen as a sterile mass, little more than a medium within which to anchor plant roots and provide the occasional home for a bug or worm that happened to be passing through, while animal manure was thought to be the direct source of all the nutrients required to feed plants. As usual, Darwin looked a little closer than everyone else. As usual, the scientific community ridiculed him. But gardeners wisely ignored the opinions of the scientists and bought his book in droves — it was a bestseller that outsold all his earlier books, including his seminal work about his Theory of Natural Selection. In time, Darwin's book about worms would lead him to be recognized, along with the aforementioned Russian geologist Vasily Dokuchaev, as the father of soil science.

Darwin noticed that it did not take very long for dead plant litter, lime, ash, chalk, and other debris to completely vanish in the soil. This led him to recognize the crucial role that earthworms play in digesting plant litter, sediments, and other raw ingredients to turn them into soil. *"I was thus led to conclude that all the vegetable mould over the whole country has passed many times through, and will again pass many times through, the intestinal canals of worms."* Darwin realized that worms were effectively plowing farmers' fields and creating humus. *"The plough is one of the most ancient and most valuable of Man's inventions; but long before he existed the land was in fact regularly ploughed, and still continues to be thus ploughed by earthworms."* Darwin's insights helped scientists recognize soil as a living skin on the Earth.

Whenever Darwin had a chance, he dug trenches all over the English countryside. He observed that in the moist temperate climate of the English countryside, forests create soil at a rate of a few inches per century. He found similar rates of soil formation all over the English countryside — one of his favorite methods was to see how much soil had accumulated to bury ancient Roman ruins by looking at the dates printed on old Roman coins found in the ruins, which provided a good estimate for how long the ruin had been abandoned. He found that most Roman ruins in Britain were covered by approximately 2.5 feet of soil.

He also obsessively weighed worm castings (a "worm casting" is the digested soil coming out of the back end of the worm) and discovered worms were producing ten to twelve *tons* of worm castings *per acre* every single year — in other words, on every single acre of land, between 125 and 185 full wheelbarrows of soil were passing through the digestive systems of earthworms each year! He noted, *"It may be doubted whether there are many other animals which have played so important a part in the history of the world, as have these lowly organized creatures."* In essence, over time, every single square inch of English soil travels through the digestive system of earthworms, over and over again. The worms' passage through the soil (and the soil's passage through the worm) also helps to aerate the soil and creates the spongey texture that helps soil absorb and hold water — this aeration provides soil microorganisms with both the water and the oxygen required for life.

Furthermore, Darwin noticed that as earthworms continually churned the soil, that soil was continually on the move — downhill. Millimeter by millimeter, billions of tons of soil were constantly on the move towards the valley bottom. And yet, despite the never-ending movement of soil from hillside to valley bottom, the rich deep soils found on forested hillsides demonstrated that, in the absence of erosion caused by farmers' plows, soil nevertheless grew faster than the rate at which earthworms moved soil downhill. Thus, he recognized that this thin skin of soil covering the planet's surface was not a fixed commodity, but a skin that continually replaced itself.

Through improper stewardship, we can deplete soil faster than it forms either by (1) increasing the rate of erosion above the natural replacement rate, (2) by unsustainable agricultural practices that deplete humus and nutrients faster than they are refilled by the tireless action of soil microorganisms, or (3) by employing land management practices that suppress the amount of biological activity in the soil, which reduces the rate at which humus and nutrients are created via biological weathering. This has been the fate of countless civilizations since the dawn of agriculture, which inadvertently put a countdown clock on the lifespan of their civilizations by using up their soils faster than earthworms and other soil microorganisms can create it.

In his book *Dirt: The Erosion of Civilizations*, geologist David Montgomery calls it "mining the soil". As he points out in his book, there are very few exceptions where civilizations learned how to build agricultural soils faster than they used them up. The hillside terraces of the Inca, the pockets of pre-Columbian *terra preta* soils in the Amazon basin, the Dutch farmland reclaimed from the sea by dikes and fertilized by copious amounts of cow dung, and the forest gardens of Tikopia Island in Melanesia stand out among the few exceptions to the rule. The rise and fall of empires can be measured by how quickly they eroded and used up their soils. As David Montgomery demonstrates in his book, exhausted soils ultimately give way to economic decline, rising rates of poverty, bitter social conflicts, war, and mass migrations as some members of society (or in some cases entire tribes) begin to migrate in search of better opportunities and more fertile soils to begin the cycle all over again. The holy grail of sustainable farming and, thus, a flourishing civilization, is learning how to build soil faster than it is consumed and how to ensure that the population does not outgrow the soil's ability to feed its people — more on that soon.

However, not all climates have the warm moist growing conditions of the English countryside that enable earthworms to thrive in such large numbers. The Great Plains of the American West are a perfect example of a semi-arid climate that plays by an altogether different set of rules. In the winter the soils are frozen, causing microbial life to come to a screeching halt. Then, after a brief interval of spring rains that temporarily allows microbial life to resume, the summer dry season sets in leaving the soil dry and parched, which once again brings microbial life to a screeching halt. Bacteria, fungi, earthworms, and other soil microorganisms living in the humus ecosystem cannot do their miracle work without warm moist soil conditions — many arid and semi-arid regions don't have any earthworms at all, only soil bacteria, fungi, and other more microscopic organisms that can lie dormant for months or even years during drought conditions. Without water, the humus ecosystem goes to sleep.

In these arid and semi-arid regions, the window of opportunity to break down plant litter and to liberate fresh nutrients from inorganic sand, silt, and clay shrinks down to a very brief period each year — outside of that window, the soil ecosystem is completely dormant in all but the deepest layers where a hint of moisture remains. This boom-and-bust cycle between rainy and dry seasons takes a terrible toll on soil microorganisms as their populations briefly surge whenever moisture returns to the soil, but then dwindle away as drought or winter sets in. At first glace, this doesn't seem like a very favorable place to build deep fertile soils. And yet, we know that the North American prairie and other grassland ecosystems have some of the richest soils in the world. So, how did these fertile humus-rich soils form in such dry climates?

The simple answer is herbivores.

Herbivores not only return nutrients to the ground as they graze across the dry plains, but they also take pressure off soil microorganisms by predigesting and partially decomposing plant litter *before* it reaches the ground so that the soil microorganisms have far less work to turn it into humus. In warm moist climates, the intestinal tracts of Darwin's beloved earthworms (along with fungi and bacteria) are the engines that efficiently transform plant debris into humus, with or without the help of livestock. But in dry climates, you cannot take the herbivore (and the herbivore's intestinal tract) out of the equation because the activity of soil microorganisms is so limited by dry soil conditions. In effect, in dry climates the herbivore is the irreplaceable engine for creating and maintaining humus.

The warm moist conditions inside the digestive tracts of herbivores allow digestive bacteria and enzymes to thrive, day in and day out, so they can continue their important work of breaking down lignin- and cellulose-rich plant fibers into humus even when the weather outside is freezing, even if the soil is too dry to support an active community of microorganisms, and even if wind and rain would otherwise wash away the plant litter before it can be transformed into humus. By the time the manure hits the ground, the plant fibers are already so thoroughly pre-digested that even the short window of the brief wet season enables soil microbes to finish their work of transforming lignin- and cellulose-rich plant residues into humus. In essence, the great herds of bison that once roamed the Great Plains were the digestive system of the prairie.

In warm moist climates, herbivores are optional, though they can dramatically speed up the creation of fertile soil beyond what even an army of earthworms and fungi can do on their own. But in dry climates, soils rapidly begin to deteriorate without regular grazing from herbivores — without herbivores, grasslands are soon reduced to deserts as is happening all over the world today, just like in the example I described earlier of Kalamalka Lake Provincial Park near my hometown in BC. You cannot build humus in dry climates without herbivores — what earthworms are able to accomplish in a well-watered English garden simply doesn't work across millions of acres of dry rangeland and deserts across Africa, Asia, and the American West.

The importance of herbivores extends far beyond their role in the nutrient cycle and their role in turning plant fibers into humus. Without them, it is also impossible to create the lush thick sod mats that protect grassland and desert soils from erosion and oxidation. Without regular grazing, grass does not thrive, and sod mats begin to break down.

We repeatedly hear that desertification is caused by overgrazing, but that's not strictly true. I've alluded to this in earlier chapters, but at this point in the story it's time to truly put grazing under the microscope. It is true that goats, sheep, cattle, and other livestock are responsible for severe soil degradation, erosion, and desertification, but once again there's a lot more to the story than simplistically dividing the world into simple categories labelled "good" vs "evil." The problem is not the animal, it is the grazing technique. After all, as bison herds roamed across the historic Great Plains (pursued by hungry wolves), their feet and their mouths were directly responsible for creating the rich prairie soils and endless grassland expanses that stretched, unbroken, from the Mexican deserts all the way to northern Alberta. And yet, when a herd of bison is kept in a large pasture that is grazed continuously (without wolves or electric fences to bunch them together to create a herd migration effect), they too will destroy the entire pasture just as fast as any cow, sheep, or goat would using this same continuous grazing technique. Grazing behaviour is everything.

As you know from cutting your lawn, grass sod requires frequent harvesting to trim away mature grass — this mowing/grazing action causes grass sod to spread out, filling the spaces between grass stalks with even more grass stalks. It's not an accident that grass co-evolved with herbivorous mammals 55 million years ago. To understand how this powerful symbiotic relationship between herbivores and grass works (and how to harness it for our benefit), we need to take a brief detour into cow psychology.

If you put a cow in a single large pasture and leave her there for the full summer, she will spend her days searching out the tastiest grasses while ignoring the less tasty stuff. Thus, the best grasses will be overgrazed as she returns to re-graze the tastiest morsels, chewing them right down to the roots every time the tiniest millimeter of fresh re-growth tries to poke its head out of the soil. This leaves their roots too small and weak to reach groundwater below the surface even as the sun beats down on soil that has been laid bare by overgrazing. And yet, even as the picky cow overgrazes these tastiest grass species (essentially destroying them), less tasty grasses and weeds are systematically ignored, which allows these less-desirable plants to reach maturity, spread out across the pasture, become fibrous and stalky, and die. To add insult to injury, this dead unpalatable grass debris will then choke out fresh regrowth from other more desirable grass species. Furthermore, the seeds from these less desirable plants will slowly colonize the rest of the pasture because the picky cow never allows the tastiest grass species to reach maturity and produce seed.

As this process is repeated year after year, the plant species in the pasture will slowly deteriorate as the tastiest grasses are entirely replaced first by less tasty grasses and then by weeds, which don't form a protective sod mat. The expanding patches of bare soil between the weeds becomes increasingly vulnerable to erosion from rain, wind, and sun, and the humus is gradually eroded and oxidized away. Thus, grasslands slowly turn to desert simply because the grazing strategy allowed the tastiest grasses to get perpetually overgrazed even as weeds and other less palatable plants were consistently *undergrazed*. This style of grazing is known as *continuous grazing*.

Sadly, continuous grazing is still the most common way of managing livestock and is undeniably one of the factors driving desertification in the world. But the problem is not livestock, it's the grazing technique. Nor can you solve the problem of desertification by removing the livestock because dry climates depend on regular grazing and manure to keep the sod mat and the soil microorganisms healthy — the desertification continues unabated even if the livestock are removed. The solution is not to remove livestock, but to change the grazing behaviour of the livestock.

Grazing behaviour completely changes when, instead of giving livestock access to a large paddock for long periods of time, livestock are bunched together in large numbers and only given access to single's day's slice of pasture each day, much like the grazing behaviour of a wild migrating herd of bison or wildebeest as they bunch themselves together out of fear of the wolves or lions stalking their flanks, and as this tightknit herd continually migrate onwards in search of fresh grass. In this style of grazing, each slice of pasture only gets grazed for a single intense day of grazing, but then it is left to rest undisturbed until the next rains produce a fresh flush of grass. This style of grazing is known as *mob grazing* or rotational grazing — it is specifically designed to mimic the grazing behaviour of a wild herd migration.

Mob grazing completely transforms the behaviour of herd animals. By bunching animals tightly together into a small slice of pasture that only has just enough grass for a single day's grazing, it creates competition between herd members as they scramble to get their share of the grass. *This change in grazing behaviour is a purely psychological effect created by the <u>illusion</u> of scarcity and by the competitive instinct that is triggered by grazing in a group.* Picky cows that once roamed aimlessly across continuously-grazed pastures in search of the tastiest morsels while ignoring weeds and less palatable grasses are thus transformed into greedy indiscriminate feeding machines, desperate to gobble up their share of the pasture before their rivals do. The picky grazing behaviour completely disappears. Everything goes down the hatch — tasty grass, weeds, old grass, new grass, it doesn't matter, it all gets vacuumed up in a greedy frenzy to keep up with the rest of the herd. Nothing gets overgrazed or undergrazed, it all just gets grazed. And then, when the grass is chewed down, the herd moves on, leaving the land to rest until the rains return and the grass has a chance to regrow. Under this style of grazing, grasses produce thick expanding sod mats that choke out weeds and cover the soil with a thick protective living skin, even as manure is evenly distributed behind the migrating herd. Soil fertility soars. Humus production soars. Erosion stops. And desertification is reversed.

Grocery stores frequently weaponize this same psychological effect among humans by creating an artificial illusion of scarcity in order to increase sales — for example, they will set up a freezer full of some product, like frozen chickens, and hang a sign above that says, "limited stock, only two per customer." Soon people who didn't even go to the store to buy chicken are putting a chicken in their grocery basket, while those who only wanted one instead come home with two. Meanwhile, there are truckloads of frozen chickens arriving at the back of the store to keep up with surging demand. And look at the run that happened on toilet paper at the start of Covid lockdowns as people ran from store to store to buy toilet paper until shelves were empty, for weeks, despite the fact that toilet paper factories never slowed down production and even ramped up production to try to meet demand. The mad scramble of Black Friday sales achieves the same psychological effect. So does standing in line at the Apple Store for the next product release. The illusion of scarcity is

Marketing 101, and it works wonders for turning grazing animals into a force for change out on the grasslands.

However, there's more to mob grazing than just the psychological effect of competition. Under continuous grazing, the "herd" acts as a loose collective of self-serving individuals. But once a herd is tightly bunched together as a migrating grazing mob, through the power of routine, through the effect of close contact, and through the security that comes with being part of a tightly knit herd, each individual herd member begins to bond with the group and begins to merge their individual identity to that of the collective herd so that the herd begins to behave as a single organism. Like teenage girls that have to go everywhere as a group, and being left behind even for a minute is a fate more painful than death, once a herd develops a herd identity, they will eat, sleep, go to water, and migrate everywhere as a group. The herd represents safety, belonging, and comfort. Sleeping, grazing, and chewing cud in the midst of the group is preferable to wandering off in search of a favorite shade tree. Either everyone wanders off together, or no-one does. This psychological shift in the herd's behaviour, which emerges during mob grazing, is part of the reason why mob grazing has an altogether different impact on the land than a scattered herd of animals in a continuous grazing scenario.

In the wild, fear of predators motivates the herd to stay tightly bunched together at all times and to migrate as a cohesive unit. Among domestic livestock, shepherds and electric fences replace the wolf to create the same herd effect. But as I noted in a previous chapter, when alpha predators are removed in the wild, like when all the wolves were killed in Yellowstone Park, the elk herds shed their herd-bound group identity, stopped grazing as cohesive migrating herds, and spread out all over the park instead, thus adopting all the destructive habits of continuous grazing. These scattered elk systematically destroyed much of the vegetation and devastated the aspen trees in the park until wolves were reintroduced. Once the wolves were reintroduced, the elk bunched up again for protection and began, once again, to graze as a migrating mob. The grasslands and aspen groves recovered even as the total wildlife carrying capacity of the park *increased* purely as a result of the beneficial impact of mob grazing.[1]

---

[1] Farquhar, B. (2023, Jun 22). Wolf Reintroduction Changes Ecosystem in Yellowstone. *Yellowstone National Park Tips.* https://www.yellowstonepark.com/things-to-do/wildlife/wolf-reintroduction-changes-ecosystem/

As the grass runs out beneath the feet of a mob of grazing animals, the herd naturally begins to move in search of fresh grass and so, the migration begins. Graze, chew cud, sleep, drink, walk to fresh grass, rinse and repeat, forever. Without wolves and other large predators to bunch the wild grazing herds together, there would not have been a bison migration to turn the Great American Desert into a lush and fertile grassland with some of the deepest and richest soils on the planet — instead continuously grazing bison would have turned the Great Plains into a barren desert long before farmers and plows arrived on the scene. Likewise, without lions there would be no million-strong wildebeest migration across the Serengeti, only a million scattered individual wildebeest gradually destroying the savannah and turning it into desert. *Migrating* herbivores create the thriving grassland skin that protects soil in dry brittle climates. Predators are nature's equivalent to shepherds and electric fencing to keep those grazing herds bunched tightly together and migrating as a cohesive group in a never-ending search for fresh grass. Continuous grazing destroys soils and turns grasslands into barren deserts. Mob grazing builds soil and turns deserts back into grasslands.

Mob grazing also has a huge beneficial impact on the soil nutrient cycle. In a continuous grazing scenario, after cows finish wandering about as scattered individuals in search of the best grass, they then return to their individual favorite shade trees to lounge and chew their cud, day after day after day. Slowly, nutrients are transported from the pasture to their favorite trees, depleting the pasture of nutrients as efficiently as any farmer's harvester would do if he never put any fertilizer back on the soil to replace the nutrients removed by the harvest. And as that manure piles up under the favorite shade trees, it builds to such extreme levels that it slowly poisons the soil around the shade tree. But a grazing mob never has access to the same slice of grass for more than a day, so there is no favorite tree to serve as a nutrient magnet because yesterday's trees are no longer accessible today. And besides, once the livestock develop a tightknit herd identity, bedding down at the center of herd is preferable to scattering about in search of the perfect individual tree. The manure stays in the pasture where it belongs so it can enrich rather than deplete the soil. And so, the soil grows more fertile over time.

Then there's the issue of soil erosion and soil compaction caused by cattle feet. In a continuous grazing scenario, picky cows use the same paths day after day. One cow follows another until deep ruts and trails of compacted soil stretch across the paddock from the water trough to the shade tree. By contrast, a grazing mob grazes side-by-side across the pasture, beds down as a herd to sleep, and then moves on as a tightly bunched group in search of the next day's fresh slice of grass. The competition for grass and the tightly bunched conditions ensure that no cow pays close attention to where it puts its feet. Instead of carving up the landscape with deeply rutted cowpaths that are pounded into oblivion day after day after day, the entire surface of the landscape is instead systematically churned by their hooves (shaped like tiny chisel-plows) as they trample plant litter and manure down through the soil surface so that soil microbes can reach it while breaking apart hard compact soils. And after a single day's grazing, the herd moves on, leaving the land to rest, and doesn't come back until the grass has had time to regrow. The thousands of hoofprints, the trampled grass, the chisel-plowed soil surface, and the long period of rest after the herd moves on all combines to loosen the sunbaked soil surface and prime it to be able to quickly absorb every drop of rain when the rains begin to fall.

As soil dries out during the drought season, soil microbes cannot survive near the soil surface, leaving dry undigested plant litter on the surface to be ravaged by sun, wind, rain, and oxygen. But indiscriminate trampling by thousands of sharp migrating cattle or bison feet mash that plant litter and manure down through the dry crust into the moist soil below where soil microbes can then resume their work of composting that debris into humus. And so, the effect of mob grazing creates the conditions to enable soil microorganisms to continue to create humus even during the dry season while the soil surface is parched and lifeless.

Reducing livestock numbers on "overgrazed" land does not fix the problem of desertification, even if it satisfies the beliefs of politicians and activists, because now there are even less mouths and less hooves to encourage vigorous grass growth, sod formation, and rapid nutrient cycling. In arid low rainfall regions, reducing livestock numbers often simply makes desertification worse. It's not the number of animals that is the problem, it's that, under our management, the grazing *behaviour* of the animals has changed, turning them from a constructive into a destructive force. When their behaviour is harnessed to bring out their constructive properties, they begin to heal the land and, in return, the land can support ever larger numbers of grazing animals.

This was the insight that led ecologist Allan Savory to develop his holistic grazing practices. He tells the story of how elephants in Zimbabwe (formerly Rhodesia) were decimating the landscape, which he and other leading scientists in the employ of the government blamed on soaring elephant numbers. Under the direction of these leading scientists, the Rhodesian government killed 40,000 elephants to try to prevent desertification in what Savory describes as "the saddest and greatest blunder" of his life. With elephant numbers reduced, desertification accelerated. It was only then that he realized that desertification was not being caused by the number of animals but by the behavioural changes triggered under our management — without large herds of migrating herbivorous grazing animals bunched together by pressure from predators, the grassland ecosystem quickly unravels.

Mr. Savory says he is the only scientist involved in that sad experiment to admit he was wrong and to learn from it — in true bureaucratic fashion, the rest of them continue to advocate the same herd culling and anti-livestock prescriptions to prevent "overgrazing", with predicably destructive results for the ecosystem. Meanwhile, Mr. Savory has been vilified by environmentalists and academics ever since, even as these dry ecosystems continue to fall apart under the oversight of central planners while healing under his. In an article about his work, Range Magazine shared a revealing story about the opposition Mr. Savory has encountered to his work: *"During a big conference in Rhodesia years ago, someone asked Savory, "Why is there such an intense, almost violent opposition to your thinking? Is it vested interest?" Savory had no answer but someone else said, 'Allan, you are up against the biggest vested interest in the world — professional people's egos.'"*[2] The whole article, linked in the bibliography, is well worth your time.

Mr. Savory has spent the rest of his life teaching these concepts, reversing desertification in communities in Africa and around the world, and practicing what he preaches on his land — the ultimate successful testing ground for his ideas. And, at 88 years old, he's still going strong!

I don't need some snarky peer-reviewed academic paper to tell me what to think about his ideas — I was first exposed to his holistic management practices and saw firsthand how successful those ideas are at reversing

---

[2] Hadley, C.J. (Fall 1999). The Wild Life of Allan Savory. *Range Magazine.*
    https://www.rangemagazine.com/archives/stories/fall99/allan_savory.htm

desertification when I apprenticed on ranch in Namibia in the late 1990s (a ranch in such a dry region of Africa that it needed more than 10x as much land as my dad had in British Columbia to support the same number of cattle). I saw what decades of poor grazing practices had done to the land, with 6-foot-deep erosion gullies criss-crossing the veldt and with former grassland savannas reduced to little more than rocky plains, and how, through the introduction of holistic grazing, the gullies were beginning to heal, the grass was slowly beginning to recolonize the barren plains, and the carrying capacity of the land was beginning to recover.

A great many environmentalists, politicians, and academics are destroying grasslands and deserts with their peer-reviewed environmental diktats that remove livestock from the land. Despite their good intentions they are quite literally destroying the very thing they are trying to protect. And they cannot see the error of their ways because they are so welded to their beliefs (and egos) that they cannot see how they have led themselves astray.

Grazing that mimics a wild herd migration is the only thing that can protect these dry grassland and desert regions — both continuous grazing and the removal of livestock are equally detrimental death sentences that slowly turn grasslands into deserts. The thin crust of cryptobiotic soils in the Canyonlands National Park in Utah are the last dying gasp of a soil on its way from grassland to barren desert. The historic grasslands that used to cover these soils once pulled vast quantities of CO_2 out of the air to store it in the soil as humus. Today, that same soil is slowly losing its humus to wind and water erosion, and to oxidization as atmospheric oxygen reacts with carbon in the bare exposed soils. These changes are not caused by "climate change", they are caused by the disappearance of the large herds of migrating herbivores and the alpha predators that once stalked their flanks, which together kept these grasslands healthy.

~

One last point about herbivores before we return to the topic of grass. As I mentioned earlier, herbivorous mammals evolved alongside grass around 55 million years ago, not long after the Cretaceous extinction. 5 million years after the first mammalian herbivores evolved, an even more efficient herbivore family emerged — ruminants. The difference between horses and cattle illustrates why ruminants are such an important evolutionary adaptation. Both are herbivores, but unlike horses, cattle have four stomach chambers and chew their cud. Consider the difference this makes: grass takes 24 to 36 hours to pass through a cow's digestive tract; among horses it passes through within 2 hours.[3] By the time digested grass comes out of the back end of a horse, most of the grass fibers are still easily identifiable. By contrast, cattle not only employ a host of fermentation microbes (especially methanogens) within their four-chambered digestive tract, but they also regurgitate and rechew their cud whenever they're not out grazing, thus grinding up the tough lignin and cellulose fibers into extremely fine particles that are mixed with digestive enzymes from their saliva before the whole mouthful is sent back down for further fermentation. This rumination process enables cattle to extract far more energy from their food than horses can. And because the plant debris is so well digested during this fermentation/cud-chewing process, the soil microbes have a much easier job of breaking down cattle manure to recycle the nutrients and turn the plant fibers into humus.

In practical evolutionary terms, this means that a horse requires much higher quality food than a cow and, in the wild, essentially has to spend every waking hour of the day grazing just to get enough calories to survive. It also means that horses are far less efficient at turning plant fibers into humus. By contrast, cattle and other ruminant species are capable of surviving on much lower quality grasses while still having time to settle down for an afternoon of leisurely cud chewing, which is why almost all of the really large wild grazing herds that once roamed across the world's grasslands are ruminants. Bison, elk, deer, antelope, moose, wildebeest, hartebeest, oryx, caribou, reindeer, and wild cattle are all ruminants. Most of the world's rich grassland soils were created by this symbiotic relationship between grass and ruminants, with a much smaller population of non-ruminant herbivores making up the difference.

~

Grass has several other evolutionary tricks up its sleeve that turn it into a far more efficient mechanism for building humus and sequestering carbon than other plant species. One of those tricks is the speed at which carbon cycles in and out of the atmosphere in a grassland environment

---

[3] "Architecture of the Equite Digestive System." (2012, Nov 1). *Practical Horseman.* https://practicalhorsemanmag.com/health/architecture-of-the-equine-digestive-system-11756/

compared to how quickly this cycle happens in a forest. A tree grows over decades, even centuries, so all the carbon they remove from the atmosphere and all the nutrients they pull up out of the soil just sits there, locked inside the tree for decades or even centuries before it is returned to the ground. Forest soils are notoriously low in nutrients. And then, when the trees die, it takes a long time to decompose those tough woody tree trunks before those nutrients can be returned to the ground and before all those tough lignin- and cellulose-rich fibers can be turned into humus. Soil builds very slowly in a forest environment, which is why forest soils typically contain far less humus than grassland soils.

Nutrients and carbon are recycled far faster in a grassland environment because the entire year's vegetative growth is fully recycled back to the soil within the same growing season. This cycle is sped up even further by the manure from herbivores. By contrast, a dense forest doesn't get a lot of animal manure because ever since the dinosaurs went extinct, there are no large herds of herbivores that eat wood — moose and elk will browse on saplings but even they can't eat a Douglas Fir tree.

A dense forest has another major limiting factor that slows growth — competition for sunlight. Once the forest canopy closes up, little if any direct sunlight reaches the forest floor. If you look at tree rings from old-growth forests (these are visible in the massive timbers used by the pioneers to build their barns and homes, some of which still stand today), the tree rings are extremely close together, which demonstrates just how slowly trees grew in dense old-growth forests — tree growth was limited by the lack of sunlight filtering down through the thick forest canopy. Tree rings from managed forests where trees are widely spaced to let in more light can achieve much faster growth rates (and have much wider growth rings), but only up to a point. Even in a managed forest, photosynthesis is mostly limited to the forest canopy as broad leaves and dense upper branches drastically limit how much sunlight is available to fuel growth. By contrast, there is *a lot* more photosynthesis happening in a meadow or grassland environment where sunlight floods every square inch of the grass — the shape of the thin narrow grass leaves maximizes how much light filters down through the entire grassland.

The differences in yields speak for themselves. An unmanaged forest produces around 1-2 tons of biomass per acre per year — in other words, it pulls 1 to 2 tons of carbon out of the atmosphere each year. A well-managed forest can produce up to 3 to 4 tons per acre per year. But a grassland can produce 7 to 10 tons of biomass per acre per year. And, unlike in a forest, all of that grassland biomass is recycled back into the soil every single year as roots die back, as fungi and earthworms digest it, and as herbivores turn it into manure to feed the microorganisms living in the soil. It's yet another factor which explains why a properly managed grassland can build soil faster than any other ecosystem in the world (and why the destruction of our grasslands has such a dramatic impact on how much carbon is building up in our atmosphere — more on that shortly). Most of the world's richest agricultural soils were created by migrating herds grazing across large open grasslands and savannahs, not by trees growing in dense forests. Most of the carbon that was once stored in the world's soils was stored in grassland rather than forest soils.

The USDA estimates that, on average, it takes about 500 years to form one inch of topsoil via the chemical weathering of rocks. In warm humid English soils, Darwin's beloved earthworms speed up that process via biological weathering to a few inches per century. Gardeners manually adding large volumes of compost and manure to their garden beds can produce humus even faster (thanks to fungi and other microbes involved in the composting process). And, similarly, humus can also be produced much faster wherever pastures and grasslands are grazed by large migrating herds of densely packed ruminants (rotational grazing practices usually divide paddocks into daily grazing slices, whereas mob grazing goes even further by dividing paddocks into even smaller slices so that the livestock are moved to fresh grass multiple times per day).

A small subset of ranchers and ecologists like Allan Savory have been leading the way in demonstrating how quickly humus can be created using these migratory grazing techniques, and academia is slowly beginning to wake up to what they have been saying. For example, a study from Nova Scotia found that soil organic matter increased by approximately 10% across multiple soil depths in a pasture rotation in a period of only 8 years.[4] A 3-year study undertaken on the steppe in Kazakhstan found rotational grazing increased available phosphorus in the soil by 46%, improved the soil structure by 14.4%, and doubled yields (with notable improvements

---

[4] Bouman, O.T., Fredeen, A.H., and Mazzocca, A.M. (2018). Soil organic matter storage in a perennial rotation pasture in north-central Nova Scotia from 2007 to 2015. *Canadian Journal of Soil Science*. 98(2). https://doi.org/10.1139/cjss-2017-0122

in the energy and protein content of forage).[5] Yet another study, this time from Arizona, found that rotational grazing significantly increased organic matter and nitrogen levels in the soil.[6] The key takeaway from these studies is not just that rotational grazing and mob grazing reverse soil erosion and rebuild humus, but that these big changes happen in an incredibly short period of time.

Back in figure 97, I showed you a photograph from the Karoo region of South Africa in which a fence runs across the plains, separating a lush grassland on one side of the fence from a barren desert composed mostly of bare soil that is interrupted by the occasional anaemic clump of vegetation. The barren desert has been managed by conventional "continuous grazing" techniques for the past 50 years. The lush grassland on the other side of the fence has been managed with Allan Savory's holistic grazing techniques over that same 50-year period. On one side of the fence, cattle are creating a lush and thriving grassland, on the other side cattle are destroying it.

Holistic grazing, mob grazing, rotational grazing, Voisin-style grazing, short-duration grazing, adaptive multi-paddock grazing (a.k.a. AMP grazing, as it's often called in academia), and so on are all variations of the same idea of managing grasslands by mimicking the grazing impact of migrating herds of herbivores. The pioneers of this style of grazing are individuals like André Voisin, Allan Savory, Stan Parsons, Joel Salatin, Gabe Brown, Allen Williams, among others. As their ideas spread, countless different schools of thought are emerging as different people experiment and adapt these ideas to suit the nuances of different climates, soil types, and ecosystems. Once again, the free market is willing to explore, experiment, observe, and ask questions that academia will only begrudgingly entertain when all other options for defending established dogma are exhausted. Anyone who expects government scientists to drive this innovation is delusional — the farmer who loses his livelihood as his soils are exhausted has skin in the game — he and his entire family has to

---

[5] Nasiyev, B., et al. (2022). Changes in the Quality of Vegetation Cover and Soil of Pastures in Semi-Deserts of West Kazakhstan, Depending on the Grazing Methods. *Journal of Ecological Engineering.* 23(10). https://doi.org/10.12911/22998993/152313

[6] Roberts, A.J. and Johnson, N.C. (2021). Effects of Mob-Grazing on Soil and Range Quality Vary with Plant Species and Season in a Semiarid Grassland. *Rangeland Ecology & Management.* 79. https://doi.org/10.1016/j.rama.2021.04.008

live with the consequences of his management decisions — whereas the government scientist can be wrong his entire life and still be rewarded with lucrative research grants, which may even increase thanks to the publicity generated by being the one working to solve an unresolved crisis.

One of the best illustrations of how these grazing techniques can restore grasslands to good health comes from a farmer I had the privilege to get to know a few years ago after he reached out to me on my cattle ranching website (*www.grass-fed-solutions.com*). Neil Dennis was a fourth-generation cattle rancher from Saskatchewan and an icon of the mob grazing movement. In the late 1990s, his farm (Sunnybrae Farms) was struggling to stay afloat. At his wife's urging, he attended a Holistic Management seminar hosted the Savory Institute. In characteristic stubborn farmer fashion, after the seminar he set out to prove the instructor wrong. He set up one small corner of his farm as a test plot for mob grazing while continuing to do what he had always done on the rest of the property. It didn't take long for the test plot to turn his world upside down.

When I spoke with him in 2014, he had just completed a series of soil tests, conducted over a 3-year interval, to track the impact that mob grazing was having on organic matter in his pastures — his results from two fields are reproduced in figure 102 below. The test for soil organic matter is quite simple — each soil sample is oven dried to remove all moisture before it is weighed and then incinerated to burn off all the carbon. The decrease in weight after incineration tells you how much organic matter was in the soil. Neil took samples from three different soil depths in each field. As you can see from figure 102, the rapid increases in soil organic matter over this brief 3-year interval are astounding in and of themselves, but even more remarkable is that these increases are not just in the topmost layers of the soil but extend over a foot down into the soil! Neil's soil tests illustrate that soil is not just built from manure and plant debris accumulating at the soil surface but also from the decomposition of roots after plants are harvested. Every time grass is grazed, a portion of the grass roots dies back in order to maintain the balance between photosynthesis in the green leafy parts above ground and the nutrient and water harvesting capabilities of the roots below.

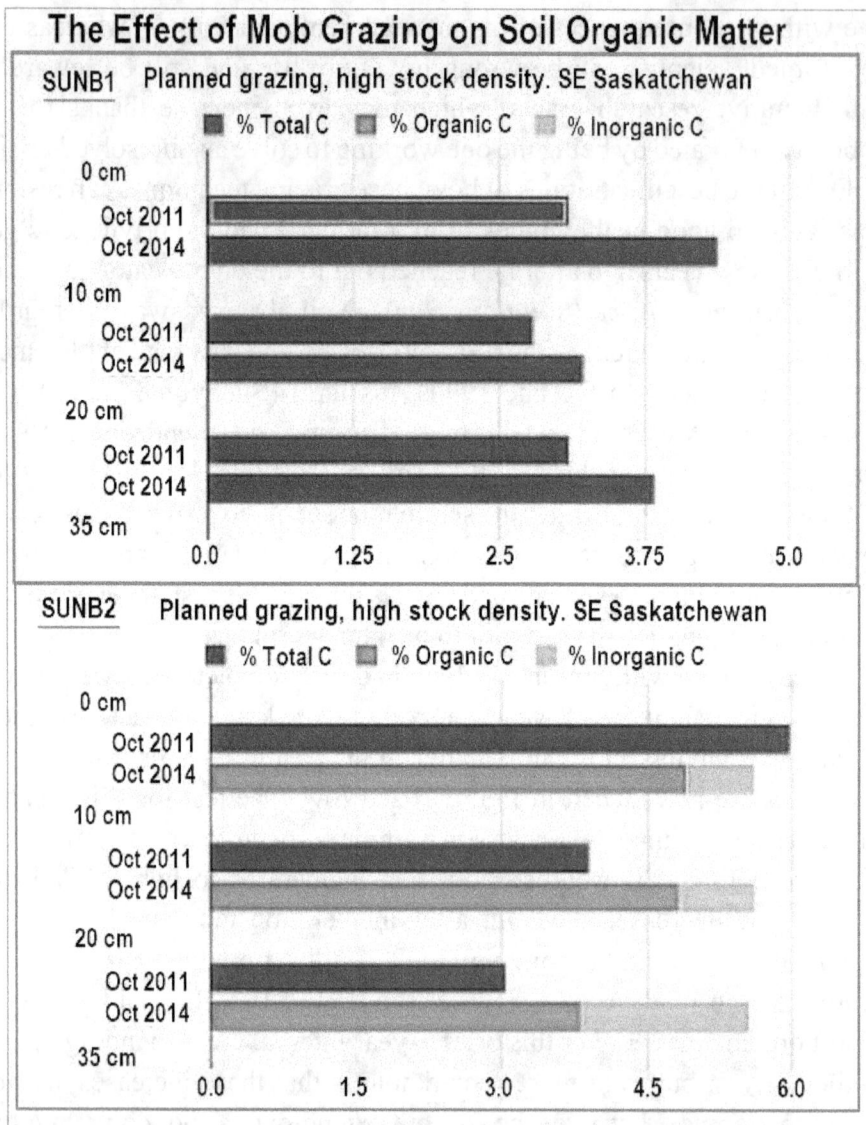

Figure 102. Soil organic matter improvements caused by mob grazing at Sunnybrae Farms.
Source: Neil Dennis (personal communications) — additional results from other farms can be viewed on the Soil Carbon Coalition website. https://web.archive.org/web/20150115093610/https:/soilcarboncoalition.org/data/challengecharts.htm

The cycle of root die-back and regrowth that follows every grazing cycle is why grasslands are such a powerful engine for building humus-rich soils in a rotational grazing system (or holistic grazing or mob grazing system, or whatever term you prefer to describe the underlying grazing philosophy). When the symbiotic relationship between grasslands and ruminants is harnessed via a holistic grazing system, livestock become an incredibly powerful force capable of pulling colossal volumes of carbon dioxide out the atmosphere and storing it in the ground.

Before Neil switched to mob grazing, his 1,170-acre farm supported 200 to 300 head of cattle. After switching to mob grazing his stocking capacity grew to 800 to 1000 head of cattle on the same 1,170 acres[7] — and it happened in a span of just over a decade and without any off-farm fertilizer inputs! These are extraordinary numbers. In effect, it's as if he had quadrupled his land, yet he did it without buying a single additional acre. Sadly, Neil passed away in 2018, but you can still hear him describe his experiences in his own words if you search his name on YouTube. I also recommend an 11-minute video called *Soil Carbon Cowboys*[8] in which Neil and two other farmers, Gabe Brown and Allen Williams, describe the transformative changes they experienced on their land when they switched from conventional to mob grazing.

In 2016, I posted an interview on my cattle farming website with Daniel Suárez — he is a commercial cattle rancher (beef and dairy) and grazing educator from the dry tropical savannah-like region near Chiapas, Mexico. He switched his ranch from conventional to Voisin-style grazing in 2010. Prior to the switch, the carrying capacity of his land was around 0.6 cows per hectare during the rainy season (while relying on outside feed during the dry season). The initial electric fence grid that he installed in order to begin divvying out fresh slices of pasture every day immediately doubled his carrying capacity during the rainy season from 0.6 to 1.3 cows per hectare — he credits this initial increase to the improved grazing efficiency and improved pasture utilization that comes from bunching cattle together to graze as a mob. And then, year after year, the land began to change as the humus began to accumulate beneath his feet. After only 6 years of using these techniques, his carrying capacity surged to 1.8 cows

---

[7] "Success Stories: Sunnybrae Farms." *Holistic Management International.* Accessed 3 Jun 2024. https://holisticmanagement.org/holistic-management/success-stories/case-studies-sunnybrae-farms/

[8] Roots So Deep (you can see the devil down there). (2020, May 14). *Soil Carbon Cowboys.* [Video]. YouTube. https://youtu.be/MDoUDLbg8tg?si=PaC6z510u9vXvavM

per hectare *averaged over the whole year, without any outside feed inputs even during the dry season,* and to over 5 cows per hectare during the rainy season! As he points out in his interview, he accomplished all of this without a drop of fertilizer or herbicide:

"Instead of using fertilizers and herbicides, we use a machine that fertilizes the grass and attacks weeds at the same time. That machine is the cow. The four legs (trampling) and the muzzle (grazing) are the weed control mechanisms of this machine. And the manure is the fertilizer mechanism of this machine. [...] There is no better or cheaper fertilizer for the soil (notice I said soil, not plant) than manure, in our case cattle manure. A 500 kg cow produces approximately 25 kg of manure every day. If we consider that we have a stocking density of 200 cows per hectare per day, we will deposit a total of 5000 kg of manure per hectare. No one else fertilizes their pastures this much. And there is no cheaper or more evenly distributed manure than this!"[9] You can read my full interview with Daniel at www.grass-fed-solutions.com/rational-grazing.html.

Based on Daniel's experiences and the experiences of other farmers using these techniques in his region, he expects his year-round carrying capacity to continue to rise to over 4 cows per hectare as organic matter continues to build up in the soil, essentially doubling even that which he has already achieved today. It's almost like he's purchased 8 additional farms without having to spend a dime — all it cost him was some electric fencing wires to be able to mimic the impact that wild herds of ruminants have on the land.

In a recent presentation hosted by the Savory Institute, Dr. Richard Teague, professor emeritus at the department of ecosystem science and management at Texas A&M University, summarized research into the effects of holistic planned grazing conducted by himself and a large number of his peers over the course of his long career.[10] They found that holistic planned grazing will, over a period of years, generally increase pasture productivity by up to a factor of 4, the equivalent of getting four

---

[9] "Voisin-style Rational Grazing – Interview with Daniel Suárez (rancher and grazing educator)." *Grass-fed Solutions.* Accessed 4 Jun 2024. https://www.grass-fed-solutions.com/rational-grazing.html

[10] Savory Institute. (2021, Apr 29). *The Science of Holistic Planned Grazing | Dr. Richard Teague.* [Video]. YouTube. https://www.youtube.com/watch?v=aVNmM5dkG-Y&ab_channel=SavoryInstitute

farms for the price of one. They also found that compared to pastures managed via continuous grazing, *pastures managed via holistic planned grazing were adding an additional 8.6 tons of carbon per hectare to the soil <u>every single year</u>*. Furthermore, parallel to the increase in carbon, *they also found a corresponding 6-fold or more increase in <u>microbial biomass</u> living within the holistically grazed soil*!

Nor were the effects limited to just carbon and biomass. Since soil carbon is the main storehouse of stored nitrogen in the soil, stored nitrogen levels rose as carbon levels increased (without the help of fertilizer amendments (!), rising to higher levels than those found in continuous grazed pastures — they even exceeded levels found in continuously grazed pastures that were fertilized with extra nitrogen fertilizers!

Dr. Teague and his colleagues also found that as the soil structure improved under holistic planned grazing, the amount of rainfall that could be captured and stored increased dramatically because water infiltrates faster and to a greater depth (measured in cm/hour) in humus-rich soil. Results will obviously vary greatly from farm to farm, but at 24:45 of his presentation[11] Dr. Teague shows results from Neil Dennis' ranch (which was included in one of his studies) —water infiltration was around 3 cm/hour under continuous grazing versus around *27 cm/hour under holistic planned grazing* — more than 9x higher! This represents a *massive* increase in how much rainwater will be captured and stored by the soil, how deep rainwater will penetrate even after a light rain shower, and how much water will be available to plants instead of running off or evaporating away immediately after a rainfall. In effect, this makes rainfall 9x more effective. Or, to put it another way, it makes the land 9x less vulnerable to drought!

By contrast, you can manufacture a drought even without reducing the amount of rainfall simply by eroding away and degassing the amount of humus stored in the soil. The amount of rainfall matters far less than how well you prepare the soil to capture and store every drop that falls, and how well you protect the soil from evaporation by how well you manage the sod cover that protects the soil from wind and sun.

Understanding this link between humus and water infiltration is of vital importance: if rainfall pools on the surface instead of being quickly absorbed into the ground, it will immediately evaporate when the sun

---

[11] Savory Institute. (2021, Apr 29).

comes back out after the rainstorm ends — the water will be gone as though the rain never happened. And, if it's big rainstorm, if that water cannot be quickly absorbed, it will begin to flow downhill into the streams and rivers, carrying precious soil along with it. Thus, as humus is eroded and degassed away, floods become increasingly common, again without any increase in rainfall, and soil erosion dramatically escalates. Flood and drought — the very things that the climate brigade keeps pointing to as evidence of "global warming" is, in fact, a consequence of soil degradation caused by poor local land management practices.

One of the leading causes of desertification is the loss of humus in the soil — it's not that there's necessarily less rain in areas suffering from desertification, it's that the rain becomes less and less effective as humus levels (and sod cover) decline. Small rains become ineffective because they aren't absorbed, and large rains become destructive because the soil lacks both the humus and the sod cover to quickly absorb that rainfall and to bind the soil together to keep the rainwater from washing away the soil. Go look at the rivers draining off the continent of Africa (and it's not the only place this is happening) — they are completely choked with sediment eroding off the continent. Yet when Africa's earliest explorers (like David Livingston) first ventured to explore the continent, most of these same rivers ran crystal clear.

On his ranch in Zimbabwe, Allan Savory demonstrates the healing effect of changing the behaviour of grazing livestock through holistic planned grazing (a.k.a. mob grazing). By restoring the grasslands on his land with his cattle herds, not only do the streams draining from his land run clear, but the vegetation is also filtering out the sediments washing downstream from other farms located uphill from his property — brown coffee-coloured water flows onto his property but leaves crystal clear. You can see the effect yourself in the documentary put out by the Savory Institute, called *Running out of Time,* available on YouTube.[12] Continuous grazing destroys continents, mob grazing restores them.

---

[12] Savory Institute. (2018, May 1). *Running out of Time | Documentary on Holistic Management.* [Video]. YouTube.
https://www.youtube.com/watch?v=q7pI7IYaJLI&t=271s&ab_channel=SavoryInstitute

Chapter 23

From Net Carbon Sink to Net Carbon Emitter

"Sometimes the things that happen the slowest are the hardest to stop because they never quite reach the level of an immediate crisis."

— David Montgomery, author of Dirt: The Erosion of Civilizations

Humus is a strange sort of carbon sink. In some ways, the carbon stored in soil is similar to coal in that photosynthesis pulls vast quantities of CO_2 out of the air and stores it in the ground after plants die. But coal just sits there, buried under tons of sediments, forever. There are individual coal seams in Wyoming that are over 250 feet thick! Peat bogs are also obvious carbon sinks — the Philippi peatland in Greece extends down over 190 meters and contains reserves of over 4,300 million m^3 of peat. You can put your finger on the carbon stored in a coal seam or peat bog and see where all the CO_2 has gone after it was scrubbed out of our atmosphere by photosynthesis. But with humus, it's not so easy.

Researchers in England have calculated that an acre of soil with 4% soil organic matter contains as much energy as 20 to 25 tons of anthracite coal (that's the highest coal grade).[1] Many prairie soils have up to 6-12% organic matter. And, because CO_2 has two oxygen atoms that are released back into the air during photosynthesis for every carbon atom that is stored in the soil, it takes 3.67 tons of carbon dioxide scrubbed out of the atmosphere to produce 1 ton of soil carbon. Multiply that carbon capture by the tens of billions of acres of soil covering our continents and you quickly get a sense of the sheer scale of this carbon sink (and the sheer scale of CO_2 that is being degassed back into the atmosphere by soil degradation and erosion). Coal beds and peat bogs only form in basins that catch and hold water so there aren't many places on our planet with the

---

[1] Sachs, P. (n.d.). Humus: Still a Mystery. *The Natural Farmer.*
https://www.humintech.com/fileadmin/content_images/agriculture/information/articles_pdf/Humus_by_Paul_Sachs.pdf

right conditions to form coal or peat. By contrast, soil forms on almost *everything*.

Nonetheless, soil is a very thin skin on the Earth — most topsoils around the world are less than a foot thick. The world's deepest soils are the chernozem soils in the Ukraine, and even they are only a few feet deep before you run into the subsoil layer that is composed primarily of inorganic gravel, sand, silt, and clay. If soil is such a powerful carbon sink, why don't soils keep getting indefinitely thicker like the coal seams that formed at the bottom of Carboniferous bogs?

Bogs are basins that slowly fill with organic matter until the basin is full. By contrast, soil is the skin of the earth that grows on flat land, hillsides, and virtually all slopes. Almost all land, even if it looks flat, is gently sloped towards the nearest stream or river — if it doesn't have any slope, it turns into a lake or bog. As Darwin pointed out during his studies of earthworms, soil is constantly on the move, downhill and downstream. A healthy soil stores organic matter, but only to a point — the soil doesn't grow indefinitely thicker because, as geologists and soil scientists have shown, even in a healthy environment that is not being eroded by human activities, the natural rate at which soil forms is in near perfect balance with the rate of natural soil erosion (from wind and rain). Agricultural techniques that build soil, like mob grazing, composting, and other regenerative agricultural strategies, can speed up the rate of soil formation. But when nature works on its own, soil formation is a very slow process that can barely keep up with the speed at which natural non-human erosion carries away that precious soil skin. So, what happens to all the carbon that is lost to erosion? The answer is not as simple as you might think.

Even in a healthy environment without human-caused erosion, whenever soil is exposed to the air, some of the carbon will get oxidized and return to the atmosphere as CO_2. But most of the erosion in a natural environment happens when rainwater washes away sediments — that process takes carbon, which began its life as soil, on a very long journey downstream until it reaches its final resting place at the bottom of a bog, lake, or ocean. And so, to understand how soil functions as a carbon sink you have to understand this long conveyor belt that is continually transporting carbon from where it is initially created until it reaches its final resting place in a permanent carbon sink further downstream. That conveyor belt ensures that a large portion of the carbon removed from the atmosphere by plants is continually being swept away and stored in carbon sinks far from where the carbon was first scrubbed out of the atmosphere by photosynthesis.

Most of the sediments that are being washed downstream are inorganic sand, silt, and clay particles, but they are also chock-full of the carbon that began its life as humus, which is now on its way to long-term storage. The largest dam in the U.S. is the Hoover Dam across the Colorado River, which created Lake Mead. The reservoir only has an expected lifespan of 100 to 400 years, depending on how fast carbon-rich sediments from the surrounding watershed accumulate at the muddy bottom of the lake. Lake Kariba in Zambia is the largest artificial reservoir in the world — in the midst of Africa's high soil erosion rates, it only has a projected lifespan of 140 years. The enormous Aswan Dam across the Nile is expected to be completely silted in within 200 years, mostly from sediments washed off the Ethiopian Highlands. The Three-Gorges Dam across the Yellow River in China has a predicted lifespan of at least 100 years.

But while some carbon ends up at the bottom of lakes and man-made reservoirs, most humus that began its life as soil ends its journey in a carbon-rich mud layer in a river delta. For example, 5,800 years ago, the Sumerian city of Ur in ancient Mesopotamia was located at the mouth of the Euphrates River. Today, Ur is more than 255 km (158 miles) inland as mud accumulating in the river delta slowly filled in the entire northeastern end of the Persian Gulf. Today's fertile marshes at the mouth of the Euphrates River are full of the carbon that began life as humus on ancient Sumerian farms and pastures. Slowly, over millennia, the fertile soils of the Fertile Crescent were washed away by erosion, leaving behind bare deserts… and muddy river deltas. In time, the weight of all these muddy layers will begin compressing the lowermost layers to become carbon-rich shales. Likewise, in the present-day United States, the Mississippi Delta is expanding by 0.8 to 3.1 km^2 *every single year*. The world's largest delta —the mangrove-covered Sundarban Delta at the mouth of the Ganges River — is likewise growing by around 2.8 km^2 per year. On the other hand, the Nile delta began shrinking after the Aswan Dam began catching all of the Nile sediments that once made their way to the Mediterranean Sea.

This never-ending conveyor belt is how soil functions as a carbon sink as billions upon billions of tons of carbon are continually transported downstream to their final resting place at the muddy bottom of a bog, lake, or ocean. It's an unstoppable process... though it can be dramatically accelerated by poor soil management.

As long as soil carbon is eroded by water, very little of it will be exposed directly to the air where it can get oxidized — humus particles floating in water are not easily oxidized, obviously. But as tillage and bad grazing practices (like continuous grazing) destroy the sod cover, not only does this make the soil more vulnerable to wind and water erosion, it also massively accelerates how much oxygen is able to come in contact with humus, causing it to degas as CO_2.

Part of that degassing is caused by oxidation at the soil surface as humus reacts with oxygen in the air. And part of that degassing is a consequence of drying out the soil by removing the vegetative debris and sod mat that once shielded the soil from the sun — as oxygen replaces moisture in the soil and circulates freely through a soil that has lost it protective cover of sod or mulch, even deeply buried humus layers will begin to oxidize and degas. In short, dry soils "burn up" humus much faster than moist soils.

I want to emphasise this point because it is so important to everything yet to come in our journey through geological time. Because oxygen easily reacts with humus whenever humus is exposed to air or whenever a soil dries out, soil moisture plays a pivotal role in preventing oxidation. While a soil is moist, moisture shields humus and other organic plant debris from reacting with oxygen. But without soil moisture, the biological processes that convert organic matter into stable humus shut down even as oxygen begins to react with the organic matter in the soil, causing it to begin to degas back into the atmosphere. Dry soil = increased CO_2 degassing. And so, simply as a consequence of drying out, a soil that once stored vast quantities of carbon begins instead to degas carbon back into the atmosphere.

Continuous grazing, tillage, cutting down trees, the loss of vegetation that would otherwise shade the soil from the hot sun, and any other process by which we remove the thin mulch of plant debris that shields the soil from drying out — all these processes that dry out the soil and mix oxygen into the soil dramatically increase CO_2 degassing even as they simultaneously interrupt with the natural biological processes that convert organic matter into humus. A USDA study demonstrated that the amount of CO_2 that degases from soil after tillage increases 1.5- to 2-fold compared to untilled soil[2] simply as a consequence of the extra oxygen that is able to penetrate the soil after tillage.

A single tillage event (i.e., a single pass with a plow) can reduce soil carbon by as much as 1 to 11%.[3] This rapid loss of carbon is caused (1) because of the extra oxygen that is added to the soil, (2) because of the loss of sod and/or plant debris that shields the soil's surface from oxygen and prevents soil from drying out, and (3) because of the disruption to the soil microorganism community living below the soil's surface (i.e. breaking apart the fungi networks, thus killing them, which causes them to also degas as CO_2 as they decompose). The more frequent the tilling, the faster soil carbon is depleted. It's hard to imagine the vast quantities of humus that were consumed (and degassed) by the incessant plowing that prepped the Great Plains for the Dust Bowl. It's hard to imagine how much humus was converted back into CO_2 by Mao's fatal agricultural reforms that forced farmers to plow the living daylights out of China's soils. All that precious humus ended up back in our atmosphere as CO_2. And those same processes are still at work all over the world today as deforestation, continuous grazing, tillage agriculture, bad forestry practices, and a whole host of other human activities continue to disrupt the biological processes that once turned soils into vast carbon sinks. Not all forms of tillage are equally destructive — cultivating and ripping leave a lot more plant debris on the surface and are therefore somewhat less damaging than the mouldboard plow, which completely turns soil over, mixes oxygen deep into the soil horizon, and buries all the plant debris.

The take-home point here is that *tillage, bad grazing habits, deforestation, and other destructive human activities have transformed the world's soil from a net carbon sink to a net carbon emitter.* As our agricultural footprint expanded to keep up with our growing population, and as tillage and continuous grazing became the basis of much of industrialized agriculture, the amount of humus that was oxidized and degassed into the atmosphere went up, and up, and up.

---

[2] Sainju, U.M., Jabro, J.D., and Stevens, W.B. (2006). Soil Carbon Dioxide Emission as Influenced by Irrigation, Tillage, Cropping System, and Nitrogen Fertilization. [Conference Paper]. *USDA. Workshop on Agricultural Air Quality: State of Science. June 5-8, 2006.*
https://www.ars.usda.gov/ARSUserFiles/35789/Sainju_198562.pdf

[3] Conant, R.T., et al. (2007). Impacts of periodic tillage on soil C stocks: A synthesis. *Soil and Tillage Research.* 95(1-2). https://doi.org/10.1016/j.still.2006.12.006

Everyone is familiar with the correlation between rising fossil fuel emissions and rising CO_2 levels. But the amount of land required for agriculture (and which is thus degassing instead of storing humus) also correlates extremely closely with rising atmospheric CO_2, as shown in figure 103 below — it complicates the story a bit, doesn't it?

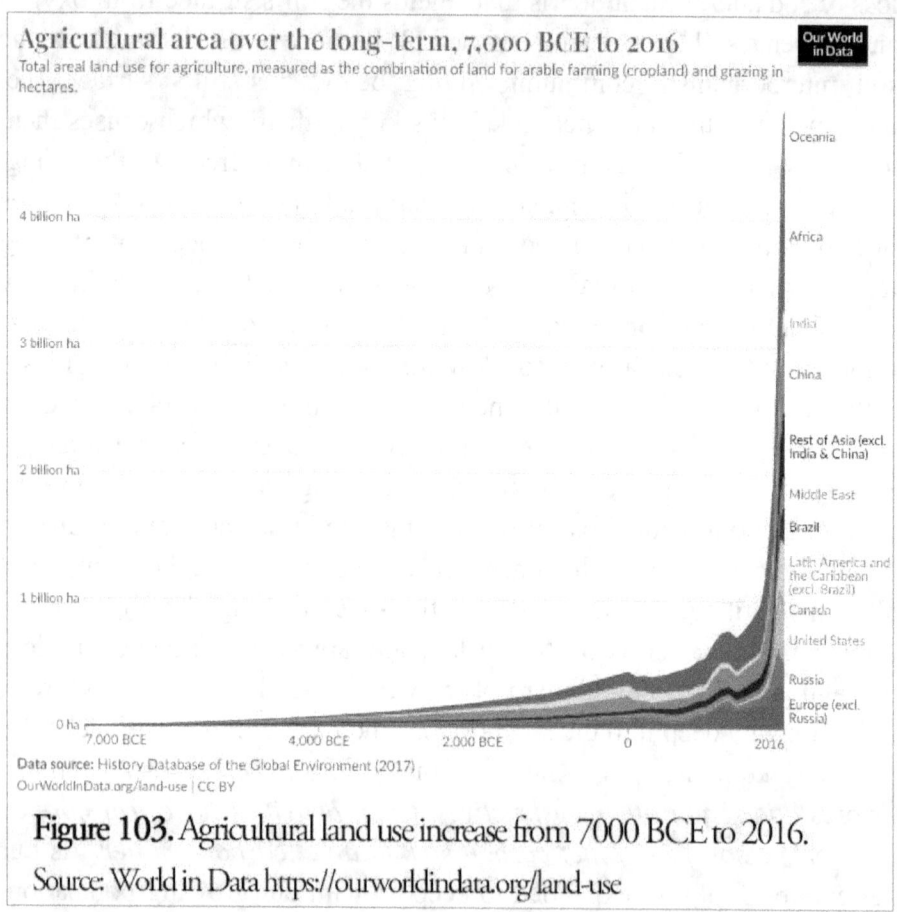

Figure 103. Agricultural land use increase from 7000 BCE to 2016. Source: World in Data https://ourworldindata.org/land-use

Dr. John Baker, a soil scientist and World Food Prize nominee in 2013 and 2014, has been studying how conventional tillage affects the soil and how its negative impacts can be reversed by switching to no-till farming. He found that *"when a farmer ploughs and cultivates a paddock it releases CO_2 into the atmosphere. The vast majority (95 percent) is released from soil with the other five percent coming from tractor exhausts. [...] The amount of CO_2 released by cultivation during reseeding can be approximately three tonnes per hectare.* "When you look at it from a global level, you realise that 15-20 per cent of the CO_2 in the world's *atmosphere comes from ploughing.*"[4] It's extraordinary to think that an acre of plowed soil degasses 19x more CO_2 into the atmosphere than the amount of CO_2 exhausted by the diesel-burning tractor that was used to plow that acre of soil!

By contrast, no-till farming (which plants seeds into the un-tilled soil and keeps the soil surface mulched with plant debris) reverses that carbon loss and once again begins to *store* carbon back in the soil, much like mob grazing does out on pastureland. A study at the University of Nebraska-Lincoln found that no-till farming was a net carbon sink, adding up to 1 ton of carbon per acre to the soil per year. The average carbon sequestration in the study area was 0.77 tons of carbon added to the soil per acre per year (equivalent to storing 2.82 tons of CO_2 in the soil per acre per year — remember it takes 3.67 tons of CO_2 to produce 1 ton of carbon).[5]

Back in chapter 12, I showed you several controversial studies (Skrable et al, 2022; Berry, 2023; Harde and Salby, 2021), which found that fossil fuel burning only accounts for around 12-15% of the rise in atmospheric CO_2, thus contradicting the consensus narrative that fossil fuels account for 78% of the rise. Soil degradation isn't just a peripheral issue in the story of atmospheric CO_2 when global soils contain around 2,500 gigatons of stored carbon (which equates to 9,175 gigatons of carbon dioxide). In other words, soils contain three times more carbon that the total carbon found in our atmosphere and four times more carbon that the amount of carbon stored in all living plants and animals combined.[6] With these kinds of numbers, a little bit of soil degradation goes a very long way towards oxidizing vast volumes of humus to turn it into CO_2 and would offer an explanation for the controversial findings in these controversial studies. After everything we've learned so far in this chapter, let's play with some

---

[4] "20 per cent of world's CO_2 from ploughing – soil scientist." (2014, Jan 15). *Farming Online*. Accessed 20 Apr 2024. https://farming.co.uk/news/20-per-cent-of-world's-co2-from-ploughing-—-soil-scientist

[5] Havens, A., Perrin, R., and Fulginiti, L. (2023, Mar 22). Carbon farming: A preliminary economic analysis of carbon credits for no-till and cover crops. *University of Nebraska-Lincoln, Institute of Agriculture and Natural Resources*. https://agecon.unl.edu/carbon-farming-preliminary-economic-analysis-carbon-credits-no-till-and-cover-crops

[6] Cho, R. (2018, Feb 21). Can Soil Help Combat Climate Change? *Columbia Climate School – State of the Planet*. https://news.climate.columbia.edu/2018/02/21/can-soil-help-combat-climate-change/

numbers to see how realistic the findings from those controversial studies might be:

Dr. John Baker's studies of conventional tillage agriculture calculated that 15 to 20% of global CO_2 emissions come from tillage alone. That's just from tillage — that doesn't include CO_2 degassing as a consequence of pasture degradation, forest degradation, desertification, conversion of forest into cropland, and so on. Only one-third of agricultural land is used for crops (11 million km^2), the other two-thirds is pastureland (37 million km^2), much of which is also in various stages of soil degradation thanks to the widespread use of continuous grazing techniques. The world's surviving grasslands and semi-arid deserts cover another 11 million km^2. And deserts cover another 40 million km^2. So, the effect of tillage is barely scratching the surface of the complete impact that soil degradation and desertification are having on atmospheric carbon dioxide levels.

If we assume another 20% of global CO_2 emissions comes from a combination of soil degradation in pastures, forests (like the soil degradation described in British Columbia's forests in chapters 3, 4, and 5), and from the conversion of virgin wilderness that is converted into cropland each year to keep up with our expanding agricultural footprint (as shown in figure 103), this puts us at the top end of the range calculated by yet another study, which found that a combination of land use and land use changes account for 28-40% of global emissions.[7] Not quite consistent with the "consensus narrative" that fossil fuel emissions are to blame for 78% of all atmospheric CO_2 increases, is it? And that study did not even include CO_2 released by desertification!

Now let's consider how much more CO_2 is released by desertification each year as once lush grasslands and forests are turned into bare deserts. The fastest expanding desert in the world is the Gobi Desert, which is growing larger by over 360,000 hectares *every single year* as the dry grasslands surrounding the desert slowly retreat. Globally, over 1.5 *billion* hectares of formerly productive land (15 million km^2) have already been lost to desertification[8] — that's almost 1.5x more land than is currently

---

[7] Houghton, R.A. (2010). How well do we know the flux of CO_2 from land-use change? *Tellus (International Meteorological Institute in Stockholm).* 62(5). https://doi.org/10.1111/j.1600-0889.2010.00473.x

[8] Howarth, T. (2023, Dec 9). Desertification: the world is losing healthy land at an astonishing rate. *Geographical.* https://geographical.co.uk/news/desertification-the-world-is-losing-healthy-land-at-an-astonishing-rate

used to grow crops on our planet — and sadly desertification is continuing to claim more than 12 million additional hectares per year.[9]

This land degradation has been matched by a corresponding increase in dust — global dust emissions have increased by 25% over the past century simply because of desertification.[10] Desertification is a particularly big net emitter of CO_2 because dry windblown soils are particularly easily oxidized once they lose their vegetative cover.[11] As noted by Project Wadi Attir, a think tank founded by Israel's Bedouin community in the Negev desert, which is dedicated to research into how to create sustainable desert communities, *"land degradation and desertification [over the last 7,800 years] has resulted in carbon dioxide emissions estimated at 450 – 500 gigatons of carbon (Ruddiman, 2003, Fig. 5; Lal, 2004),* **which corresponds to more than the total amount of CO_2 emitted from fossil fuel combustion so far** *[my emphasis]."*[12] Take a moment to let that sink in.

One research study calculated that desertification alone is responsible for at least 4% of global CO_2 emissions.[13] Another study places that number at a much higher 20%[14] but included deforestation and dryland degradation in their calculations so for the purposes of this thought exercise, I'll stick with the lower number of 4%. As you'll soon see, even these conservative estimates quickly add up to obliterate the consensus narrative about what's causing the sharp rise in atmospheric CO_2 levels

---

[9] United Nations. (2019, Jul 27). *Every Year, 12 Million Hectares of Productive Land Lost, Secretary-General Tells Desertification Forum, Calls for Scaled-up Restoration Efforts, Smart Policies.* [Press release]. https://press.un.org/en/2019/sgsm19680.doc.htm

[10] Stanelle, T., et al. (2014). Anthropogenically induced changes in twentieth century mineral dust burden and the associated impact on radiative forcing. *JGR Atmospheres.* 119(23). https://doi.org/10.1002/2014JD022062

[11] McSweeney, R. (2019, Aug 6). Explainer: 'Desertification' and the role of climate change. *Carbon Brief.* https://www.carbonbrief.org/explainer-desertification-and-the-role-of-climate-change/

[12] "Desertification and Climate Change." *Project Wadi Attir.* Accessed 17 May 2024. https://sustainabilitylabs.org/ecosystem-restoration/desertification/

[13] "Desertification – 7. Is there a link between desertification, global climate change, and biodiversity loss? *Green Facts.* Accessed 27 May 2024. https://www.greenfacts.org/en/desertification/l-3/7-climate-change-biodiversity-loss.htm

[14] McSweeney, R. (2019, Aug 6).

while giving credence to the controversial findings by Skrable et al (2022) and their peers.

Another non-fossil-fuel CO_2 source of emissions is dredge-net fishing, which releases at least one gigaton of carbon dioxide into the atmosphere each year as the nets get dragged through the carbon-rich muds settling on the seafloor — this is equivalent to how much CO_2 is released each year by the global aviation industry (2% of global fossil fuel emissions!).[15]

If we add up the CO_2 emissions from these sources (40% from agricultural land use and land use changes, another 4% from desertification, and 2% from dredge-net fishing), we're already at 46% of global CO_2 emissions coming from non-fossil-fuel sources— the consensus narrative is crumbling fast.

However, as I've shown over course of this book, atmospheric CO_2 levels are not static but are instead a balance of carbon that is removed from the atmosphere by plants, carbon that is added to the atmosphere by oxygen breathers, carbon that is permanently lost from the system as it gets trapped in carbon sinks, and fresh carbon that is added to the atmosphere from natural sources, like the CO_2 released by volcanic eruptions, seafloor spreading ridges, rift zones, degassing peat bogs, thawing permafrost, and even by the rocks themselves as they react with oxygen in the atmosphere. The popular idea of a circular carbon cycle creates the oversimplified impression that there is a permanent supply of carbon continually being recycled in and out of the atmospheric by plants and oxygen breathers. Thus, whenever atmospheric CO_2 levels rise, we expect any increase to be the consequence of some entirely new source of CO_2 (like fossil fuel emissions). However, because volcanoes and other natural sources are continually adding carbon to the atmosphere, even if the rate of volcanic emissions and rock weathering remains unchanged and even if there are no other "fresh" sources of CO_2, *you can still increase the amount of carbon accumulating in the atmosphere simply by decreasing how much carbon is scrubbed from the atmosphere by photosynthesis before it is stored in permanent carbon sinks.* Biomass matters because it affects how much plant debris is available to get trapped in long-term carbon sinks.

---

[15] Agence France-Press. (2021, Mar 18). Fishing boats that dredge their nets release carbon equivalent to aviation. *Firstpost*.
https://www.firstpost.com/tech/science/fishing-boats-that-dredge-their-nets-release-carbon-equivalent-to-aviation-9433101.html

And so, by transforming soils from net carbon sinks into net carbon emitters, we also have to account for how much of the CO_2 released by volcanoes, rock weathering, and other natural sources of fresh CO_2 is now building up in the atmosphere instead of accumulating in the soil as humus.

We've already seen this concept in action during previous mass extinctions, like the Great Dying, when the collapse of the global biomass led atmospheric CO_2 levels to surge because there simply wasn't enough plant debris available to get trapped in long-term carbon sinks in order to offset the amount of fresh CO_2 released by volcanoes and rock weathering. Although, contrary to Greta Thunberg's tweets, we are not living through a mass extinction today, our activities have nevertheless triggered a massive reduction in the total biomass of the planet thanks to deforestation, desertification, and soil degradation. And so, our global biomass is no longer producing as much plant debris as it once did and thus it no longer sustains the equilibrium that once existed between carbon consumed by photosynthesis and carbon sinks versus the amount of CO_2 being added to the atmosphere each year by volcanoes, seafloor spreading ridges, and other natural sources of fresh CO_2.

Perhaps a good analogy to illustrate this concept is the image of a bathtub with a running faucet and an open drain. As long as the water from the faucet matches the amount lost down the drain, the water level remains roughly balanced. There are two ways to make the water in the bathtub begin to rise. One is to increase the amount of water being added to the tub (the image of smokestacks belching out CO_2 comes to mind). But the other is to simply reduce the size of the drain (i.e. reduce the global biomass or turn soils from carbon sinks into carbon emitters), which causes water levels in the tub to begin to rise *even without increasing the amount of water coming from the faucet.*

The normal state of affairs across geological history is that CO_2 levels have ever so slowly declined across time because the drain of carbon sinks is slightly bigger than the faucet that continually adds fresh volcanic gasses into our atmosphere. However, as tillage, continuous grazing, deforestation, desertification, and other human activities converted soils from net carbon sinks into net carbon emitters, and as deforestation and desertification reduces the total biomass on the planet, we have effectively reduced the size of the drain in the bathtub, causing water levels in the bathtub to begin to rise.

The consensus climate narrative explicitly claims that the volcanic gases continually being released by 3.4 million submarine volcanoes and by over 80,000 km of mid-oceanic rift zones are not relevant to calculating rising CO_2 levels because CO_2 levels have fallen steadily over the last 66 million years even as these eruptions were ongoing. In other words, they claim that because these volcanic emissions were already part of the carbon equilibrium before the industrial revolution, these volcanic emissions are already accounted for in the balance between CO_2-in vs CO_2-out. But as I've just shown with the bathtub analogy, their logic is fundamentally flawed because our human activities have transformed a large portion of the planet's soils from carbon sinks to carbon emitters even as we have severely reduced the total biomass on the planet, and so we have severely reduced our planet's ability to offset these volcanic emissions.

But how do we put a number on this change? We'll deal with affects of deforestation and desertification on the global biomass in a moment — for now let's focus on the consequences of turning soils from net carbon sinks into net carbon emitters. Let's assume that no-till agriculture represents the natural rate at which carbon was sequestered in the soil before we turned those prime soils into farmland for tillage agriculture. Based on the aforementioned study by the University of Nebraska-Lincoln, which showed that no-till farming stores an average of 0.77 tons per acre of carbon each year, let's assume in addition to the amount of degassing that is caused by tillage agriculture, that the introduction of tillage agriculture also took an additional 0.77 tons per acre of carbon storage offline (equivalent to 2.83 tons of CO_2 — remember, 1 ton of carbon is equivalent to 3.67 tons of carbon dioxide).

Since we know that tillage agriculture is directly responsible for 20% of global CO_2 emissions, we can use the ratio between Dr. John Baker's numbers (tillage releases 3 tons of CO_2 per acre per year) and the University of Nebraska's numbers (no-till sequesters the equivalent of 2.83 tons of CO_2) to guestimate that in addition to the CO_2 directly released by tillage, *tillage is also responsible for an additional 18% of the rise in atmospheric CO_2 levels simply because these farm soils are no longer scrubbing of CO_2 from the atmosphere to keep up with volcanic emissions.* By leaving out the part of the story that shows how much of the planet's carbon storage capacity has been taken offline by soil degradation, activist-scientists have been able to scapegoat fossil fuels as the primary source of CO_2 increases in our atmosphere in order to suit their biases and agenda.

The same hidden loss of carbon storage is also at work in pastures, forests, and deserts, though it's a little more difficult to put a number on how much the drain on atmospheric CO_2 has been reduced in these other ecosystems as a consequence of our poor land management. Cropland only represents around 10% of our planet's habitable land (the rest being forests (38%), pastures and grasslands (35%), scrubland (16%), and urban sprawls (1.4%)), but since I couldn't find any semi-reliable numbers to work with to guestimate how much carbon sequestration has been lost in these other ecosystems as they too have been turned from net carbon sinks into net carbon emitters or as their sequestration rates have declined under suboptimum land management, let's be extremely conservative and pretend that they haven't changed at all. Let's only add the additional 18% from cropland to our running total. This obviously vastly underestimates the total reduction in the bathtub drain, but it still gives us something to work with for our thought experiment. By adding that 18% to our previous running total of 46%, we're now at 64% of the rise in global atmospheric CO_2 levels being caused by non-fossil-fuel sources. The "narrative" keeps unravelling. And I'm not done yet.

Now, let's tackle another "inconvenient truth" — deforestation fires — these are not part of the regular wildfire cycle but are fires used to clear land in order to convert forests into farmland and to turn wood into charcoal as a cheap source of energy. These deforestation fires are mostly happening in South America, Asia, and Africa — poor countries with rapidly expanding populations where many people are trapped in a self-reinforcing Malthusian cycle of poverty, lack of economic opportunities, and limited access to cheap fossil fuel energy, which condemns billions of people to rely on subsistence agriculture even as their countries run out of land to keep up with their growing populations. In theory, we already accounted for deforestation fires — they fall into the category of land use and land use changes. However, a study in 2022 found that emissions from these deforestation fires are underestimated by a factor of 200-300% and that deforestation fires in Brazil and Indonesia alone accounted for 7%

and 3% of global greenhouse gas emissions in 2019 and 2020.[16] That's 10% of global emissions caused by deforestation fires in only 2 countries! And it just keeps growing year by year as their populations keep growing. Adding these numbers from only these two countries (while ignoring how much deforestation fires are underestimated everywhere else) brings our running total up to 74% of the rise in global atmospheric CO_2 levels being caused by non-fossil-fuel sources.

Now let's look at Africa, nicknamed the "fire continent" because of how much of it burns each year — more than 4.9 million square kilometers (490 million hectares or 1.2 billion acres) burned in Africa in 2016[17] — this is 25x more acreage than was burned in the whole of Canada and the USA combined during the 2023 wildfire season (Canada's worst fire season in decades at 45 million acres, plus another 2.6 million acres in the USA). For context, the land mass of Canada plus the USA is 2/3rds the size of Africa. Our wildfires are a drop in the bucket compared to what is happening in Africa every year.

Africa's population surged from 140 million people in 1900 to 1.5 billion in 2024, of which only 300 million are middle class — the other 1.2 billion are desperately poor and rely on woodburning and charcoal stoves for cooking and are relentlessly burning down more forest each year to make room for more subsistence farms keep up with the rapidly growing population. In other words, the annual amount of forest that Africa's growing population burns each year as fuel for cooking and to clear land for subsistence farming has increased by a factor of 8.5x since 1900 — all that carbon is ending up back in the atmosphere instead of decomposing in the soil as it once did when these areas were still wilderness.

A study in 2021 found that the impact from these countless small African fires on global emissions is underestimated by at least 31% to

---

[16] Datta, A. and Krishnamoorti, R. (2022). Understanding the Greenhouse Gas Impact of Deforestation Fires in Indonesia and Brazil in 2019 and 2020. *Frontiers in Climate*. 4. https://doi.org/10.3389/fclim.2022.799632

[17] Petesch, C. (2019, Aug 28). Africa is the 'fire continent' but blazes differ from Amazon. *PHYS.org*. https://phys.org/news/2019-08-africa-continent-blazes-differ-amazon.html

101% and that Africa's fires account for a full 14% of global emissions.[18] Before 1900, back when Africa only had 140 million people, 1/8th of those fires would still have happened, so it's safe to assume that 7/8th of those fire-emissions (12% of global emissions) *are directly linked to Africa's rapidly exploding population during our modern era*. And this doesn't even factor in the fact that as these forests are cut down, they stop acting as net carbon sinks (thus further reducing the size of the drain in our bathtub analogy).

When we add Africa's small fires to our running total, *we're now at 86% of the rise in global atmospheric CO_2 levels being caused by non-fossil-fuel sources* — we're right in the ballpark of the levels calculated by Skrable et al. in their controversial 2022 study, which only attributed 12% of the rise in atmospheric CO_2 to fossil fuels, with the other 88% coming from non-fossil-fuel sources. And, we haven't even calculated the impact of deforestation fires in Asia where 60% of the world's population lives. Hmmm. Even without knowing anything about the accuracy of the mathematical calculations used by Skrable et al (2022), here we have an entirely independent way of guestimating non-fossil-fuel emissions using all peer-reviewed studies, and we still arrive in the same ballpark as Skrable et al (2022).

All of this raises the question: If global soils were restored to good health (especially in the developing world where soil degradation is happening the fastest) and if the developing world was not being held back by green colonialism (so that population numbers would stabilize and reverse the pressure to gobble up forests and wildlands to make room for more subsistence farms), would thriving forests, grasslands, deserts, pastures, and farmland be capable of absorbing all of our fossil fuel emissions? In other words, are soil degradation and the reduction of biomass the primary culprits behind the rise in atmospheric CO_2 levels?

~

Almost every significant rise in CO_2 levels in our planet's history happened during times when photosynthesis was dialed back to the point where it was unable to keep up with volcanic CO_2 emissions, and then

---

[18] Ramo, R. et al. (2021). African burned area and fire carbon emissions are strongly impacted by small fires undetected by coarse resolution satellite data. *PNAS*. 118(9). https://doi.org/10.1073/pnas.2011160118

began to fall again once photosynthesis recovered to previous levels. Why should this time be different?

During the Ordovician Glaciation, CO_2 levels rose as long as the entire planet was encased in a giant snowball, which all but extinguished photosynthesis for millions of years, but began falling again as soon as the ice melted. The sharp rise in CO_2 during the Great Dying (Permian extinction) happened at a time when photosynthesis again almost came to a halt; once again CO_2 levels began to fall as soon as plant growth recovered. Likewise, the sharp drop in biomass after the Cretaceous extinction played a big role in why CO_2 was able to rise, albeit briefly, until vigorous plant growth resumed. Atmospheric CO_2 levels are a never-ending balancing act between how much new CO_2 is added to the atmosphere from various sources versus how much is pulled back out by photosynthesis to be stored in long-term carbon sinks. You simply cannot understand the atmospheric carbon dioxide balance without taking account of changes in global biomass.

Global biomass has been on a steady decline since the beginning of the Industrial Era. And thanks to yet another heretical geologist, we can put some numbers to that biomass decline and test the controversial results from Skrable et al. (2022) from a different angle. The results will blow your mind.

The Link Between Deforestation and Aridity

A geologist by the name of Timothy Casey decided to approach the question of rising atmospheric CO_2 levels from a completely different angle by looking at how much the planet's declining photosynthesising biomass (caused by indiscriminate deforestation) has reduced the planet's ability to self-regulate the amount of carbon dioxide in the atmosphere.[19] Before I show you his conclusions, let me point out a few things to explain the underlying assumptions of his calculations because, once again, those assumptions reveal some surprising insights about the forces that shape our climate.

Deforestation has some obvious effects — by stripping away vegetation you reduce our planet's capacity to remove carbon from the atmosphere via photosynthesis and store that carbon in soil. However,

---

[19] Casey, T. (n.d.). Deforestation & Carbon Emission. *Geologist-1011.com* Accessed 8 Jun 2024. http://deforestation.geologist-1011.net/

deforestation also has another less obvious and more indirect impact on the amount of biomass that our planet can sustain because deforestation makes the air drier, reduces rainfall, and increases the soil's vulnerability to drought, all of which reduces the total biomass living on our planet. This link between deforestation and aridity is one of the most important drivers of desertification.

Forests pump large volumes of moisture into the air via a process called evapotranspiration as moisture escapes the pores of leaves as those pores open up to absorb CO_2. Research has demonstrated that at least 40% of rainfall on land comes from moisture escaping into the atmosphere from trees (in the Amazon it's even higher with evapotranspiration accounting for more than 70% of rainfall).[20] Furthermore, researchers found that *a mere 10% increase in humidity levels increases rainfall by two to three times*.[21] The same relationship holds true in the opposite direction — small decreases in humidity produce large decreases in rainfall. In other words, the moisture that's pumped into the atmosphere by forests is instrumental for creating abundant rains — the fastest way to tip a local ecosystem into drought is to cut down all the trees and strip the soil of its vegetation. This crucial point is going to come back and haunt us in upcoming chapters.

Trees also help shield the soil from direct sun, even as leaf litter and vegetation build up on the forest floor to prevent the soil from drying out — thus, cutting down the forests rapidly dries out the exposed soils that are left behind.

Then there is the effect of vegetation on air temperature. Photosynthesis absorbs large amounts of solar energy as plants consume sunlight in order to turn CO_2 into food, which helps cool the air above a forest and, by consequence, helps trigger the condensation required for clouds and raindrops to form when moist vapour passes overhead. Deforestation eliminates that cooling effect as photosynthesis declines.

Furthermore, the bare ground left behind after deforestation reflects solar energy, which further raises air temperatures directly above the soil and reduces the ability for water vapor to condense as clouds and raindrops. *Researchers have found that as much as 18% of warming on*

---

[20] Ellison, D. et al. (2017). Trees, forests and water: Cool insights for a hot world. *Global Environmental Change*. 43. https://doi.org/10.1016/j.gloenvcha.2017.01.002

[21] Ellison, D. et al. (2017).

land comes just from the effect of deforestation.[22] If that doesn't throw a massive wrench into all our assumptions about recent warming, I don't know what does...

In sum, deforestation creates a noticeably drier climate. And a drier climate can't support as much plant growth as plants are starved for moisture, which puts further downward pressure on biomass. That's a true "feedback effect". Less biomass leads to less carbon dioxide being scrubbed out of the atmosphere and less carbon being stored in the soil.

The effect of the Amazon creating its own weather and the drop in rainfall that happens when the rainforest is cut down are well known. But the consequences of deforestation on humidity (and on local weather patterns) are illustrated by another equally poignant example that has been systematically abused to propagate the global warming narrative. Climate change activists like Al Gore tell us that Mt. Kilimanjaro's shrinking glaciers are caused by rising temperatures and reduced rainfall caused by "climate change". Once again, a half-truth is a whole lie. A series of studies (Mason in 2003[23], Kaser in 2004,[24] and then Pepin in 2010[25]) obliterated these activist claims by demonstrating that Kilimanjaro's shrinking glaciers are caused by deforestation on the flanks and perimeter of the mountain, which has caused the air flowing up and over the mountain to become significantly drier as the forests around the mountain were chopped down. Less water vapour in the air as it passes over the mountain causes less snow to fall at the top and increases the intensity of the sunshine radiating down on the glaciers whenever the sun is shining.

In short, Kilimanjaro's shrinking glaciers are caused by a change in *local* weather patterns triggered by deforestation, not by global climate shifts triggered by CO_2 emissions. That didn't stop Al Gore from shamelessly using Kilimanjaro as a fear-inducing example of global warming in his 2006 documentary, *An Inconvenient Truth*. His shameless

---

[22] Ellison, D. et al. (2017).

[23] Mason, B. (2003). African ice under wraps. *Nature*. https://doi.org/10.1038/news031117-8

[24] Kaser, G. et al. (2004). Modern glacier retreat on Kilimanjaro as evidence of climate change: observations and facts. *International Journal of Climatology*. 24(3). https://doi.org/10.1002/joc.1008

[25] Rastetter, C. (2010, Oct 5). Deforestation May Lead to Kilimanjaro Glacial Melt, Study. *Circle of blue*. https://www.circleofblue.org/2010/world/deforestation-may-lead-to-kilimanjaro-glacial-melt-study/

deception is reminiscent of another scandal that broke in 2019 about claims made by Sir David Attenborough in the Netflix documentary *"Our Planet."* The documentary showed a scene in which a large number of walruses fell to their deaths from a high cliff — Attenborough's narration blamed their deaths on overcrowding due to a lack of sea ice caused by climate change. But then video footage from a Russian photographer emerged showing a group of polar bears driving the walruses off the same cliff on the same day. The scene and Attenborough's claims were created by omitting polar bears from the story.[26] As Mark Twain once said, *"Never let the truth get in the way of a good story."*

As geologist Timothy Casey points out, while deforestation directly adds 2.3 gigatons of carbon to the atmosphere each year (from carbon released by burning), the biggest impact of deforestation is the loss of photosynthesizing biomass on our planet, which has shrunk from 696 gigatons of biomass in 1850 down to only 540 tons of biomass in 2000, *thus depriving the planet of over 156 gigatons of photosynthesising biomass that existed in 1850 but no longer exists today.* As that deficit grows larger each year, the amount of carbon that is scrubbed out of the atmosphere by plants each year shrinks accordingly. By 2000, the annual deficit in the amount of carbon scrubbed out of the atmosphere by photosynthesis was 34 gigatons less than in 1850! And that annual deficit is growing larger by an additional 0.5 gigatons each year.

In other words, *deforestation has reduced photosynthesis on our planet by 22% since 1850.* Our planet has lost 1/5th of its total photosynthesis capacity since 1850 just because of deforestation — in other words, the drain on the atmospheric bathtub has shrunk by 1/5th! Ouch.

According to a biomass census cited by Greenpeace, since the dawn of agriculture our human activities have reduced the total global biomass by a colossal 50%[27] (from 1,000 gigatons to 545 gigatons), which make Timothy Casey's numbers look conservative. The Greenpeace numbers also serve as a reminder that the loss of biomass began long before the Industrial Revolution when we first began clearing the forests to build

---

[26] "New footage reveals Netflix fakes walrus climate deaths." (2020, Nov 19). *Polar Bear Science.* https://polarbearscience.com/2020/11/19/new-footage-reveals-netflix-faked-walrus-climate-deaths/

[27] Weyler, R. (2018, Jul 18). How much of Earth's biomass is affected by humans? *Greenpeace.* https://www.greenpeace.org/international/story/17788/how-much-of-earths-biomass-is-affected-by-humans/

civilizations and make room for the plow, and that this biomass loss only accelerated with the population explosion that was triggered by the Industrial Revolution.

In 2007, the IPCC stated that fossil fuel burning contributed 7.8 gigatons of carbon to the atmosphere each year. But how do those fossil fuel numbers stack up to the amount of CO_2 that is no longer being removed from the atmospheric bathtub because of our colossal reduction in photosynthesis caused by deforestation? Based on Timothy Casey's biomass calculations, we know that by 2007 photosynthesis consumed *37.5 gigatons less carbon each year than it would have consumed back in 1850*. That annual biomass loss, in combination with the additional 2.3 gigatons of CO_2 released into the atmosphere by deforestation fires, is *5x larger than the total amount of CO_2 released by fossil fuel burning each year*.

You can read about Timothy Casey's calculations and find all the links to the peer-reviewed studies that he used for his calculations at *http://deforestation.geologist-1011.net/*. He concludes that: "**Carbon emissions due to fossil fuel combustion represent less than 20% of the total human impact on atmospheric carbon levels.** *Deforestation not only contributes a relatively minor one-off carbon emission of some 2.3 gigatons of carbon to the atmosphere, but an ongoing loss of photosynthetic carbon sequestration to around 38 gigatons per annum that is growing at the rate of 500 megatons every year. It is clear from the fact that this amount dwarfs the present 7.8 gigaton fossil fuel combustion contribution (IPCC, 2007), that the cessation of fossil fuel combustion will not halt the rise of atmospheric carbon dioxide because the loss of photosynthesising biota and the corresponding fall in photosynthesis is so much greater. The current focus on fossil fuel combustion to the exclusion of ongoing impacts of deforestation only serves to blind the public to the consequences of excessive land clearance and the fact that deforestation and consequent soil deflation are the simplest explanation for the unprecedented rise in global aridity during a warming phase.*"

In other words, even if we stopped all fossil fuel use today, CO_2 levels would continue to rise nearly as fast as before simply because of the accumulated reduction in photosynthesis caused by deforestation. In fact, if we phased out all fossil fuel (remember UN's Secretary-General António Guterres' words from 2023: "*Not reduce. Not abate. Phaseout.*"), we might even cause CO_2 levels to *increase* because without access to cheap energy, a lot more people are going to be forced into subsistence lifestyles all over the world, which would cause deforestation and roaming village goats to become even more widespread.

Timothy Casey's last point about *"the unprecedented rise in global aridity during a warming phase"* is also an extremely important observation. Even as the planet is getting slightly warmer during our current epoch, many regions of our planet are also getting noticeably drier. This is the opposite of what you would expect to happen as the climate gets warmer. Once again, we stumble across another paradox. A warmer climate increases evaporation rates, both on land and over the oceans. However, 71% of the planet's surface is covered in oceans so as a climate warms, the increased evaporation on land is more than offset by increased evaporation over the oceans, which raises global humidity levels and increases rainfall over land as clouds bring moisture from the oceans up onto land. Individual rain shadows on the leeward side of mountains may get drier, but overall, the global climate gets wetter whenever the world gets warmer. If the small increases in local humidity pumped into the atmosphere by local forests can trigger massive increases in rainfall (as I mentioned a few paragraphs ago, a mere 10% increase in local humidity levels causes local rainfall to increase by two to three times), imagine what increased evaporation over the global oceans can do to increase *global* humidity levels and increase global rainfall amounts!

The geologic record is absolutely clear on this point — the global climate gets more humid every time global temperatures rise, while the global climate gets much drier during cold periods. The wettest periods in our planet's history were always when it was hottest while the driest periods invariably happened at the coldest points of the most brutal ice ages. This relationship holds true throughout geologic history: hot and humid, cold and dry.

There are individual exceptions that prove the rule, like during the warm Cretaceous period at the end of the Age of the Dinosaurs when the newly forming Rocky Mountains and other new mountain chains began scrubbing moisture out of the clouds, causing the climate east of these mountain chains to get increasingly dry. But even then, the Cretaceous did not get drier because of hotter temperatures, but because the rising mountain ranges changed global weather patterns by forcing moisture-laden clouds to dump their moisture on the windward side of these growing mountain ranges before traveling further across the continents.

The ice core data from Vostok, Antarctica, shown back in figure 74, clearly demonstrates this relationship between temperature and aridity. The uppermost chart in figure 74 shows temperature while the bottom chart shows windblown atmospheric dust trapped inside the ice cores. As predictable as clockwork, you can see that as temperatures decline, dust levels rise. And as temperatures rise, dust levels decline. Hot and humid, cold and dry.

And that leads us to another paradox. As global humidity increases as a consequence of a warming climate, that extra humidity helps trap heat from radiating out into space at night while reflecting incoming radiation from the Sun during the daytime. Increasing global humidity causes nights and winters to become milder even as extreme heat waves in the summer become less frequent. And that's precisely what we see happening today, as demonstrated by the charts I shared back in chapter 8 and 9 from the Fourth National Climate Assessment, which showed that heat waves are getting less extreme (figure 44), summer highs are essentially unchanged and winter lows have gotten warmer over the past century (figure 61), even as overall precipitation has increased slightly (figure 66). What is happening is precisely what you would expect from a slightly warming climate that increases moisture evaporation over the oceans: warmer winters, less frequent cold spells, warmer nights, less frequent and less extreme summer heat waves, and a slight increase in overall precipitation.

And yet, activist-scientists, media, and corrupt organizations like the United Nations continually tell us that drought and desertification today are caused by rising temperatures, especially in tropical regions. This corrupted logic is the justification for why the United Nations predicts that there will be around 1.2 *billion* climate refugees by 2050.[28] Yet the evidence provided by the geologic record illustrates that it is not possible for a warming climate to tip the world into drought and desertification — on the contrary, the opposite should be the case as evaporation over the oceans increases global humidity levels and rainfall on land.

Indeed, it is worth bearing in mind that 8,000 years ago, when temperatures were around 4 °C warmer than they are today, the Fertile Crescent in the Middle East (where the first large civilizations emerged)

---

[28] Elton, C. (2023, Jun 20). World Refugee Day: How will climate change force us to rethink attitudes to mass migration. *Euronews*.
https://www.euronews.com/green/2023/06/20/climate-change-will-displace-millions-of-people

was still covered by vast grasslands and lush forests, like the prized cedar forests of Lebanon. Ancient Greek historian Herodotus described the climate of Assyria, in modern-day Iraq, as too wet to grow vineyards! Today, the humus from those civilizations is at the bottom of the Persian Gulf and the region has become a desert. The perennial drought of the Middle East wasn't caused by a warming climate, it was caused by deforestation and soil degradation. Full stop.

Many regions are undeniably getting drier today as desertification continues to get worse, but these drying trends have local rather than global cases. Once again, the mainstream "consensus" narrative is using half-truths to tell a whole lie — in reality local climates are getting drier *despite recent warming* — the fault lies with local deforestation, with vegetation stripped from soils, with forests cut down or burned to make room for expanding subsistence farms and plantation-scale commodity agriculture, with deforestation fires for cooking and making charcoal, with roaming goats and subsistence farmers trying to scrape a living from tiny and steep plots of land, with overpopulated developing countries that have exceeded their food production capacities, with unsustainable tillage practices and continuous grazing, and with rampant soil erosion and widespread soil degradation, all of which are making local climates drier despite increasing evaporation over the oceans.

And yet, in so many ways, a large portion of the blame for this deforestation and for the slow-rolling aridification that follows in its wake lies squarely on the shoulders of industrialized nations and academic institutions because of their arrogant green colonialism, because of climate policies that hinder development and restrict access to cheap energy in developing nations, because of their endless wars to impose their vision of neo-liberalism by force, because of the foreign aid that keeps strongmen in power and populations in poverty, and because our politicized self-serving Western scientific institutions have completely blinded themselves and everyone who listens to them to the real causes underlying these environmental changes even as they weaponize fear as a strategy to raid the public treasury to keep their research grants flowing. Once again, like so many times before, our academic intelligentsia are leading the charge into madness.

The United Nations may yet be right that there will be over 1.2 *billion* "climate" refugees fleeing developing countries by 2050, but if that happens it will be the result of local desertification (aridity) caused by rampant local deforestation and rampant local soil degradation, not because of "global warming". Those refugees most certainly will not be fleeing local climates destabilized by fossil fuel emissions from rich industrialized countries. This slow-rolling ecosystem collapse only has the appearance of being global because almost everyone is using the same ecosystem-destroying land management practices in their local backyards. Curbing CO_2 emissions won't stop a single tree from being cut down, it won't stop a single molecule of soil from eroding away, and it most certainly won't do anything to stop the relentless aridification that is spreading around the world in the wake of rampant deforestation.

Rising CO_2 levels are merely signalling just how many local jurisdictions are suffering from the same soil-wrecking and biomass-destroying habits. Even if we magically stabilized CO_2 levels with elaborate carbon capture machines capable of pulling billions of tons of carbon out of the air, even if Bill Gates' harebrained schemes to dim the Sun could magically stabilize temperatures exactly where they are at today, even if George Monbiot got his wish to remove all livestock from the land, and even if cloudseeding programs could magically trigger timely rains, absolutely nothing would change. None of these "solutions" would fix the ongoing deforestation and soil degradation unfolding all over the world — desertification and aridification would continue at the exact same pace as our destructive land management processes continue to hollow out entire ecosystems.

Degraded soils cannot absorb or store water and are extremely vulnerable to erosion, drought, and floods. Deforestation and soil erosion dry out the local climate. Falling soil humus levels eat away at agricultural yields — even fertilizer amendments are merely a temporary band-aid because they cannot slow erosion and they cannot prevent a region from becoming more vulnerable to drought as a consequence of declining humus levels. This is the ongoing story of the century, happening right beneath our noses even as the experts look everywhere except at what matters. Rising CO_2 levels are merely evidence of this complex story. The fault for this slow-rolling ecosystem collapse can be laid squarely at the feet of climate science for leading a credulous world around by the nose hairs, and at the feet of the political architecture that incentivized our scientific institutions to abandon the principles of truth-seeking in their pursuit of government contracts.

And yet we, the voting public, paved the way for this madness, over generations, through our appetite for a Ministry for Everything — through our appetite for a benevolent shepherd to run our lives with carrots and sticks — because we fell for the illusion that the progressive administrative state would remain our servant instead of becoming our master, and because we naively believed that the ever-growing government trough wouldn't deteriorate into a gigantic self-serving racket. Now all that stands between us and an end to the madness is a $5 trillion dollar "green" economy that has been built upon that madness, which is now financially incentivized to keep that fictitious narrative intact. If you back a rat into a corner, it will bite you — those whose reputations, careers, and livelihoods have been built upon this climate narrative are not going to go down without a tremendous fight.

Even Dr. Frankenstein eventually realized he had created a monster he could not control. But how do you dismantle a hideous creation that develops a mind of its own and an instinct for survival?

Chapter 24

The Slow Slide into the Icehouse

"Geological processes are very slow, but they are very persistent and have all the time in the world."

— Anonymous.

At this point the broad outlines of the "story of the century" have become visible. However, what we have touched upon so far is only the beginning — there are still darker insights waiting for us up ahead as we come face to face with what our system has become. But for now, it is time to finish our stroll through the final 66 million years of our planet's geological history to prepare us for turbulent Age of Man that lies ahead.

By now you are familiar with all the key biological, chemical, physical, and geological forces that shape our planet — photosynthesis, methanogens, oxygen breathers, fungi, earthworms, herbivores, ruminants, solar cycles that bombard our planet with solar radiation and magnetically shield our planet from cosmic radiation, ocean currents that transport heat from the tropics to the poles, wobbles in our planet's orbit around the sun (Milankovitch cycles), our solar system's circular path through the spiral arms of our galaxy where higher densities of cosmic radiation lurk, the many and complex processes that drive cloud formation, the back-and-forth fluctuation between Icehouse and Hothouse climates, the central role of water vapour in controlling temperature on the planet's surface, the drift of our continents on tectonic plates, the mountain chains that drain moisture out of the air as clouds pass overhead, the importance of sod and plant debris to shield soil from sun, wind, and rain, the origins and reactive nature of oxygen in our atmosphere, the never-ending processes that continually scrub carbon dioxide from our atmosphere, the link between deforestation and aridity, the conveyor belt that links soil to carbon sinks at the bottom of bogs, lakes, and oceans, the degassing of soils through tillage and exposure to the atmosphere, the processes that create and destroy humus, and the effect of reducing biomass on the amount of CO_2 accumulating in our atmosphere. Each of these complex processes is part of a large, interconnected, and dynamic puzzle that continually shapes and reshapes our planet. As we watch that puzzle play out over the course of our own era of geologic history, brace for a few more surprises...

We'll pick up our story at the time when dinosaurs went extinct at the end of the Cretaceous period 66 million years ago, clearing the way for mammals and flowering plants to take over our planet. Figure 104 below zooms in on the temperature record over these last 66 million years so you can follow along as the climate slowly shifted from the Hothouse enjoyed by the dinosaurs to the carbon-depleted Icehouse in which we live today.[1] As always, there are many different things that have to line up just right in order to tip the Earth into a full-blown Icehouse climate.

---

[1] "Cenozoic deep sea temperature." *Wikimedia Commons*. 28 Oct 2013.
   https://commons.wikimedia.org/wiki/File:Cenozoic_deep_sea_temperature.jpg

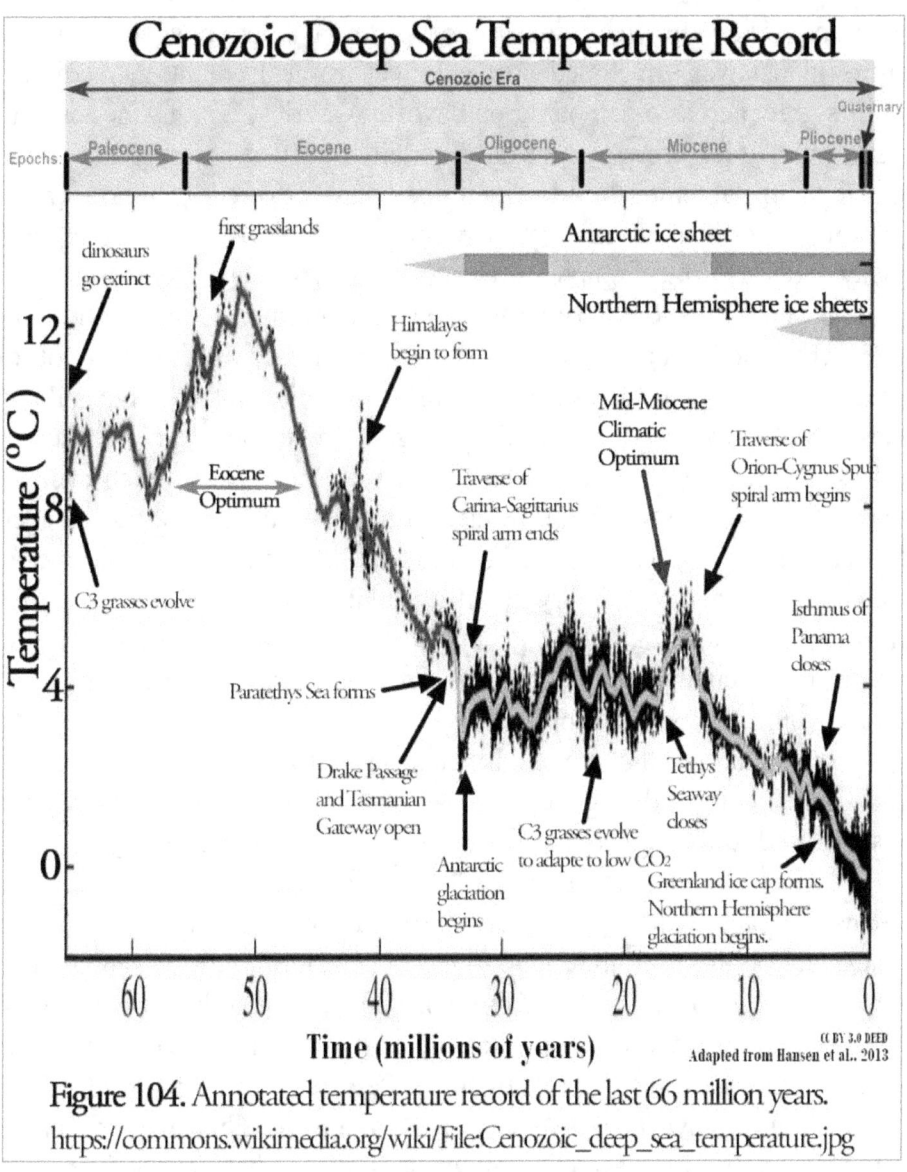

Figure 104. Annotated temperature record of the last 66 million years.
https://commons.wikimedia.org/wiki/File:Cenozoic_deep_sea_temperature.jpg

Beginning around 76 million years ago (approximately 10 million years before the asteroid that wiped out the dinosaurs), our solar system began to traverse the Carina-Sagittarius spiral arm of our galaxy, matching the beginning of the planet-wide cooling that began in the late Cretaceous period.[2] However, that gradual cooling and drying trend was interrupted around 56 million years ago by an abrupt and intense multi-million-year

---

[2] Sankaran, A. V. (2008). Galactic triggering of geologic events in earth's history. *Current Science*, 95(6), 714–716. http://www.jstor.org/stable/24102576

warming period known as the Eocene Optimum, a period when nearly the entire planet was enveloped by warm wet tropical conditions. Tropical forests grew across much of Europe, North America, South America, and Asia, palm trees grew in Alaska, crocodiles thrived north of the Arctic Circle, and Greenland and Antarctica were covered by sub-tropical rainforests. Global surface temperatures were 12–14 °C warmer than today and sea surface temperatures were up to 16 °C warmer than today.[3]

This "climate event", called the Eocene Optimum, is one of the most intensely studied geologic periods in our history. There is a lot of speculation about what caused this brief warming interval but still no definitive answers — explanations range from wobbles in our planet's orbit on time scales that are even longer than the Milankovitch cycles I discussed previously, to the usual outlandish claims of an alleged "greenhouse effect" caused by a burst of CO_2 released by volcanic activity, to a heavily criticized theory about a possible catastrophic release of a massive greenhouse-inducing pocket of methane from the bottom of the ocean, to speculation about a large comet impact, to changes in ocean circulation patterns, and more.

The explanation that makes the most sense to me (and which has the most solid geological evidence to back it up) is that this rapid climate shift was caused by changes to the ocean currents triggered by a well-documented and intense period of tectonic activity (accompanied by intense volcanic eruptions), which cut off the North Sea from the Atlantic Ocean at that time.[4] In the runup to the Eocene Optimum, Greenland had already begun breaking away from Northern Europe to begin opening up the North Atlantic Ocean — this being the last major rupture that tore apart what remained of the supercontinent of Pangaea, which began breaking apart 175 million years ago. As Greenland and Europe began to drift apart, a narrow and shallow seaway (known as Barents Shelf) formed between them, which allowed water from the deep North Atlantic to flow up into the North Sea and then onwards into the Barents Sea and Arctic Ocean, somewhat akin to the pattern of the Gulf Stream that helps equalize

---

[3] Bird, H. (2023, Jun 27). 56 million-year-old Eocene global warming may indicate a wetter future. *PHYS.org*. https://phys.org/news/2023-06-million-year-old-eocene-global-wetter-future.html

[4] Jones, M.T., et al. (2023). Tracing North Atlantic volcanism and seaway connectivity across the Paleocene-Eocene Thermal Maximum (PETM). [preprint]. *EGUsphere*. https://doi.org/10.5194/egusphere-2023-36

temperatures on our planet today, though in those days the Atlantic Ocean was still much smaller. However, around 56 million years ago a large plume of molten rock pushed up underneath the tectonic plates beneath the shallow Barents Shelf. This triggered widespread volcanic eruptions all through the region (including the eruptions that formed the famous Giant's Causeway in Ireland) and uplifted large portions of this shallow sea out of the water (the magmatic hotspot underneath Iceland is a remnant of that giant magma plume) — this tectonic uplift is also what caused England, Ireland, and large portions of Scandinavia to rise up above sea level.

With the North Sea cut off from the North Atlantic by this uplifting, warm waters accumulated in the North Atlantic instead of flowing northwards up past Greenland and Europe to mix with Arctic waters. With the Atlantic Ocean unable to vent its accumulating heat into the Arctic, the Atlantic began to warm up like a gigantic warm bathtub. The heat didn't just build up in the ocean — with nowhere to vent the heat accumulating in equatorial waters, the warming Atlantic Ocean effectively pumped vast quantities of humidity into the atmosphere, creating a humidity-induced global greenhouse effect that produced the warm and humid Eocene Optimum — basically Florida's climate on steroids, exported to the whole world as soaring humidity levels spread all around the world.

But the same tectonic forces that created this unique moment in geological history soon unravelled it (by "soon" I mean on a geologic time scale — an 8-million-year climate event is many times longer than the entire span of human history but to a geologist it is a mere blink of an eye). As the continental plates continued to spread apart along the Mid-Atlantic Ridge, the connection between the North Atlantic and the Barents Sea was "unblocked", allowing ocean currents to once again begin venting heat to the Arctic. The widening Mid-Atlantic Ridge also served to deepen the Atlantic Ocean, which helped create a much colder deep-water reservoir at the bottom of the Atlantic, leading to a colder global climate and an accompanying drop in global humidity. Shallow seas are warm; deep oceans are much, much colder — as the Atlantic Ocean widened and deepened, this too helped cool the global climate.

The net effect of all these changes is that global temperatures dropped back down, the humidity that had accumulated in the atmosphere during the Eocene Optimum began to disperse, and the long-term cooling trend that had begun 76 million years ago resumed.

The very first fossils of grasses emerged right around the time the dinosaurs went extinct. By 55 million years ago, near the peak of the Eocene Optimum, the first grasslands began to aggressively spread out in the dry rain shadows of the largest mountain ranges, like in Patagonia and in the rain shadow of the Rocky Mountains. As the planet rapidly cooled and became markedly drier after the end of the Eocene Optimum, forests around the world went into rapid retreat and grasslands quickly colonized much of the planet. Herds of large herbivorous mammals co-evolved with the grasslands, creating the symbiotic relationship I described in earlier chapters.

Back in chapter 13, I told the story of how plumes of molten rock erupting onto our planet's cooling crust created the first continents 2.5 billion years ago, contributing to the planetary cooling that triggered the Huronian Glaciation — the large, bare, rocky continents emerging from the oceans reflected sunlight back into space since rocks are reflective and do not store heat, unlike the dark oceans that absorb and store heat. The retreat of forests and the expansion of grasslands 55 million years ago had a similar cooling effect on global temperatures because light-colored grasslands reflect far more sunlight back into space than dark-coloured forests.[5] And, because grasses pump far less moisture back into the atmosphere than trees do (grasses are better at conserving moisture and thus produce much less evapotranspiration), the expansion of grasslands contributed to a significant drying of the global climate, in the same way that replacing Amazon rainforest with soybean fields and pasturelands reduces rainfall in the region. This reduction in global evapotranspiration also triggered yet another feedback mechanism because in a drying climate there's less water vapour in the air to insulate the planet, which further contributes to cooling.

As you can see in figure 104, temperatures once again stopped falling around 30 million years ago, ushering in yet another temporary warming period that began during the Oligocene epoch and peaked with what is known as the Mid-Miocene Climate Optimum (another warm period during which forests expanded and recolonized many areas that had previously been colonized by grasslands). Although temperatures had

---

[5] Retallack, G. (2013). Global Cooling by Grassland Soils of the Geological Past and Near Future. *Annual Review of Earth and Planetary Sciences.* 41. https://doi.org/10.1146/annurev-earth-050212-124001

already fallen by more than 8 °C since the peak of the Eocene Optimum, temperatures during this Mid-Miocene Climate Optimum were still approximately 4 °C warmer than they are today (this is despite the fact that by this time CO_2 levels had already fallen to present-day levels of around 300 to 500 ppm — oops!)[6]. Furthermore, studies have shown that *"late Oligocene warming [was] coincident with declining atmospheric CO_2"*.[7] In other words, this warming period happened against a backdrop of *falling* CO_2 — yet another oops for the climate modellers who keep trying to explain everything with CO_2-centered climate models.

Once again, the reasons for this Oligocene-Miocene pause/warming are still very poorly understood, though continental drift and its effect on ocean currents and global humidity levels undoubtedly played the biggest role, which I will explain in a moment. However, one of the contributing factors to the brief warming period of the Mid-Miocene Climate Optimum is that our solar system exited the main trunk of the Carina-Sagittarius spiral arm around 35 million years ago, giving our solar system a 15-million-year break from the cosmic radiation inside these spiral arms, only to begin traversing the Orion-Cygnus Spur (a side branch off of the main Carina-Sagittarius spiral arm) starting around 20 million years ago (right at the end of the Mid-Miocene Climate Optimum).[8] Even today, we are still somewhere inside this Orion-Cygnus Spur. But back to the main story of the impact of plate tectonics on the climate before, during, and after this Mid-Miocene Climate Optimum.

Earlier in the book I told the story of how the Pangaean supercontinent began to break apart around 175 million years ago when Jurassic dinosaurs still ruled the Earth (that's when South America and Africa began to unzip). Most of the world's heavily forested landmasses were all located near the equator at that time. But as Pangaea split apart, the North American and Eurasian continents began to drift northwards even as they began to spread apart, while Antarctica and Australia drifted south towards the South Pole, leading to a general cooling climate on all these continents

---

[6] Rae, J.W.B., et al. (2021). Atmospheric CO_2 over the past 66 Million Years from Marine Archives. *Annual Review of Earth and Planetary Sciences.* 49. https://doi.org/10.1146/annurev-earth-082420-063026

[7] O'Brien, C.L. et al. (2020). The enigma of Oligocene climate and global surface temperature evolution. *Proc. Natl Acad Sci U S A.* 117(41).

[8] Sankaran, A. V. (2008). Galactic triggering of geologic events in earth's history. *Current Science*, 95(6), 714–716. http://www.jstor.org/stable/24102576

as they drifted further from the equator. As the continents' positions changed, they also increasingly disrupted and re-routed the ocean currents that were responsible for efficiently transferring warm water from the tropics to polar regions, which further reinforced the cooling trend.

By 100 million years ago, southern Africa, Antarctica, India, and Australia were all still joined, as shown in figure 105.[9] The only direct connection between the Atlantic and the Indian Ocean was via a very narrow passage between Europe and Africa. Instead, the Indian Ocean was surrounded by land on three sides, effectively creating a large seaway known as Tethys Sea (although, as you can see in the map in figure 105, it's not actually correct to call it the Indian Ocean at that time since India is still on the wrong side of it — Tethys Sea was between India and Asia, but as India moved north, Tethys Sea slowly closed while opening up the Indian Ocean behind it). By the Mid-Miocene Climate Optimum, India's northward march had shrunk Tethys Seaway so that, once again, we have a giant shallow bathtub full of warm water straddling the equator without efficient ocean currents to vent warm water to the poles, which helped create a warmer and much more humid climate during the Mid-Miocene Climate Optimum.

---

[9] "Tethys Ocean." *Wikipedia*. 9 Apr 2024. https://en.wikipedia.org/wiki/Tethys_Ocean

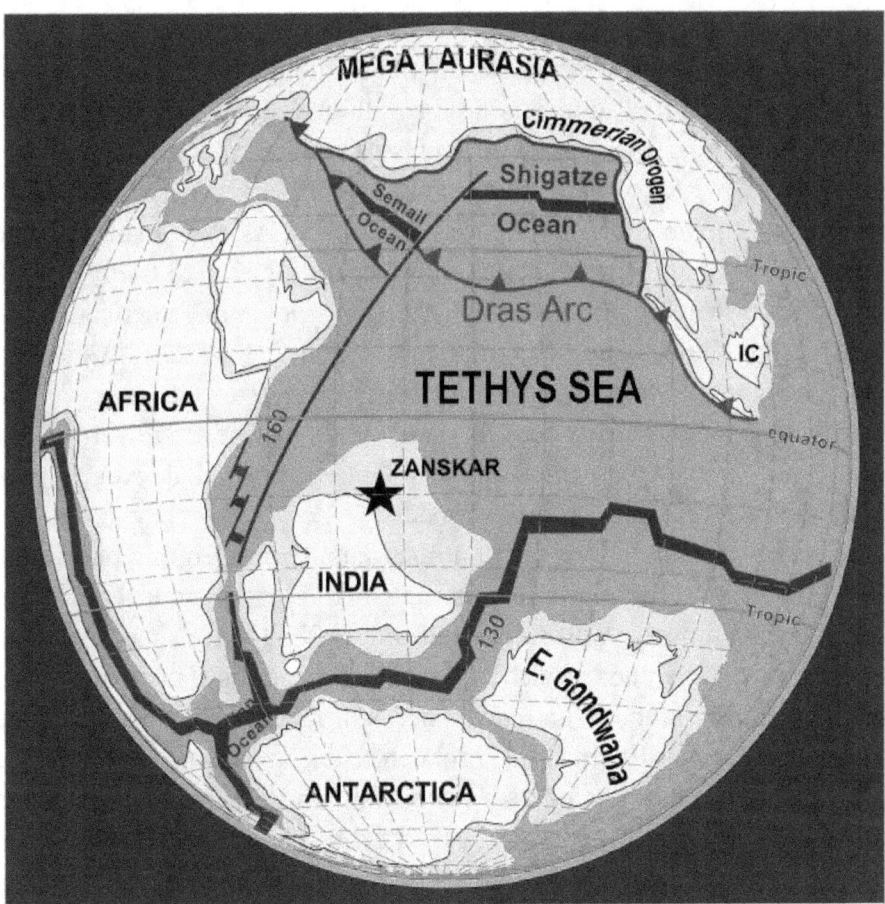

Pierre Dèzes 1999, "Tectonic and metamorphic Evolution of the Central Himalayan Domain in Southeast Zanskar (Kashmir, India)". *Mémoires de Géologie (Lausanne)* No. 32, ISSN 1015-3578

Figure 105. Plate tectonic reconstruction of Tethys Sea at 100 mya. South Africa, Antarctica, and India were all connected at that time.

Source: https://en.wikipedia.org/wiki/Tethys_Ocean

But by the end of the Mid-Miocene Climate Optimum (around 18 to 14 mya), this shrinking warmwater bathtub finally closed (known as the closing of Tethys Seaway), bringing an end to the warm humid global climate. The long-term cooling trend resumed. To fully understand the impact that Tethys Sea had on the global climate (and the sharp cooling that happened once Tethys Sea closed and the Indian Ocean opened up), we need to have a closer look at how continental drift was reshaping the southern hemisphere since the breakup of the Pangaean supercontinent.

Antarctica reached its current position at the South Pole around 70 million years ago, and then stayed there, though it remained warm, forested, and ice-free until around 34 million years ago when the land bridge connecting the Antarctic Peninsula to the Andes mountains in South America broke apart, thus slowly opening a gap between the two continents, known today as the Drake Passage. As the Drake Passage opened, cold ocean waters increasingly began to circulate freely around Antarctica instead of being pushed north along the South American coast, thus creating the Antarctic Circumpolar Current that insulates Antarctica and the Southern Ocean from warmer ocean waters to the north.[10] With cold waters accumulating near the South Pole instead of flowing north along the South American continent, Antarctica cooled even as the continents closer to the equator warmed.

And so, intermittently at first but then rapidly accelerating during the Mid-Miocene Climate Optimum as the Drake Passage widened and the Antarctic Circumpolar Current became more efficient, ice sheets began to spread across Antarctica, which in turn reflected sunlight back into space and cooled the continent (and the planet) still further. The opening of the Drake Passage was the straw that broke the camel's back — it plunged the entire Antarctic continent into a bitter 34-million-year-long ice age that is still ongoing today. As ice was locked away in Antarctica, falling sea levels further contributed to a global cooling by reducing the surface area of the oceans (where heat is stored) while simultaneously increasing the surface area of the land masses (which reflect heat back into space).

By around 45 million years ago, the subcontinent of India (which had been steadily drifting north across the Tethys Sea) begun colliding with Asia and, as India plowed forward, a great crumpling happened at the leading edge of this collision, forcing the Himalayan mountains to begin to rise (even today they are still rising at around 5 mm per year as India continues to plow north into Asia). The top of Mount Everest is covered in ancient limestones that were once at the bottom of Tethys Sea but have now been crumpled and thrust up onto the roof of the world.

The emerging Himalayas disrupted the monsoon winds that carry moisture from the Indian Ocean into Siberia, creating a much warmer and

---

[10] Davies, B. (2020, Oct 5). Antarctic Peninsula Ice Sheet evolution. *AntarcticGlaciers.org*. https://www.antarcticglaciers.org/glacial-geology/antarctic-ice-sheet/icesheet_evolution/

wetter climate on the windward side of the Himalayas and a giant rain shadow on the leeward side. This rain shadow is why, even as Europe and North America were eventually covered by vast ice sheets (up to 2.4 kilometers thick over what is now New York City), there was so little snow north of the Himalayas that Eastern Siberia remained ice-free throughout all the many ice ages that followed. Even today, despite the bitterly cold Siberian winters, Siberia gets very, very little winter snow thanks to the rain shadow cast by the Himalayas.

There were several other large tectonic "events" that completely changed the circulation of the global ocean currents during this period and, by extension, changed the global climate. For example, the Tasmanian Gateway opened up as Australia began to split away from Antarctica around 33.5 million years ago, enabling ocean waters to circulate between them. Furthermore, both the Andes in South America and the mountains ranges in western North America were getting higher, contributing to a drying of both the North American Great Plains and the Pampas region of South America. Likewise, in Europe the Alps began to form as the northward-moving African plate began to collide with the Eurasian plate. And, in Western Siberia, the vast shallow inland sea (shown on the upper map in figure 106 as the interconnected Peri-Tethys and Western Siberian basin), which had once stretched across the Eurasian continent from the Arctic Ocean to the Tethys Sea and which had helped moderate the Eurasian climate, also dried up, leaving behind only the much smaller Paratethys Sea, as shown in the lower map in figure 106 below.[11]

---

[11] "Tethys Ocean." *Wikipedia.* 9 Apr 2024. https://en.wikipedia.org/wiki/Tethys_Ocean

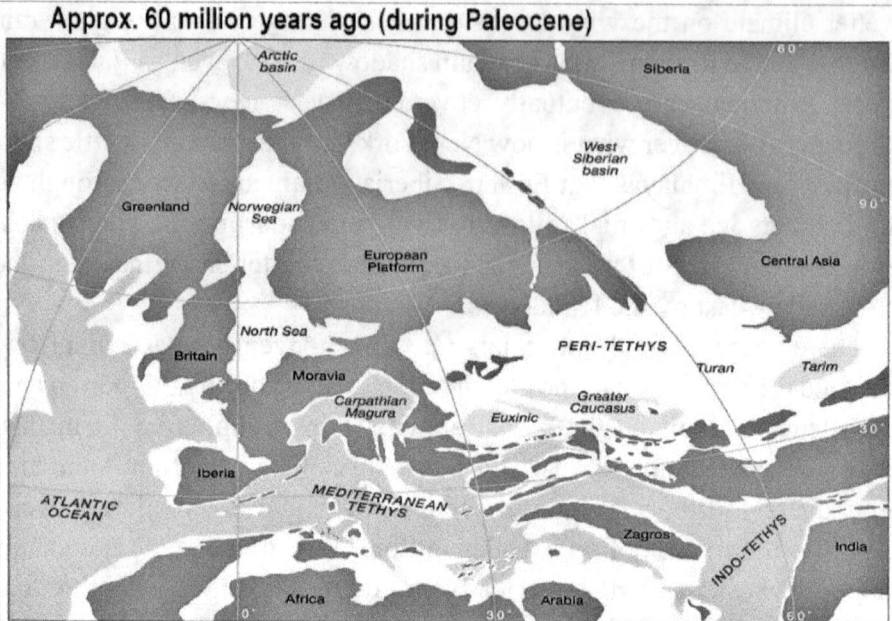
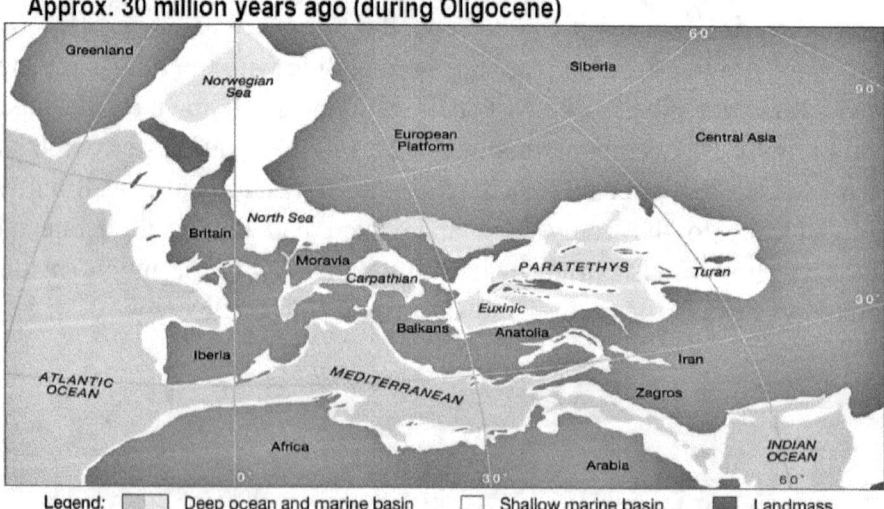

Figure 106. Vast regions of Europe and west-central Asia were still covered by a contiguous Tethys at the start of the Eocene that stretched all across Western Siberia (top image). But by the Oligocene most of this had dried out (bottom image). Thus, the Arctic Ocean was completely cut off from the Indian Ocean, and the deepwater connection was severed between the Indian Ocean and the Mediterranean Sea.
https://en.wikipedia.org/wiki/Tethys_Ocean

By the time the Mid-Miocene Climatic Optimum came to an end and global cooling ramped up again, the Drake Passage and the waters surrounding Antarctica had become so deep that the ice-encased southern pole and the waters surrounding it became a major global reservoir of cold water that helped chill the entire global climate as deep-sea currents known as "Antarctic bottom water" spread northwards along the bottom of the ocean to cool the Pacific, Atlantic, and Indian Oceans from below.

As you can see, the study of plate tectonics isn't only a geological 3-D jigsaw puzzle to work out the path of drifting continents. It is also constantly rearranging ocean currents, interfering with the Earth's ability to vent heat from equatorial regions to the poles, creating mountain ranges that change the balance of humidity and aridity all over the world, creating and destroying shallow seas that contribute to global humidity levels, and deepening reservoirs of cold water at the bottom the ocean. You cannot understand the climate without understanding continental drift, plate tectonics, ocean currents, and the effect that these have on global humidity levels and global temperatures.

The C$_4$ Revolution

Even as Antarctica froze 34 million years ago, the rest of the globe was still many millions of years away from joining Antarctica in an ice age. But the relentless scrubbing of CO_2 from the atmosphere was starting to take its toll on CO_2 starved plants. The grasses that evolved around the time that dinosaurs went extinct are part of a family of flowering plants known as C_3 plants — the C_3 refers to the fact that the first (simplest) carbon compound produced by these plants during photosynthesis contains three carbon atoms. C_3 plants are extremely efficient at converting CO_2 into plant fibers... but only when atmospheric CO_2 levels are high. As CO_2 levels fall, yields from C_3 plants collapse as CO_2 fertilization turns into CO_2 starvation.

Furthermore, as CO_2 levels fall, C_3 plants have to open their pores wider to be able to absorb enough CO_2 from the air, which allows a lot of moisture to escape their pores. And so, as CO_2 levels continued their long slow multi-million-year decline, C_3 plants became increasingly vulnerable to drought in the increasingly dry global climate. Equally important (and equally challenging to C_3 plants), because C_3 plants have to open their pores so wide at low CO_2 levels, these open pores also let in large quantities of oxygen molecules, which interfere with CO_2 absorption.[12]

---

[12] Osborne, C.P. and Beerling, D.J. (2006). Nature's green revolution: the remarkable evolutionary rise of C$_4$ plants. *Philos Trans R Soc Lond B Biol Sci.* 361(1465).173-94. https://doi.org/10.1098%2Frstb.2005.1737

By the Oligocene epoch (33.7 to 23.8 million years ago), low CO_2 levels were becoming a serious obstacle to C_3 photosynthesis, particularly in arid regions where water is the main obstacle to plant survival.

Figure 107 below traces the evolution of atmospheric CO_2 from the time the dinosaurs went extinct until the modern era. Although the climate was still much warmer and more humid during the Oligocene than that it is today, by that time CO_2 levels were already hovering below 400 ppm (below today's levels). But as always, life adapts.

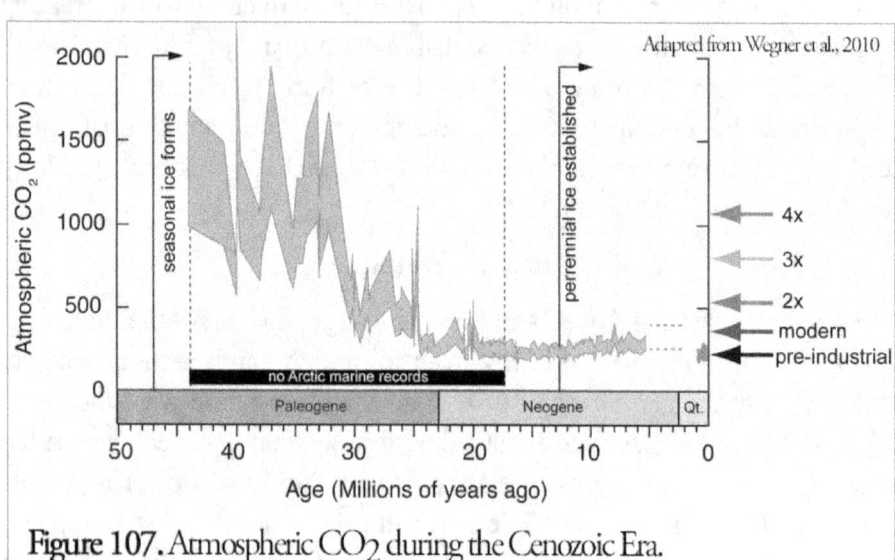

Figure 107. Atmospheric CO_2 during the Cenozoic Era.

Source: Wegner et al. 2010. *Arctic in Rapid Transition (ART) Science Plan*, Arctic Ocean Sciences Board / International Arctic Science Committee (AOSB/IASC), 34 pp. https://epic.awi.de/id/eprint/23791/1/Weg2010e.pdf

A revolutionary new innovation in photosynthesis emerged during this period, which is much more efficient at binding carbon dioxide at low CO_2 levels — these plants are called C_4 plants because they produce a four-carbon compound at the first stage of photosynthesis. And because they are more efficient at absorbing CO_2 from the atmosphere at low CO_2 levels, they don't have to open their pores as widely. This allows them to conserve *a lot* of water compared to their C_3 cousins. A C_3 plant loses approximately 833 molecules of water for every CO_2 molecule that it absorbs from the atmosphere. By contrast, C_4 plants only lose 277 water molecules, which means they only lose one-third as much water as C_3

plants.[13] This gave C_4 plants a major evolutionary advantage in dry climates.

Figure 108 below[14] shows how C_3 grasses (top) and C_4 grasses (bottom) respond to different levels of CO_2 in the air. There is less of a CO_2 fertilization effect among C_4 plants — they don't mind rising CO_2 levels but can't take advantage of the extra CO_2 to the same degree at their C_3 cousins. Whereas yields among C_3 plants will double with a doubling of CO_2 from current levels, C_4 plants will only experience up to a 40% increase in yields as CO_2 levels double.[15] But on the other end of the scale, C_4 plants continue to thrive at CO_2 levels that are fatally low to C_3 plants.

---

[13] "C₄ Carbon fixation." *Wikipedia.* 8 May 2024. https://en.wikipedia.org/wiki/C4_carbon_fixation

[14] von Caemmerer, S., Quick, W.P., and Furbank, R. (2012). The Development of C4 Rice: Current Progress and Future Challenges. *Science.* 336(6089). https://doi.org/10.1126/science.1220177

[15] Gerhart, L.M. and Ward, J.K. (2010). Plant responses to low [CO₂] of the past. *New Phytologist.* 188(3). https://doi.org/10.1111/j.1469-8137.2010.03441.x

Figure 108. Photosynthesis at various CO_2 levels – C3 vs C4 plants. Source: von Caemmerer et al., 2012. *The Development of C4 Rice*, Science 336(6089): 1671-2.
https://www.science.org/doi/10.1126/science.1220177

Over 40 evolutionary groups of plants independently evolved this same C4 genetic adaptation during this time period, demonstrating that low CO_2 levels had become a serious challenge to photosynthesis, especially in dry climates. The C4 genetic adaptation is a prime example of convergent evolution, which is when multiple species independently evolve similar features without having a common ancestor.[16]

Rice, wheat, soybeans, cotton, tobacco, bamboo, and almost all trees are C3 plants, whereas the C4 newcomers include corn, sorghum, sugarcane, and millet, along with most warm season grasses (the grasses that thrive at the hottest and thus the driest time of year). Today, C4 plants dominate grasslands and tropical savannahs, they account for 30% of global terrestrial carbon fixation, and they represent 30% of global

---

[16] Osborne, C.P. and Beerling, D.J. (2006). Nature's green revolution: the remarkable evolutionary rise of C4 plants. *Philos Trans R Soc Lond B Biol Sci.* 361(1465).173-94. https://doi.org/10.1098%2Frstb.2005.1737

agricultural grain production.[17] By 5 to 8 million years ago, C₄ plants had come to dominate terrestrial ecosystems —you only have to look at falling atmospheric CO_2 levels in figure 107 to see why C₄ plants are winning the race. Compound that with the reduced moisture evaporation from the oceans as temperatures fell and the increased aridity that came to dominate the world as shallow seas dried up and as emerging mountain ranges produced ever larger rain shadows, and it becomes easy to understand why C₄ plants had so much success in colonizing the world. Even plant evolution is signalling that we are currently living in a CO_2-starved world.

I previously stated that the increasingly dry climate of the last 66 million years pushed forests into retreat and opened the door for grasslands to colonize deforested regions. But the story of a drying climate is only half of the puzzle to explain the rise of large grasslands replacing the forests. Falling CO_2 levels are equally important in explaining this widespread transition from forests to grasslands — if CO_2 levels had not fallen to such low levels, aridity would have posed far less of a challenge to trees (which are all C₃ plants) because they wouldn't have had to open their pores so wide to absorb enough CO_2. Falling CO_2 levels are thus equally to blame for the slow but steady retreat of forests and the relentless expansion of grasslands, which has been unfolding throughout our own geologic era and which began even before the end of the reign of the dinosaurs.

From a geological perspective, we are living in an age when low CO_2 levels are completely transforming what kinds of plants dominate our planet. C₃ plants are being challenged by low CO_2 levels, C₄ plants are emerging to fill the void. Natural selection favors the plants that can thrive in a CO_2-starved world. Any scientist, journalist, or politician who claims that 413 ppm CO_2 in the atmosphere is a lot is willfully, deliberately, shamelessly misleading you.

A study published in the New Phytologist in 2010[18] summarized the body of scientific research on the effects of low CO_2 on the biosphere. They concluded that, *"low [CO2] of the past 21,000 years favored grasslands and tundra over the forests currently located at temperate and*

---

[17] Osborne, C.P. and Beerling, D.J. (2006). Nature's green revolution: the remarkable evolutionary rise of C₄ plants. *Philos Trans R Soc Lond B Biol Sci.* 361(1465).173-94. https://doi.org/10.1098%2Frstb.2005.1737

[18] Osborne, C.P. and Beerling, D.J. (2006).

boreal latitudes" and that, since grasslands pump less water back into the atmosphere than forests, "*[t]hese changes would have greatly affected evapotranspiration and possibly the entire water cycle of the region.*" In plain English, this study is pointing out that low CO_2 levels are causing the planet to get drier by turning forests into grasslands as trees become increasingly drought stressed as they are forced to open their pores ever wider to compensate for falling CO_2. The same study also pointed out that "*the low CO_2 effects on vegetation were the dominant factor*" in reducing the total biomass on the planet during the coldest part of the last ice advance, even ahead of the effect of the planet becoming increasingly cooler and drier as the ice sheets advanced. In other words, CO_2 starvation during the coldest phase of the last ice age had a bigger impact on global biomass than cold temperatures!

So, it turns out that CO_2 does have a major impact on the climate in that it indirectly affects <u>humidity</u>, not temperature, because it affects what plant species can survive in low CO_2 conditions. As grasslands and tundra replaced CO_2-starved forests (which reduced global evapotranspiration rates), falling CO_2 levels indirectly triggered a major drying effect on the global climate as a consequence of the vegetation changes caused by falling CO_2. Our world is full of complex surprises.

The Isthmus of Panama and the Start of the Quaternary Glaciation

Until around 3 million years ago, North and South America were still separated by the Central American Seaway, which allowed ocean waters to circulate freely between the Pacific and the Atlantic Ocean through the gap between these two continents. But, once again, plate tectonics intervened to change the world and added yet another serious chill to the global climate.

Beneath the Central American Seaway, the Cocos tectonic plate began to slide under the Caribbean plate, creating an arc of volcanoes over the collision zone, which began to fill in the gap between the two continents. As the Cocos plate slid under the Caribbean plate, parts of the Caribbean plate were also forced upwards above sea level and so, around 3 million years ago, the land bridge known as the Isthmus of Panama closed off this gap between the two continents. The ancestors of opossums, armadillos, and porcupines migrated north across the land bridge, while the ancestors of bears, cats, dogs, horses, llamas, and raccoons trekked south into South America. Ocean currents that had previously circulated between the two oceans came to an abrupt halt as warm tropical waters were rerouted northwards to establish the global ocean circulation patterns we see today — in the Pacific these warm waters are carried north along the coast of the Pacific Northwest, while in the Atlantic this gave rise to the Gulf Stream, which circulates warm Caribbean waters up into northern Europe.

Paradoxically, these warmwater currents pushing north accelerated the formation of continental ice sheets in the northern hemisphere. While the Himalayan rain shadow still prevented ice sheets from forming in dry Siberia and Alaska, the warm ocean currents carrying moisture northwards after the closing of the Isthmus of Panama triggered much larger snowfalls in Europe, Greenland, and up both sides of the North American continent. And so, as summer melting failed to keep up with these enhanced winter snowfalls, the first ice sheets began to form in Greenland, Europe, and North America even as eastern Siberia and Alaska stayed ice-free. Ice sheets began growing on Greenland just over 3 million years ago, which further cooled the Arctic, and by 2.58 million years ago the entire northern hemisphere was plunged into the Quaternary Glaciation. Even today, we are still in that Quaternary Glaciation — we are merely in a temporary warm interglacial period between long glacial advances. The closure of the Isthmus of Panama was the final tipping point that triggered our ongoing Quaternary Glaciation.

Ice Age Struggles

Figure 109 zooms in on the temperatures of the last 5 million years to illustrate the gradual long-term cooling and drying trend, which began even before the extinction of the dinosaurs 66 million years ago and continues unbroken even today.[19] By now we're zooming in so much that the individual temperature swings caused by wobbles in the Earth's orbit (Milankovitch cycles) become visible on the chart — those orbital wobbles caused temperatures to constantly rise and fall within a tightly constrained range even as the long-term trendline of global temperatures continued to decline throughout this 5-million-year period. For context, the 800,000-year chart published by National Geographic in 1976 (shown in figure 41 back in chapter 8) only covers the final eight glacial advances and retreats — the small box in the top right corner of figure 109 below

---

[19] "Mid-Pleistocene Transition." *Wikipedia.* 11 Jun 2024.
  https://en.wikipedia.org/wiki/Mid-Pleistocene_Transition

shows the period that overlaps with the National Geographic chart back in figure 41.

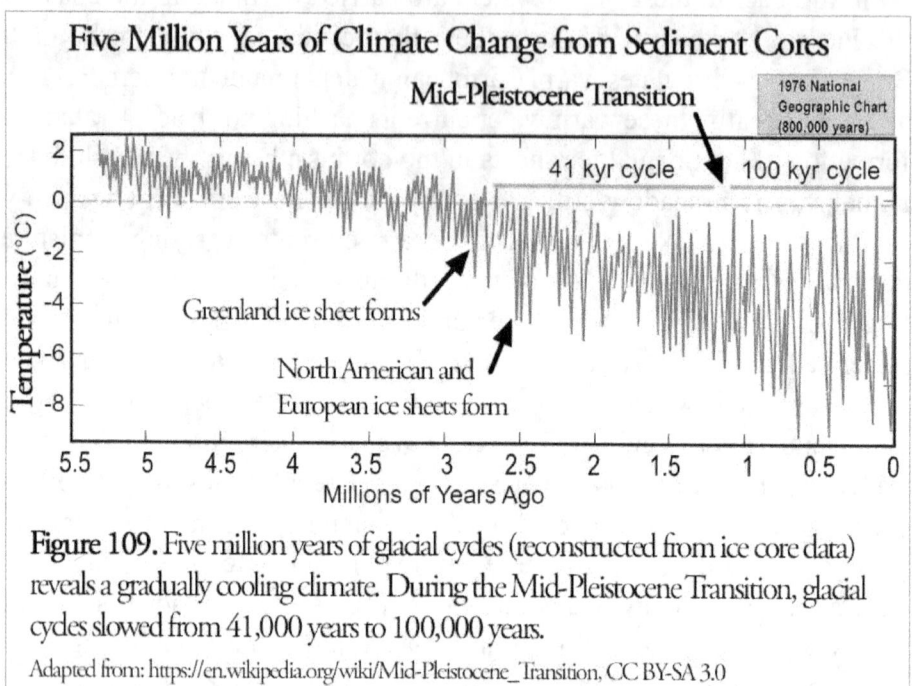

Figure 109. Five million years of glacial cycles (reconstructed from ice core data) reveals a gradually cooling climate. During the Mid-Pleistocene Transition, glacial cycles slowed from 41,000 years to 100,000 years.

Adapted from: https://en.wikipedia.org/wiki/Mid-Pleistocene_Transition, CC BY-SA 3.0

You can see from figure 109 that the onset of glaciation in Greenland, North America, and Europe slowed down the rapidly vibrating temperatures while also increasing the range of each temperature swing as cold spells got markedly colder (which makes sense when you think of the extra chill caused by covering entire continents with ice). The ice sheets served as a buffer to smooth out the weaker temperature swings caused by the lesser solar cycles. From 3 million until 1.2 million years ago, the shorter 41,000-year Milankovitch cycle caused by wobbles in the Earth's tilt dominated the cycle of advancing and retreating ice sheets. But then, from around 1.2 million years ago until 800,000 years ago, the cycle of advancing and retreating ice sheets gradually began to slow down even more (with much longer and colder glacial periods and only very short interglacial periods) — this transition period is known as the Mid-Pleistocene Transition. After this transition, the cyclical advance and retreat of the ice sheets (and the corresponding global temperature fluctuations) shifted permanently to match the 100,000-year Milankovitch cycle caused by the stretching and shrinking of Earth's orbit. But what caused this change from 41,000- to 100,000-year cycles?

As usual, some climate modelers blame CO_2 (the magical control knob for everything), though the alleged mechanism of how CO_2 is supposed to affect this cycle shift escapes me. The CO_2 explanation makes absolutely no sense because, apart from its inability to meaningfully affect temperatures, there also weren't any significant changes to global CO_2 levels across this transition period —they always bottomed out around 180 ppm at the coldest extent of each ice advance and peaked just under 300 ppm during each warm interglacial period both before, during, and after the transition.[20]

Others have proposed that changes in the behaviour of the ice sheets themselves could have caused this change, suggesting earlier ice sheet advances scraped away the loose rock and sediments that would have made it easier for ice sheets to slide (thus making these earlier ice sheets thinner), but once the ice sheets began sliding on bare rock the increased friction would have slowed the advance of the ice sheets and made them thicker. Still others are searching for a tectonic trigger as a mechanism to alter the global temperature balance or alter ocean circulation pattern, but so far none have been found.

And yet, in my opinion, the obvious answer is staring us in the face when we look at the long-term cooling trend in figure 104, which has been ongoing for at least 76 million years. As temperatures continued to cool, the amount of heat required to melt away the ice sheets after each glacial advance necessarily increases with each passing cycle. And so, as the planet continued to cool, at some point (around 1.2 million years ago) the modest warming produced by changes in the Earth's tilt (the 41,000-year cycle) was no longer sufficient to melt the ice after a glacial advance; instead the advancing ice sheets persisted until the warmest part of the stronger 100,000-year cycle began as Earth's orbit was stretched to its most elliptical point (which floods the planet with 23% more solar radiation during the summer compared to when the orbit is more circular). In other words, against the backdrop of a long-established cooling trend, as the cooling continued it necessarily would take more heat to melt the ice sheets, hence the shift to the more powerful 100,000-year cycle

---

[20] Da, J. et al. (2019). Low CO_2 levels of the entire Pleistocene epoch. *Nature Communications*. 10(4342). https://doi.org/10.1038/s41467-019-12357-5

Thus, no new mechanism is required to produce this transition from the 41,000-year cycle to the 100,000-year cycle — it's just the inevitable continuation of all the diverse interconnected factors that have been gradually cooling our planet since the dinosaurs went extinct:

- Starting 76 million years ago, plate tectonics caused the continents to drift ever further away from the equator and rerouted ocean currents to create a general cooling of the climate.
- Our solar system's traverse through the Carina-Sagittarius Spiral Arm of our galaxy and then through the Orion-Cygnus Spur magnified that general cooling trend by increasing cloud cover over our planet.
- A cooler planet is inevitably going to be drier, but a drier planet also is cooler because the greenhouse effect produced by water vapour gets weaker as humidity falls. Remember, humidity causes the most important greenhouse effect, by a wide margin — unlike CO_2, water vapour is a greenhouse gas that actually matters to global temperatures. So, as global humidity levels declined, global temperatures declined in lockstep. Northern regions and high elevations are especially sensitive to this cooling triggered by falling global humidity levels.
- Ongoing tectonic forces were fueling the rise of mountain ranges (even today the Himalayas, Andes, Rocky Mountains, and Alps are all still growing higher), which also contributed to a drier (and thus cooler) global climate.
- Plate tectonics also caused many of the planet's warm, shallow, inland seas to shrink, which had once acted as large reservoirs of warm water to help buffer local climates — all that remains today of the enormous Paratethys are tiny remnants like the Black Sea, the Caspian Sea, and the Aral Sea.
- A cooler and drier climate also caused forests to retreat and grasslands to expand. But since grasslands reflect more solar radiation back into space than forests do, this too contributed to a general cooling of the climate.
- In a self-reinforcing feedback effect, as forests shrunk and grasslands expanded, this further decreased water vapour in the air (and thus decreased the greenhouse effect of water vapour)

because grasses produce much less evapotranspiration than trees.
- As falling CO_2 levels triggered the shift from C_3 to C_4 plants, this further contributed to a drier global climate with less greenhouse-inducing water vapour in the air because C_4 plants lose much less moisture through their pores than C_3 plants. As C_4 plants spread, evapotranspiration declined and global humidity levels fell, thus reducing the greenhouse effect from water vapour still further.

A supporting piece of evidence for this explanation comes from windblown dust deposits, which provide a window into the aridity of past climates. Dust deposits demonstrate that the climate continued to get drier throughout the Mid-Pleistocene Transition — dust is a good proxy for both aridification and the increased reflectiveness (a.k.a. albedo) of land surfaces.[21] It's not a smoking gun, but it certainly indicates that the complex mix of dynamic of processes that have been cooling and drying the global climate for over 76 million years continued unbroken right through this transition period. At some point it was inevitable that the amount of extra warmth produced by the weaker 41,000-year Milankovitch cycle would be insufficient to melt the ice sheets and that the ice age cycle would thus necessarily have to switch to the much stronger 100,000-year Milankovitch cycle to trigger a glacial retreat.

Nor has this long-term cooling and drying trend stopped. Have another closer look at the ice core data from Vostok, Antarctica, shown back in figure 74. Obviously, dust rises and falls as temperature moves between cold glacial and warm interglacial periods. However, if you look at the amount of dust at the bottom of each cold glacial low, you can see that even across this most recent 400,000-year period, each cold period is demonstrably dustier than the last — the long-term drying trend continues unabated towards an ever-drier climate. Furthermore, figure 74 also shows that although the last four interglacial warm periods were slightly warmer than earlier periods, the coldest temperatures reached during each glacial advance have nevertheless consistently continued to get colder. At this

---

[21] Herbert, T. (2023). The Mid-Pleistocene Climate Transition. *Annual Review of Earth and Planetary Sciences*. 51. 389-418. https://doi.org/10.1146/annurev-earth-032320-104209

scale, the long-term cooling trend that has been in place for the last 76 million years is clearly still ongoing, which would explain why the 41,000-year cycle became insufficient to end glacial periods and why the extra warmth from the stronger 100,000-year cycle was required to turn the tide. Presumably, unless something changes, the next glacial advance will be even colder and drier than the last.

The only thing that hasn't changed across the Mid-Pleistocene Transition, nor over the multiple glacial advances and retreats that have happened over the subsequent 800,000 years is CO_2, which has consistently continued to cycle back and forth between ~180 ppm and ~280 ppm as the ice sheets advanced and retreated. Why did it stabilize in this range instead of continuing to fall to new lows? —more on that in a moment.

All this raises the question, if this long-term multi-million-year cooling trend continues, will future glacial advances continue to be interrupted every 100,000 years by warm interglacial periods (like the one we're enjoying today)? Or will the 100,000-year cycle also eventually break down, possibly being replaced by the 400,000-year cycle (an even longer Milankovitch cycle that I haven't discussed yet)? Or worse? After all, many of the really severe glaciations from earlier in our planet's geological history did not have these frequent, short bursts of glacial advances and retreats — many of those earlier ice ages were marked by individual uninterrupted glacial advances that each lasted for *tens of millions* of years at a time.

~

The cooling and aridification of our planet over the past 76 million years had yet another unanticipated impact on the global jigsaw puzzle of interconnected parts — oxygen levels and biomass — both of which have significant implications for understanding the rise in CO_2 levels we are experiencing today. Back in figure 101, I showed you how atmospheric oxygen levels changed over the course of geologic time. A brief recap is in order before I show you the surprising thing that happened to oxygen levels over the last 800,000 years and explain the implications for the total biomass on our planet.

In the carbon/oxygen cycle, oxygen and carbon follow inverse paths as plants add oxygen to the atmosphere even as oxygen breathers pull it back out. Because some carbon is continually lost to permanent carbon sinks before it can be oxidized back into CO_2, there's always some extra oxygen left over in this circular exchange so that oxygen levels gradually rise whenever there's vigorous photosynthesis happening on the planet — the best example comes from the rapid rise in oxygen during the Carboniferous period when most of the world's coal beds were formed.

However, oxygen-breathers aren't the only processes that consume oxygen — oxygen also weathers silicate rocks, oxidizes various minerals like pyrite, and is consumed as long-term carbon sinks like humus are exposed to the atmosphere. And so, whenever biomass declines because of a cooling climate, increased aridity, deforestation, desertification, or because of a mass extinction, at some point this reduces the amount of photosynthesis to such a low level that plant growth cannot produce enough oxygen to completely replace all the oxygen that is continually being consumed by all these processes. Once global biomass falls below a certain point, rising oxygen levels stall and then begin to fall.

In figure 110 below, I've combined three graphs (biomass,[22] extinctions,[23] and oxygen levels[24]) to show you how oxygen levels consistently fall every time there's a severe decline in biomass, like during the biomass decline at the beginning of the Devonian Period when the Age of Fish began, during the Great Dying at the end of the Permian period, during the Triassic-Jurassic extinction, and briefly at the end of the Cretaceous period when the dinosaurs went extinct. You can see that whenever biomass falls (which causes a decline in global photosynthesis), oxygen also falls. And whenever biomass increases, oxygen levels begin to climb.

---

[22] Svensmark, H. (2022). Supernova Rates and Burial of Organic Matter [burial fraction of organic matter adapted from figure 4.] *Geophysical Research Letters.* 49(1). https://doi.org/10.1029/2021GL096376

[23] "Extinction." *Wikipedia.* 7 Jun 2024. https://en.wikipedia.org/wiki/Extinction

[24] Payne, J.L. et al. (2010). The evolutionary consequences of oxygenic photosynthesis: a body size perspective. *Photosynthesis Research.* 107. https://doi.org/10.1007/s11120-010-9593-1

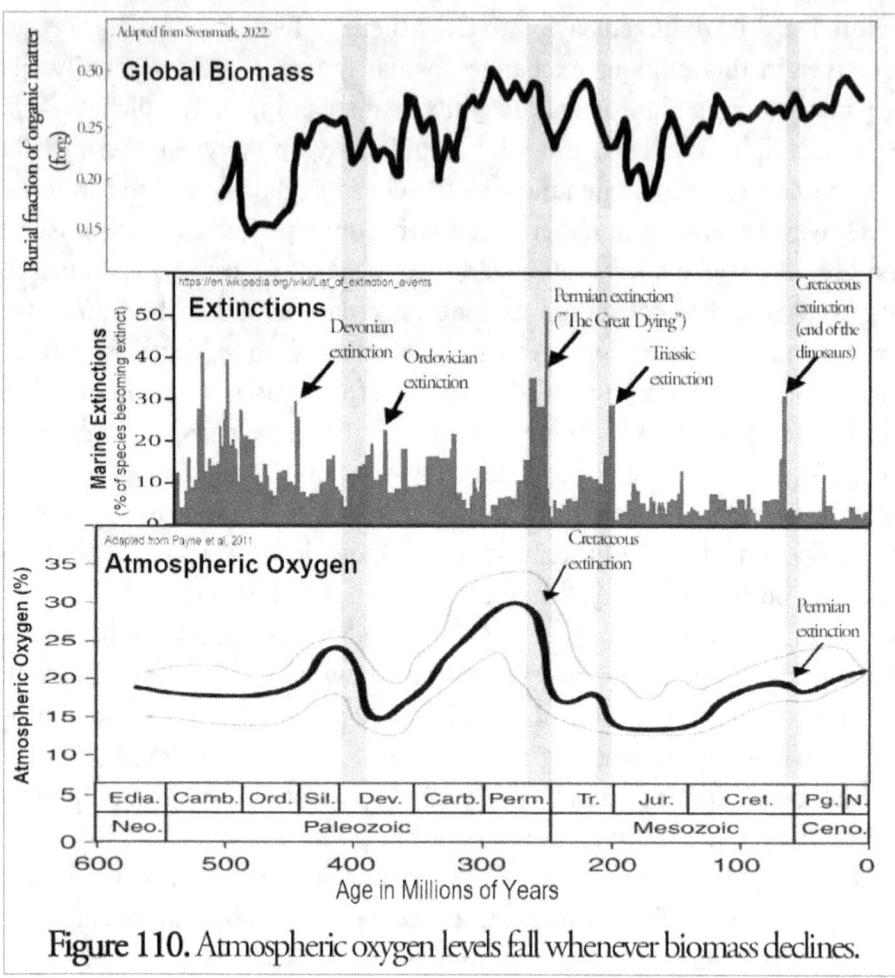

Figure 110. Atmospheric oxygen levels fall whenever biomass declines.

Here's why this matters: before the Mid-Pleistocene Transition (when ice advances and retreats were controlled by the shorter and weaker 41,000-year Milankovitch cycle), oxygen levels had been steadily rising for almost 200 million years (with the exception of a temporary dip when the dinosaurs went extinct), just as you would expect while there's a flourishing biomass and lots of vigorous photosynthesis. But 800,000 years ago, midway into the cycle of bitterly cold Quaternary glaciations and right around the time of the Mid Pleistocene Transition when the 41,000-year glacial cycles slowed down to be replaced by 100,000-year glacial cycles, oxygen levels stalled and then slowly began to fall. For the first time in 65 million years, oxygen consumption exceeded oxygen production. The obvious implication is that by 800,000 years ago the combined effect of a cooling climate, falling CO_2 levels, and the increasingly dry conditions on our planet *had caused biomass to shrink to such low levels that photosynthesis no longer could keep up with the rate at which oxygen is depleted from our atmosphere.*

Over the past 800,000 years and persisting throughout at least eight major glacial advances and retreats until just before the Industrial Era began, global atmospheric oxygen levels steadily decreased by 0.7%,[25] which works out to an average annual decline of 0.00875 ppm per year. This 0.7% decline over 800,000 years may seem insignificant, but this switch from rising to falling oxygen levels is a monumental shift from the long-established atmospheric conditions that came before the Mid-Pleistocene Transition.

Increased erosion/weathering rates in a more arid climate and the long-term cooling of the oceans (which absorb more oxygen as they get colder) may have played a contributing role in causing oxygen levels to flip from rising to falling. However, oxygen levels continued to fall even while large portions of the continents were protected from weathering by thick ice sheets. And they also continued to decline despite the fact that the last 4 interglacial peaks were warmer than earlier peaks (as shown by the Antarctic ice core data). This leaves only one possible explanation — falling global biomass.

As the authors of the study point out and as is confirmed by the Antarctic ice core data, CO_2 levels have not changed throughout this 800,000-year period, remaining within a tightly controlled range between ~180 ppm and ~280 ppm. Indeed, if you look at the CO_2 reconstruction shown back in figure 107, *CO_2 levels have essentially flatlined for the past 25 million years.* This is an important observation because it means that as the global climate cooled and got drier, biomass (and thus photosynthesis) declined to the point where the amount of new CO_2 added to the atmosphere by volcanic gases is just able to keep up with the amount of carbon that is lost from the carbon cycle as it gets trapped in carbon sinks. That can only happen if there's either a major increase in volcanic emissions (for which there's no evidence) or if global biomass (and photosynthesis) has been severely reduced. In short, biomass on our planet today is severely depressed by historic geological standards — the global

---

[25] Stolper, D.A., et al. (2016). A Pleistocene ice core record of atmospheric O_2 concentrations. *Science.* 353(6306). https://doi.org/10.1126/science.aaf5445

biomass hasn't been this depressed since the dinosaurs went extinct at the end of the Cretaceous period!

Since the Industrial Era began, atmospheric oxygen depletion has accelerated, from 0.00875 to 4 ppm per year. Although fossil fuel burning, increased rock weathering (as a consequence of all our earth moving and tillage), increased deforestation fires, and the rapid oxidation of humus in our soils are all playing a role in using up oxygen, the biggest factor is that through all our activities we have reduced global biomass by over 22% since the 1850s (as highlighted by geologist Timothy Casey's study featured in the previous chapter) and (if we are to believe Greenpeace's numbers) by about 50% over the last 5,000 years.[26]

The take-home message is that, even before our earliest human ancestors evolved, our cold, dry, CO_2-depleted world was much reduced from the thriving biosphere it once was when the planet was warmer, wetter, and more CO_2-enriched. This was a world already primed to begin accumulating atmospheric CO_2 if (1) some new "event" causes the planet's total biomass (and photosynthesis) to decline still further, (2) if some new "process" releases CO_2 from carbon sinks back into the atmosphere, or (3) both.

And, right on cue, along comes a big-brained primate with opposable thumbs, an appetite for clearing forests and plowing the soil, a tendency to cause massive soil degradation through his land management practices, and a habit of digging up fossil fuels to set them on fire. And so, the tumultuous human chapter of history begins....

---

[26] Weyler, R. (2018, Jul 18). How much of Earth's biomass is affected by humans? *Greenpeace.* https://www.greenpeace.org/international/story/17788/how-much-of-earths-biomass-is-affected-by-humans/

Chapter 25

Mining the Soil in My Father's Fields

"Read books and observe nature. If the two disagree, throw away the book."

— Dr. William A. Abrecht, soil scientist and the foremost authority on the relationship between soil fertility and human health.

My father's farming career began with a raw piece of virgin land. Some parts had previously served as open rangeland for a large local cattle ranch, some was covered by forests, and a lot of it was a waterlogged swamp. He built fences for his cows, cut down trees and pulled the stumps to make room for fields, trapped the beavers that were flooding the valley bottom, and bought crateloads of dynamite (!) to blast ditches through the sections of bog that were too wet to support a bulldozer (in those days farmers could still freely buy dynamite for farm purposes — as novelist L.P. Hartley once wrote, "*the past is a foreign country: they do things differently there*"). And yes, it sometimes went wrong — at one point he overdid one of his blast charges and ended up with a giant soggy sod mat hanging off one of the high voltage transmission lines that cross the farm on their way to providing electricity for our local community —the powerline survived unscathed nonetheless, as did my father, though I'm sure it caused a few lights to flicker in Lumby.

By the time I returned from university with my geology degree, his fields had only seen a mere twenty years of continuous tillage since he drained the swamps and busted the virgin sod with his plow. Upon my return, I did a series of soil tests on his pastures and fields that fundamentally changed my view of the world.

My geology degree had given me the context of the long view of geological history, which helped me see through the growing hysteria about CO_2-caused global warming that was emerging in those days. When Al Gore's climate film, *An Inconvenient Truth,* was released a few years later, I never imagined any serious scientist would take its fear-ridden message seriously. Nevertheless, I initially "trusted" the central story about fossil fuels being the main source of rising CO_2 levels. But then I did those soil tests, which we'll dive into shortly. In addition to the lessons that they taught me about how we were managing the soils on our farm, they also planted the first heretical seeds of doubt about the simplistic "consensus" fossil-fuel explanation for rising CO_2 levels.

Many farmers today take advantage of the free soil-testing services offered by fertilizer companies in order to plan their fertilizer program for the coming year. So did my family throughout my childhood. However, I had stumbled across Dr. William Albrecht's work on soil fertility while browsing through the book catalogue at *Acres USA* and decided I wanted a "second opinion" from a soil lab using Dr. Albrecht's technique (known today as the Albrecht-Kinsey method).

With degrees in biology, agriculture science, and botany, Dr. Albrecht served as chairman of the Department of Soils at the University of Missouri from 1938 until 1959. Dismayed by America's declining soil fertility, the loss of organic matter from the soil, the growing deficiency of nutrients and trace minerals, and the rapid chemicalization of farming that followed WWII, he became a vocal critic of conventional soil tests that hyper-focused on soil pH and nitrogen-phosphorus-potassium formulas at the expense of a more rounded view of soil fertility. Through his subsequent work, he became a foremost authority on the relationship between soil fertility and human and animal health and, by extension, the complex relationship between soil organic matter (a.k.a. humus) and soil nutrients.

The Albrecht-Kinsey soil testing method resonated with me because my geology education had taught me to view ecosystems as a comprehensive biological-geological *system* (just as I have laid out here in this book). The conventional soil testing approach focuses on measuring absolute levels of soluble fertilizer elements in the soil. Dr. Albrecht (and others who have followed in his footsteps) was more concerned with nutrient *ratios* because he believed that healthy mineral ratios combined with adequate levels of organic matter would enable soil microorganisms to thrive so that bacteria, fungi, and Darwin's beloved earthworms can do the hard work of turning inorganic sediments into *biologically-available* fertilizer elements — from this the concept of "insoluble yet biologically available" emerged. (If you are a gardener or farmer, a good technical primer to dive deeper into Dr. Albrecht's soil fertility research is *Hands-On Agronomy*, by Neal Kinsey and Charles Walters.)

In essence, Dr. Albrecht was pointing out that if you "feed the soil" and take good care of your humus, the soil will provide the necessary nutrients that plants need, which in turn will provide all the nutrients that animals and humans need. You can see why his views resonate in the regenerative agriculture community. In every possible way, this view of agriculture is the polar opposite of the hydroponics systems and factory mega-farms emerging at the other end of the modern agricultural spectrum, which treat plant growth like a factory-scale laboratory experiment that is hyper-focused on maximizing plant yields — after all, food is sold by the pound, not by mineral and vitamin content or nutritional value.

The point I am driving at here is that *there is a direct link between humus degradation and declining human health*. The way we are mismanaging our land isn't just turning our precious humus into CO_2 and making our soils more vulnerable to drought and flood, the far-reaching consequences of how we are mismanaging our land reach right into biological machinery at work inside every single one of our bodies' cells. Even if you aren't a farmer, the story of the century affects us all.

Humus is the central building block in a long chain of biological and chemical processes that lead directly to the nutrient levels in our food. But instead of taking care of our humus so it can take care of us, we're degrading it and degassing it, turning humus into CO_2, or simply eroding it away. Without humus, the minerals that were once stored within it are also gone, washed away by wind and rain, leaving soils depleted and exhausted. As Dr. Albrecht noted, *"If the soils that have lost their organic matter are to be restored,* **the loss of minerals, which has probably been fully as great, must be taken into account** *[my emphasis], and provision must be made to restore these mineral deficiencies before attempting to grow crops for the sake of adding organic matter."*[1] Dr. Albrecht's soil tests weren't about maximizing plant yields, they were intended to restore the nutrient balance of the soil *in tandem with rebuilding soil organic matter*, in order to fix the complete soil-nutrient-plant-animal ecosystem. Same fertilizer inputs, but with a completely different goal in mind.

Just as the mouldboard plow out on the Great Plains set forces in motion that ultimately unravelled the prairie ecosystem by making it

---

[1] Albrecht, W.A. (1938). Loss of Soil Organic Matter and Its Restoration. *U.S Dept. of Agric.* – *"Soils and Men, Yearbook of Agriculture 1938)*. pp 347-360. https://soilandhealth.org/wp-content/uploads/01aglibrary/010120albrecht.usdayrbk/lsom.html

extremely vulnerable to drought, decades of soil degradation and humus losses have also set forces in motion that are now playing out in terms of plant, animal, and human health. Soil degradation isn't just putting us on the path to desertification and ecological destruction, our health and the health of our animals are also in play. As I showed back in figure 23 (chapter 5), the average mineral content of calcium, magnesium, and iron in our food has dropped by a staggering 80-90% between 1914 and 2018. Even National Geographic has written at length about the significant 70-year-long documented nutritional decline of our food, highlighting in particular the declines in protein, calcium, phosphorus, iron, riboflavin, and vitamin. For example, wheat saw a 23% decline in protein levels from 1955 to 2016, while iron levels dropped by 30 to 50% in crops like sweet corn, red-skinned potatoes, and green beans.[2]

Health depends on a wide range of essential macronutrients, micronutrients, and trace minerals to optimize our health and productivity — not only in absolute amounts but also in correct ratios. And yet, the conventional approach to soil testing focuses primarily on the macronutrients that control yield, leading to chronic deficiencies in lesser nutrients that are nevertheless vital to animal and human health, with pharmacare and the supplement industry ready to sell you a remedy when those chronic deficiencies begin to express themselves as disease. Dr. Albrecht referred to these chronic deficiencies as a "hidden hunger." He wrote that, "*A declining soil fertility, due to a lack of organic material, major elements, and trace minerals, is responsible for poor crops and in turn for pathological conditions in animals fed deficient foods from such soils, and that mankind is no exception. [...] NPK formulas, (nitrogen, phosphorus, potassium) as legislated and enforced by State Departments of Agriculture, mean malnutrition, attack by insects, bacteria and fungi, weed takeover, crop loss in dry weather, and general loss of mental acuity in the population, leading to degenerative metabolic disease and early death.*"[3]

---

[2] Colino, S. (2022, May 3). Fruits and vegetables are less nutritious than they used to be. *National Geographic*. https://www.nationalgeographic.co.uk/environment-and-conservation/2022/05/fruits-and-vegetables-are-less-nutritious-than-they-used-to-be

[3] Schuurman, D. (2018, Aug 8). Healthy Soils, Healthy People: The legacy of William Albrecht by John Ikerd. *Biologix.co.nz* https://biologix.co.nz/blogs/news/healthy-soils-healthy-people-the-legacy-of-william-albrecht-by-john-ikerd

It has become common today to try to link every health issue to some agrifood chemical, pharmaceutical product, or pollutant — some of those concerns are valid, especially in an era when regulatory agencies seem to behave like marketing arms for political and corporate entities instead of upholding the scientific principles upon which they were founded. But as any farmer can tell you from decades of managing the health of their livestock, chronic nutrient deficiencies are every bit as damaging to health as unnatural things that are in food but shouldn't be, perhaps even more so because chronic deficiencies seriously impair the body's ability to regulate its own health when it is challenged by pests, parasites, toxins, and disease. Ask any farmer what a chronic selenium deficiency will do to his cows' fertility, or what a chronic copper deficiency will do to his cattle's health and productivity (incidentally, copper deficiency is recognized as one of the most common disease problems in beef cattle in California, causing diarrhea, poor weight gains, infertility, anemia, swollen joints, and reduced immune function, among other symptoms[4]). Why anyone would think that humans are exempt from the health consequences caused by nutrient deficiencies that are already affecting the health of many of our livestock is yet another of the unexplained mysteries of human irrationality. As Dr. Albrecht clearly demonstrated, soil erosion, humus degradation, nutrient deficiencies, and human and animal health are all inextricably linked. Everything is connected.

In keeping with the broader theme of this book, a short anecdote about the critical role of calcium in the soil will suffice to explain the differences between the conventional approach to soil fertility and Dr. Albrecht's approach. In the conventional soil fertility approach, soil pH is used as the measure to determine whether a soil is calcium deficient. By contrast, Dr. Albrecht focused on the ratio of calcium to magnesium, and on their combined ratio in relation other soil nutrients. Focusing on pH as a measure of calcium leads to scenarios where a soil may have the perfect soil pH, yet fields are nevertheless calcium deficient because magnesium can play the same role as calcium in controlling soil pH. That undiagnosed calcium deficiency will nevertheless cause the soil to become rock hard, the hard soil will struggle to absorb rainwater, various weeds linked to

---

[4] Maas, J. (2007, Oct). Methods to prevent copper deficiency in beef cattle. *California Cattlemen's Magazine – UCD Vet Views.*
https://vetext.vetmed.ucdavis.edu/sites/g/files/dgvnsk5616/files/local_resources/pdfs/pdfs_beef/cca0710-copper-defic.pdf

calcium-deficient soils will begin to thrive, and health issues related to calcium deficiency will begin to crop up in the livestock. For example, beef and especially dairy farmers often have to deal with cows that come down with milk fever (calcium deficiency) or grass tetany (magnesium deficiency) shortly after calving. Both conditions are extremely common, especially in the dairy industry, which often has to supplement their cows with calcium and/or magnesium to avoid these debilitating conditions. Both conditions make the cows extremely excitable, they begin to get tremors, then they go into paralysis. Both are ultimately fatal conditions if left untreated (usually requiring an IV infusion of calcium and/or magnesium, depending on which deficiency caused the problem). And it happens really fast — from onset to death can take as little as 12 hours. Grass tetany is a particularly memorable (and scary) condition because it also often transforms the most calm, docile cow into a wildly aggressive monster, which makes it especially difficult to bring sweet ol' Daisy back to the barn for treatment without getting yourself killed. But there's no *visible* sign of the "hidden hunger" until the cow's body begins to prepare for lactation, which dramatically increases her calcium requirements, at which point all hell breaks lose. As her calcium requirements suddenly spike, an already calcium-deficient cow feeding on a calcium-deficient diet will exhaust the calcium reserves in her blood *and in her bones* within a matter of hours, leading to the sudden calcium (or magnesium) crisis. The deficiency was already there; lactation exposed it. And it all began with a calcium deficiency in the soil.

The inorganic sand, silt, and clay in our soils have almost all the fertilizer elements a plant needs, but they are locked away inside minerals that are inaccessible to the plant. But a humus-rich soil supporting a thriving microorganism community will "digest" those sediments to release these fertilizer elements, which are then stored in the soil in insoluble yet biologically available forms until plants need them. But when humus degasses and erodes away, nature's self-improving soil-nutrient system breaks down along with it, leaving us ever more dependent on artificial sources of fertilizers and other agrochemicals, and ever more vulnerable to health issues as the "hidden hunger" takes its toll.

The Carbon-Nitrogen Connection

It's only possible to understand the full implications of what happens to soil as it loses its humus once you understand the central role that soil carbon plays in the nutrient cycles of other minerals. As I explained earlier, humus is the end product of breaking the complex carbon chains found in plant and animal tissues into their smallest elemental form. Humus is approximately 50% elemental carbon, 35% oxygen, 8% hydrogen, 5% nitrogen, 2% sulfur, plus a bunch of other trace elements. These tiny carbon particles behave a lot like ultra-microscopic inorganic clay particles (known to soil scientists as *colloidal clay*), which electrostatically bond with chemical nutrients in order to store those nutrients in the soil (this is why biochar is such a powerful soil amendment because the carbon that it adds to the soil dramatically boosts the soil's nutrient-storage capacity). And like those colloidal clay particles, soil carbon plays a vital role in facilitating the exchange of nutrients between soil and plant roots. However, as you will soon see, there are things that carbon can do that clay cannot.

The electrostatic bonds that colloidal clay and carbon particles use to capture, store, and exchange nutrients with plant roots work like the static electric charges that are formed by rubbing a balloon on your hair, which create weak but stable bonds between these colloidal clay or carbon particles and various fertilizer elements, like calcium, magnesium, phosphorus, potassium, and so on. Colloidal clay particles are negatively charged, so they bond with positively charged mineral ions until root acids break those weak electrostatic bonds to absorb those nutrients into the roots of the plant. This is what it means for a fertilizer element to be insoluble, yet biologically available.

The remarkable thing about soil organic carbon is that it is 5x more effective than clay particles at attracting and storing nutrients. That's one of the primary reasons why eroding humus causes such catastrophic declines in soil fertility. Furthermore, unlike colloidal clay, *these tiny carbon particles have patches of both positive <u>and</u> negative charges on the same particle*, so they can attract and bond with some mineral ions that clay particles cannot bond with. One of those nutrients is nitrogen.

The problem with nitrogen (i.e. nitrate fertilizers) is that they are negatively charged, so they are unable to bond with clay particles (like charges repel each other). So, if the plant doesn't absorb these nitrate fertilizers immediately after they are added to the soil, they are either washed away by water trickling through the soil or they evaporate back into the air. That's where soil carbon comes in because, unlike colloidal clay, the positively charged carbon particles are able to bond with nitrates, essentially turning soluble nitrogen fertilizers into insoluble yet biologically-available nitrogen compounds that can be stored in the soil in a stable form for later use.

Humus typically has a carbon-to-nitrogen ratio of around 10:1 (it ranges from around 9:1 to 11:1). Humus is quite literally the soil's nitrogen storage system. But soil degradation destroys that nitrogen storage capacity because, as the soil carbon is oxidized or eroded away, the nitrates that were previously bonded to the carbon particles are released back into the soil and are then washed away by rainwater, leached away by groundwater, or simply evaporate back into the atmosphere. Without humus, soil cannot store nitrogen.

And where does nitrogen come from in the first place when it isn't being delivered to the field by a fertilizer spreader or dropping out of the back end of a cow? Again, Dr. Albrecht sums it up best: *"Soil organic matter is the source of nitrogen. In the later stages of decay of most kinds of organic matter, nitrogen is liberated as ammonia and subsequently converted into the soluble or nitrate form. The level of crop production is often dependent on the capacity of the soil to produce and accumulate this form of readily usable nitrogen."*[5] In other words, decomposing plant and animal tissues are the original source of nitrogen, and so nitrogen is released into the soil as those tissues break down, where it promptly bonds with carbon in humus so it can be stored for later use by plant roots. If you deplete your humus, you better keep your fertilizer company on speed dial.

This is why dry climates are utterly dependent upon livestock — their role in "digesting" and trampling plant debris (along with their manure and urine) ensures that nitrogen from plant tissues is recycled back into the soil instead of simply degassing away. When livestock are removed from an arid ecosystem, microbes in the dry soil are not able to efficiently recycle nitrogen on their own. Without grazing herds of livestock, the fragile nitrogen cycle breaks down in arid climates. Without nitrogen, plants wither away and dry ecosystems quickly turn to desert. Removing livestock from dry ecosystems is every bit as misguided and ecosystem-destroying as Mao's attempt to eradicate the sparrow. Everything is connected.

---

[5] Albrecht, W.A. (1938). Loss of Soil Organic Matter and Its Restoration. *U.S Dept. of Agric.* – *"Soils and Men, Yearbook of Agriculture 1938).* pp 347-360. https://soilandhealth.org/wp-content/uploads/01aglibrary/010120albrecht.usdayrbk/lsom.html

Humus and animal manures aren't the only natural sources of nitrogen in the soil. Legumes have a symbiotic relationship with nitrogen-fixing bacteria (Rhizobium bacteria) that live on nodules on the legume roots — these bacteria convert nitrogen gas from the atmosphere into ammonia fertilizer. As long as there is enough carbon in the soil to capture and hold onto the ammonia produced by these bacteria, that ammonia will also be stored on the tiny soil carbon particles until root acids from growing plants tap into those nitrogen reserves. This is why many crop farmers include legumes in their crop rotations in order to increase nitrogen levels in the soil. But, once again, you need a rich storehouse of humus in the soil to be able to capture and hang onto the nitrogen produced by these nitrogen-fixing bacteria, otherwise those nitrogen reserves will be leached away by rain and groundwater or evaporate away before the plants can get around to using it.

There are also a number of other less known free-living nitrogen-fixing bacteria (like Azotobacter, Cyanobacteria, Nitrosomas and Nitrobacter) that also fix nitrogen directly from the air. These bacteria don't need legumes to help them do their work; they live directly in the soil. These free-living nitrogen-fixing bacteria were disregarded for a long time because, rather than creating ammonia, they create much larger molecules of nitrogen-rich amino acids by bonding nitrates to sugar molecules. However, there's increasing evidence that root acids are able to absorb these amino acids as a source of nitrogen, and even prefer them to other forms of nitrogen.[6] But once again, the ability to create and store these nitrogen-rich amino acids depends entirely on the amount of carbon in the soil.[7] If there's a biological keystone to Life on Earth, it is soil organic carbon — humus.

There's another curious phenomenon we need to discuss before we look at the results from the soil tests in my father's fields. It would be wonderful if we could create humus simply by mixing lots of plant debris

---

[6] Leu, A. (2012, Aug). Soil Organic Matter: Tips for Responsible Nitrogen Management. *AcresUSA*. https://www.ecofarmingdaily.com/build-soil/soil-inputs/minerals-nutrients/soil-organic-matter-tips-nitrogen-management/

[7] Cao, X., et al. (2016). Elevational Variation in Soil Amino Acid and Inorganic Nitrogen Concentrations in Taibai Mountain, China. *PLoS One*. 11(6). https://doi.org/10.1371%2Fjournal.pone.0157979

into the soil — if only it were that easy. Unfortunately, it is well documented that when plant debris is plowed under or buried in the soil, surprisingly little of it is converted into humus — most is "digested" by microbes living in the soil and quickly degases back into the air as CO_2. In other words, instead of being turned into humus, carbon-rich plant debris that gets buried in the soil simply becomes food for bacteria, gets digested, and floats away as CO_2 in the same way that most of the carbon that we ingest when we eat a meal is converted to CO_2 and gets exhaled back into the air.

Even more upsetting is that when you add nitrogen fertilizers to a soil that is full of buried plant debris, that extra nitrogen will rev up microbial activity, triggering a bacterial bloom that will dramatically accelerate how quickly buried plant debris will be digested and degassed. As a 2021 study from Illinois points out, *"The problem is that when microbes have a high nitrogen supply, they also have a high demand for carbon as an energy source.* **With high nitrogen rates their demand may exceed the carbon supply in residues, which may cause them to attack stable organic matter** *[my emphasis]."*[8]

In other words, highly soluble nitrogen fertilizers like those provided by the chemical fertilizer industry can stimulate soil bacteria so much that they not only completely digest and degas the plant residues in the soil (preventing plant debris from being converted into stable organic matter), but the bacterial bloom can even begin to attack and "burn up" the stable carbon in your humus. As another 2021 study notes, *"five decades of synthetic [nitrogen] fertilization led to a net decline in [...] storage of SOC [soil organic carbon][...], which is consistent with similar evidence from numerous other long-term cropping experiments throughout the world."*[9] Nitrate fertilizers are a temperamental beast.

There's an additional dimension to this. Without enough nitrogen fertilizer in the soil, something equally frustrating happens when you bury plant debris. Since bacteria feed on complex carbon chains, when fresh

---

[8] Quinn, L. (2021, Dec 15). Study clarifies nitrogen's impact on soil carbon sequestration. *University of Illinois Urbana-Champaign – College of Agricultural, Consumer and Environmental Sciences.* https://aces.illinois.edu/news/study-clarifies-nitrogens-impact-soil-carbon-sequestration

[9] Jesmin, T., Mitchel, D.T., and Mulvaney, R.L. (Short-Term Effect of Nitrogen Fertilization on Carbon Mineralization during Corn Residue Decomposition in Soil. *Nitrogen.* 2(4). 444-460. https://doi.org/10.3390/nitrogen2040030

plant debris is added to the soil, these bacteria immediately go to work to digest that new food source. But since they need nitrogen to do their work, they rapidly gobble up and consume all the readily available nitrogen in the soil, *including stripping nitrogen off the humus particles,* which temporarily depletes the soil of all its available nitrogen and leaves plants starved for nitrogen. Many novice gardeners make this mistake when they mix a bunch of plant residues into the garden soil in the spring to try to improve their soil's organic matter only to discover yellowing leaves symptomatic of a nitrogen deficiency or to see plant growth stall altogether. Even a "green manure" crop can trigger this nitrogen starvation phenomenon if the timing to plow it under is wrong, though the problem is most noticeable with fibrous lignin- and cellulose-rich crop residues that are more energy-intensive to digest, which is why many gardening books recommend never to mix wood chips into the soil (only leave them scattered on the soil surface) and why it is better to add compost rather than raw plant debris to the soil. Composting not only breaks down the raw organic matter into humus *before* it is added to the soil, but the composting process also ensures that any nitrogen released by the decomposing organic matter during the composting process will be captured and bonded to carbon particles instead of feeding humus-destroying bacterial blooms in the soil.

~

As a side note, successful composting requires a combination of green material (as a nitrogen source to fuel the digestive process of composting bacteria), brown material (as a source of carbon to serve as food for the bacteria), and oxygen (to give these bacteria the oxygen they need to drive the chemical reactions that break complex carbon chains into simpler carbon molecules). Without enough oxygen, the bacteria die even as molds, methanogens, and anerobic bacteria take over the compost heap. Without enough nitrogen (green material), the bacteria responsible for composting stay dormant and once again molds and fungi take over the compost heap. Without enough carbon (brown matter), the green material rots through bacterial action and is completely broken down into CO_2 without turning any of the carbon into humus. And without enough moisture, everything goes to sleep and the compost heap turns into a moldy bale of hay.

The solution to the nitrogen-starvation paradox comes from the Mulch Queen herself, famous gardener Ruth Stout, who advocated spreading plant mulch *on top of the soil*, rather than digging it in (she typically used old hay as mulch).[10] When old hay or other plant debris is spread on top of the soil surface rather than being mixed in, that debris not only keeps the soil covered and reduces moisture evaporation, it also allows the mulch to slowly compost at the soil interface, producing vast quantities of humus-rich soil at the mulch-soil interface without triggering any nitrogen deficiencies in the lower soil horizons. The work of worms and insects continually churning the soil then carries that freshly created humus down into deeper soil horizons.

By this simple mulching technique, the composting process happens at the soil surface rather than down in the soil, which prevents soil bacteria from revving up in the lower soil horizons where the roots live and, by consequence, the slowly composting mulch layer doesn't deplete nitrogen stores deep down in the soil. In essence, by spreading the mulch on top of the soil rather than mixing it in, Ruth Stout was allowing her old hay to compost *before* it became part of the soil, while simultaneously using that old hay as a mulch to keep the soil moist. Win-win. Plus, she didn't need to spend time or money turning a compost heap.

Perhaps it shouldn't surprise us that the Mulch Queen's approach works so well — it mimics the way that leaves and dead plant debris slowly break down on the forest floor. Once again, farming and gardening techniques that mimic nature work best. And yet, as I described in chapter 5, Bill Gates and his peers are funding research to bury tree trunks as carbon sinks, which will not only deprive the forest floor of the debris that is able to slowly compost to create fresh soil, but will provide bacteria, mold, and fungi with a giant food source underground that they will simply "digest" and turn back into CO_2. In order to build humus and replenish the soil nitrogen cycle, a healthy soil needs to be covered by rotting plant debris, including rotting tree trunks. If only Bill Gates and his peers got a little garden dirt under their fingernails from time to time, maybe they wouldn't pour their money (and ours) into such hare-brained schemes.

---

[10] Stout, R. (2021). *The Ruth Stout No-Work Garden Book: Secrets of the Famous Year Round Mulch Method.* 12 Sirens. ISBN: 978-1927458365

Mob grazing essentially creates humus in the same way — the livestock spread manure and trample plant debris at the surface but don't bury it deeply, which allows the plant debris to compost in the uppermost layer of the soil, just deep enough so the soil bacteria have enough soil moisture to survive, yet without depleting nitrogen stores at deeper soil levels from which the roots draw most of their nutrients. Once again, mob grazing successfully builds humus without starving either the soil or the plants of nitrogen, even as the manure paddies and trampled debris serve as a bit of a mulch to reduce soil moisture evaporation.

This is also the secret behind the success of no-till farming, which leaves plant debris to accumulate as a mulch at the soil surface to protect the soil from erosion, choke out weeds, and conserve moisture, even as that plant debris slowly composts to build humus, just like in Ruth Stout's garden. Win-win-win. Under no-till farming, fertilizers are also typically injected into the soil directly next to the seed rather than being broadcast across the entire soil surface, which reduces fertilizer leaching, reduces bacterial blooms, reduces how much fertilizer is needed to feed the crop, and starves weeds growing between the rows even as the crop gets the fertilizer it needs. Another win. Research consistently shows that no-till farming builds soil organic matter whereas, in most cases, tillage farming depletes it — now you know why.

All this also serves to highlight the importance of the traditional model of the mixed family farm in organic production systems (a model that combines crops and livestock on the same farm) because no-till farming doesn't have any easy answers for how to control weeds without pesticides — the mulching effect of plant debris and targeted fertilizer injections both help reduce weeds, but it's not a foolproof weed control system because the crop residues are nowhere near thick enough to fully suppress weeds in a corn or wheat field. Within the mixed-farm model, fields rotate back and forth between being used for tillage crop production and pastures — tillage can be used to suppress weeds during the crop production part of the cycle, but then the land is "rested" for a while by reseeding it to pasture so that the symbiotic relationship between livestock and grass can be used to rebuild the humus that was depleted by tillage. No-till builds humus slowly and consistently, whereas the mixed farm model alternates between rapid humus drawdowns during the tillage part of the cycle followed by rapid humus rebuilding phases during the pastureland part of the cycle.

However, a word of caution is required here. There is no one-size-fits-all solution — once you understand the overarching principles, it then becomes necessary to adapt those principles to different local climates (i.e. arid vs moist), different soil types, different seasonal rainfall patterns, different crop types, different topography (i.e. hillside vs flat land), etc. A thriving farm ecosystem is intensely local — never let an academic, politician, bureaucrat, or activist tell you otherwise as they try to impose blanket farm practices on all farmers.

What's right for Lumby's semi-arid Douglas fir grasslands isn't even necessarily right for Kelowna's arid pine forests, let alone England's wet climate. The mulching techniques of the Mulch Queen work well in moist climates, but in arid climates, even mulching is no guarantee that the garden will retain enough moisture to sustain the composting process at the peak of the dry season. Without enough moisture, composting stops and plant debris is instead slowly broken down by sunlight and oxidation as atmospheric oxygen slowly converts plant debris back into CO_2. You can see this in dry climates where there isn't enough moisture to rot dead plant debris — it slowly turns grey and becomes increasingly brittle and then, ever so gradually, it completely disintegrates as oxygen oxidizes the plant debris until it has all been turned back into CO_2. In dry climates, mulching isn't enough; you have to enlist the help of ruminant animals to recycle plant debris through their moist digestive tracts on the way to creating humus.

As Allan Savory has shown, the only way to effectively prep a garden plot in his arid local Zimbabwean region without relying on chemical fertilizers is to congregate cattle and other grazing animals over the plot where you want to plant the next season's garden. In that region, the cattle need protection from lions and other predators at night, so they build temporary movable corrals made of plastic sheeting to enclose the cattle at night. This concentrates their manure and tramples plant debris within the corral enclosure — that corral becomes the next year's fertile vegetable plot.[11] As you will see in chapter 27, traditional farming techniques in the dry regions of Africa kept moving their garden plots around to follow the cattle manure and to allow the grassland to reclaim

---

[11] Savory Institute. (2018, May 1). *Running out of Time | Documentary on Holistic Management.* [Video]. YouTube.
https://www.youtube.com/watch?v=q7pI7IYaJLI&t=271s&ab_channel=SavoryInstitute

and heal soils depleted by a period of tillage. In the dry regions of Africa, the cow isn't just a pastoral beast, until Europeans tried to impose their agricultural techniques onto the continent, the cow not only fertilized the cropland but was also absolutely essential to the process of rejuvenating exhausted cropland by turning that cropland back into grassland pastures.

By contrast, in the moist climate of the British Isles, earthworms, microbes, and mulch can build fertile soils without the help of livestock — livestock are an added bonus that will speed up the process of building humus in the soil, but they are optional because the humid climate and abundant rainfall allows plant debris to successfully compost right at the soil surface even without the help of the ruminant digestive system. And so, in the British Isles, simply resting the land (rewilding) by removing livestock and pausing tillage for a time period will slowly return the land to forest while enriching the soil. Likewise, leaving a field idle for a season (fallowing) works well in the moist European climate. But when fallowing or rewilding are used in arid climates, they are a disaster — removing livestock from the dry grasslands of the American Southwest or from the Zimbabwean savannah will rapidly degrade those arid grasslands into barren desert. In dry climates, even a single season without hungry mouths, sharp trampling hooves, and an ample supply of manure will interrupt the process that produces humus and recycles nitrogen, and thus will begin the slow deterioration of the ecosystem from grassland to weedy scrubland to barren soil.

The same is true of no-till farming — it is also not a one-size-fits-all solution. No-till works great in the dry climate and light soils of the semi-arid Great Plains. But in the wet humid climate and heavy clay-rich soils of the Netherlands or Germany, tillage is essential to oxygenate the soils, otherwise the heavy waterlogged soils become anaerobic, and crops won't thrive.

Earlier I showed you that soil erosion is much lower in Germany and in the Netherlands than almost anywhere else in the world despite the fact that they don't use no-till farming techniques. The flat topography reduces the risk of erosion by water, the moist heavy clay-rich soils are particularly resistant to wind erosion, and those same heavy wet soils (and humid climate) also reduce the risk of humus oxidation when the soil is aerated because the moist climate ensures that the soil stays wet enough to prevent oxygen from easily oxidizing carbon in the soil. In this wet climate, tillage adds just enough air to these soils, but not too much. As long as soil is fully saturated with water, there's little opportunity for oxygen to start reacting with the soil's humus.

By contrast, the dry soils out on the arid Great Plains (where the sodbusters plowed up the plains in the lead up to the Dust Bowl) are made of the fine windblown dust and silt that was left over after the glaciers retreated (this lightweight glacial silt is called loess) — this type of soil is extremely lightweight, so it is extremely easily eroded by tillage, by rain, and especially by wind. This lightweight soil does not need tillage to inject additional air into the soil, it already has plenty. The combination of lightweight soil and dry climate turn tillage into a surefire recipe for erosion and for rapid humus oxidation, but it's the ideal environment for no-till farming. Mao's deadly experiment with deep plowing likewise took place on the dry silty loess soils of north-central China, with similarly deadly results.

Even fire is either a friend or a foe depending on the environment in which it is used and on how hot it burns. In semi-arid regions that lack sufficient grazing animals, low-intensity fires are an essential alternative tool to burn away dead plant debris to prevent it from choking out the vulnerable grass shoots that emerge at the base of the sod clump and to quickly return nutrients and carbon to the soil before oxygen in the atmosphere oxides them away. Furthermore, while fire during the dry season is more likely to burn up the carbon in the dry soil, an intentionally lit low-intensity fire in early spring or late fall when the ground is wet can burn across the landscape without any risk to the humus below. In dry grassland climates, you either need livestock or low intensity fires to maintain a healthy grass sod — either will help add carbon to the soil, though livestock will rebuild humus faster while low-intensity fires do it slower. However, if the fuel load is allowed to build up, this will lead to high-intensity wildfires that scorch the underlying soil and burn away the humus, leaving a sterile humus-depleted wasteland behind — the hottest fires bake the soil like a pottery kiln leaving behind a hard reddish baked soil that has been stripped of all its carbon and nitrogen.

Livestock, low-intensity fires, no-till farming, mulch, mixed farming, earthworms, crop rotations, tillage, and so on are all merely tools in a large toolbox. The farmer's job is to understand the principles behind each one and then figure out a *system* that combines these tools in a way that is appropriate for each individual parcel of his land — in many ways, every field is its own ecosystem. Land management is as much an art as it is a science — observe, think, place things in context — that's why academia can inform and test, but should never have the ultimate say over how a farmer designs the system with which he manages his own land. In German there is saying that *"the eye of the farmer fattens the cow."* It's worth noting that the proverb does not say that 'the eye of the academic, politician, or bureaucrat fattens the cow'.

The Soil Test that Changed My View of the World

The thing that struck me when the soil tests came back from my father's fields so many years ago was the difference in organic matter between my dad's prime crop fields versus the marginal pastureland on the adjacent steep hillsides. Two adjacent pieces stood out in particular. The crop field had beautiful deep loam soils and was flat as a pancake. It lay at the foot of a steep adjacent hillside pasture (which had been grazed in a continuous grazing system, not via mob grazing). This hillside pasture had a wafer-thin soil skin covering the surface of hundreds of feet of coarse gravel and sand left over from an ancient glacial river delta that once flowed through our valley at the end of the last ice age — if you disturb the thin sod skin holding that gravelly soil in place on that steep hillside pasture, that soil is *gone!* Presumably, twenty years earlier when my dad first plowed the virgin soil in that flat, rich, fertile loam field on the valley bottom, that loamy soil must have had *at least* as much soil organic matter as the thin soil skin on the adjacent gravelly hillside pasture, if not higher. Was I ever in for a surprise.

The hillside came back at a respectable 4% soil organic matter. The prime cropland came back at meagre 2.5%. We'd been "mining the soil", as geologist David Montgomery would say. Although presumably the prime soil at the valley bottom probably once far exceeded the amount of humus on that steep hillside, let's assume that both plots started out at the same level of organic matter (around 4%. Over a period of only around twenty years we'd used up at least 38% of the humus in our prime cropland despite dutifully following all the government-recommended conventional agricultural best practices. This was the ultimate betrayal of "trust the science".

That loamy field is as flat as a pancake. Water erosion could not have carried away so much humus so quickly — the water in the swampy ditches draining the margins of this field were crystal clear. So where did all that carbon go?

The only explanation is that most of that carbon must have gone up into the air. Some was probably carried away by the wind after tillage during the dry summer heat. A lot of it was probably oxidized away by atmospheric oxygen once the sod mat was peeled off the virgin soil and especially each time another round of tillage mixed a fresh flush of oxygen into the soil. And some was probably digested and degassed by soil microbes after heavy nitrogen fertilization applications. For twenty years, my dad's valuable humus had been degassing away between our feet, slipping away as CO_2 despite following all the expert guidelines. I was gobsmacked. A farm is supposed to be a multigenerational "forever" business. And yet, that soil test spelled out clear as day that if we continued business as usual, our family's stewardship of the land would only last until the humus runs out. Tick tock.

When a representative from a fertilizer company presents soil test results at your kitchen table, the eyeball tends to hyperfocus on the big red dollar signs next to the big-ticket items like nitrogen, phosphorus, potassium, and sometimes calcium. Sometimes a trace mineral even gets a little attention if it is particularly deficient. But carbon doesn't get delivered to the farm on a fertilizer truck, so it tends to fade into the background. And the soil degradation happens so slowly that you won't notice any significant changes in the soil carbon numbers from one year to the next — the slowest changes are the hardest to recognize.

If I hadn't done the soil sampling myself that year, I wouldn't have noticed, much less understood the implications. The eye can only see what the mind is prepared to comprehend. By pushing the probe into the soil myself, by seeing the soil as I put the samples into the collection bags, and by seeing the topography where each sample was collected, I was primed to expect certain results. And so, when the numbers came back, the soil organic matter numbers jumped straight out at me because they contradicted what I expected to see. This shattered my confidence in the viability of the recommended best-management practices we were diligently following. It was turning out that "trusting the experts" wasn't the winning strategy that it purported to be. Perhaps if my father had come from a family that lived through the nonsensical advice spewed by government advisors in the lead-up to the Dust Bowl, he might not have been quite so trusting.

Let's put some hard numbers to those humus losses based on calculations provided by the University of Washington's Center for

Sustaining Agriculture and Natural Resources (CSANR).[12] A decline from 4% to 2.5% soil organic matter works out to 30,000 lbs of carbon lost *per acre* (that's 15 tons per acre!). Converting that carbon to CO_2 using the 3.67:1 conversion ratio works out to a loss of *55 tons of CO_2 per acre,* most of which was lost back into the atmosphere because of oxidation. In a period of less than 20 years. These are colossal numbers.

Most cropland around the world has been under agricultural production for a lot longer than the fields on my father's farm. Add in humus loses caused by pasture degradation, deforestation, desertification, and so on, and it only takes a little imagination and a simple pocket calculator to begin to question the simple story about fossil fuels being the overwhelming driver of CO_2 increases in the atmosphere. Figure 111 below shows a photo from a ranch where I apprenticed in Namibia many years ago that shows the effects of erosion triggered by continuous grazing in the arid Namibian savannah over a century of ranching. It doesn't take a rocket scientist to recognize the colossal amounts of carbon that have been eroded and degassed with such extreme amounts of soil erosion.

Figure 111. Soil loss on a cattle ranch in Namibia caused by poor grazing management. Author's photo.

---

[12] McGuire, A. (2023, Nov 7). Putting Numbers to the Difficult Task of Increasing Soil Organic Matter. *Washington State University – Center for Sustaining Agriculture and Natural Resources.* https://csanr.wsu.edu/putting-numbers-to-the-difficult-task-of-increasing-soil-organic-matter/

Globally, there are over 48 million km² of agricultural land (cropland and pastures combined), 40 million km² of forests (many of which are regularly harvested for timber), 14 million km² of scrubland in various states of soil degradation, and another 20 million km² of barren land (parts of which only recently degraded into desert as our planet's deserts continue to expand). If my dad's well-managed cropland, which followed all the recommended best-management practices, can lose that much carbon from the soil in only 20 years, how much carbon is degassing everywhere else? Factor in the massive reduction in photosynthesis caused by global biomass reductions (thanks to deforestation, desertification, and soil degradation), and you quickly arrive at an alternate explanation for why atmospheric CO_2 has been rising so sharply over the past century.

Now you understand why my ears perked up when I ran across the 2022 study from Skrable et al., which showed that fossil fuels only accounted for around 12% of the increase in atmospheric carbon dioxide instead of the 78% promoted by the climate brigade. Perhaps the suspicions raised by the soil tests in my father's fields were not entirely unfounded. It prompted me to dig deeper.

Now consider the 10:1 ratio of carbon to nitrogen stored in humus. A loss of 30,000 lbs of soil organic matter per acre represents a loss of 3,000 lbs per acre of stored nitrogen that has also disappeared. Consider how much extra off-farm fertilizer has to be purchased to make up for these soil nitrogen losses in order to sustain annual crop yields! The humus degradation is small enough from one year to the next that farmers don't notice, but over time the costs creep up and it becomes a recipe for bankruptcy as the cost of inputs overwhelm the farmer's bottom line. The same is true for all the other minerals and fertilizer elements that were once stored in that humus before it eroded and degassed away.

And what does it take to rebuild that lost humus? Again, the CSANR numbers provide some startling insights. On average, you have to compost 10 tons of raw plant material to produce 1 ton of soil organic matter. So, if you were trying to use composting to increase soil organic matter from 2.5% back up to 4% soil organic matter, you would have to compost 300,000 lbs of plant debris *per acre* (150 tons per acre) to fix this. What works in a small garden setting simply isn't feasible at this scale of agricultural production.

What about switching to no-till farming, since no-till farming is a net carbon sink? Using the numbers from the University of Nebraska-Lincoln, which found that no-till farming added, on average, around 0.77 tons per acre of carbon to the soil per year, it would take around 19 years to rebuild the humus that was lost from my father's field — roughly the same amount of time as it took to lose it in the first place.

By contrast, based on Dr. Richard Teague's numbers that I quoted back in chapter 22, which found that holistic planned grazing can rebuild soil organic matter at up to 8.6 tons per acre per year (on average), returning this cropland to pastureland and managing it under an intensive mob grazing program could, theoretically, rebuild all the lost carbon in as little as two years!

These numbers highlight why before the era of hyper-specialization, chemical fertilizers, and shareholder meetings, the sustainability of the traditional family farm in bygone eras was utterly dependent on maintaining a herd of livestock alongside their cropland. The livestock were more than just a source of manure to feed the plants and fertilize the soils, they were also the tool that enabled the farmer to rotate fields between crops and pastureland in order to rebuild humus degraded by tillage in order to prevent the slow degradation of the farm ecosystem. The crops provided higher value sources of income, but the livestock and pastures ensured that future generations would inherit a farm that still had humus upon which to grow crops. Ancient historians from both Roman and medieval eras record that soil degradation was worst during eras when social factors (like high tax rates, aggressive feudal tithes, or plantation-style ownership focused on growing monoculture commodity crops) incentivized farmers to reduce their livestock numbers to put a greater portion of their land into crop production to squeeze higher profits out of their land.

That's also why we need family farmers, not agricultural conglomerates, out there working the land — the independent farmer has future generations in mind and will work out the subtle details of how to rebuild humus and perfect crop rotations so that there's still a farm left for his grandchildren to inherit (as long as politicians don't take away all the farmer's tools in their crusade to "save" the planet or squeeze the farmer so hard that he has to sacrifice tomorrow's humus to pay his bills today). The incentives of the multigenerational family farm encourage long-term thinking far beyond what employees and shareholders ever think about.

The employees that work for the mega-corporations, like the slaves on Roman-era plantations, have no incentive to incorporate that kind of long-term thinking into their day-to-day activities since they get no benefit from them. If anything, they are rewarded for maximizing this year's profits and punished if this year's profits are sacrificed for something that takes generations to bear fruit. Likewise, the shareholders of agricultural corporations are focused only on their return on investment — yields, not soils. The kind of long-term thinking required to keep farm ecosystems healthy requires people to not only live on the land, but to own it, and to own it for generations.

Whichever route you choose to repair damaged soils (grazing, composting, mulching, no-till farming, etc.), once you do rebuild the humus, you also have to replenish all the minerals and fertilizer elements that were lost. You can either wait for soil microbes, root acids, and earthworms to do their work of breaking down sand, silt, and clay particles to release fresh minerals locked up inside of these rock particles (a process which happens relatively quickly under holistically managed pastures) or, as Dr. Albrecht visualized, in a crop setting you can do it with the help of a fertilizer truck (which may be the only option if you need to keep harvesting crops to pay the bills or if the farm isn't set up with the infrastructure to manage livestock). Fertilizers are nothing but one tool among many — banning certain fertilizers, as our politicians would like to do, merely cripples a farmers' ability to repair nutrient losses from the past and encourages more mining of the soil. Knowledge makes the world a better place, diktats and central planners destroy it.

Today, a little over twenty-five years after I took those soil tests on my father's farm, that same loamy field is hovering just over 2% organic matter. I don't know at what level it bottomed out in the intervening years since I left the farm, but it is now rebuilding soil organic matter under an aggressive manure and compost spreading program, combined with various crop rotations as befits the overall farm plan. Knowledge is power.

But if you think those soils are in safe hands now, just wait until you get to the Epilogue of this book where I tell you about the latest government policies to throw a wrench into the future of cattle farming in British Columbia. The Ministry for Everything is at it again — and what they have recently done to cattle farmers in B.C. makes my blood boil.

Chapter 26

From Stone Age Farmers to Agri-Food Corporations

"A great civilization is not conquered from without until it has destroyed itself from within."

— Will Durant.

The first evidence that humans began to fundamentally change carbon dioxide levels in the atmosphere does not come from the era of rapid industrialization that began in the 1950s, nor from the era following the invention of the diesel engine in 1893, nor even from the era following the invention of the steam engine back in 1712. Once again, we have to look at CO_2 records preserved in the Greenland ice core data to see the first evidence of Man's impact on the atmosphere — that's the ice core data I showed you in figure 55 back in chapter 8.

To recap, the Greenland ice core data shows that after the ice age ended, temperatures peaked during the Holocene Optimum around 8,000 years ago, with trees growing hundreds of miles north of where they grow today. Despite the many wild temperature swings that happened between then and now, the overall temperature trend since the Holocene Optimum has been down, albeit slowly. The current warming is merely a counter-trend rally in the longer-term downtrend. The Minoan Warm Period, the Roman Warm Period, the Medieval Warm Period, and the Modern Warm Period were each successively cooler than the warm periods that preceded it.

During previous interglacial periods, CO_2 levels always peaked shortly after the ice sheets retreated, slightly lagging temperatures that also peaked within the first few thousand years after the ice sheets melted. That early peak marks the high point during each warm interglacial — from that point onwards temperatures always gradually began to cool, and CO_2 levels always continued to gradually decline during the remaining millennia of the warm interglacial period until, around 13,000 years or so after the ice sheets initially melted, both temperature and CO_2 levels suddenly dropped off a cliff, marking the plunge into the next glacial advance. For example, during the warm interglacial period 800,000 years ago, the gradual decline in temperatures was mirrored by CO_2 levels, which declined by 17 ppm during the 8,000 years that followed the initial peak.[1]

The Greenland ice core data shows that in the early years of our own warm interglacial period, CO_2 levels again initially began following that same well-established pattern as both temperature and CO_2 both slowly declined from their peak after the ice sheets melted. But then, starting 7,000 years ago, CO_2 broke with the established pattern. Despite temperatures continuing to cool, CO_2 unexpectedly began a slow but steady climb that continued uninterrupted right through into modern times. Something new happened 7,000 years ago to break the well-established pattern that had happened during every other previous interglacial period.

According to the Greenland ice core data shown in figure 55, from the point when this trend reversal began around 7,000 years ago until just before the beginning of the Industrial Era in 1750, instead of continuing to fall like we would expect from previous interglacial periods, CO_2 reversed course and instead rose around 25 ppm from 258 to 283 ppm. According to a 2003 study by professor of paleoclimatology William Ruddiman from the University of Virginia, the CO_2 increase over those 7,000 years was even larger than that shown by the Greenland ice core data in figure 55, rising by 40 ppm, not 25.[2] This decoupling of direction tells us that something monumental happened 7,000 years ago to disturb the balance between all the long-standing drivers that control the amount of CO_2 in the atmosphere — some new force had entered the equation that did not exist during previous interglacial cycles.

From everything we have learned over the course of this book, there are only three things that can raise CO_2 levels against the backdrop of a cooling trend: (1) a surge in volcanic activity (for which there is no evidence), (2) a decrease in biomass (less biomass = less photosynthesis),

---

[1] Stanley, S. (2016, Jan 19). Early Agriculture Has Kept Earth Warm for Millenia. *Eos Magazine.* https://eos.org/research-spotlights/early-agriculture-has-kept-earth-warm-for-millennia

[2] Stanley, S. (2016, Jan 19). Early Agriculture Has Kept Earth Warm for Millenia. *Eos Magazine.* https://eos.org/research-spotlights/early-agriculture-has-kept-earth-warm-for-millennia

or (3) an increase in some other form of CO_2 emissions (something other than volcanic emissions, like an uptick in fire, humus oxidation, deforestation, and/or desertification).

7,000 years ago also marks the point in time when the planet's first large civilizations emerged — in Mesopotamia, in Egypt, in China, and in the Indus Valley. It's hard not to suspect that our ancestors' fingerprints are all over this.

The first hesitant steps towards domesticating plants and animals began even earlier, around 10,000 years ago, but it took thousands of years to evolve the complex social structures to compel people to accept living under the dominion of a distant centralized authority. It's quite a leap to transform a small close-knit kinship-based tribal society into an obedient densely-populated civilization that accepts the authority (and tax demands) of some leader that most people have never even seen in person — a leader who acquires his status purely based on his hereditary status rather than on a track record of competency and yet can nonetheless project the authority of his state onto the lives, property, and labours of his subjects, even imposing conscription to force people he has never met and to whom he is not related by blood to fight and die for his military ambitions if he so desires. In other words, although we began by domesticating cattle, sheep, goats, chickens, grain, and vegetables, it just took a little longer to develop the social and cultural innovations to enable us to domesticate ourselves.

As always, changes in politics are inextricably linked to changes in land use. The leap from tribal farmers living in small rural villages to the establishment of large urban civilizations led to soaring population numbers and a corresponding dramatic intensification of agriculture to support these large, rapidly growing urban populations. The earliest evidence of the plow dates to 6,000 years ago in Mesopotamian and Egypt and to 4,500 years ago in the Indus Valley. In China, large-scale rice cultivation and dry-land agriculture emerged around 7,000 years ago. As agriculture expanded its footprint and intensified its impact on the land, soil degradation accelerated dramatically. And ever larger tracts of forest were cut down to harvest timber and clear space for farms in order to keep up with the demands of a rapidly expanding civilization. And so, CO_2 began to accumulate in the atmosphere under the combined effects of accelerating deforestation, accelerating soil degradation, the draining of wetlands, and biomass reduction. Pollen grains trapped in sediments in

Europe and archaeological data from China provide a record of vegetation changes across this time period and reveal the accelerating deforestation that accompanied our expanding footprint throughout the pre-industrial era.[3]

By around 5,000 years ago, atmospheric methane levels also began increasing.[4] Since we know methane doesn't last long in the atmosphere (only 12 years), this indicates that some new sustained force began actively producing large quantities of methane at that time, year after year, continuously over thousands of years, all of which was soon converted to become CO_2. The obvious suspects to trigger this rise in methane are the widespread adoption of irrigated rice paddies in Asia (which began during that era), the drainage of wetlands (which releases large amounts of methane), and accelerating erosion (the carbon-rich soil that erodes off fields and mountains slopes accumulates in lakes, bogs, and river deltas where it begins to decompose in anaerobic mud deposits where methanogens thrive — for example, filling in the northern end of the Persian Gulf with billions of tons of soil eroded from Sumerian, Babylonian, Mesopotamian, and Assyrian hills and fields inevitably produced a massive surge in methane percolating out of the accumulating mud at the bottom of the Persian Gulf).

I'm far from the first to be struck by the fact that CO_2 began to rise around the same time as the first large-scale civilizations emerged — Professor Ruddiman first drew attention to this correlation in 2003. Back in chapter 23 in the discussion about desertification, I included the following quote from Project Wadi Attir, "*land degradation and desertification [over the last 7,800 years] has resulted in carbon dioxide emissions estimated at* **450 – 500 gigatons of carbon** *(Ruddiman, 2003, Fig. 5; Lal, 2004),* **which corresponds to more than the total amount of CO_2 emitted from fossil fuel combustion so far**." According to Our Word in Data, from the start of the Industrial Era in 1751 until 2003 (the year of Professor Ruddiman's study) the world emitted 1.13 trillion tonnes of CO_2

---

[3] Stanley, S. (2016, Jan 19). Early Agriculture Has Kept Earth Warm for Millenia. *Eos Magazine*. https://eos.org/research-spotlights/early-agriculture-has-kept-earth-warm-for-millennia

[4] Hance, J. (2008, Sep 3). Did prehistoric farmers drive early global warming? *Mongabay News*. https://news.mongabay.com/2008/09/did-prehistoric-farmers-drive-early-global-warming/

from burning fossil fuels[5] — using the conversion factor of 3.67 to 1 to convert from CO_2 to carbon, this works out to only 308 gigatons of carbon released by the burning of fossil fuels compared to 450 – 500 gigatons of carbon released by land degradation and desertification! This rise in CO_2 levels beginning 7,000 years ago, caused by deforestation, soil erosion, the draining of wetlands, and rice farming delivers yet another fatal blow to the already fatally destroyed consensus climate narrative.

If the destructive processes triggered by our expanding footprint on the planet raised atmospheric carbon dioxide by 40 ppm during the first 7,000 years of civilization (before we started burning fossil fuels), those same destructive processes clearly also dramatically accelerated starting in the late 19th century when humanity began a rapid 8-fold explosion in the global population (see figure 112 below) and a corresponding 4-fold expansion in the agricultural footprint (see figure 103 back in chapter 23).

---

[5] Ritchie, H. (2019, Oct 1). Who has contributed most to global CO_2 emissions? *Our World in Data.* https://ourworldindata.org/contributed-most-global-co2

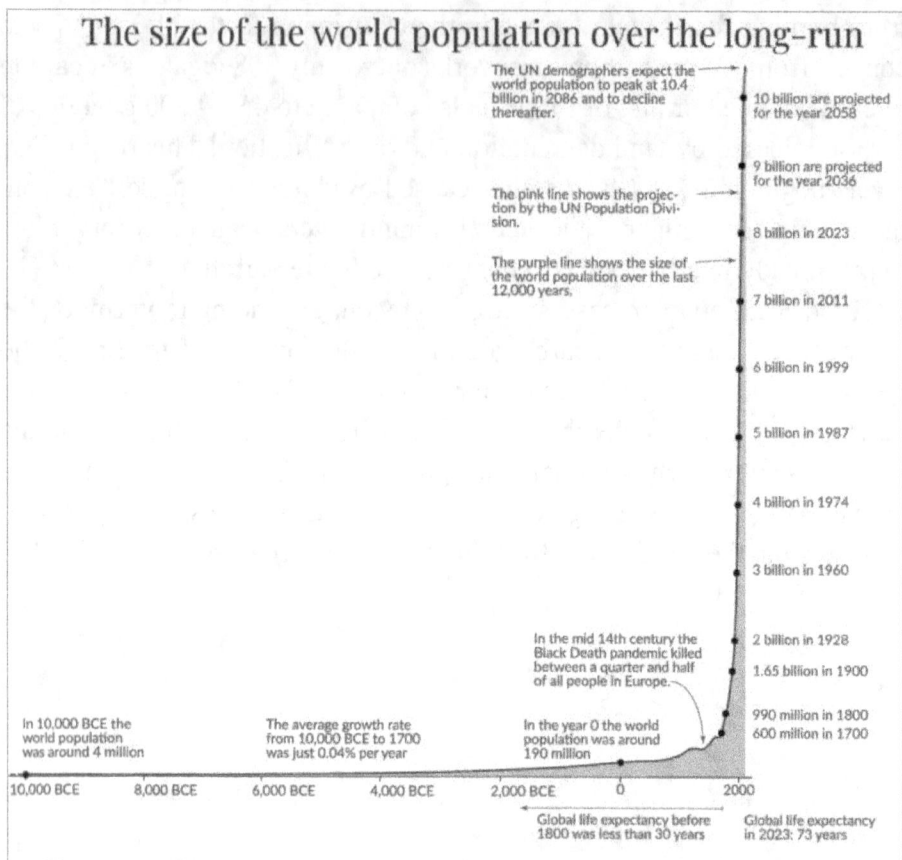

Figure 112. Population growth over the last 10,000 years. Population has expanded 8-fold since the late 19th century.

https://en.wikipedia.org/wiki/File:Annual-World-Population-since-10-thousand-BCE-1-768x724.png
CC BY 4.0 DEED

Since the end of the 19th century, CO_2 levels have increased by roughly 140 ppm. As I mentioned previously, the "consensus" climate narrative claims that fossil fuels are allegedly responsible for 78% of this increase (roughly 110 ppm of the 140-ppm increase), which leaves a mere 30 ppm to other non-fossil-fuel processes. It's a patently ridiculous assertion. For that to be true we would need to believe that although deforestation, soil degradation, desertification, wetland drainage, and rice farming during the pre-industrial era (committed with hand axes and horse-drawn plows) was responsible for raising atmospheric CO_2 by 40 ppm, we are nevertheless supposed to believe that those same destructive non-fossil-fuel processes after the start of the Industrial era was only able to contribute a much smaller additional 30-ppm increase despite an eight-fold population increase, despite a four-fold expansion of our agricultural footprint, despite vastly more powerful machines that allowed us to plow much deeper and disturb far more soil than ever before, and despite the fact that the Industrial Era marked the point when we learned to extend our exploitation of the planet's resources to almost every corner of our globe in order to cut down enough trees, till enough soil, drag-net enough oceans, and drain enough swamps to harvest enough resources to feed humanity's growing appetites.

Are we to believe that oxen, wooden plows, and hand-operated garden hoes in the pre-industrial era caused more CO_2 degassing from our soils than deep tillage with modern tractors pulling steel-bottom plows? Were deforestation fires more destructive in the pre-industrial era despite the fact that our pre-industrial ancestors only cleared less than a quarter of the forest acreage that has been cleared since the Industrial Era began? Was the traditional mixed-farm model of yesteryear that regularly tried to rebuild humus using manure and crop rotations more destructive than the monoculture plantations of today, which cover vastly larger areas, and which largely ignore declining humus levels because we have easy access to chemical fertilizers to make up for nutrient deficiencies as soil organic matter declines? Did the thriving grasslands of yesteryear release more CO_2 into the atmosphere before the Industrial Era killed off most of the world's large migratory herds and before soil degradation and desertification spiralled out of control in these grassland ecosystems? Did shovels, pickaxes, and handsaws disturb more soils and than bulldozers, excavators, and clearcut logging? Did wars waged with spears and arrows create more wastelands than our modern industrial-scale warfare that leaves entire ecosystems and entire countries in ruin? Did an impoverished feudal society with less than $1/8^{th}$ of the population consume more raw commodities sourced from the land than the global consumer society of today? We know the answer to these questions, but the activist-scientists have been successfully gaslighting us to ignore these facts, to ignore the story told by the Greenland ice core data, and to ignore the implications of Professor Ruddiman's calculations.

Nothing over the past 7,000 years before the start of the Industrial Era comes anywhere close to the scale of environmental degradation that has happened over the past 120 years. So, if anything, we would expect the CO_2 released by these destructive processes during the Industrial Era to be orders of magnitude greater than the initial 40 ppm increase calculated by Professor Ruddiman, not less. And so, Professor Ruddiman's efforts to reconstruct CO_2 and methane levels during the pre-industrial era provide us with yet another strong line of evidence to demonstrate that fossil fuel burning is a mere side-show compared to the main drivers of rising CO_2.

Professor Ruddiman took his research one step further — this is where our conclusions part company. He proposed that the greenhouse effect from that 40-ppm human-triggered pre-industrial CO_2 increase (from the time of the first large civilizations 7,000 years ago until the end of the pre-industrial era) is the reason why the world hasn't already slipped back into another ice age. According to his theory, the Canadian High Arctic would already be under permanent snow cover, with the first seeds of ice sheets beginning to grow, were it not for the CO_2 released by farming and deforestation over the past 7,000 years. That's hardly likely considering everything we've learned over the past chapters of this book about how insignificant the greenhouse effect of CO_2 is on global temperatures. Furthermore, it ignores the fact that previous warm interglacial periods lasted around 13,000 years (on average) so there's no reason to suspect that this one would have ended so much sooner. Nevertheless, it is a theory that is perfectly suited to our algorithm-driven era in which everything is viewed through the lens of CO_2 as the control knob on the climate. If your trusted peers assign a high value to CO_2 feedback effects in climate models, this is precisely the kind of conclusion you would expect to emerge from a climate model.

The Erosion of Civilizations

Professor Ruddiman's calculations about how much CO_2 was released into the atmosphere by pre-industrial agriculture and by deforestation over the past 7,000 years is well supported by archaeological and geological evidence illustrating the scale of erosion and soil degradation that accompanied the rise of increasingly populous civilizations across those long millennia. In his book, *Dirt: the Erosion of Civilizations*,[6] geologist David Montgomery meticulously documents soil degradation since the dawn of civilization; it's a sobering read to see how, with few exceptions, you can follow the rise and fall of civilizations based on how fast they used up their soils. Obviously, there are lots of factors that come into play

---

[6] Montgomery, D. R. (2012). *Dirt: the Erosion of Civilizations*. University of California Press.

to tip a civilization over the edge, but as soils become exhausted, the wealth of once-flowering civilizations begins to decline, population pressures mount, famine and malnutrition increases, soils become more vulnerable to drought and other natural climate cycles, it becomes harder to sustain large armies, emigration and colonization accelerates as the struggling population begins to yearn for greener pastures, and the home front increasingly deteriorates into political turmoil as the ailing civilization turns on itself. As Will Durant once said, *"a great civilization is not conquered from without until it has destroyed itself from within."* What is true about rotting political institutions and rotting cultures that lose a sense of themselves is equally true about the soils that sustain nations — everything is connected.

Although there are many other factors that influence the lifespan of civilizations and although different civilizations consumed their soils at different speeds depending on what kinds of agricultural techniques they used, a common theme that emerged during David Montgomery's research is that a large number of completely unrelated civilizations, across multiple continents, all have lifespans of around 1,000 years before they collapse. A few unravel faster if they consumed their soils faster, but only a handful of civilizations ever managed to sustain their soils for longer than that. We know that these civilizational collapses did not happen as a consequence of large climate shifts because even as older civilizations collapsed into drought, chaos, and famine, their younger neighboring civilizations thrived. As Montgomery so clearly demonstrates, the story of that internal deterioration perfectly mirrors how quickly each civilization runs down the clock on their soils — when the clock runs out, civilization collapses.

Each civilizational collapse is followed by a long dark age with drastically diminished populations during which the soils slowly begin to rebuild their humus, thus laying the groundwork for a reinvigorated agricultural base that can sustain a renewed flowering of civilization... only to repeat the same cycle of erosion, decline, and collapse all over again. For example, Montgomery describes the Argolid Peninsula in Greece (part of the Peloponnese peninsula), which used up its soils during the Bronze Age — the Bronze Age Collapse was followed by a long dark age, until, in time, classical Greek farmers emerged to farm the land once more until it collapsed again, but then in the late Roman period the region flourished once more before it was depopulated once again, with another flourishing emerging in the 7th century AD, followed by yet another collapse. As he writes in his book, *"About fifteen inches of soil are estimated to have been lost from the Argolid uplands since the start of Bronze Age agriculture. As much as three feet of soil may have been stripped from some lowland slopes."*

He summed up the process as follows, *"Initially, agriculture in fertile valley bottoms allowed populations to grow to the point where they came to rely on farming sloped land. Geologically rapid erosion of hillslope soils followed when vegetation clearing and sustained tilling exposed bare soil to rainfall and runoff. During subsequent centuries, nutrient depletion or soil loss from increasingly intensive farming stressed local populations as crop yields declined and new land was unavailable. Eventually, soil degradation translated into inadequate agricultural capacity to support a burgeoning population, predisposing whole civilizations to failure."*

Archaeologists have long argued about the causes of the Bronze Age Collapse during the 11th century BC, when all the flowering Bronze Age civilizations around the Mediterranean all collapsed within the span of only 50 years. A variety of explanations were proposed for the collapse based on archaeological evidence — war, disease, drought, volcanic eruptions, invasions from the Sea Peoples, changes in military technology, and so on, each backed by compelling archaeological evidence, yet unsatisfactory nonetheless as there was no clear unifying link between them. And yet, writers in Ancient Greece writing only a few centuries after the Bronze Age Collapse had a more nuanced tale to tell. Plato, Aristotle, and Theophrastus were all convinced that soil degradation was at the root of the problem — a consequence of civilizations mining their soils until the land stopped supporting their populations, leading to war, famine, drought, and massive migrations of displaced peoples.

In our own era, geologists have teased apart the subtle evidence to confirm the explanations offered by these ancient Greek thinkers, clearly demonstrating that the flowering civilizations of the Bronze Age succumbed because they used up their soils. To geologists, the flourishing warm period during the Bronze Age was called the Minoan Warm Period — the Greenland ice core data back in figure 55 shows it was significantly warmer than temperatures today. The Bronze Age Collapse happened as that warm period abruptly ended and the globe was plunged into a prolonged period of cold and dry weather. Sediments record a period of drought at the time of the Bronze Age Collapse triggered by a sharp cooling of Mediterranean Sea surface temperatures, which reduced evaporation from the sea and, by consequence, reduced the amount of rain falling over adjacent lands.[7] And yet, geologists also discovered that the climate during the Bronze Age Collapse was, in fact, no drier than during previous dry cycles that had occurred in the region since the end of the last ice age. However, before the region was populated by farmers, previous dry cycles did not suffer the widespread soil erosion that characterized the Bronze Age Collapse — the ecosystem of the region weathered these dry cycles just fine before the start of large-scale civilizations. But during the Bronze Age Collapse, things completely fell apart.

The sedimentary record tells a cautionary tale of the consequences of centuries of deforestation and soil erosion stripping the Mediterranean region of humus and vegetation. When the drier and cooler weather began at the end of the Minoan Warm Period, the degraded landscape was incapable of withstanding the drier conditions. And so, a perfectly natural dry phase in the regional climate instead triggered an ecosystem collapse not unlike that of the 1930s Dust Bowl out on the Great Plains, which in turn lead to widespread crop failures, displaced peoples, malnourishment, raiders from the sea and invading tribes on land, and ultimately war. Countless cities were laid to waste, leaving only ash and ruin in their wake. Over 90% of all small cities and towns on the Peloponnese peninsula in southern Greece were abandoned, suggesting a massive depopulation of the region. Furthermore, countless temples were burned throughout the region, never to be rebuilt, suggesting a massive cultural upheaval and a wholesale rejection of the traditional belief systems upon which these communities were built — the gods had failed them, the sacrifices had been made in vain, and so the gods were punished by being condemned to the ash heap of history.

Ancient Rome faired no better. Ancient Roman historians documented drastic declines in agricultural yields across the centuries until many farms were utterly abandoned, with land increasingly flowing into the hands of large landholders even as their tenants slid into chronic debt servitude. Over time, Rome became utterly dependant on grain imports from Egypt

---

[7] Drake, B.D. (2012). The influence of climatic change on the Late Bronze Age Collapse and the Greek Dark Ages. *Journal of Archaeological Science*. 39(6). https://doi.org/10.1016/j.jas.2012.01.029

(which was able to sustain its bountiful harvests thanks to the annual Nile floods that replenished the soil each year with humus washed off the Ethiopian Highlands during the rainy season, which was deposited in fresh layers on Egyptian fields as the floodwaters receded). It is questionable whether Rome would have survived as long as it did without Egyptian grain — whoever controlled Egypt controlled the Mediterranean. Indeed, the fresh soil deposited on Egyptian fields by the annual floods is the *only* reason why Egypt was able to sustain a continuous civilization across thousands of years even as other great civilizations came and went. Egypt's farm soils only began to decline after the Aswan Dam was built in the 1960s, which stopped the annual flooding as the mud that had previously fertilized farmers' fields instead settled to the bottom of the reservoir, and as the introduction of modern irrigation techniques raised the water table, which began causing massive problems with salinity building up in the soils of Egyptian fields.

As the Roman empire grew larger, its original small-holder farmers were replaced by ever larger landowners using tenant farmers and slaves to grow plantation-style commodity crops destined for export to distant cities. Flooding a country with cheap labor (either through slaves captured in war, by impoverishing local citizens, or by taking advantage of desperate migrant labour) creates the cheap labour required for plantation-style agriculture while destroying the time-tested model of independent multigenerational farmer who actually care about the quality of the soil that their grandchildren will inherit. Roman writers like Columella and Pliny described in detail how damaging slave labour was to the soil because the watchful eye of the autonomous farmer was replaced by distant city-dwelling landlords and by hired overseers telling slaves what to do. Few who work someone else's land notice the slow but steady degradation of the soil; fewer still care while labouring for someone else's benefit. Stewardship of the soil requires a multigenerational vested interest in creating an inheritance for the next generation.

Columella, writing in the 1st century AD, wrote that, *"So I am decidedly of the opinion that repeated letting [renting out] of a place is a bad thing, but that a worse thing is the farmer who lives in town and prefers to till the land through his slaves rather than by his own hand. [...] it is better for every kind of land to be under free farmers than under slave overseers [...]. To such land a tenant farmer can do no great harm [...] while slaves do it tremendous damage."*[8] Today's ideological bureaucrats and politicians would do well to heed Columella's wise words as they try to micromanage farmers from distant halls of power as they introduce one policy after another that undermines the model of the traditional family farm to the benefit of large shareholder-owned corporations and absentee landlords like Bill Gates, who view farmland as investment opportunities (Bill Gates is the single largest landowner in the United States, owning over 270,000 acres![9]).

As Ancient Rome continued to decline, the migration of impoverished tenant farmers fleeing from the countryside to the cities reversed as countless lower-class plebians fled the crumbling cities in search of land for subsistence farming out in the countryside. By then soils in the countryside around Rome were so degraded that the slave-run plantation-style farming system was breaking down and was slowly reverting back to the mixed-farm model of subsistence farms, out of necessity, because slave labour was completely unsuitable for managing the complicated system of animal husbandry and crop rotations required to sustain a mix of livestock and crops in a mixed-farming system. Furthermore, with over-extended Rome no longer conquering fresh territory, the supply of cheap slaves had dried up and so the farm model had to change. Landowners began offering mutually beneficial leases on tenant farm plots to these desperate city dwellers in exchange for a share of the yields (sharecropping). Many freed slaves were also offered tenant plots in order to encourage the transition to this style of farming.

However, as the land continued to deteriorate, this fresh wave of tenant farmers was also pushed ever further into debt and dependency as they were unable to meet their tax obligations. This produced a new kind of tenant farmer who signed a contract to work for the landowner indefinitely in exchange for the landowner taking over all the tax obligations. But even that was not enough to take the pressure off the poor tenant farmers — by

---

[8] Columella. (1941). *De Re Rustica.* Loeb Classical Library Edition. University of Chicago.
https://penelope.uchicago.edu/Thayer/e/roman/texts/columella/de_re_rustica/1*.html

[9] Orf, D. (2023, Jan 18). The Truth About Why Bill Gates Keeps Buying Up So Much Farmland. *Popular Mechanics.*
https://www.popularmechanics.com/science/environment/a42543527/why-is-bill-gates-buying-so-much-farmland/

the end of the third century AD, the farms had become so degraded that desperate tenant farmers were abandoning their leases in ever greater numbers as they walked off the land until Emperor Diocletian passed a series of financial reforms that made debts heritable, limited freedom of movement, and prevented peasants from escaping their tax obligations by fleeing their land — the sum of these laws effectively bound tenant farmers to their land, in perpetuity.

Things continued to deteriorate until, in 332 AD, Emperor Constantine passed another series of laws that allowed landlords to keep their tenant farmers in chains if they suspected they might try to flee. Although the landlord could not strip the tenant farmer of his land as long as the tenant farmer continued to service his debts and obligations, the tenant farmers were forbidden to sue their landlords for any reason other than if the landlord raised the rent, and the tenant farmer was forbidden to transfer property without the consent of the landlord. The sum of all these laws meant that tenant farmers could no longer leave the land or change profession, and if their leased land was sold, the tenant farmers were effectively transferred into the dominion of the new property owner. Tenant farmers were also forbidden to marry outside of their status, and their children were tied to the land on which they were raised and automatically bound by the same terms as their parents. By the late Roman Empire most of the big farms, which had once used slave labour to grow monoculture crops for the commodity market, had been reduced to little more than self-sufficient estates that were being farmed by these captive tenant farmers who were barely able to scrape a subsistence living from the land to which they were bound. After the Western Roman Empire fell, these Roman laws became the legal basis of the feudal system.

Throughout the Roman era, soil degradation and deforestation caused a massive and well-documented increase in aridity, leading to countless cities and towns being abandoned altogether. For example, in the Syrian Desert there is a ruin of a Roman city called Ilandarin, which covered about one square mile and had a population of between 20,000 and 100,000 citizens. But as soil degradation and deforestation took its toll, the wells dried up and the once-productive farmland turned to desert, incapable of supporting even a single inhabitant.[10] Meanwhile, soil eroding off Rome's upland farms on the Italian peninsula clogged up the marshes so that, by 200 BC, malaria had become a major problem in the region as mosquito populations flourished.

Writing about the decline of Roman agriculture, V. G. Simkhovitch, Professor of Economic History at Columbia University observed that:

> *"The steady shrinkage of population in the ancient world did not follow, curiously enough, in the wake of its bloodiest wars, but in times of complete peace. The fearful losses of Rome's greatest wars on the other hand, losses for instance occasioned by the Punic Wars, were rapidly made up, and in spite of further wars the population was steadily increasing. The same was true of the temporary decrease of population occasioned by a plague. Different was the situation in the period under discussion [that is, the period of the steady decline of Rome]. Losses occasioned by war and plagues were never made up, and during the longest and profoundest peace that Rome ever enjoyed the Roman population was steadily shrinking and its national strength steadily melting away."*[11]

The slow shift towards ever more centralized and authoritarian government during the Roman era mirrored the slow degradation of the land. This trend towards authoritarianism accelerated during periods when natural climate cycles intensified the stress on the population. Ellsworth Huntington, professor of geography at Yale University, wrote about the impact of soil degradation and climate change during the Roman Empire:[12]

> *"In the fourth century B.C., Italy appears to have been favored with so fine a climate that less than five acres was enough to support an average family. Cultivation was highly intensive so that the most advanced methods of agriculture were developed. Failures of the*

---

[10] Huntington, E. (1917). Climatic Change and Agricultural Exhaustion as Elements in the Fall of Rome. *The Quarterly Journal of Economics.* 31(2). https://doi.org/10.2307/1883908

[11] Huntington, E. (1917). Climatic Change and Agricultural Exhaustion as Elements in the Fall of Rome. *The Quarterly Journal of Economics.* 31(2). https://doi.org/10.2307/1883908

[12] Huntington, E. (1917).

crops were rare, and general prosperity prevailed. The farmers lived in comfort on their little farms and asked nothing of anyone, and the towns reflected their condition. Then when the spring and summer rains diminished — to speak by hypothesis — a small tract of land was not enough to furnish a living for the farmer and his family. Crops that had previously been profitable ceased to be worthwhile, the farmers ran into debt, and their lands gradually fell into the hands of large landowners. Since crops were no longer profitable the land was used for grazing purposes, as classical writers often point out. This was bad in two respects. In the first place, sheep and goats eat not only grass, but seedling trees, and thus prevent the growth of new forests. Where they pasture in abundance, the soil is badly trampled, and is no longer held in place by roots. Hence it is washed away by the winter rain, leaving the hillsides barren and ruining the fields in the lowlands. In the second place, sheep-raising and cattle-raising demand large areas. Hence, they increase the tendency toward the concentration of land in the hands of a few individuals. During the Augustan Age the farmers apparently recovered somewhat, and presumably were better off than in the second century B.C. Then came renewed climatic stress at the end of the first century A.D., and later the long deadening decline that culminated in the seventh century. In those days the Roman farmer was in circumstances as discouraging as those of [a] banker [reduced to] a mechanic's income.

Such economic changes must inevitably produce political results. One of the first and most obvious is a disturbance of the system of taxation. Theoretically, taxes ought to be proportioned to the income of the people who pay them. Practically the adjustment is most imperfect and has a disagreeable way of remaining fixed when other conditions change. When crops are bad the expenses of the government do not diminish. A tax which was easily paid from a full grain bin becomes oppressive when the grain bin is half empty. It is not surprising that the people were discontented, and agrarian reforms were needed in the days of the Gracchi. At that time Rome apparently suffered from climatic conditions more unfavorable than at any other period previous to about 300 A.D. **Under such circumstances the poverty and discouragement of the many almost** *inevitably favor the concentration of power in the hands of a few. Hence democracy suffers and a plutocratic form of government is superimposed upon the old framework. [..]* Not only Rome itself, but the provinces were suffering, and it is not strange that their discontent was finally an important element in the break-up of the Roman Empire. *[my emphasis]"*

In his book, David Montgomery tells detailed stories of innumerable civilizations across thousands of years of history, each following a similar cycle of soil degradation and decline. He doesn't just showcase what happened in the big well-known civilizations like Babylonia, Sumer, Assyria, Greece, and Rome, he also demonstrates that this cycle of deforestation, soil degradation, and desertification happened in virtually every civilization on every continent and on every inhabited island on the planet.

For example, consider the abandoned ruins of the Jordanian city of Petra, in the Arabian Desert, a once-flowering civilization powerful enough to build an amphitheatre capable of seating thousands and carve elaborate temples into the rock walls of its canyons. Yet all that remains of the now long-abandoned farms that once sustained this long-since dispersed civilization is scant evidence of farm terraces cut into the hillsides, ruined beyond recovery as erosion slowly stripped those hillsides down to bedrock.

Another example comes from Iceland. When the Vikings first settled Iceland, over 40% of the island was covered by birch forests. Once the early settlers cut down the forests to serve as building material and fuel, the very existence of those forests completely faded from cultural memory as the centuries passed. Once the forests were stripped off the hillsides, winds began to carry away the soil. Most shocking for me, having travelled around the island many years ago, was to discover from David Montgomery's book that the barren rocky interior highlands had once been lush pastureland for sheep. Today all that remains is a barren rocky plain that has been completely abandoned by farmers — sparse tufts of moss and windswept boulder plains are all that remain in that region of the island today.

The destruction of Iceland's soils continues unabated even today as continuously grazed sheep tear holes in the remaining pasturelands, which allows the wind to begin hollowing out the lightweight volcanic ash soils below the thin sod skin. Once there is a hole in the sod, the relentless wind gouges out the lightweight volcanic ash soils below (anywhere between 1 and 10 feet thick) until the scar reaches all the way down to the barren rock below the bottommost layer of those lightweight volcanic soils. And then these holes begin to expand, like a band of small cliffs, eating their way across the landscape at anywhere between half an inch to a foot and a half per year. Iceland is losing approximately 0.2 to 0.5% of its remaining soils every single year, yet it is happening so slowly that most people, including most of its citizens, don't even notice, even as Iceland's scientists predict that the island will completely run out of soil within the next few centuries.

The indigenous agricultural communities that emerged in the Americas long before the first Europeans landed on the continent suffered similar fates as their humus degassed and their soils eroded away. In America's Southwest, Chaco Canyon was the epicenter of the now long-lost Pueblo culture. But under pressure to grow crops, harvest timber, and provide firewood for its burgeoning populations, the canyon's once thriving ponderosa forests were reduced to scrubland as soils depleted by farming and by the increased aridity triggered by deforestation turned normal climate fluctuations into an existential and ultimately terminal threat. During a dry episode in the region's natural climate cycles, drought consumed the Pueblo culture, leading to the wholesale abandonment of entire cities and a mass migration out of the region. The climate cycles are perfectly normal, but the soil degradation, deforestation, and destruction of humus turns these perfectly normal climate cycles into existential crises. Again and again, we see the same pattern of slow creeping soil degradation (unnoticed during periods of abundant rains) until at some point the increasingly arid landscape and depleted soils are suddenly pushed into catastrophic collapse when the next natural drought cycle begins.

One of the most tragic stories is the story of Easter Island, which once supported a thriving Polynesian farming culture of almost 10,000 people, along with a thick cover of trees. But once again deforestation and soil erosion took their toll. After the last trees were cut down (thus preventing any escape by building canoes), the eroding soils gradually exhausted their capacity to raise chickens and grow sweet potatoes until, in time, the Easter Island population descended into perpetual warfare and cannibalism. By the time Europeans discovered the island, the population of Easter Island had shrunk by 80% to around 2,000 warring cannibals, and so many generations had passed since the last tree had been cut down that no-one knew the island had once been treed. With the memory of trees long forgotten, not a single islander knew how the giant carved stone figures that dot the Easter Island landscape had been moved into place by their ancestors. Things that happen slowly are the most difficult to recognize, and to stop.

Culture evolves slowly and unconsciously to respond to the societal pressures that emerge when a population outgrows the land's carrying capacity or as soil degradation eats away at the land's productivity. Thomas Malthus wrote about famine, yet this is only the most acute example of what happens when land catastrophically fails to feed its people. But in the absence of some extreme crisis triggered by a pronounced drought that tips already stressed soils into sudden ecosystem collapse, soil degradation happens slowly enough that it begins to produce cultural "adaptations" among the increasingly stressed population: colonial emigration, a culture of war, feudal laws to tie farmers to the land, human sacrifice, and even cannibalism (in Easter Island's case) — these are but five of countless cultural adaptations that emerge as soil degradation leads to overcrowding, dwindling opportunities, and intensifies the struggle to feed one's family.

One of the few examples in David Montgomery's book where a society figured out how to sustain its soils and stabilize its population before soil degradation went into terminal decline comes from Tikopia, a tiny volcanic island of only 5 square kilometers out in the middle of the Pacific among the Solomon Islands. The beginning of Tikopia's 2,400-year history reads much like that of other islands, with rapid soil erosion, deforestation, and pigs rooting up the vegetation. But then their agricultural practices changed, and soil degradation began to reverse as they turned their entire island into a kind of forest garden, with coconuts and fruit growing overhead and yams and taro growing beneath the canopy. They also completely eradicated all pigs from the island to stop the destructive uprooting of the fragile volcanic soils. That still left the islanders with the constant threat of overpopulation, which they solved through a combination of radical cultural adaptations that included strictly enforced contraception, celibacy, abortion, infanticide, and even waves of forced emigration whereby a portion of the island's population was periodically given canoes and told to "go explore the Pacific to colonize new islands, never to return" — in view of Tikopia's remote location in the midst of the Pacific, for many this was probably a death sentence. This combination of sustainable agriculture and population control allowed the island to prosper for thousands of years even as other surrounding Pacific islands devolved into war, cannibalism, famine, extreme population declines, irreversible soil losses, and complete ecosystem collapse.

The story of colonialism in North America is typically told through the lens of pilgrims escaping religious persecution and monarchs seeking to conquer new territory to secure their name in the history books as the founders of colonial empires. And yet, as documented in Mr. Montgomery's book, from the first tentative pilgrims to the later waves of emigration from England, Ireland, Germany, France, Italy, Eastern Europe, and beyond, each fresh wave of emigration began as populations in their home nations exceeded the carrying capacity of their exhausted soils. At its core, colonialism was driven by the need to import food and export excess mouths from the degraded home country. As Mr. Montgomery points out in his book, *"By 1900, Britain imported four-fifths of its grain, three-quarters of its dairy products, and almost half of its meat. Imported food pouring into Europe mined soil fertility on distant continents to further the growth of industrializing economies."*

17th-, 18th-, and 19th-century colonialism was not so different from the ancient Greek colonies that sprang up around the Mediterranean as Greece's population outgrew the capacity of its deteriorating soils to feed its people. As ancient Greek soils declined, Greek civilization expanded by sending excess citizens to set up fresh Greek colonies throughout the Mediterranean — the rulers dreamt of conquest, empire, and an abundant supply of commodities sent back to the capital from its colonies, while the colonists dreamt of virgin soils and breathing space from heavy-handed rulers. But in the end, Ancient Greek colonialism wasn't enough to stem the decline of Greek civilization. Degraded soils at the core of the empire were soon joined by degrading soils in the colonies, which ate away at the empire's revenues and manpower, thus overtaxing the empire's ability to project power out over its provinces and fend off younger and better-fed rival empires emerging outside of its borders, even as resentment within the colonies surged as the tax demands of the empire continued to grow despite falling agricultural productivity in the colonies.

In time, the cycle would repeat as Greek colonial expansion was replaced by Roman colonial expansion as Greece's successor in the

Mediterranean likewise sought to subsidize itself by "acquiring" resource-rich provinces, until Roman decay also became so widespread that collapse set in once again. The story repeats itself in later centuries with the spread of Islamic caliphates and then the Ottoman Empire, only to be followed by European colonization during the Colonial Era as European soils depleted by centuries of soil degradation, humus oxidation, and deforestation forced Europe's nations to turn their gaze towards distant shores. The solution is always the same: ship excess people abroad and then force colonists to send all of their commodity products back to the home country while using tariffs and shipping costs to capture all of the resentful colonists' profits. The names and the faces change during each chapter of history, but in the end, the story is the same one.

In our global economy, there's no place left to colonize. But as long as everyone is fixated on carbon emissions, the most important story of the century is going on mostly unnoticed beneath our feet even as top-down, centrally planned climate policies (like efforts to forcibly reduce livestock numbers) are only making things worse by taking away the most important tools that farmers have for rapidly rebuilding humus in exhausted soils and for creating a healthy sod cover to protect soils in arid grassland climates.

America's early colonists found an entire continent overflowing with rich and abundant soils just waiting for the farmer's plow. And yet, by the 19th century, those same soils in the original colonies were exhausted and heavily eroded after centuries of farming and plantation-style agriculture (tobacco is especially hard on soils). Soil degradation in the original colonies was a major factor in fueling the appetite for ever further westward expansion, out of necessity, in search of virgin soil to feed hungry families. Soil degradation along the eastern seaboard was at the heart of the desire to open up the Great Plains for settlement and the reason why so many people were so eager to believe that the rains would follow the plow out onto the prairie. Our willingness to believe the great fictions of our eras arise because, at some level, they provide relief — an illusion of hope — to rescue us from the forces eating away at the foundations of our civilizations. Buried beneath the follies of each age are poorly understood yet very real pressures that shape the irrational beliefs that lure civilizations into madness, decay, and conflict. Even the sparsely populated North American continent in the pre-colonial era was a never-ending running battle between different warrior cultures waging war that evolved in response to the non-stop competition for limited hunting territory.

All this begs the question of how much more violent our world would be today if chemical fertilizers were not compensating for degrading soils and forestalling the hunger that normally follows in the wake of soil degradation — the alternative to starvation has always been conquest of new lands and the subjugation of other peoples. Historically, empathy for others and a respect for their rights and liberties are luxuries only afforded to cultures where bellies are full. We would be wise to understand the lessons of our age and recognize the temporary nature of the relief offered by chemical fertilizers if we continue to ignore and/or misdiagnose the relentless erosion, oxidation, and nutrient-depletion of our soils and if we continue to ignore the true causes of why carbon is piling up in our atmosphere instead of building ever richer soils beneath our feet.

Classical liberalism was built on a foundation of individual human rights, property rights, and the idea that every individual has moral worth, regardless of social rank — noble ideas that quickly evaporate and are replaced by age-old tribal rivalries and feudal impulses when civilizations run out of fertile soil to sustain themselves within their own borders. Here in the wealthy West, we live surrounded by abundance despite our own slowly deteriorating humus levels as technology masks and prolongs the decline. But the clock is ticking much faster in the rest of the world, in Africa, in South and Central America, in Asia, and in the Middle East as they buckle under the combined pressures of soil erosion, humus depletion, continuous grazing, hillside farming, deforestation, desertification, war, and rapidly expanding populations driven into the cities in search of work as their subsistence farms become unsustainable. It is the story of Rome's decline, repeating itself on a much grander scale.

Soil degradation isn't just a farmer's problem; the pressures it creates (along with the accompanying health consequences) quickly morph into social and political problems. Understand these processes and you understand what is driving wave upon wave of migrants to flood out of impoverished countries where rivers run brown with eroded soil, where sunbaked soils are turning to dust, and where the forests are in rapid decline to make way for subsistence farms and provide fuel for cooking. The climate brigade offers the illusion of "easy answers" ("fossil fuels did this"), a ready scapegoat to blame and plunder ("those dirty industrialized capitalist nations"), and emotionally attractive solutions (fancy technology, punitive taxes, and socialist redistribution of wealth). And yet, none of those things offer real solutions (and arguably only make everything immeasurably worse) even as efforts to blame these scapegoats distract a population from recognizing the real root causes underlying their decline. Ultimately, the political, academic, and corporate peddlers of these fairy tales are nothing more than parasites feeding on a stressed society.

Even the oppressor-oppressed victimhood narrative — that poisonous gift that Marx and Lenin gave to the world — is rapidly grafting itself onto the fictitious climate change narrative. *Scientific American* recently spilled copious amounts of ink to contemplate "the racial implications of climate anxiety" and explain to readers why climate change is a social justice issue that demands "collective liberation."[13] And a chilling recent BBC article that is absolutely dripping with the seeds of resentment and division, entitled "why climate change is inherently racist", went as far as to declare that *"There is a stark divide between who has caused climate change and who is suffering its effects. People of colour across the Global South are those who will be most affected by the climate crisis, even though their carbon footprints are generally very low. Similar racial divides exist within nations too, due to profound structural inequalities laid down by a long legacy of unequal power relationships. [...] For some, it can be disconcerting to hear terms such as "racism" and "white supremacy" used in discussions about climate change. [...] 'If you want to understand why 40 years of climate diplomacy has failed to bend the curve on temperature rises, you have to go back and understand racialized capitalism – how race is codified to justify the exploitation and subjugation of people.' [...] 'The nations of the Global North have effectively colonised the atmospheric commons. They've enriched themselves as a result, but with devastating consequences for the rest of the world and for all of life on Earth.'"*[14] Judging by the fires being stoked by the media, industrialized nations are spiralling towards a full-blown self-inflicted self-hating cultural revolution.

---

[13] Ray, S.J. (2021, Mar 21). The Unbearable Whiteness of Climate Anxiety. *Scientific American*. https://www.scientificamerican.com/article/the-unbearable-whiteness-of-climate-anxiety/

[14] Williams, J. (2022, Jan 26). Why climate change is inherently racist. *BBC News*. https://www.bbc.com/future/article/20220125-why-climate-change-is-inherently-racist

At some level, the paternalistic West (the politicians, the scientific institutions, and the public at large) recognizes that not all is well in the world. And yet, through its ideological beliefs and the perverse incentives created by our top-heavy kleptocratic governance, the West has completely blinded itself to the true underlying causes of the desertification and the environmental degradation happening in so many regions of the word. And, by exporting those erroneous beliefs to the developing world, they are also undermining the ability of the poorest countries to recognize, let alone solve their problems. The ideological blindness of the world's "experts" has become the greatest obstacle to recognizing and solving the centrifugal forces tear society apart even as those same "experts" are actively fueling the most vile and detestable tribal hatreds, most especially within the West itself.

Chapter 27

Wards of the State — the Story of the Century

"The forest keeps shrinking, yet the trees keep voting for the axe. For the axe is clever and convinced the trees that he is one of them, because his handle is made of wood."

— Turkish Proverb

One of the most important realizations to come from this book is that despite the fact that soil erosion, deforestation, and soil degradation are happening in almost every single corner of the world, they are in fact an endless series of *local* crises that can only be understood (and solved) within the context of those local cultures, local climates, and local decision-making processes. We cannot solve this by painting the world with a single brush — that's how we got into this mess in the first place. And so, in this chapter, I am going to use three case studies (the Sahel region of Africa, Australia, and Haiti) to illustrate the complex political dynamics that have inadvertently changed the incentives in how the land is managed, which in turn is at the core of all the soil erosion, deforestation, and humus destruction happening all over our world. Politics and land management are inextricably linked.

Mob grazing and no-till farming on their own cannot fix what is happening if we do not understand the destructive incentives created by our centralized, institutionalized, debt-fueled, corporate-dominated, transnational political architecture and how this system is obstructing the decision-making processes that can lead us out of this mess and back towards thriving local ecosystems capable of sustaining their local populations. And so, the final pieces of "the story of the century" fall into place as we explore the link between the land and the decision-making processes by which we govern ourselves. A fish rots from the head — everything is connected.

Case Study # 1:
The Sahel, a Systemic Unravelling

"In the matter of reforming things, as distinct from deforming them, there is one plain and simple principle; a principle which will probably be called a paradox. There exists in such a case a certain institution or law; let us say, for the sake of simplicity, a fence or gate erected across a road. The more modern type of reformer goes gaily up to it and says, "I don't see the use of this; let us clear it away." To which the more intelligent type of reformer will do well to answer: "If you don't see the use of it, I certainly won't let you clear it away. Go away and think. Then, when you can come back and tell me that you do see the use of it, I may allow you to destroy it.""

— G.K. Chesterton, the Parable of the Fence

The 20-year drought and deadly famines that gripped the Sahel region of sub-Saharan Africa from the late 1960s until the early 1980s were a mild preview of what is to come if we continue on the present course. The region, commonly known as Africa's "Hunger Belt", is a perfect case study of how the destructive processes of deforestation, soil degradation, and desertification play out against the backdrop of natural climate variations after the breakdown of the traditional cultural and economic systems that once enabled populations to thrive in this brittle climate.

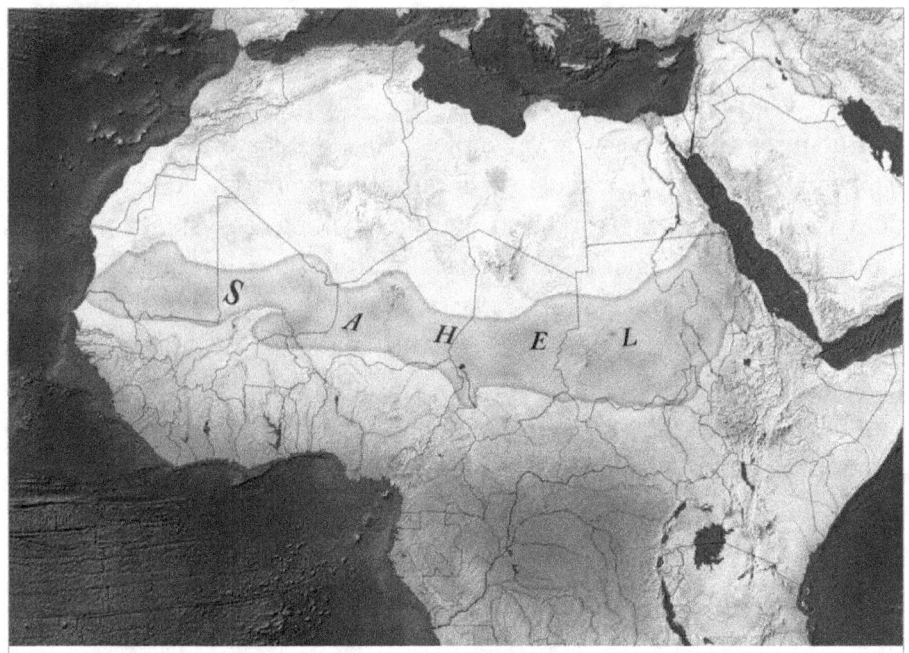

Figure 113. Sahel region of Sub-Saharan Africa
https://commons.wikimedia.org/wiki/File:Map_of_the_Sahel.png CC BY-SA 3.0 DEED

The Sahel is the one of the poorest regions in the world. It also has some of the world's highest birthrates. And it is the epicenter of the migrant exodus out of Africa to Europe (and also, increasingly to the USA) as people flee their broken countries in search of greener pastures.

As I write this today, the Sahel is once again facing one of the fastest-growing displacement crises in the world, partially because of severe drought and partially because of violent inter-tribal conflicts and terrorist militias that are rapidly spreading through the region. In early 2024 there were already over 3.7 million internally displaced people in the Sahel region, over 1 million refugees and asylum-seekers, over 7,800 closed schools, and over 30 *million* people in need of assistance and protection across the region.[1] The entire region is teetering on the brink, yet most people outside of the region have no idea any of this is happening — instead the media is focusing our attention on pseudo-crises like burping

---

[1] INTERSOS Humanitarian Aid. (2022, _, _). *Sahel, epicenter of a fast-growing humanitarian crisis.* [Press release]. INTERSOS in partnership with the UN Office for Coordination of Humanitarian Affairs.
https://www.intersos.org/en/sahel-epicenter-of-a-fast-growing-humanitarian-crisis/

cows, microaggressions, and pronouns, while, as per usual, the United Nations is busy rallying support for a *"global phase-out of fossil fuels"* as the answer to everything.[2]

But just because very few people in the West are aware of what's happening in the Sahel and just because everyone is focused on CO_2 doesn't change the fact that the current slow-rolling ecosystem collapse of the region, which began in 2010, has already reached Europe's doorstep as a key contributor to Europe's massive migrant crisis — the ecological has become political; the dominoes have begun to move. What's happening in the Sahel has also begun to affect the USA as a growing number of African migrants are joining the sea of people flooding across America's open southern border. The destructive destabilizing consequences of more than century of Wilson's liberal international world order are coming home to roost.

Despite the UN's disingenuous focus on "global warming", switching to electric cars and banning red meat in Europe and North America cannot fix this unfolding humanitarian disaster, nor defuse any of the tectonic political forces that have been set in motion by these mass migrations. Nor will these high-tech pseudo-solutions do a thing to stop rising CO_2 levels caused by the wholesale soil degradation, deforestation, and desertification happening all across Africa and elsewhere in the world. If anything, their latest buzzword — "degrowth" — will only make things worse by accelerating the poverty spiral. The only thing that can fix this is dismantle the mad crusade against CO_2 so people can recognize the real underlying causes of the Sahel's troubles.

The Sahel region spans right across the African continent along the southern edge of the Sahara, from Senegal on the Atlantic coast to Eritrea and the Ethiopian Highlands on the Red Sea coast. The region has been described as a "giant free-range pasture" where cattle, goats, and sheep roam aimlessly between scattered villages and degraded subsistence farm plots — there is no mob grazing or no-till farming being practiced here. Virtually all of the land across this vast region is at least moderately degraded. A colossal 650,000 square kilometers of land (over 20% of the Sahel region) that was once suitable for pasture or crop production has

---

[2] United Nations. (2024, Feb 13). *Climate Action Can Help Fight Hunger, Avoid Conflicts, Official Tells Security Council, Urging Greater Investment in Adaptation, Resilience, Clean Energy.* [Press release]. UN Security Council. https://press.un.org/en/2024/sc15589.doc.htm

already been completely lost to desertification over the last 50 years.[3] Over 90% of the soils are sorely deficient of important soil nutrients like phosphorus.[4] The traditional fallow periods that local farmers once incorporated into their crop rotations to rest their land have been shortened to try to squeeze more productivity from the land, which has only accelerated nutrient depletion and eroded crop yields — in Nigeria alone over 20% of the farmland has already been completely destroyed by soil erosion.[5]

Years of soil degradation and deforestation have taken their toll. The bare, hard, sun-baked, humus-depleted soil is incapable of quickly absorbing the rain and so, with the water cycle impaired, the rainwater begins to pool and flood across the land — every time it rains, it floods. This increased local flooding has nothing to do with global climate change, nor does it represent a change in weather patterns and storm intensity — it is nothing more and nothing less than the consequence of soil degradation impairing the ability of rainwater to soak into the soil.

And because bare, degraded, sun-baked, humus-depleted soils have become incapable of efficiently absorbing the rains, the land returns to drought almost as soon as the floodwaters subside. Soil degradation simultaneously leads to both flooding and drought even when historic rainfall patterns remain unchanged.

The flash floods that now follow every rainstorm are carrying away billions of tons of topsoil each year from the region — rivers that once ran clear have turned into brown, sediment-filled conveyor belts that carry topsoil away to lakes, wetlands, and oceans. Mud-clogged wetlands and the pools of floodwater that puddle on the hard sunbaked ground also become massive breeding grounds for billions of malaria-infested mosquitoes, just like what happened in ancient Rome, thus fueling the malaria epidemic in Africa. And the combined effect of inefficient

---

[3] Norman, S. (n.d.). *Environmental issues in the Sahel [Geo Factsheet Number 16]*. The Geographer Online.
https://www.thegeographeronline.net/uploads/2/6/6/2/26629356/a116_sahel.pdf

[4] Norman, S. (n.d.).

[5] Norman, S. (n.d.).

irrigation and pooling rainwater is creating massive problems with salt build-up in the soil, further damaging crops.[6]

An extraordinary statistic from the Orange River in South Africa gives a sense of the scale of the erosion happening all across the African continent: the amount of mud carried by the Orange River has increased tenfold during the modern era compared to the average of the last 10,000 years *while soil erosion rates within the catchment basin were found to be 100x higher today compared to the average of the last 10,000 years!*[7] Furthermore, erosion in the pre-industrial era was largely confined to the steep slopes of the mountains, but today it is predominantly coming from the flat intensively cultivated lands and, to a lesser extent, from the grazing areas within the catchment basin.[8] This story repeats, continent by continent, all over the world. According to a 2006 meta-study, the world is losing its soils 10-40 times faster than the natural replenishment rate and, as a result, 30% of the world's arable land has already become unproductive in the past 40 years.[9] What's happening in Africa is happening everywhere, but in the driest regions things unravel the fastest, which makes the process easier to see — it's part of the reason why I chose the Sahel as one of the case studies for this chapter.

According to the United Nations, there are more than 4 *trillion* tonnes of carbon stored in soils around the world[10] (for comparison, the woody biomass in forests only contains around 360 billion tonnes of carbon and the entire global atmosphere only contains around 258 billion tonnes of

---

[6] Norman, S. (n.d.). *Environmental issues in the Sahel [Geo Factsheet Number 16]*. The Geographer Online. https://www.thegeographeronline.net/uploads/2/6/6/2/26629356/a116_sahel.pdf

[7] Compton, J.S., Herbert, C.T., and Hoffman, M.T. (2010). A tenfold increase in the Orange River mean Holocene mud flux: implications for soil erosion in South Africa. *The Holocene*. 20(1). http://dx.doi.org/10.1177/0959683609348860

[8] Compton, J.S., Herbert, C.T., and Hoffman, M.T. (2010).

[9] Lang, A. (2021, Mar 1). Conservation Agriculture: Necessary for the Future, and Incentives Are Needed. *No-Till Farmer*. https://www.no-tillfarmer.com/blogs/1-covering-no-till/post/10430-conservation-agriculture-necessary-for-the-future-and-incentives-are-needed

[10] "Soil Fertility and Erosion." *Global Agriculture – Agriculture at a Crossroads*. Accessed 29 Apr 2024. https://www.globalagriculture.org/report-topics/soil-fertility-and-erosion.html

carbon[11]). In other words, there is 15x more carbon stored in soil than there is carbon in the entire global atmosphere — even small increases in soil erosion and deforestation have big impacts on the atmosphere. The United Nations estimates that, globally, we are losing 24 billion tonnes of fertile soil per year due to erosion, which works out to an average of *3.4 tonnes of soil lost per person per year for every single person alive on the planet today!*[12] The fastest erosion is happening in Africa and Asia.

The most sediment-laden river on Earth is the Yellow River in China. And yet, the Yellow River and many of its tributaries, like so many other muddy rivers all around the world, had clear water until 700 years ago when the population in the region began to increase, land was cleared to make way for more farms, and large parts of the region were deforested to provide fuel for the expanding population. However, the most rapid acceleration of soil erosion in the Yellow River basin began in the 1900s — today more than 70% of the plateau that serves as the main catchment basin for the Yellow River is dissected by a dense network of erosion gullies.[13] Some of the carbon from those billions of tons of eroded soil ends up in lakes, bogs, and oceans, but a large portion also directly reacts with oxygen in the atmosphere and turns into CO_2. Likewise, as deforestation and soil degradation take their toll on the total amount of vegetation that can be supported by the exhausted land, the region's total biomass is plunging.

In the Sahel region, deforestation and soil degradation have likewise led to a dramatic reduction in biomass — one study found that dry-matter biomass in the vicinity of a borehole in Nigeria had declined from 100 kg/hectare to just 8 kg/hectare in a span of only 8 years.[14] Of course, as biomass declines, less solar energy is absorbed by photosynthesis, which means that more solar energy heats up and bakes the bare ground, which

---

[11] "Carbon dioxide in Earth's atmosphere." *Wikipedia.* 11 Jun 2024.
https://en.wikipedia.org/wiki/Carbon_dioxide_in_Earth%27s_atmosphere

[12] "Soil Fertility and Erosion." *Global Agriculture*

[13] Dotterweich, M. (2013). The history of human-induced soil erosion: Geomorphic legacies, early descriptions and research, and the development of soil conservation–A global synopsis. *Geomorphology.* 201.
https://doi.org/10.1016/j.geomorph.2013.07.021

[14] Norman, S. (n.d.). *Environmental issues in the Sahel [Geo Factsheet Number 16].* The Geographer Online.
https://www.thegeographeronline.net/uploads/2/6/6/2/26629356/a116_sahel.pdf

warms up the air directly above the ground surface and impairs moisture condensation in clouds passing overhead. But activist-scientists point to all of it as evidence of global warming.

Forests in the region are in a death spiral of overexploitation. In Burkina Faso, deforestation has been so extensive that, going forward, if woodcutting was restricted only to a sustainable harvest capable of maintaining the currently remaining forest without further deforestation, those remaining forests would only be able to provide enough firewood for cooking fuel for around 1% of the country's population.[15] In Ethiopia at least 66% of the country was covered by forests as recently as 150 years ago; by 1982 only 3% of its forests remained. As Ethiopian forests were cut down, the rivers and streams that once flowed through them year-round all dried up and the ground quickly became parched (water evaporates off bare soil extremely quickly once the soil loses its ground cover).[16] In Sudan, the city of Khartoum has almost no trees left within a 90 km radius of the city — they have all been cut down for fuel and to make space for farming.[17] As firewood supplies dwindle away in the region, dried cattle dung has become a common replacement fuel, which further robs the land of fertilizer and organic matter that would otherwise be returned to the soil. Despite the claims made by the United Nations, rising CO_2 didn't do any of this. Nor can net-zero policies do a thing to stop this.

What's happening today in the Sahel is a repeat of the Dust Bowl of the 1930s, only it's happening on a slower timescale and on a much bigger scale (the Sahel region alone is approximate half the size of the continental United States, and the same process is underway across much of the rest of Africa). During the Dust Bowl, the Black Sunday storm that I wrote about in chapter one carried away 300,000 tons of topsoil; dust storms in the Sahel region carry away 182 *million* tons of dust every single year[18]

---

[15] Norman, S. (n.d.).

[16] Leaf of Life. (2024, Jan 21). *How They Transformed Desert Into Fertile Farmland and Forest.* [Video]. YouTube. https://www.youtube.com/watch?v=RBP2uRQk5pQ&ab_channel=LeafofLife

[17] Norman, S. (n.d.).

[18] Touray, K.S. (2022, Feb 25). A step from hell. The seasonal dust storms of the Sahel are a reminder that we are one step from hell. *The Standard.* https://standard.gm/a-step-from-hell-the-seasonal-dust-storms-of-the-sahel-are-a-reminder-that-we-are-one-step-from-hell/

— some of that dust originates in the Sahara but the rest is stripped from pastures and farmland in the Sahel. Before human populations exploded, the region was a giant fertile savannah grassland with seasonal migrations of vast herds of oryx, gazelles, buffalo, and hartebeest covering vast migration routes. Those migrations are gone now.

Likewise, before modern borders and the population explosion, the region had been continuously inhabited by nomadic pastoralists for thousands of years — overgrazing was not a problem in those days because the land was grazed by herds of migrating livestock, not by continuously-grazed livestock scattered randomly all over the region.[19] This point is important to understand — the positive impact of livestock on the land doesn't just come from giving the animals a fresh slice of grass each day. Mob grazing has its most positive impact on the land when cattle are *bunched* tightly together as a compact herd, whether by shepherds, fear of predators, or electric fences. This "bunching effect" is what changes the herd's impact on the land so that plants are grazed evenly, manure is spread evenly, and soil and vegetation is systematically trampled to prevent remaining dead plant debris from being able to choke out the next year's regrowth — it is this combined impact that essentially "preps" the land to receive the next rains, stimulates the nutrient cycle, and optimizes conditions for grass to thrive in the next growing season.

The nomadic prehistoric pastoralists of the region instinctively bunched their cattle tightly together to protect them from predators and to be able to control the herd as a single manageable unit. Furthermore, they combined their livestock to create huge migrating communal herds to keep the cattle safe rather than every cattle owner doing off to do their own thing. In every way, the nomadic pastoralists managed their livestock to mimic the behaviour of the wild grazing herds. Most modern farmers do not.

At the heart of the Sahel region lies the hidden jewel of Central Africa — Lake Chad. During the 19th century it was the sixth largest lake in the world at over 28,000 square kilometers. But as settled farmers displaced nomadic pastoralists over the last century, the amount of water flowing into the lake decreased as water was diverted for irrigating farmland and

---

[19] Norman, S. (n.d.). *Environmental issues in the Sahel [Geo Factsheet Number 16]*. The Geographer Online. https://www.thegeographeronline.net/uploads/2/6/6/2/26629356/a116_sahel.pdf

as the combined effects of deforestation, vegetation loss, and soil degradation took their toll on the land — once the bare soil was exposed and sunbaked, a lot of the rainwater simply evaporated or flooded away after the storms instead of refilling the groundwater table, which caused aquifers, springs, streams and rivers across the region to dry up. Reduced rainfall is only a small part of the problem — reduced soil moisture *absorption* is at the core of the issue. Starting during the droughts of the 1960s, the lake began to shrink and has fluctuated between a mere 2,000 and 5,000 square kilometers ever since (between 7 and 17% of its former size).

At the end of the 1960s, natural climate cycles caused rainfall in the region to plunge by over 30% compared to 1950s levels — and it stayed that low for two decades. The 20-year drought from the 1960s to the 1980s was not unusual for the region as long drought cycles lasting multiple decades (and even multiple centuries) are part of the normal cyclical nature of the climate in the Sahel region. Long brutal cyclical droughts have always been a part of this ecosystem, yet famine was not.

The unbroken grasslands of the region were Nature's defense against soil degradation and moisture evaporation during these long droughts. But all the soil degradation and deforestation that I described above, which are still happening today, were already at work in the decades leading up to the late 1960s famine (albeit on a smaller scale because the population was much smaller then). As the weather turned drier in the 1960s, the deteriorating ecosystem that had been held together by the high rainfall of the previous decades simply fell apart. With less trees to pump water into the air, with less humus in the soil to absorb limited rains, with less vegetation to shield the bare soil from the sun, and with less sod to prevent the wind from carrying away the parched soil, the ecosystem had been stripped of all its natural defenses to help carry it through the dry part of the region's climate cycles. Famine killed over 100,000 people, left another 750,000 dependent on food aid, and affected all of the region's 50 million inhabitants. One third of the region's livestock died. The scale of the catastrophe was so immense that it became known as the Great Sahelian Drought.

The reasons for these recurring cyclical Sahelian droughts are still hotly debated, however, the timing of the 1970s drought suggests that *cooler global weather* (and, by consequence, reduced evaporation from the oceans) probably played a significant contributing role in reducing humidity and rainfall in the region. The same thing happened during the cooler weather of the Little Ice Age — the Sahel region experienced a crippling multi-century drought that lasted from 1400 to 1750, essentially straddling the full span of the Little Ice Age. Nor is evidence for drought during the Little Ice Age restricted to Africa — the entire tropics from the western Pacific to the Arabian Sea, continental Asia, and even the tropical parts of South America were all much drier during the cold centuries of the Little Ice Age.[20]

Today, as the Sahel's climate begins to turn drier once more, there are almost 7x more people living in the region than at the start of the 1960s (today the region has over 333 million people). And there have been another 40 years of accumulated soil degradation and deforestation since the Great Sahelian drought. It is a ticking time bomb of colossal proportions.

Sadly, neither Allan Savory's holistic grazing techniques nor Ruth Stout's mulching and no-till gardening are at the forefront of the efforts to educate farmers in the region. And I don't need to remind you of the Malthusian Trap created by the paternalistic West's green colonialism, which is obsessed with reducing greenhouse gas emissions instead of encouraging economic development and cheap fossil fuel energy in the Sahel, which would ultimately lead to lower birthrates (and take pressure off the forests).

~

In 2007 the African Union launched an initiative called the Great Green Wall. It is a plan to combat desertification and stop the expansion of the Sahara Desert by planting a wall of trees from Senegal to Djibouti, all across the northern frontier of the Sahel where the Sahel meets the Sahara. You may have seen the emotionally compelling videos on YouTube of the half-moon crescents of soil that are being hand-dug across the region to capture rainwater and loosen the soil so that trees and vegetation can be planted in the crescents. Like all centrally planned ideas that emerge from the optimistic minds of well-meaning bureaucrats, in theory it seemed like a wonderful idea. In practice the results have been... mixed.

---

[20] "Droughts in the Sahel." *Wikipedia*. 14 Feb 2024.
https://en.wikipedia.org/wiki/Droughts_in_the_Sahel

Some of the best results have been in Senegal where crop plantations were incorporated into the designs to stimulate the local farming economy. Anecdotal reports emerged of villages that had previously been depleted of all their young people as they fled to Europe in search of work, which were starting to see some of those young people return because there was a way to making a living again in the village.[21] However, despite some success stories, overall, things are not really going according to plan. By 2020, halfway through the planned schedule, only 4% of the Great Green Wall's planned area had been planted, and even in Senegal (the posterchild of the program's success) grave doubts were being raised about the survival of the 12 million trees that had already been planted.[22]

As reported by Science magazine in 2021, *"In recent years, research by ecologists, economists, and social scientists has shown that many forestry projects around the world have failed because they didn't adequately address fundamental social and ecological issues. Project leaders often didn't ask communities what kinds of trees they wanted, planted species in places where they didn't belong, and did little to help the saplings survive. 'Tree planting is often viewed as the simple act of digging a hole,' forest scientists Pedro Brancalion of the University of São Paulo, Piracicaba, and Karen Holl of the University of California, Santa Cruz, noted last year in a review of agroforestry projects in the Journal of Applied Ecology. 'But this short-term, naïve view has resulted in large quantities of money being spent on ... efforts that have failed almost entirely.'"*[23]

The Science article goes on to highlight that these reforestation programs frequently don't have buy-in from local communities because the plantings conflict with other priorities, they frequently plant fast-growing species that are not native to the area and which in some cases

---

[21] Millison, A. (2024, Feb 18). *How the UN is Holding Back the Sahara Desert.* [Video]. YouTube. https://www.youtube.com/watch?v=WCli0gyNwL0&ab_channel=AndrewMillison

[22] Watts, J. (2020, Sep 7). Africa's Great Green Wall just 4% complete halfway through schedule. *The Guardian.* https://www.theguardian.com/environment/2020/sep/07/africa-great-green-wall-just-4-complete-over-halfway-through-schedule

[23] Cernansky, R. (2021, Feb 11). Taking Root. *Science.* https://www.science.org/content/article/great-green-wall-could-save-africa-can-massive-forestry-effort-learn-past-mistakes

turn into invasive weeds that compete with native plants for water and space, and they often have unrealistic expectations about costs, available infrastructure, and the suitability of the land being set aside for reforestation.

One of the problems with community projects is that when everyone owns something, in reality no-one owns it. As the project begins to fail, everyone blames everyone else while no-one is truly responsible, and the "experts" become more authoritarian as they begin to call for greater control and greater technical oversight. Unlike a private initiative in which those who have skin in the game are forced to change course or suffer the consequence, in these public "communal" projects, failure simply feeds the beast as ever greater sums of money are poured down the same rabbit hole and as an every larger "economy" of NGOs, international organizations, and well-meaning outsiders impose themselves onto the target communities. It may only take one dollar to plant a tree, but it takes a lot more than a dollar to maintain it — who has time to nurse someone else's tree when they have their own crops to plant, livestock to take care of, and a family to feed. Lack of ownership and lack of direct economic benefits leads to a lack of incentives once the original enthusiasm wears off. If only the experts knew how to listen as well as they know how to talk and write grant proposals.

Furthermore, as many critics of the Great Green Wall pointed out, the one-size-fits-all approach to this centrally-planned initiative ignored the fact that the project often tried to reforest low-rainfall areas that have historically always been grasslands, not forests — the deck was already stacked against the seedlings. In many of these historic grasslands, helping communities rekindle herd migrations (mob grazing) would likely be a far more effective way to revegetate the landscape and rebuild soils. But of course, that's a tough pill to swallow for central planners and ideologues like the United Nations who have come to view livestock as Public Enemy #1 and who view themselves as the unassailable experts with all the answers (Remember when the UN declared: "*We own the science, and we think that the world should know it.*"[24]) People with that attitude aren't going to listen to anyone.

---

[24] Bernstein, B. (2022, Oct 4). U.N Communications Official Touts Google Search Partnership: 'We Own the Science.' *National Review.* https://www.nationalreview.com/news/u-n-communications-official-touts-google-search-partnership-we-own-the-science/

None of these failures has deterred enthusiasm and fresh infusions of Western taxpayer money for the Great Green Wall. The UN is even funding an expensive new campaign called "Growing a Wonder World" designed to "raise awareness about the Great Green Wall"[25] — it describes its campaign as a "Great Green Wall Accelerator"[26].

With all this money sloshing about and with campaigns being funded solely to bring awareness to other campaigns, I am reminded of when Bono's ONE non-profit foundation came under fire in 2010, as reported by the Daily Mail, when it was revealed that of the almost £9.6 million collected in donations (most of which came from the Bill & Melinda Gates Foundation), only £118,000 was paid out to good causes (1.2 per cent) while £5.1 million was consumed by the salaries of the foundation staff. In essence, the foundation was a lobbying firm designed to liaise with politicians and UN members, and to organize high-profile, celebrity-supported events. As the One foundation themselves point out: *"We don't provide programmes on the ground. We're an advocacy and campaigning organisation."*[27] Right… because the world needs more people with a saviour complex telling everyone else how to live. A vast economy comprised of aid organizations, charities, NGOs, consultancy firms, environmental groups, lobbying groups, and intranational organizations has been built on the misery of developing countries, which are now wholly dependent on a continuation of the "established" narrative to continue to justify their own continued existence.

A far more successful but much smaller initiative than the Great Green Wall is being run by an organization called "Trees for the Future", which provides hands-on agronomy training to local farmers to teach them how to plant forest gardens on part of their land, in effect giving farmers the skills and tools to create their own viable economic systems based around

---

[25] "What is the Great Green Wall?" *United Nations*. Accessed 17 May 2024. https://www.unccd.int/our-work/ggwi

[26] "The Great Green Wall." *The Great Green Wall*. Accessed 17 May 2024. https://thegreatgreenwall.org/about-great-green-wall

[27] "Bono's ONE foundation under fire for giving little over 1% of funds to charity." (2010, Sept 23). *Mail Online*. https://www.dailymail.co.uk/news/article-1314543/Bonos-ONE-foundation-giving-tiny-percentage-funds-charity.html

their forest gardens and creating an economic model for other local farmers to mimic.[28]

~

All this raises an important question — why didn't the Sahel ecosystem fall apart under nomadic pastoralism during the pre-colonial era? Although there is some documented evidence of soil erosion in Africa before the colonial era, there is no evidence anywhere on the continent for the development of the kinds of extensive badlands that emerged after the arrival of Europeans.[29]

As I've pointed out before, land management and politics are inextricably linked. It's easy to explain the difference between the pre-colonial and the modern eras from a standpoint of the impact of the animals on the grasslands, but it's only a partial explanation because it ignores the fact that land management is deeply embedded within our culture and within the social and economic systems that incentivize our behaviour. I know that sounds like a dry mouthful, so let's have a closer look at what destabilized the Sahel ecosystem as tradition gave way to modernity.

Prior to the colonial era, the vast grasslands of the Sahel had no tillage agriculture. Crop farming was restricted to the southern portion of the region where rainfall rates are higher, and where the forests begin. Nomadic pastoralism and settled farmers growing crops create completely different societies. If you invest time and effort into planting and cultivating crops, you need ownership of your little plot so that someone else doesn't run off with the fruits of your labour. That requires some form of individual private property ownership tied to the land. Nomadic pastoralism is the opposite — the grasslands belonged to the tribe (and were defended by the tribe), whereas the cattle were owned by individuals. The herdsman's effort goes into managing his animals — removing ticks,

---

[28] Millison, A. (2023, Nov 15). *How 8,000 Food Forests Grew Africa's Great Green Wall.* [Video]. YouTube. https://www.youtube.com/watch?v=1LCTVO_Y5Rs&ab_channel=AndrewMillison

[29] Dotterweich, M. (2013). The history of human-induced soil erosion: Geomorphic legacies, early descriptions and research, and the development of soil conservation–A global synopsis. *Geomorphology.* 201. https://doi.org/10.1016/j.geomorph.2013.07.021

providing protection from predators, maintaining water sources, finding mineral licks, and participating in battles against rival tribes vying for the same land. If the rains failed or the soil degraded, nomadic tribes were extremely mobile as they migrated elsewhere in search of better grass — these nomadic migrations prevented famine but also produced the never-ending tribal wars that helped kept population numbers in check.

Much of Africa was polygamous prior to the arrival of Christianity on the continent. If each man had three or four wives, that means there were *a lot* of men that didn't have any wives. These "excess men" were consumed by the never-ending intertribal wars. Polygamy also reduced population growth because although a man might have many children, each woman only had a small number of children. In a pastoralist or agriculturalist pre-colonial society, a man could only take as many wives as he could provide for, each wife had her own hut, and there were harsh social stigmas levied against any man who failed to provide for his family. Even the practice of demanding a bride price (dowry) before marrying off the family's daughters served as a social mechanism to force a man to prove to the bride's family that he had the skill to be able to provide for her and her children. Nor was this the only cultural strategy to limit population numbers. Anthropologists studying pastoral camel-herding tribes in Kenya[30] observed that they *"intentionally control their own fertility in order to adapt the growth of the human population to growth rates of camel herds."* Those rules included *"infanticide, post-partem taboos, and cultural prescribed delayed marriages of women"*, which had a *"sizeable impact on population and subsequently sets a limit on the rate of population growth."* It's only recently, in the modern era, that the livestock-to-person ratio has become disconnected as the reliance on pastoralism as a way of life has declined and as social stigmas against those who father more children that can be provided fall away. Irregardless of the noble purposes claimed by liberal feminists, the net result of their efforts has been the triumph of irresponsible playboys and the loose morals of the emancipated nymphs who follow in their wake, who together impose the costs and consequences of their hedonism on everybody else while the administrative state, driven by understandable

---

[30] Bollig, M., and Lang, H. (1999). Demographic Growth and Resource Exploitation in Two Pastoral Communities. *Nomadic Peoples.* 3(2), 16–34. http://www.jstor.org/stable/43124086

empathy, forces taxpayers to pick up the tab. Cultural mores can provide a measure of restraint that regulation alone cannot.

A Kenyan recently shared her ancestral family tree with me, which demonstrated how these cultural changes played out in her family. Her tribe migrated from the Sahelian grasslands to Kenya with their cattle around 300 years ago at around the same time as the Maasai arrived in Kenya — the Maasai retained their pastoral lifestyle as they moved onto the grasslands of the Serengeti, whereas her tribe settled in the Lake Victoria region where they combined pastoralism with farming and fishing. Her father was the genealogist in his family and could recite, from memory, 13 generations of their family tree going back 300 years to the time when their ancestors began their southward migration. Generation after generation, each wife in the family tree only had a small number of children... until Christianity arrived, monogamy replaced polygamy, and the number of children born to each wife soared, marking the tipping point when Africa's population suddenly went into overdrive. But then, as the children of farmers and herdsmen moved off-farm to take city jobs as engineers, doctors, and so on in the modern economy, birthrates quickly plummeted again as they do in every country where standards of living begin to rise. Prosperity, not the UN's new obsession with "degrowth", is the key to the overpopulation problem — if only Thomas Malthus has taken that last step to study which parts of society have more and which parts of society have less children, he might have figured out that prosperity, rather than hard-hearted poor laws and grinding toil, were the solution to the Malthusian Trap he had identified with his statistics and mathematical formulas. By failing to see the complete picture, Malthus created beliefs that continue to haunt us to this day.

Traditional pastoralism was more than just a way of grazing cattle. Pastoral culture evolved over many generations to combine grazing habits, marriage rites, animal and human fertility, and the productivity of the land into a comprehensive socio-economic system to keep people, land, and animals in balance. Break even one small part and the entire system spins out of balance unless a complete new economic model is able to replace the traditional way of life. Economic development in Africa and other overcrowded countries is vital to bringing birthrates back in line with their nation's sustainable agricultural carrying capacity.

In a subsistence lifestyle divorced from its historic pastoral and agricultural traditions, children serve two purposes: they are a retirement strategy for aging parents, and they are a source of free labour on farms (many hands make short work), so having lots of them made good economic sense for the family even if it had disastrous long-term consequences to the future of the community as a whole. But as wealth increases, savings become the dominant retirement strategy. And as soon as people moved off-farm to take jobs in the modern economy and began buying their food from the grocery store instead of growing it themselves, children turned from an asset into a liability as the cost of education, extracurricular activities, food, clothing, babysitting, etc. began to add up. Thus, cheap child labour turned into expensive mouths to feed. The only way to escape the Malthusian birthrate trap is for countries to get rich. Dismantling ancient traditions and ancient lifestyles causes entire nations to spiral into famine and overpopulation unless nations are able to complete the full transition to become a modern fully developed economy.

But thanks to the climate crusade, green colonialism (fear of fossil fuels) is interrupting that economic transition. If you step back far enough, all these threads lead back to Woodrow Wilson's legacy of the masterful administrative state surrounding itself with scores of statistics and "expert planners" as they set out to make plans for a better world. Wilson's masterful administrative state was mirrored in European nations by an equally aggressive push to centralize power and resources during that same time period (Wilson merely represents the American head of a global appetite for centralization that infected the Western world during that period). This centralization spawned colossal levels of corruption as everyone began whispering in the ears of the central gatekeepers for their "fair share" of the tax grab, which ultimately rotted out the core of academic institutions who have traded their intellectual curiosity for a government contract — these government-funded institutions will keep spreading whatever message is required to keep their funding taps wide open. But those same funding-hungry institutions set the agenda and shape the beliefs of the entire world — they have become the legitimizing force behind the "green" crusade against carbon dioxide, which has emerged as one of the principal obstacles to lifting poor countries out of poverty. Everything is connected.

The colonial era had a profound impact on traditional agriculturalist societies in Africa. In a research paper published in 2013 by Markus Dotterweich of the Institute of Geography at the Johannes Guttenberg

University in Mainz, Germany,[31] he describes how long-established sustainable farming techniques completely fell apart in southern Malawi as an unintended consequence of societal forces that were set in motion during the colonial era, triggering the extreme levels of erosion that have plagued the continent ever since:

> *"[...] during the pre-colonial period, the semi-sedentary Africans had a very good knowledge about soil fertility and soil management. They used many different local conservation practices such as general tilling before planting, raised mounds, mixed cropping, and planting protective grasses to prevent runoff and soil erosion. According to oral traditions, they were aware that undertaking conservation helped to prevent soil erosion, improved soil fertility and therefore increased yields.*
>
> *Elders knew of different strategies like placing grass or tree branches on the top of the outer ridge or planting sugar cane or other plants and fruits on the outer ridge of their gardens. In addition, various types of grasses were used for livestock feeding. In hilly areas, the plantations were set along the contour lines. Peasants only used basic tools like hoes and axes and non-labor-intensive practices. In some areas, stone walls were built across the slopes in order to prevent runoff and soil erosion. One other method was to divert a stream from its natural course above a garden. The water was sometimes directed into a furrow and/or stored in a pond behind an earthen dam to irrigate the fields [...].*
>
> *This type of soil and water management was widely used until a significant population increase caused by immigration of different African groups and European settlers in the late nineteenth century CE occurred. Various social and power relations competed against the traditional systems. Slash-and burn practices and shifting cultivation became more popular on the scarce land resulting in an increase of soil erosion and environmental degradation.*

---

[31] Dotterweich, M. (2013). The history of human-induced soil erosion: Geomorphic legacies, early descriptions and research, and the development of soil conservation–A global synopsis. *Geomorphology.* 201. https://doi.org/10.1016/j.geomorph.2013.07.021,

In the early colonial period, African production systems and conservation practices increasingly came under attack from colonial state officials and some missionaries who sought to control access to land and labor resources. The Colonists did not realize the intrinsic causes of land degradation and the historic practices; instead they believed that Africans had no idea of manuring and effective management of land resources. The European idea of soil conservation was to settle permanently in a certain area with careful manuring allowing a constant rise of crop yields.

Soon after a declaration in the 1890s, a series of restrictive measures were introduced resulting in increasing rates of deforestation, wildlife depletion, and soil erosion [...]. A report on soil erosion on private estates from 1930 indicates that sheet and gully erosion had been taking place to an alarming degree [...]. After the 1930s, the colonial state began to realize the importance of soil erosion control, but state interventions bifurcated between rural and private lands. Harsh conservation campaigns were spread over the rural areas, while on the mainly European-owned private estates they came much later and were not adequately enforced. On the whole, estate owners did not use many methods to control soil erosion. But in the early 1940s, with the help of well-educated soil conservationists, there was a new focus on what might be called integrated land use management in modern terminology. This was the birth of Malawi's modern soil conservation network."

There is a bitter irony about all this — by disrupting the traditional agricultural techniques, those techniques were forgotten. And now these same broken countries have to rely on Western institutions to help them relearn those very same skills. And so, we get the Great Green Wall and armies of virtuous institutional experts demanding vast amounts of funding to study and find solutions to the problems while locals like Allan Savory who actually grew up on the land and understands his local community structures are vilified by those same international "experts".

As modern borders were established, as fences were built, as populations expanded, and as the general chaos of the colonial era displaced many tribes who all got mixed together within those new borders, long-established traditions broke down even further and pushed crop farmers ever deeper into the fragile grasslands of the region.

However, the biggest changes are far more subtle, yet had the most far-reaching consequences on the land. In the pre-colonial era, the Sahel region was home to several large kingdoms. Professor of Geography Bill Mosely at Macalester College in Minnesota describes how these pre-colonial kingdoms collected a portion of the harvest as a tax, which was used (in part) to build social safety nets by creating community grain stores to avert famine in the dry years. All that fell apart during the colonial era.[32]

The colonial regimes switched to a head tax (much like a feudal tax) as a way of setting production quotas to force farmers to produce much higher yields. The head tax served *"as a tool for forcing local farmers to grow more cash crops and store less surplus grain. Some herders were also encouraged to abandon their animals in favour of farming, or to develop fenced-off ranches."*[33] Of course, the cash and crops collected under the head tax system did not stay in the region but was sent off to Europe (or to Egypt, since the Eastern Sahel was colonized by the Khedivate state of Egypt, which was a tributary state of the Ottoman Empire). After all, the main point of colonialism was to send money and food back to the home country, not to improve life in the colony. As a consequence, the social safety nets designed to insulate people from famine during drought years collapsed. Furthermore, while vast extended kinship networks in the pre-colonial era had helped avoid famines during drought periods, the head tax system essentially kept people chained to their land to try to squeeze what they could out of the parched soils to avoid defaulting on their tax obligations and losing their land.

The post-colonial system is arguably no better. In chapter two I described the Green Revolution that revolutionized the agricultural industry with chemical fertilizers, new high yielding seed varieties, pesticides, irrigation, increasing mechanization, and western agricultural advisors. That Green Revolution also reached Africa where it is known as the New Green Revolution for Africa. At first glace this would seem like a wonderful thing. But once again, these technologies are not just tools, they are part of an economic system imposed from the outside rather than

---

[32] Moseley, W.G. (2022, Feb 15). The trouble with drought as an explanation for famine in the Horn and Sahel of Africa. *Conversation Media Group.* — *United Nations Office for Disaster Risk Reduction.* https://www.preventionweb.net/news/trouble-drought-explanation-famine-horn-and-sahel-africa

[33] Moseley, W.G. (2022, Feb 15).

developing organically from within, and so, by consequence, it further destabilized the African agricultural system.

The New Green Revolution in Africa led to a widespread shift to monoculture planting, even in the dry regions of Africa. The plow arrived to break the grassland sod. Irrigation and fertilizer replaced careful soil conservation efforts designed to reduce moisture loses from the soil (like mulching). And large monoculture plantations left the soil bare for long periods between crops (which increased soil moisture evaporation, humus degassing, sunbaking, and soil erosion when it rains). And, of course, new technologies don't come cheap, and so the debt-fueled serfdom of the modern economy also came to Africa. Professor Bill Mosely writes: *"These new systems imperil farmers as they often purchase inputs on credit, become indebted when harvests are less than ideal, or must sell their crops to settle bills, a practice that undermines any effort to hang on to surplus production for a future year."*[34] As we can see even in our Western economies, easy access to debt quickly incentivises society to live hand-to-mouth. In 2023, CTV News reported that even in allegedly wealthy Canada, *"more than half of Canadians say they are $200 away or less from not being able to pay all of their bills at the end of the month."*[35] Reliance on the state has replaced the savings account as easy access to debt has completely destroyed the culture of savings. The modern creditor is the new feudal master or colonial head tax collector, and the worst part is that this debt slavery was entered into voluntarily.

Debt pushes everyone to focus on yield at the expense of everything else — subtle considerations like soil humus, moisture evaporation rates, and the root causes of bare soil between crops pale in significance to the pressure to make ends meet to be able to hold on to the farm. And opting out often isn't an option — those who take on debt to expand their operations quickly dominate markets while those who do not are gobbled up by the bigger players. The poor get poorer, and the rich get richer as

---

[34] Moseley, W.G. (2022, Feb 15). The trouble with drought as an explanation for famine in the Horn and Sahel of Africa. *Conversation Media Group. — United Nations Office for Disaster Risk Reduction.* https://www.preventionweb.net/news/trouble-drought-explanation-famine-horn-and-sahel-africa

[35] "More than half of Canadians $200 away or less from not being able to pay all of their bills." (2023, Jul 10). *CTV News.* https://www.ctvnews.ca/business/more-than-half-of-canadians-200-away-or-less-from-not-being-able-to-pay-all-of-their-bills-1.6473939

the corporate model slowly imposes itself onto the value chain, pushing everyone else to the margins and gutting the socio-economic incentives of that once sustained independent farmers and craftsmen. Even the United States is losing farms at a rate of around 142,000 farms per 5-year interval[36] (the interval of U.S. ag censuses) as small farms are gobbled up and consolidated in the hands of bigger (and often investor-funded) players. The Epoch Times recently reported that the U.S. is losing its independent cattle ranches at a rate of over 20,000 ranches per year, with more than 650,000 ranches lost since the 1980s![37]

Compounding all of this is the financial architecture of the modern nation state. I've already mentioned that in the United States, the purchasing power of the U.S. dollar has declined by over 97% since the introduction of the Federal Reserve in 1913, effectively eroding all savings (and eroding the culture of savings) while incentivising debt-fueled risk-taking. It's wonderful for asset holders who see their assets rise in value and who see their debts get easier to pay off as price inflation makes it easier to pay off their debts. But for the smaller, weaker players (like independent family farms), it is a recipe for getting trapped on a treadmill with no way off as inflation eats away at savings and wages, which slowly tips small farmers into perpetual debt slavery and forces them to prioritize yields at the expense of everything else (including rebuilding humus) until someone scoops up their land in a distressed land sale. Likewise, financial reforms like the Frank-Dodd Act (passed in the wake of the 2008 Financial Crisis) was meant to clamp down on excessive risk-taking, but it had the unintended effect of lowering borrowing costs for the largest Wall Street companies (who are ultimately backstopped by the US taxpayer because they are "too big to fail"), while significantly *raising* borrowing costs for small farmers, family-run operations, and independent farm businesses (which, by definition, were higher risk enterprises due to their smaller size and lack of a Wall Street connections). By making loans more expensive for family farms, small farm bankruptcies increased and the ongoing transfer of U.S. farms from the

---

[36] Baethge, J. (2024, Feb 13). U.S. lost 142,000 farms according to ag census. *Farm Progress*. https://www.farmprogress.com/farm-policy/u-s-lost-142-000-farms-according-to-ag-census

[37] Stocklin, K. (2024, Aug 1). America's Looming Cattle Crisis. *The Epoch Times*. https://www.theepochtimes.com/article/americas-looming-cattle-crisis-5697192

hands of independent family farmers into the hands of large agribusinesses and Wall Street investors only accelerated still further.

But inflation in the West is a mere fraction of what it is in the developing world. Printing presses run by corrupt kleptocratic banana-republic dictators certainly deserve a lot of the blame. But, as usual, there's more to the story. President Franklin D. Roosevelt took the U.S. off the gold standard in 1933; before then (since the U.S. Civil War) every dollar was tied to an equivalent amount of gold stored in the vaults of the United States Bullion Depository in Fort Knox, Kentucky. After 1933, the Federal Reserve could flood the system with as many dollar bills as it wanted in order to "stimulate" its domestic economy and to fund its military and corporate adventures around the world. But because the U.S. dollar is the reserve currency for other central banks around the world and because it is the currency used to conduct international trade, the U.S. could print vast quantities of dollars without triggering runaway inflation at home — the rest of the world wanted as many dollars as the U.S. was willing to print. In effect, all these U.S. dollars floating around the rest of the world have exported America's inflationary money printing to other countries while incentivizing other nations to optimize their local economies to produce commodities for export back to the U.S. in order to try to get a hold of more of those precious U.S. dollars. This retooling of local economies to optimize exports for world markets has completely hollowed out prior nation-centric agricultural systems that were adapted to take account of local ecological circumstances and insulate local communities from the punitive measures imposed by foreign creditors.

As this vicious cycle takes its toll, Western institutions, like the IMF, step up to help struggling nations with debt relief, but that debt relief is frequently structured to privilege export agriculture, which further reshapes local institutions in these struggling countries to promote commodity-oriented monoculture practices that push small farmers into debt and soil degradation. And "assistance" from abroad comes with pressure to dismantle all trade barriers in order to open up the local economy to competition from abroad, which leaves local small farmers competing with plantation-scale monocultures in other countries. As local businesses gain access to the lowest-cost commodities imported from somewhere else in the world, small local producers are effectively shut out of the market economy, leaving them as subsistence farmers on the margins of society even as the big corporate farms gobble up the land.

It's the same effect as the Walmart Effect — at first it seems to uplift the poor by providing basic needs at the lowest price, until one day you discover that the local economy is completely hollowed out and everyone has been forced to become a serf to a global mega-corporation because the local market economy has been entirely replaced by Wall Street dominated globalism. The economy of independent business owners is slowly being replaced by cashiers, Amazon delivery workers, and distant factory workers. In time, even the family farms and independent woodlot owners give way to shareholder-owned farming and forestry conglomerates. The long view of the farmer securing his grandchildren's inheritance is replaced by the short-term thinking of CEOs maximizing this year's share price and doing whatever is necessary, regardless of the long-term cost, to reduce the cost of production below those of other global competitors.

The global marketplace is optimized, by design, to favor mega-corporations, monoculture production, and corporate-scale farm production — the entire value change is financed and retooled for the benefit of the big players with the closest relations to central banks. Meanwhile, smallholder farmers are left scrambling to hawk their produce outside of the established system, like through farmer's markets, farm gate sales, etc. They are effectively shut out from participating in the conventional corporate-dominated value chain, which leaves small farmers having to build entirely independent marketing systems outside of the established system. This entire debt-fueled economic model is ill-suited to smallholder farmers living in an ecosystem that experiences dramatic decade-long drought cycles, like the farmers in the Sahel. Resolving the destructive incentives created by this system is going to take a lot more re-thinking than just planting some trees.

I'll leave the last word on the Sahel to Professor Bill Mosely: *"And last but not least is people's limited power or agency to shape their own food systems. From the colonial period forward, this has led to an unsustainable mix of livelihood practices that are constantly threatened by drought."*[38]

---

[38] Moseley, W.G. (2022, Feb 15). The trouble with drought as an explanation for famine in the Horn and Sahel of Africa. *Conversation Media Group.* — United Nations Office for Disaster Risk Reduction. https://www.preventionweb.net/news/trouble-drought-explanation-famine-horn-and-sahel-africa

Case Study # 2:
Australia, a Continent on Fire

"It is hard to imagine a more stupid or more dangerous way of making decisions than by putting those decisions in the hands of people who pay no price for being wrong."

— Thomas Sowell

One-third of the world's forests have been lost since the beginning of human civilization — an area twice the size of the United States. Half of that forest loss happened since 1900. That is already a significant story unto itself. But two-thirds of the world's landmass is covered by arid and semi-arid ecosystems — these are brittle, low-rainfall ecosystems that are either covered by grasslands, scrubland, or desert. And those grasslands are disappearing even faster than the forests as they are either converted into farmland or deteriorate into scrubland and desert. Only 10% of the planet's historic grasslands are still intact.[1]

Australia is the world's second driest continent (Antarctica is the driest), it has the least fertile soils, and its historic climate cycles (since long before colonization) alternate between extreme drought and extreme floods — it is a land of extremes. Today 40% of Australia is covered by dunes.[2] Yet the first Europeans to arrive in the 17th, 18th, and early 19th century described an open landscape with very few trees but covered by a vast sea of grass.[3] Writing in 1770, Botanist Sir Joseph Banks reported that Australia had *"very few tree species, but every place was covered with*

---

[1] Sullivan, K. (2019, Jun 5). Only 10 percent of the world's grasslands are intact. *Popular Science.* https://www.popsci.com/grasslands-disappearing-chart/

[2] Sullivan, R. (2010, Feb 10). Great Sandy Land. *ABC Science.* https://www.abc.net.au/science/articles/2010/02/10/2807917.htm

[3] Forbes, V. (2022, Feb 15). The destruction of Australia's grasslands. *Spectator Australia.* https://www.spectator.com.au/2022/02/the-destruction-of-australias-grasslands/

vast quantities of grass".[4] In 1827, Henry Hellyer (a surveyor) described *"an excellent country, consisting of gently rising, dry, grassy hills [...] They resemble English enclosures in many respects, being bounded by brooks between each, with belts of beautiful shrubs in every vale."*[5]

The last 200 years have not been kind to Australia historic ecosystems. 99% of Australia's temperate lowland grasslands have been destroyed, along with 83% of Tasmania's grasslands and grassy woodlands, 90% of Victoria's grasslands, 95% of Queensland's brigalow scrub, 90% of Australia's temperate woodlands, 60% of coastal wetlands in southern and eastern Australia, 75% of its rainforests, and nearly 50% of all of Australia's forests.[6] What caused this colossal ecological destruction?

I've already described the vital link between large migrating grazing herds and healthy grasslands and what happens when that link is severed in arid climates by either plowing up the grasslands or replacing herd migrations with continuous grazing. However, herds of kangaroos, emus, and other grazing animals were only partially responsible for creating Australia's vast historic grasslands. The other half of the equation in pre-colonial Australia was fire.

In order to sustain any healthy grassland, the large volume of plant material that emerges during the growing season *must* be removed before the next growing season begins in order to prevent it from choking out fresh grass shoots. Like the indigenous peoples in my home province of British Columbia, Canada, the Aboriginal people of Australia routinely lit fires all over to continent to remove that dead grass. In most of the world's grasslands, planned fires played a secondary role compared to the grazing impact from vast herds of migrating ruminant animals. But in Australia, fire played the central role in maintaining the grassland ecosystem.

In the drier regions of Australia, fire kept the grasslands healthy and kept scrubland from invading the open landscape. In the wetter regions, it kept the rainforest at bay. For example, northwest Tasmania today is a

---

[4] Forbes, V. (2022, Feb 15).

[5] Fletcher, M-S. (2020, May 11). This rainforest was once a grassland savanna maintained by Aboriginal people – until colonisation. *The Conversation.* https://theconversation.com/this-rainforest-was-once-a-grassland-savanna-maintained-by-aboriginal-people-until-colonisation-138289

[6] "What's happening to Australia's biodiversity?" (2023, Jul 24). *Australian Museum.* https://australian.museum/learn/science/biodiversity/whats-happening-to-australias-biodiversity/

temperate rainforest dominated by gigantic trees, yet as recently as two hundred years ago this entire region was a "poa tussock" grassland region, kept open by Aboriginal fires.[7] Archaeologist Rhys Jons coined the term "fire-stick farming" to describe the indigenous planned cool season cultural burning that was regularly practiced by Aboriginals to clear away dead grass, remove brush, concentrate kangaroos to make them easier to hunt, and reduce the risk of high-intensity wildfires during the peak dry season.

Professor Michael-Shawn Fletcher, professor of biogeography at the University of Melbourne, uses the example of the temperate rainforests that now cover northwest Tasmania to illustrate the profound impact that "fire-stick farming" had on Australia's historic grasslands: *"... the speed at which rainforest invaded and captured this Indigenous constructed landscape [after the practice of planned burning was banned by modern politicians] shows the enormous workload Aboriginal people invested in holding back rainforest."*[8]

As grass matures, it slowly yellows even as nutrients are diverted from the leaves to the seed head. In a wet climate, the high humidity and high rainfall soon causes dead grass to fall over and rot. But in a dry climate, there isn't enough moisture to rot the grass. If the grass is not grazed by animals shortly after it dies, it slowly turns grey and becomes increasingly brittle, until eventually it simply crumbles away. That destructive process is oxidation — oxygen is slowly reacting with the carbon in the grass to turn it back into CO_2. In a dry climate, that slow oxidation process can take over 40 years during which the dead material chokes out fresh shoots trying to grow from the base of the stem (this is why weeds take over when dead grass is not regularly removed by either grazing or fire). Compounding the problem is the fact that sun and rain leach nutrients out of the dead grass — by the end of winter (or by the end of the dry season in dry climates) any leftover grass that did not get grazed will become so

---

[7] Fletcher, M-S. (2020, May 11). This rainforest was once a grassland savanna maintained by Aboriginal people – until colonisation. *The Conversation.* https://theconversation.com/this-rainforest-was-once-a-grassland-savanna-maintained-by-aboriginal-people-until-colonisation-138289

[8] Fletcher, M-S. (2020, May 11). This rainforest was once a grassland savanna maintained by Aboriginal people – until colonisation. *The Conversation.* https://theconversation.com/this-rainforest-was-once-a-grassland-savanna-maintained-by-aboriginal-people-until-colonisation-138289

nutrient deficient that it takes more energy to digest it than what animals can extract from it — the animals can literally starve to death trying to eat it. So, if this dead grass is not removed before it loses all nutritional value, it isn't just damaging to new shoots and doesn't just encourage the spread of weeds, it also turns a fertile grassland into a food desert — the animals either migrate away or starve.

Fire is fast oxidation whereas leaving dead grass to waste away is slow oxidation. Fire looks more dramatic, but either way the same amount of CO_2 will ultimately end up back in the atmosphere. In theory, if animal numbers were maximized so that every single speck of old grass was trampled or eaten by the time the next growing season begins, then you wouldn't need fire. But climate is *not* stable, it is cyclical, especially in a dry place like Australia. If animal numbers were always maximized to the total volume of available grass, vast numbers of wildlife would starve every time there is a dry year. And so, planned fire lit by hunter-gatherers became an essential part of maintaining a healthy ecosystem in arid climates where dead vegetation doesn't simply rot way. During years with poor harvests (drought years), there would be little grass to burn by the start of the next growing season, whereas during years with abundant growth, more grass would need to be burned before the next growing season began. And so, in arid climates indigenous fire management became an essential part of the ecosystem balance to keep the grasslands healthy while also helping to prevent starvation among the wild herds by creating a stable balance between population numbers and the amount of food available to them a fire-managed landscape.

If you read through the academic literature on wildfires, there is no shortage of "peer-reviewed evidence" claiming that all fires are bad for the soil, including all low-intensity fires. That evidence is just plain wrong. It is true that during the dry season, even a low intensity fire will put humus, sod, and even roots at risk of being consumed by fire because there is no moisture in the soil to prevent the fire burning down into the organic matter in the soil. However, a planned fire during the cooler months (winter in Australia, early spring in Canada) will not only burn much cooler, but the damp ground will protect the soil from being scorched by the low-intensity flames — planned grass fires conducted when the soil is moist pose no threat whatsoever to the sod and humus. On the contrary, as long as the low-intensity fire happens while the ground is moist, something magical happens…

Fire turns plant debris into a combination of carbon (ash and charcoal) and CO_2. As long as the ground is moist as the fire passes overhead, soil microorganisms are active right up to the surface of the wet soil so the ash and charcoal that fall to the ground after a planned fire are immediately consumed by those soil microorganisms and carried down into the lower soil horizons instead of simply sitting on the surface. Soil scientists have demonstrated that soil microorganisms rapidly transport all types of organic matter from the surface down into the deeper soil layers. Planned burning while the ground is moist ensures that the charcoal left behind by a fire is quickly added to the carbon inventory in the soil. And so, fire becomes a tool for enriching the soil and building humus.

Back in chapter 24, I explained how microscopic carbon particles act as a storehouse for plant nutrients, which electrostatically bond with the tiny carbon particles. Planned, low-intensity fires release nutrients from dead plant debris, which falls to the ground as ash, while simultaneously releasing a flood of fresh carbon particles that can electrostatically bond with those nutrients — thus the carbon released by burning immediately serves as the storage mechanism to capture and hold onto all the other nutrients that are also released by burning. Win-win. By contrast, a fire during the dry season causes the ash to settle on parched soils, only for the wind (or subsequent rainfall) to carry it away because its too dry for soil microbes to efficiently incorporate those nutrient-rich ashes into the soil.

Back in chapter 17, I described the *terra preta* soils that were created by indigenous peoples in the Amazon jungle by adding charcoal to soil to transform the otherwise extremely nutrient deficient Amazonian soils into some of the most fertile soils on the planet. Likewise, the historic, planned, low-intensity grass fires lit by Aboriginal Australians, just like the grass and brush fires once used by indigenous peoples throughout western North America, were essential for creating fertile carbon-rich soils in these arid climates. Without these planned fires, the grasslands and open forests of these arid regions would have turned to desert long ago. Fires, like grazing animals, are merely tools — at their worst they can be destructive beyond our wildest imaginations but at their best they are indispensable for building fertile, sustainable, drought- and erosion-resistant landscapes.

Early settlers in Australia kept up the seasonal burning to some degree. However, they also introduced all the modern tillage practices that I have written about throughout this book (remember at the end of chapter one when the British thought that the rains would follow the plow into the interior of Australia more than 660 km north of where crops are grown today, back when it was still a grass-covered plain). And in areas not considered suitable for cultivation, the settlers scattered sheep and cattle across the landscape, managed via the destructive practice of continuous grazing (if someone doesn't know about the importance of mob grazing, why would they bunch their livestock together in a grazing mob in a country where there are no large grassland predators to threat livestock, apart from the occasional crocodiles lurking beneath the surface of waterholes). Today, over two-thirds of Australia's agricultural land has been degraded by water erosion, wind erosion, loss of vegetation, and salinity.[9] The New South Wales government estimates that the natural rate of soil formation in its region is between 0.04 to 0.4 tonnes per hectare per year, but soil erosion on grazed pasturelands in New South Wales is happening at a rate of around 1 tonne per hectare per year while on cultivated land soil erosion ranges from around 1 to 50 tonnes per hectare per year! Tick tock, the clock is ticking on Australian civilization.

However, some of the worst damage to the grasslands and the worst destruction of soil organic matter has been done by ideological environmentalists intent on suppressing any and all wildfires.[10] Part of the resistance to planned fires was motivated by their simplistic fear of greenhouse gases, part of it was motivated by concern for small wildlife caught up in the flames, and part of it was motivated by shortsighted academic research that claimed all wildfires destroy sod and humus. These environmental concerns infiltrated the political class in Australia, which worked to gradually extinguished the practice of planned low-intensity grassland fires, effectively tying the hands of local communities who know better.

As low-intensity wildfires were suppressed, dead grass built up, bare ground emerged between the unhealthy grass tussocks, and weeds and brush crowded into the spaces once occupied by healthy grass. And then, when lightening, a cigarette, a spark from a hot engine, or some arsonist with a match set fire to this accumulating fuel (usually at the worst time

---

[9] "An Overview of Land Degradation in Australia." (2022, Nov 25). *KG2.com.au – Australia's largest independent farmer database*. https://kg2.com.au/an-overview-of-land-degradation-in-australia/

[10] Forbes, V. (2022, Feb 15). The destruction of Australia's grasslands. *Spectator Australia*. https://www.spectator.com.au/2022/02/the-destruction-of-australias-grasslands/

of year when conditions are hottest and driest), enormously destructive, unplanned, high-intensity wildfires began sweeping across the Australian landscape, burning through the sod, through the humus, and through the roots below to leave behind an utterly decimated landscape fit only for weeds. The Black Summer of 2019-2020 consumed more than 46 million acres. The worst season on record was the 1974-1975 fire season, which consumed a colossal 290 million acres, leaving behind vast tracts of scorched soil! By discouraging planned low-intensity grassfires, well-meaning fire bans essentially created the conditions for a whole continent's worth of soil to go up in smoke.

Noxious weeds quickly take over the scorched soil left behind by these high-intensity fires, woody brush invades former grasslands in the absence of regular low-intensity grassland burning, and the emerging scrubland provides habitat for invasive non-native species like feral camels, rabbits, foxes, wild dogs, cats, deer, and pigs, which have decimated the native vegetation and the native wildlife. There are more than 1 million feral camels, 200 million feral rabbits, up to 25 million feral pigs, 2 million feral deer, and 6 million feral cats roaming wild across Australia!

By removing planned wildfires from the Australian landscape, the ecosystem has spiralled into decline. Once again, righteous do-gooders intent on controlling their unruly neighbors and overspecialized scientists who lost sight of the big picture convinced themselves of their own wisdom and encouraged know-nothing central planners to fiddle with complex natural systems they don't understand while ignoring the voices of those with practical experience who lack the shiny credentials to make themselves heard in the halls of power. And so, reality and politics live in two separate worlds as environmentalists, the media, the IPCC, and the United Nations all continue to beat their self-serving drums about climate change as the alleged key driver of Australia's destructive wildfires and desertification while the rest of us look on in frustrated exasperation. Local conservation efforts are slowly trying to reintroduce traditional Aboriginal fire management practices, and holistic planned grazing is slowly making headway among Australia's farmers, but their efforts continue to face stiff headwinds from establishment voices defending the dogmatic carbon dioxide narrative and vilifying fire and livestock in all their forms.

As long as our political system is centralized, our world will continue to be dominated by these centrally funded "experts" who are increasingly using political diktats and state-enforced censorship to settle scientific debates while denying anyone from outside of their elite circles from having any voice in shaping the policies that rule our lives. The halls of power are so distant that it is virtually impossible for anyone outside of the echo chamber to grab the decision-makers by the hand, drag them out into a pasture, and show them what is going on. They probably wouldn't believe their own eyes anyway unless there's some "peer reviewed study" to back up the claims, and even if there is, that study will be dismissed if those with priority access to the microphone insist that these claims contradict the "consensus" narrative. It's the perfect echo chamber designed to insulate itself against all dissenting voices.

Even if you could convince someone who allegedly has some authority to make changes, decision-making in the modern institutionalized system has become so diffuse and distant from the people who have to live with the consequences of those decisions, responsibility has been dispersed across so many different agencies, and our institutions are staffed by such a vast army of risk-averse bureaucrats all intent not to provoke the wrath of those who have influence over their career prospects that the entire decision-making process has essentially become an exercise of looking left and right to judge the opinions of their peers before putting their stamp on anything. Perhaps it is unfair to blame anyone working within this mad system — there's no reward and only a world of hurt for anyone trying to swim against the tide. Mortgages have to be paid, careers are on the line, and it doesn't pay to think outside the echo chamber. Even among the "experts", it pays to suspend critical thinking skills and simply defer to the other experts in order to lead a frictionless life.

In a decentralized world with an institutional apparatus stripped down to its barest essentials, it would only take one dissenting voice in local leadership to create one small local example of how it could be done differently to break the echo chamber. Decentralization changes the incentives to encourage people to lead by example and encourages conversations between local leaders and local populations; centralization encourages everyone to keep their heads down and run with the herd to avoid a fruitless confrontation with the dogmatic establishment.

Case Study # 3:
Haiti, a Long Descent into Hell

"Thieves of private property pass their lives in chains; thieves of public property in riches and luxury."

— Cato the Elder (234-149 BC)

If there is a poster child for the story of the century, it is Haiti, where all the destructive political and ecological forces that I have written about throughout this book come together in a perfect storm to completely wreck a country and utterly devastate its soils. There's nothing subtle about Haiti's story — it exposes in gritty detail the destructive link between centralized decision-making and rampant environmental deterioration. The one billion inhabitants who live in industrialized nations are experiencing their own problems with deforestation, soil degradation, and desertification, but what is happening in countries like Haiti where the other seven billion people live is orders of magnitude worse. Once you understand the toxic cocktail of political, economic, and environmental forces tearing Haiti apart, you will be able to see echoes of those same forces at work all around the world, including in our supposedly prosperous industrialized nations. What's happening to Haiti's soils reflects what's happening to all of civilization.

In the colonial era, Haiti was the most lucrative colony in the French empire and perhaps the richest colony in the world.[1] In the 1780s, 60 percent of all the coffee and 40 percent of all the sugar consumed in Europe came from this one small colony![2] Even as late as 1884, almost a century after the Haitian Revolution that threw off France's yoke, Sir

---

[1] "Economy – Remember Haiti." *The John Carter Brown Library.* Accessed 9 Jun 2024.
https://www.brown.edu/Facilities/John_Carter_Brown_Library/exhibitions/remember_haiti/economy.php

[2] Henley, J. (2010, Jan 14). Haiti: a long descent to hell. *The Guardian.*
https://www.theguardian.com/world/2010/jan/14/haiti-history-earthquake-disaster

Spenser St. John wrote glowingly about Haiti's rich soils: "*And yet in all the wide world there is not a country more suited to agriculture than Hayti; not one where the returns for labour are more magnificent; a rich, well-watered soil, with a sun which actually appears to draw vegetation towards itself with such energetic force that the growth of plants, though not actually visible to the eye, may be almost daily measured.*"[3]

Today, Haiti's per capita GDP has plunged to 163rd out of 189 countries in the world. Soil erosion and deforestation in Haiti are so extensive that they are visible from space, as shown in figure 114 below. There are barely any trees left on the Haitian side of the island, and its sharp gullies and ridges reveal the rapid erosion compared to the much softer hills and gullies on the Dominican side of the island.

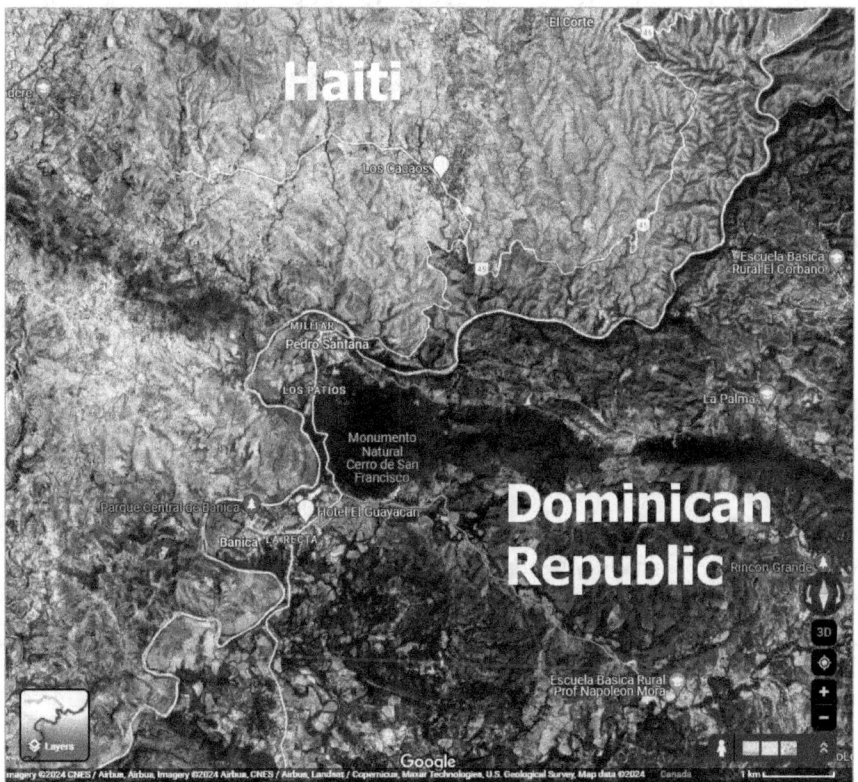

Figure 114. Deforestation and soil erosion in Haiti is so severe that it is visible from space by the sharp line that divides Haiti from its neighbor, the Dominican Republic.

---

[3] St. John, S. (1884). *Hayti; or, The black republic.* Smith, Elder, & Co. https://www.gutenberg.org/cache/epub/68592/pg68592-images.html

Less than 2% of Haiti's natural forests remain and the deforestation continues at a rate of around 15 to 20 million trees per year.[4] Haiti also has one of the highest rates of soil erosion in the world. Its neighbor, the Dominican Republic, is losing soil at an already disastrous rate of just under 25 tons per hectare per year, but Haiti's erosion is more than 3x faster at over 75 tons per hectare per year[5], which works out to more than 36 million tons of topsoil lost from the island nation each year.[6] Over 50% of its farmland has degraded past the point of no return,[7] and erosion is continuing to shrink Haiti's agricultural base by an additional 37,000 acres per year.[8] You can imagine the destruction that hurricanes leave behind every time they pass over this denuded island, which they do on a semi-regular basis. And the absence of timber means that most buildings are made of concrete instead of timber-framed houses, which is a recipe for massive infrastructure destruction every time Haiti is shaken by an earthquake (it sits on a major fault zone between the North American and Caribbean tectonic plates).

At the end of the Haitian Revolution in 1804, Haiti had a population of around half a million people. Since then, Haiti's population has steadily risen to over 11 million people today. As its soils have deteriorated and its farms have eroded away, Haiti has become wholly reliant on food aid as it lost its ability to feed itself. Over 50 percent of its food and over 80 percent of its main staple food (rice) is imported today.[9] In effect, Haiti is in the iron jaws of the Malthusian Trap, with little hope of reprieve as the nation's wealth is continually destroyed by one thing or another and as

---

[4] Araujo, S. (22013, Jan 29). *Investing in Haiti's Rural Community.* [Press release]. Relief Web. https://reliefweb.int/report/haiti/investing-haiti's-rural-community

[5] Dror, I., Yaron, B., and Berkowitz, B. (2022). The Human Impact on All Soil-Forming Factors during the Anthropocene. *ACS Environ Au.* 2(1). https://doi.org/10.1021%2Facsenvironau.1c00010

[6] Araujo, S. (22013, Jan 29).

[7] Montgomery, D. R. (2012). *Dirt: the Erosion of Civilizations.* University of California Press.

[8] Araujo, S. (22013, Jan 29). *Investing in Haiti's Rural Community.* [Press release]. Relief Web. https://reliefweb.int/report/haiti/investing-haiti's-rural-community

[9] Coello, B., Oseni, G., Savrimootoo, T., and Weiss, E. (2014, Sept). Rural development in Haiti. *The World Bank.* https://documents1.worldbank.org/curated/en/976171468032682306/pdf/955400WP0Box390portunities00PUBLIC0.pdf

politics continually imposes itself as the principal obstacle to good land stewardship.

Over half of Haitians live in rural areas, spread out over 1 million farms with an average size of only 0.93 hectares (2.3 acres). Almost all of these farms are subsistence farms, many of which are located on steep hillsides.[10] Very few farms exceed 12 hectares (30 acres).[11] Declining soil fertility is slowly undermining productivity as harvest yields per acre fall by between 0.5% to 1.2% annually.[12] As Haiti's small farms become increasingly eroded, degraded, and unproductive, millions have crowded into the big cities in search of work, creating vast slums. More than two-thirds of the entire population of Haiti's capital city, Port-au-Prince, live in shantytowns. 60 percent of Haitians depend on charcoal produced on the island for their domestic energy needs, sourced from Haiti's rapidly dwindling forest reserves — urban residents in particular depend on this charcoal because they don't have access to wooded areas to cut their own firewood.

Famed Peruvian economist Hernando de Soto dedicated his life to bringing property rights to the world's poor (which made him a target of numerous Marxist terrorist bombings and assassination threats).[13] In Haiti, he found that *"the total assets of the poor are more than one hundred fifty times greater than all the foreign investments received since Haiti's independence from France in 1804"*, but that those assets do not help the people because a lack of clear property rights means that they cannot leverage their assets to access loans, buy or sell property, get insurance, or raise capital to invest in their businesses.[14] No-one in their right mind invests time or money into improving land if they fear it could be taken away from them. No-one will allow you to use your house as collateral for

---

[10] "Farmer's Organizations for Africa, Caribbean, and Pacific (FO4ACP)." *Food and Agriculture Organization of the United Nations.* Accessed 11 May 2024. https://www.fao.org/in-action/farmers-organization-africa-caribbean-pacific/country-activities/haiti/en

[11] "Haiti – Agriculture." *Nations Encyclopedia.* Accessed 4 Jun 2024. https://www.nationsencyclopedia.com/Americas/Haiti-AGRICULTURE.html

[12] Coello, B., Oseni, G., Savrimootoo, T., and Weiss, E. (2014, Sept).

[13] Healy, G. (2004, Apr 14). Dangerous Minds: Hernando de Soto. *Cato Institute.* https://www.cato.org/commentary/dangerous-minds-hernando-de-soto

[14] de Soto, H. (2000). *The Mystery of Capital: Why Capitalism Triumphs in the West and Fails Everywhere Else.* Basic Books, New York. https://archive.org/details/mysteryofcapital00soto

a loan if it's unclear who owns the land under your house. And the legal system is decidedly two-tier, leaving the poor completely unable to defend themselves or their assets through the courts.

In Haiti, de Soto found that it took 19 years and 176 bureaucratic steps to legalize the purchase of private land — even securing a 5-year lease of farmland from the government took a little over two years and 65 bureaucratic steps, on average.[15] 68% of city dwellers and 97% of people in the countryside live in housing in which <u>nobody</u> has a clear title to the property. Even as late as 2010, prior to the 2010 Earthquake that made everything so much worse, only 40% of Haitian landowners had some kind of title or transaction receipt to prove ownership of their land.[16] In his book, *The Mystery of Capital: Why Capitalism Triumphs in the West and Fails Everywhere Else*, de Soto wrote: "*Imagine a country where nobody can identify who owns what, addresses cannot be verified and the rules that govern property vary from neighborhood to neighborhood, or even from street to street.*" In developing countries, he often found that the most reliable way to know who owned what property was to see whose dog barked as you crossed from one property to the next.

It's easy to write Haiti's problems off as the toxic product of corruption, gang wars, a self-inflicted broken culture, and a legacy of colonialism, but as this story unfolds, you'll soon see that it is much more complicated than that. Once again, the fingerprints of yesteryear's "nation building" and present-day "liberal internationalism" are all over this destructive cycle in many subtle and not-so-subtle ways. Even the full gamut of U.S. presidents make an appearance in this story, from Woodrow Wilson to Franklin D. Roosevelt to Bill Clinton, as we see what liberal internationalism and central banking really means for the world.

Tree planting and soil conservation programs can slow Haiti's environmental destruction, but to fix this requires reforms that cut right to the heart of the political and financial system — and by that I don't just mean Haiti's broken internal political system but also to the very heart of the global neo-liberal international "rules-based" order. However, you will be "happy" to know that the United Nations has managed to coax

---

[15] de Soto, H. (2000).

[16] USAid. (2010, Jan). *Land Tenure and Property Rights in Haiti*. [USAid Issue Brief]. U.S. Agency for International Development. https://pdf.usaid.gov/pdf_docs/PA00J75K.pdf

Haiti into a firm commitment to reduce its greenhouse gas emissions by at least 6.32% by 2030.[17] Nero fiddles while Rome burns.

In the late 18th century, Haiti alone accounted for one-third of the *entire* Atlantic slave trade, importing more than 40,000 slaves per year. In other words, Haiti was a meat grinder — the average life expectancy of a slave was only 21 years.[18] The bitter and bloody 12-year Haitian Revolution put an end to French rule in 1804 but left the island's infrastructure in shambles. That was only the start of Haiti's troubles. The Haitian Revolution was the first slave rebellion in the Western Hemisphere and so, out of fear of triggering slave rebellions in their own countries, the United States and other western nations refused to recognize Haiti as a sovereign nation and imposed a trade embargo on the island (the U.S. embargo lasted until 1862). This was meant to prevent Haiti from attracting foreign investment in order to push Haiti into economic failure as a dangerous example designed to discourage any further slave revolts in other colonies in the Western Hemisphere.

In 1825, France sent a fleet to reconquer the island. France's gunboat diplomacy paid off handsomely since it was clear that Haiti would not win a war against France and so, as an alternative to war, Haiti negotiated a deal that it would pay reparations (US$21 billion in today's dollars) to compensate France for its lost colony in exchange for recognition as a sovereign nation. And so, Haiti was saddled with a crippling debt that took until 1947 to pay off — by 1900, over 80% of Haiti's budget was going towards these reparations, effectively exporting all of Haiti's wealth out of the country and further impoverishing the nation as it replaced colonial masters with foreign creditors.

As Haiti continued to toil in servitude to its debt to France, its former visible chains of slavery were replaced by the invisible shackles of debt. Financial serfdom replaced the yoke. Haiti's example is merely a more extreme version of what faces many communities throughout the world today during the golden age of globalism as mom-and-pop businesses are systematically replaced by employment in transnational corporations, as institutional investors take over the housing market, as multinational

---

[17] United Nations. (2022, Jun). Haiti: Latin America & the Caribbean. *United Nations Development Programme — Climate Promise.*
https://climatepromise.undp.org/what-we-do/where-we-work/haiti

[18] Henley, J. (2010, Jan 14). Haiti: a long descent to hell. *The Guardian.*
https://www.theguardian.com/world/2010/jan/14/haiti-history-earthquake-disaster

banks crowd out local lenders in the mortgage market, as Wall Street converts farmland into "yield-generating assets", and as taxes funnel the fruits of a community's labours into distant government coffers. When only crumbs stay in a community because everything is "financialized" and everyone is but a "minion" running for their life on someone else's treadmill, everything is hollowed out — the land, the community, and the soul of a civilization.

France's recognition of Haiti as a sovereign nation didn't open the floodgates to international investment. The United States continued to refuse to recognize Haiti's sovereignty and maintained its embargo until slavery was abolished in the United States. But when Lincoln finally reversed this long-standing policy, U.S. financial institutions quickly positioned themselves to become Haiti's most important creditors, ostensibly for development loans and to help pay off Haiti's debt to France, but in reality their sudden interest in the island was driven by a deliberate U.S. policy to leverage connections in the banking industry to pull Haiti into the American orbit.

If the bulk of a nation's wealth is sent abroad to pay off loans and nothing ever gets built with those loans, it's a surefire way to trap a nation in perpetual poverty. But debt and a lack of outside investment were only a small part of Haiti's problems. After the Haitian Revolution ended in 1804, sharp geographical and cultural differences between the more peasant-dominated rural North and the more outward-looking urban South (where most of Haiti's elites lived) fueled a never-ending power struggle as these two separate regions began vying for control of the National Assembly as each side longed to use Haiti's top-down autocratic system of government to loot the other half of the nation, setting in motion Haiti's never-ending cycle of civil wars, gang wars, and coups. The stronger the centralization of government, the greater the prize. Tyranny from without was replaced by tyranny from within.

The words of Haiti's Declaration of Independence were directly modelled after the American Declaration of Independence, yet Haiti's example after Independence demonstrates that culture is what gives words meaning, not the other way around. Despite the American influence on their Declaration of Independence, after the Haitian Revolution ended the political culture that emerged was decidedly *not* American, but rather it mimicked the French Imperial system with absolute power concentrated in the hands of the island's ruler. When Haiti's Revolution began,

America's second president, John Adams, and his Secretary of the Treasury, Alexander Hamilton, were sympathetic to Haiti's bid for independence and were eager to establish closer economic ties with the island. But Haiti's rebel leader, Toussaint Louverture, rejected them, wanting to maintain close ties with France instead.

As the rebellion rumbled on, Louverture was eventually captured and deported to France where he died in prison. Jean-Jacques Dessalines (who had helped orchestrate Louverture's capture) became the island's leader and, as the rebellion came to a successful end, Dessalines ordered the genocide of Haiti's remaining white population, including many whites who had supported the Haitian revolution.[19] He then sent a letter to President Thomas Jefferson calling for closer ties between the two nations. In view of the massacre and the intense fear it provoked in the Antebellum South where slaves outnumbered whites in several states, American political support for Haiti's independence evaporated and led to Jefferson's decades-long policy of embargoing the island. Historian Kevin Julius wrote that *"As abolitionists loudly proclaimed that "All men are created equal", echoes of armed slave insurrections and racial genocide sounded in Southern ears. Much of their resentment towards the abolitionists can be seen as a reaction to the events in Haiti."*[20] If ever there was an example of a nation shooting itself in the foot, Haiti stands as the quintessential example. One is also left to wonder whether America's path towards the abolition of slavery might have taken a gentler course and happened sooner, and whether America might have been spared its bitter Civil War, if Toussant Louverture hadn't slaughtered Haiti's white population after the Haitian Revolution.

Starting with Dessalines, Haiti's early rulers fashioned themselves and their government upon France's rulers, even going as far as creating a Haitian nobility. When Napoleon crowned himself Emperor in December of 1804, Dessalines quickly followed suit by declaring himself as Haiti's first Emperor, establishing a precedent for those who succeeded him. Sir Spenser St. John, in his controversial 1884 book *Hayti; or, the black republic* describes Dessalines in the following words:

---

[19] "Jean-Jacques Dessalines." *Wikipedia.* 22 May 2024.
    https://en.wikipedia.org/wiki/Jean-Jacques_Dessalines
[20] "1804 Haitian massacre." *Wikipedia.* 13 May 2024.
    https://en.wikipedia.org/wiki/1804_Haitian_massacre

> "Dessalines, like most of those who surrounded him, was in every way corrupt; he is said to have spared no man in his anger or woman in his lust. He was avaricious, but at the same time he permitted his friends to share in the public income by every illicit means. His government was indeed so corrupt, that even the native historians allow that the administration was distinguished "for plunder, theft, cheating, and smuggling." Dessalines, when he appointed an employé, used to say, "Plumez la poule, mais prenez garde qu'elle ne crie," [translation: pluck the chicken, but take care that she does not cry out] — the rule by which the Government service is still regulated. The tyranny exercised by Dessalines and his generals on all classes made even the former slaves feel that they had changed for the worse. There were no courts to mitigate the cruelty of the hard taskmasters, who on the slightest pretext would order a man or woman to be beaten to death."[21]

After the revolution, Dessalines and his successors used military force to keep newly "freed" slaves on their plantations in a legal arrangement mimicking European feudalism — they were allowed to keep one-quarter of what they produced but were forced to hand over the rest to Haiti's elite plantation owners and to the government. They could not choose not to work and they could not legally leave the plantation they were attached to.[22] King Henry Christophe, Dessalines successor, even used forced labour to construct a new citadel, rounding up hundreds of thousands of people as involuntary workers and killing an estimated 20,000 people in the process. Once the formal debt was established in 1825 to pay reparations to France, President Jean-Pierre Boyer restricted agricultural workers rights still further and then enacted the system of *corvée*, which used police and government authority to force residents to work (unpaid) to build roads.[23] Forced labor, plantation-style agriculture, and subsistence farmers clinging to the margins of society without clear ownership of their land and without a functioning legal system to defend their interests

---

[21] St. John, S. (1884). *Hayti; or, The black republic*. Smith, Elder, & Co. https://www.gutenberg.org/cache/epub/68592/pg68592-images.html

[22] "Slavery in Haiti." *Wikipedia.* 23 May 2024. https://en.wikipedia.org/wiki/Slavery_in_Haiti

[23] "Slavery in Haiti." *Wikipedia.*

became a recipe for rampant and accelerating erosion, deforestation, and exploitation.

And so, Haiti lurched from one tyrannical government and from one civil war to the next, following the precedent set by its first ruler. Constitutions are mere pieces of paper — culture evolves by precedent and gives the words of a constitution their meaning. Haiti's political culture emerged from the mindset of a captive population held in bondage who embraced strong central leadership in the hope of replacing tyrannical rulers with more benevolent leaders who could rescue them from their desperate plight. In essence, Haiti adopted the same top-down progressive concept of Liberty, inspired by the French Revolution, which idolized a powerful state acting as society's shepherd while guillotines waited in the wings to deal with any leader who failed to deliver on their promises of prosperity. A destitute population wholly dependent on its leaders for survival does not yearn for freedom; out of necessity it yearns for a saviour to set the world right because it does not yet have the means to stand on its own two feet. And that's the situation where Haiti has gotten stuck ever since.

By contrast, the American Revolution that cast off its colonial masters emerged from the bottom-up mindset of self-sufficient and independent farmers and craftsmen, already positioned to stand on their own feet, who rose up to defend their own liberty and property from encroachment by a distant tyrant — they did not crave a saviour, they craved to be allowed to continue to provide for themselves — to be left alone — which is an altogether different mindset. Tellingly, George Washington chose to go home to his farm after the American Revolution rather than to crown himself king. His peers ultimately convinced him to run for president, but in keeping with the ethos that inspired the American Revolution, after two terms he relinquished power and returned back home to his beloved farm, leaving behind a constitution perfectly crafted to serve a nation of independent farmers and craftsmen. His example and the constitution he signed into law were the product of the unique circumstances of America's pre-revolutionary history. By voluntarily returning to his farm after serving two terms as president, he not only set the precedent for the peaceful transfer of power in America, but he also set the standard that presidents peacefully go home after two terms, which every subsequent president adhered to (though not for a lack of trying stay in power, but because the American public denied them re-election after two terms). It wasn't until the 1940s when Franklin D. Roosevelt broke the unspoken two-term rule (with four terms — he died in office), which finally prompted America to make the two-term limit official via the 22nd Amendment in 1951.

Haiti had a different history and so it made a different choice. A slave with nothing has already been accustomed to the idea of a powerful master and, even if freed, doesn't have a penny to his name and thus is wholly dependent on a generous patron. Thus, a freed slave depends on either the generosity of his former masters or the generosity of the state to help him back onto his feet following emancipation. In short, Haiti's pre-revolutionary conditions ensured that Haiti's destitute and dependent population was in need of a powerful and progressive post-revolutionary administrative state from day one, out of necessity. Circumstances led it to model itself on the example of the paternalistic French Emperor and the centralized French administrative state. And so, Haiti established a different precedent and created a different political culture that has haunted it ever since.

In recent years, it has become fashionable to criticize US Founding Father (and third US president) Thomas Jefferson for the fact that he owned many slaves throughout his lifetime. And yet, upon closer inspection of history we find that he clearly understood both the extreme vulnerability of any slave granted his freedom as well as the vulnerabilities of the indebted slave owner whose financial solvency was tied up in human capital — he spent the rest of his life trying (unsuccessfully) to find a way to solve this dilemma. In 1779, he advocated for the gradual emancipation, training, and integration of slaves rather than immediate manumission, believing that releasing unprepared persons with no place to go and no means to support themselves would lead to disaster.[24] In 1785, Jefferson observed that slavery corrupted both masters and slaves alike. And in 1824, he proposed a plan to end slavery (which was rejected) by getting the federal government to buy all slave children for $12.50 and then train them in the occupations of freemen. Sadly, all his efforts to push America's system to evolve its way out of slavery were rejected at every turn. And so, nothing changed until everything changed through bloody revolution.

---

[24] "Thomas Jefferson and slavery." *Wikipedia.* 22 May 2024.
    https://en.wikipedia.org/wiki/Thomas_Jefferson_and_slavery

In the end, Jefferson's grim predictions of disaster following emancipation proved to be right because the classical laissez-faire system set out by their Constitution — perfectly designed for independent farmers and craftsmen who just want to be left alone — was ill-suited to the sudden emancipation of millions of slaves with no assets, no skills, and no means to provide for themselves. After the U.S. Civil War ended in 1865, four million emancipated slaves suddenly found themselves cast off plantations and left utterly destitute with no job, no land, no assets, and few skills other than picking cotton, and no-one to come and save them. In his book *Sick from Freedom*, American historian Jim Downs documented how, as the plantation economy collapsed in the immediate aftermath of emancipation, hundreds of thousands of freed slaves starved to death and millions experienced destitution, malnutrition, and disease. Mr. Downs describes it as "the largest biological crisis of the 19th century."[25] Desperation ultimately drove millions back onto the very same monoculture plantations from whence they came as they signed sharecropping contracts to pick cotton, which reduced them to little more than serfs — a feudal serf actually had more legal protections (and more incentive to invest in his land) than a sharecropper as a feudal serf could not legally be deprived of his land unless he failed to fulfill his tithe obligations.

Considering Haiti's plantation economy before the Haitian Revolution and the desperate circumstances of its labouring classes in the immediate aftermath of the Revolution, it should not surprise us that Haiti turned towards the top-down French model of government, which views the state as society's benevolent shepherd, rather than adopt the bottom-up laissez-faire model of American government. But then, as the benevolent shepherd failed to deliver, the one wearing the crown was soon deposed and replaced by another. Welcome to Haiti's never-ending cycle of revolution.

The great flaw in modern liberal thinking is the idea that democracy leads to freedom when, in reality, the opposite seems to emerge as people turn to the ballot box to ease the burdens of life. The first step towards freedom is economic, not political. You can only protect your freedom at

---

[25] Harris, P. (2012, Jun 16). How the end of slavery led to starvation and death for million of black Americans. *The Guardian*.
https://www.theguardian.com/world/2012/jun/16/slavery-starvation-civil-war

the ballot box as long as the bulk of society has both the skills and the financial reserves to stand unassisted on their own two feet and does not live according to the whims of a paymaster — in short, the preservation of liberty and autonomy requires a vibrant and thriving self-employed middle class. Even that is rarely enough, as America's gradual drift from classical liberty to hyperregulated bureaucracy so vividly demonstrates. Freedom imposes the burdens that accompany autonomy (fear of failing, fear of not making ends meet, fear of not being up to the task of providing for oneself and one's family in a hypercompetitive world) — and so, many people prefer the comfort and ease of being swept along within someone else's structure. In 1941, just after being elected to his third of four terms, President Franklin D. Roosevelt explicitly referred to the "freedom from want" and the "freedom from fear" in his famous Four Freedoms speech, in which he laid out his vision of the state's ever-expanding responsibility as society's benevolent shepherd. His four consecutive terms tell us the degree to which his message resonated in America's evolving society.

But as Carl Sagan once observed, *"Once you give a charlatan power over you, you almost never get it back."* Or, as tongue-in-cheek observers of socialism like to say, *"you can vote your way into socialism, but you have to shoot your way out."*

Haiti is merely leading the way down the path that every other "progressive" neo-liberal nation is also following into Hell. Woodrow Wilson and his progressive peers gradually introduced Europe's concept of the top-down administrative state to America as America's appetite for a paternalistic shepherd emerged in a population increasingly labouring in the employ of factories and investor-owned corporations rather than for themselves on family farms and in independent workshops. It's a perfect feedback loop: the rat race heightens everyone's sense of vulnerability and fuels the appetite for centralized government, even as expanding centralized government disempowers the individual and the family business while accelerating the rise of an economy dominated by the mega-corporation, which in turn fuels the conditions that plunge everyone ever deeper into the rat race. Even our centrally controlled education system today is no longer tailored to suit the needs of educated and free-thinking freemen; it is preparing society's young for a future in the tax farm as technicians and labourers.

Name a single country today in our globalized corporate-dominated world that doesn't idolize the state as the vehicle of progress and doesn't embrace some version of the progressive values of Wilson's America — there isn't one — we are all moving towards the same destination, albeit at different speeds. The culture wars consuming Western society today are merely a never-ending argument about who is the most deserving beneficiary of the state's largess and an argument about whose preferred leader will be the better dictator. In a centralized system of governance in which vast power and resources are concentrated in the hands of a small number of gatekeepers, the culture wars that emerge are ultimately little more than a bitter rivalry about who gets to choose the winners and losers of this bloated hyper-regulated redistributionist system — from pronoun wars to climate change conferences, in the end these are all just part of the never-ending word games designed to manipulate the gatekeepers. U.S. Founding Father, John Adams, once observed in a letter to his wife that *"A Constitution of Government once changed from Freedom, can never be restored. Liberty, once lost, is lost forever."* I hope one day we can prove him wrong but so far his words stand uncontested.

President Abraham Lincoln finally recognized Haiti as a sovereign nation in late 1862 in the midst of the politics of the U.S. Civil War. Lincoln's successor, President Andrew Jackson, promptly began the process of formally annexing Haiti. Though Jackson never completed the formal annexation of the island, he did set a precedent of establishing a military and financial presence on the island "to secure a U.S. defensive and economic state in the West Indies."[26] By 1890, following U.S. involvement in yet another leadership revolt on the island, U.S. President Benjamin Harrison used gunboat diplomacy in a failed attempt to convince Haiti's newest president to lease the commune of Môle Saint-Nicolas to the U.S. government for the purpose of building a U.S. naval base on the island. By 1910, U.S. President William Howard Taft had introduced American banks to Haiti with the goal of dislodging the European presence on the island. The American banks gave Haiti loans to help service its debt to France, once again revealing the clandestine partnership between Wall Street's corporate business activities and U.S. foreign (and domestic) policy, which persists to the present day. From 19th century railroad barons and property developers on the Great Plains, to 20th century investment banks, oil barons, and the military industrial

---

[26] "United States occupation of Haiti." *Wikipedia.* 20 May 2024.
    https://en.wikipedia.org/wiki/United_States_occupation_of_Haiti

complex, to 21st century tech firms and pharmaceutical companies, the cozy collaboration between big government and big corporations are a mutually beneficial match made in heaven, consummated at the expense of individual liberty and local autonomy, and intentionally designed to facilitate America's pursuit of its national interests.

In the early 1900s, the U.S. military became increasingly concerned about the German influence on Haiti. A rapidly growing German community was bypassing the prohibition on foreign land ownership on the island by intermarrying with the most prominent mulatto families — by that time Germans controlled about 80% of the country's international commerce and had become the principal financiers of the island's competing political factions. To limit German influence, the U.S. Department of State backed a group of American investors operating as the National City Bank of New York (the earlier incarnation of today's Citibank) to acquire complete control over Haiti's national bank and treasury.

As soon as the National City Bank of New York achieve this goal, it began lobbying U.S. lawmakers to occupy the island as concern grew among the bank's investors that Haiti might default on its debts (during that period, the island went through six presidents in only four years in a series of political assassinations and coups). In 1914, the U.S. military walked into Haiti's national bank and seized all of Haiti's gold reserves at gunpoint ($13 million in 2021 dollars) on behalf of the National City Bank of New York as assurance that Haiti's debts would continue to be serviced. By 1920, National City Bank's share price was soaring as *25% of all of Haiti's revenue* landed in the National City Bank's coffers in service of Haiti's debts.[27]

When Haitian president Vilbrun Guillaume Sam (who was friendly towards the United States) was lynched in 1915 by an angry crowd, President Woodrow Wilson responded by sending the U.S. military to occupy the island, justifying the invasion partly to restore order, partly to prevent Haiti from defaulting on U.S. loans and to protect U.S. investments, and partly to exclude other European powers from establishing any further strongholds on the island.

---

[27] "United States occupation of Haiti." *Wikipedia*. 20 May 2024.
    https://en.wikipedia.org/wiki/United_States_occupation_of_Haiti

By then WWI was underway, with the U.S. involved as financier but not yet as an active belligerent (J.P. Morgan & Co. of New York was the major American financier for the Allies[28]). As Major General Smedley D. Butler later wrote in his book, *War is a Racket*:[29]

> "Woodrow Wilson was re-elected president in 1916 on a platform that he had "kept us out of war" and on the implied promise that he would "keep us out of war." Yet, five months later he asked Congress to declare war on Germany.
>
> In that five-month interval the people had not been asked whether they had changed their minds. The 4,000,000 young men who put on uniforms and marched or sailed away were not asked whether they wanted to go forth to suffer and to die.
>
> Then what caused our government to change its mind so suddenly?
>
> Money.
>
> An allied commission, it may be recalled, came over shortly before the war declaration and called on the President. The President summoned a group of advisers. The head of the commission spoke. Stripped of its diplomatic language, this is what he told the President and his group:
>
> There is no use kidding ourselves any longer. The cause of the allies is lost. We now owe you (American bankers, American munitions makers, American manufacturers, American speculators, American exporters) five or six billion dollars.
>
> If we lose (and without the help of the United States we must lose) we, England, France and Italy, cannot pay back this money . . . and Germany won't.
>
> So . . .

---

[28] "*Economic history of World War I.*" Wikipedia. 26 Jul 2024.
https://en.wikipedia.org/wiki/Economic_history_of_World_War_I

[29] Butler, S. D. (1935). *War is a Racket.* Round Table Press.

Had secrecy been outlawed as far as war negotiations were concerned, and had the press been invited to be present at that conference, or had the radio been available to broadcast the proceedings, America never would have entered the World War. But this conference, like all war discussions, was shrouded in the utmost secrecy.

When our boys were sent off to war, they were told it was a "war to make the world safe for democracy" and a "war to end all wars."

Well, eighteen years after, the world has less of a democracy than it had then. Besides, what business is it of ours whether Russia or Germany or England or France or Italy or Austria live under democracies or monarchies? Whether they are Fascists or Communists? Our problem is to preserve our own democracy."

What happened in Haiti under U.S. occupation gives us a glimpse of what Wilson's "benevolent" liberal internationalism really meant to those living outside of the United States — it was a precursor to the global liberal world order that emerged in the post-WWII era, led by Wall Street and the CIA. When the U.S. occupied the island in 1915, a much younger Franklin D. Roosevelt was the Under-Secretary for the Navy — he personally wrote a new liberal constitution for Haiti that removed all impediments to foreign land ownership on the island (those restrictions on foreign land ownership had been the most essential component of Haitian law in order to preserve some semblance of Haitian autonomy). When Haitian deputies refused to ratify this new U.S.-imposed constitution and tried to introduce a different constitution that preserved the restriction on foreign land ownership, U.S. troops forcibly dissolved the National Assembly at gunpoint and then held a public referendum. Despite the fact that only 5% of Haiti's population turned out for the vote and despite the fact that 97% of the population was illiterate, Roosevelt's imposed constitution was ratified.

U.S. business interests promptly flooded to the island to snap up land and thousands of Haitian peasants were forcibly removed from their properties to make way for American agricultural businesses setting up large plantations to produce crops for export, like sisal, sugarcane, and cotton. Shortly after the U.S. entry into WWI, Haiti's puppet leader declared war on Germany in order to confiscate all the assets held by

Haiti's German community. He also imprisoned large segments of the German community in concentration camps and then deported many of them when the war ended (without compensation for their seized assets, which were never returned).

The U.S. ruled Haiti as a brutal and violent military regime under a constant state of martial law for over 19 years. They exercised absolute control over the press to quash criticism and committed a litany of human rights abuses against Haiti's population including censorship, concentration camps, forced labor, racial segregation, religious persecution, torture, bombing of entire villages, and widespread extrajudicial killings.[30] A guerilla group called the Cacos emerged in revolt of the U.S. occupation, which prompted the U.S. to step up its use of forced labour by reinstituting the corvée system of forced unpaid labour in order to build a network of roads to make it easier for the military to violently crush the resistance and hunt down resistance fighters hiding out in remote mountains. Peasants were shackled together in chain gangs to build these roads, with widespread beatings, abuse, and murder of forced labourers to enforce discipline. Food and wages promised to these forced labourers often failed to materialize.

Backlash led the U.S. to officially abolish their corvée system in 1918 but the practice continued illegally, and no-one ever faced punishment for continuing the practice. With the corvée system officially withdrawn, the U.S. turned to prison labour to fill the forced labour gap, going as far as making false arrests when too few prisoners were available to populate the chain gangs.[31] The total death toll during the U.S. occupation is unknown, but some historians think at least 5,500 Haitians died in the labour camps, up to 3,250 rebels were killed, and as many as 15,000 Haitians in total may have been killed during the occupation.

In a masterstroke of blatant hypocrisy, at the Paris Peace Conference of 1919 Woodrow Wilson reaffirmed his support for the liberal principle of self-determination of all peoples. The U.S. occupation of Haiti continued nonetheless.

U.S. civil rights activist James Weldon Johnson visited the island in 1920 and wrote a blistering report criticizing the occupation:

---

[30] "Slavery in Haiti." *Wikipedia.* 23 May 2024.
https://en.wikipedia.org/wiki/Slavery_in_Haiti

[31] "Slavery in Haiti." *Wikipedia.*

> "Military camps have been built throughout the island. The property of natives has been taken for military use. Haitians carrying a gun were for a time shot on sight. Machine guns have been turned on crowds of unarmed natives, and United States Marines have, by accounts which several of them gave me in casual conversation, not troubled to investigate how many were killed or wounded. [...] [T]o understand why the United States landed and has for five years maintained military forces in that country, why some three thousand Haitian men, women, and children have been shot down by American rifles and machine guns, it is necessary, among other things, to know that the National City Bank of New York is very much interested in Haiti. It is necessary to know that the National City Bank controls the National Bank of Haiti and is the depository for all of the Haitian national funds that are being collected by American officials, and that Mr. R. L. Farnham, vice-president of the National City Bank, is virtually the representative of the State Department in matters relating to the island republic."[32]

Following Johnson's report, which produced a massive outcry in the United States, the U.S. government launched a congressional investigation to investigate itself, which predictably concluded that the occupation needed to continue because of the *"chronic revolution, anarchy, barbarism, and ruin"* that would befall Haiti if the United States withdrew.[33] In 1922, National City Bank bought out and completely absorbed the Bank of the Republic of Haiti, and the stolen gold became a mere entry in the National City Bank ledger.[34] The practice of forced labour expanded further in order to boost the economy and build more infrastructure. In 1929, a peaceful protest on the island was gunned down by the U.S. military, which triggered U.S. President Herbert Hoover to appoint two commissions to figure out how and when to withdraw from Haiti. But the occupation continued for another five years, under increasing tensions, until the Great Depression completely hollowed out Haiti's export-dependent economy, at which point President Franklin D.

---

[32] "United States occupation of Haiti." *Wikipedia.* 20 May 2024.
    https://en.wikipedia.org/wiki/United_States_occupation_of_Haiti

[33] "United States occupation of Haiti." *Wikipedia.*

[34] "United States occupation of Haiti." *Wikipedia.*

Roosevelt announced he was in favor of a "Good Neighbor policy" and, in 1934, finally ended the occupation.

The legacy of the U.S. occupation only further exacerbated regional tensions on the island — the urban south benefited from the export-driven commercial agricultural economy during the U.S. occupation and wanted to dominate political control over the island while the rural north resented the occupation and the impact on their livelihoods. The U.S. occupation also exacerbated racial resentments on the island as mixed-raced Haitians were consistently prioritized in leadership positions during American rule, reviving the colorism that had existed on the island in the early years after the Haitian Revolution (colorism was on the decline prior to the U.S. occupation). These tensions inspired a black nationalism movement on the island that ultimately swept Haitian dictator François Duvalier to power in the 1950s.

Export agriculture had certainly boosted the economy during the U.S. occupation, but it had come at the cost of dramatically lowering Haiti's domestic food production as land was gobbled up by plantations, leaving the population increasingly dependent on foreign imports for its everyday needs. When the U.S. finally withdrew from Haiti in 1934, most of the infrastructure left behind was in very poor condition. According to a New York Times report, the small gang of American officers who controlled Haiti during the occupation *"spent more on their own salaries than on the public health budget for two million Haitians."*[35] After the U.S. military withdrawal, U.S. banks continued to collect about 20% of Haiti's total annual revenue to service Haiti's debts — by the time the debts were paid off in 1947, 90% of the population lived "close to starvation levels."[36] As long as the U.S. government and the coercive might of the U.S. military is used to shield creditors from the consequences of making bad investments, there's no incentive to stop predatory lending practices and to stop saddling debtors with wholly unrealistic terms.

In 1935, one of the U.S. commanders who was posted to the island during the occupation, the aforementioned Major General Smedley D.

---

[35] "United States occupation of Haiti." *Wikipedia*.
[36] "United States occupation of Haiti." *Wikipedia*.

Butler, wrote a book called *War Is a Racket* in which he condemned the profit motive behind America's never-ending wars.[37] In his book he wrote:

> *"War is a racket. It always has been. It is possibly the oldest, easily the most profitable, surely the most vicious. It is the only one international in scope. It is the only one in which the profits are reckoned in dollars and the losses in lives. A racket is best described, I believe, as something that is not what it seems to the majority of the people. Only a small 'inside' group knows what it is about. It is conducted for the benefit of the very few, at the expense of the very many. Out of war a few people make huge fortunes. [...]*

> *"I spent 33 years and four months in active military service and during that period I spent most of my time as a high-class muscle man for Big Business, for Wall Street and the bankers. In short, I was a racketeer; a gangster for capitalism. I helped make Mexico and especially Tampico safe for American oil interests in 1914. I helped make Haiti and Cuba a decent place for the National City Bank boys to collect revenues in. I helped in the raping of half a dozen Central American republics for the benefit of Wall Street. I helped purify Nicaragua for the International Banking House of Brown Brothers in 1902–1912. I brought light to the Dominican Republic for the American sugar interests in 1916. I helped make Honduras right for the American fruit companies in 1903. In China in 1927 I helped see to it that Standard Oil went on its way unmolested. Looking back on it, I might have given Al Capone a few hints. The best he could do was to operate his racket in three districts. I operated on three continents."*

Although the U.S. largely left the island alone for the remainder of the 1930s after its withdrawal from the island, brutal repression continued under the next home-grown dictator, Sténio Vincent. He set the precedent of politicizing the Haitian military in the aftermath of an attempt by the neighboring Dominican Republic to buy influence inside the Haitian military in a bid to try to conquer the Haitian side of the island — up to 40,000 Haitians were butchered in the Dominican attempt to seize Haiti.

---

[37] "Smedley Butler." *Wikipedia.* 19 May 2024.
    https://en.wikipedia.org/wiki/Smedley_Butler

But as the world divided into Allied vs Axis powers during WWII and then Capitalists vs Marxists during the Cold War, the U.S. once again stepped up its involvement in the island by aggressively supporting whichever leader expressed pro-US and anti-Marxist sympathies, leading to a series of dictators, military juntas, and coups that each promised the moon only to see their regimes succumb to their personal ambitions, to rampant corruption, and ultimately to their violent ouster. In a system dominated by absolute centralized power, the incentives remain the same no matter who rises to power. Rinse and repeat.

During WWII under the rule of President Élie Lescot, large rubber plantations were established on the island under contract with the U.S. military. The program was funded by the Export-Import Bank in Washington (a U.S. federal agency established by President Franklin D. Roosevelt whose benign-sounding official goal was to "aid in financing and to facilitate exports and imports and the exchange of commodities between the United States and other Nations or the agencies or nationals thereof.") The resulting US-Haitian partnership was called the Société Haïtiano-Américane de Développement Agricole (SHADA) and was managed by an American agronomist. The project began by forcibly clearing peasants off their lands to claim over 100,000 hectares of Haiti's most productive farmland. Countless peasant homes were razed to the ground and over a million fruit-bearing trees were cut down to make way for the rubber plantations.[38] Eventually the program was cancelled when a U.S. military report concluded that the project did not have the support of the Haitian people and was a drain on American taxpayers. 90,000 labourers lost their jobs when the program was terminated. Lescot's dictatorial rule provided fertile ground for Marxists and black nationalists. Rather than let things spiral any further out of control, Haiti's military stepped in to force Lescot out of power.

In 1957, a doctor known as François Duvalier (Papa Doc) rose to power on the black nationalist platform, beginning a reign of terror that lasted until 1971 under his rule and until 1986 under the rule of his equally tyrannical son, Jean-Claude Duvalier (Baby Doc). Together, theirs were among the most repressive and corrupt regimes to have ever existed in the Western Hemisphere. Up to 60,000 Haitians were murdered by their regimes and countless others were systematically raped, beaten, and

---

[38] "Élie Lescot." *Wikipedia*. 27 Apr 2024. https://en.wikipedia.org/wiki/Élie_Lescot

tortured even as the two dictators systematically plundered the country while pocketing up to 80% of the aid money sent to Haiti (that aid money represented a substantial part of the Haitian budget). They also saddled the country with mountains of fresh debts that they had no intention of repaying. Because he distrusted the military, Papa Doc set up a rural militia to act as his secret police force and serve as a counterweight to the military (his militia grew to outnumber the military 2:1). But since militiamen had no official salary, they made their living through crime and extortion. Papa Doc even compensated them by regularly confiscating peasant landholdings to give to militia members.[39]

The tyranny continued under the son, Baby Doc, who used government coffers as a personal slush fund. President Nixon restarted American aid programs to the island during Baby Doc's reign, which Baby Doc and his wife promptly pilfered to the tune of over $540 million dollars. All the aid money in the world cannot make up for a lack of property rights and good limited governance. HIV/AIDS during Baby Doc' reign caused tourism to the island to collapse, which accelerated hunger and malnutrition on the island still further. Throughout it all, the U.S. tolerated the tyrannical regimes of the two Docs because of their staunch anti-communist stance, which provided a regional counterbalance to neighboring communist Cuba. President Ronald Reagan finally forced Baby Doc into exile in 1986; Baby Doc's parting gift to his country was that he cleaned out much of what was left in government coffers on his way out the door.

The next chapters in Haiti's history don't look much better as the endless cycle of coups, massacres, civil wars, foreign interventions, drug trafficking, prostitution, child slavery, human trafficking, and economic servitude continue to condemn the people and the land to exploitation. Until this point in the story, what was imposed on Haiti by outside forces was largely shaped by the hard-nosed pre-WWII-era colonial mindset. What comes next is the "softer" liberal international order of the post-WWII and post-Cold-War eras, where sharing is caring, where cooperation and kindness are all the rage, where the world's nations hold hands to build a kinder, gentler UN-led global community, and where everyone is lifted up by foreign aid and by mutually beneficial "strategic" investments.

---

[39] "Jean-Claude Duvalier." *Wikipedia*. 16 May 2024. https://en.wikipedia.org/wiki/Jean-Claude_Duvalier

It was meant to be a new dawn overseen by a benevolent UN-led peace corps, shaped by philanthropy and by wisely administered International Monetary Fund (IMF) loans, with prosperity delivered to one and all via open markets and free trade, and with the United Nations empowered to bind everyone together in a never-ending series of sustainability goals imposed from above like some 5-year communist development plan while allegedly giving all "stakeholders" a voice at the decision-making table. Kumbaya. As Haiti was about to find out, beneath the fluff the new world order looked an awful lot like the old world order — it's nothing more than fresh lipstick to dress up the same old pig. And it's the exact same lipstick-adorned pig that is rampaging through every other community around the world today, foreign and domestic, as the marriage between big government and Wall Street consumes us all in a giant global free-for-all.

In 1987, as the War on Drugs kicked off in the United States, the CIA set up, financed, and trained the Haitian National Intelligence Service, which quickly turned its new skills towards persecuting political opponents and trafficking drugs. Already in 1987, it massacred up to 300 voters on election day in order to cancel the election and retain control of the government. Yet the CIA knowingly continued to fund them for another 3 years, by which time the body count had risen to over 5,000 people. The New York Times even revealed that Emannuel Constant, a Haitian terrorist and founder of the paramilitary death squad FRAPH, was on the CIA's payroll throughout this period and that the CIA and U.S. military planners had been directly involved in the planning to set up this terror organization.[40] The Clinton administration denied all allegations.

In 1994, the U.S. military invaded the island again to re-install its chosen leader, Jean-Bertrand Aristide, who had previously been deposed in a 1991 coup. A decade later the CIA was involved in another coup, this time to oust Aristide from power. In 2021, President Jovenel Moïse was assassinated, and it later turned out that the assassins worked for a US-based security firm and had ties to the U.S. establishment. At least one was a US-trained Columbian commando trained in the art of kidnapping and conducting assassination operations, and two more had links to the FBI, CIA, or DEA — the U.S. Justice Department promptly sealed the

---

[40] Via, D. Jr. (2024, Mar 14). Haiti is Among the First and Latest Example of US-Led Destabilization. *The Free Thought Project.*
https://thefreethoughtproject.com/foreign-affairs/haiti-is-among-the-first-and-latest-examples-of-us-led-destabilization

file.[41] Moïse's successor, Ariel Henry, who was ousted in March of 2024, had the full-throated support of the Biden administration despite his widespread unpopularity on the island and despite suspicions that Henry had been involved in planning his predecessor's assassination and had participated in the subsequent cover-up.[42]

Throughout the years, the United Nations stationed peacekeepers on the island to crack down on gang violence. But reports surfaced of widespread sexual abuse of Haitian women and children by UN peacekeepers and a massive increase in sex trafficking in order to supply the peacekeepers with what they wanted — an accusation that has plagued UN peacekeeping missions wherever they have operated around the world. No-one ever faced repercussions. In the aftermath of the 2010 earthquake, the UN also sent aid workers to the island. They accidentally brought cholera with them, killing thousands.[43]

On the economic front, the liberal international order hasn't been any kinder to the island. Since the 1970s, the U.S. and the IMF pressured Haiti to dismantle all tariffs and trade barriers and open up its economy to world trade — it is one of the most open economies in the region. And then the U.S. flooded Haiti with food aid, decade after decade. Food aid after a temporary disaster like a hurricane or earthquake can fill a gap, but because this aid was sustained for decades, it completely destroyed the local rice farming economy, which had until then been quite capable of growing all of Haiti's rice needs. Today, with its local rice production in tatters, Haiti imports more than 80% of its rice.

Here's how the scheme works: the U.S. subsidizes American farmers to the tune of billions of dollars not only through direct subsidies but also by buying up surplus crops via USAid (the U.S. government agency for international development) to donate those surpluses to other countries as "food aid". This food aid program creates new overseas markets for

---

[41] Via, D. Jr. (2024, Mar 14).

[42] Rivers, M., Dupain, E., and Gallón, N. (2022, Feb 8). Haitian Prime Minister involved in planning the President's assassination, says judge who oversaw case. *CNN News.* https://www.cnn.com/2022/02/08/americas/haiti-assassination-investigation-prime-minister-intl-cmd-latam/index.html

[43] Via, D. Jr. (2024, Mar 14). Haiti is Among the First and Latest Example of US-Led Destabilization. *The Free Thought Project.* https://thefreethoughtproject.com/foreign-affairs/haiti-is-among-the-first-and-latest-examples-of-us-led-destabilization

American crops while simultaneously subsidizing American farmers, but it also completely hollowed out Haiti's local agricultural industry because, thanks to the USAid food program, American rice is sold in Haiti at well below the true cost of production, which leaves local growers unable to compete with the heavily subsidized American rice. Haiti's importers benefited enormously but Haiti's domestic rice industry collapsed, many thousands of farmers lost their farms, entire rural regions were impoverished, and countless more people flooded to the city slums in a desperate search for factory jobs after their farms were bankrupted by the flood of subsidized American rice. In effect, food aid is the more extreme version of the Wal-Mart Effect, which hollows out the local economy.

As Haitian farmers tried to point out, long-term food aid is the wrong long-term solution to Haiti's economic problems. In addition to the ongoing unsolved property rights issue, what Haitian farmers desperately need is to update outdated equipment, improve irrigation systems, build drainage ditches to protect crops from flooding, gain access to fertilizer, acquire modern rice cleaning and de-hulling equipment (most Haitian-grown rice is still hulled by hand), and get access to credit to invest in their farms. Haitian farm advocate Camille Chalmers sums up the food aid issue as follows: *"It creates food dependence and food insecurity [...] And it's a problem also because a lot of money leaves the country when it could return to help our national economy."*[44] But who listens to locals these days when we have so many enlightened "experts" populating our intranational organizations who know what's best for everyone else.

Instead, planners in the U.S. and at the United Nations decided that the best way to "help" Haiti would be to push Haiti towards industrialization by collapsing the subsistence farming economy while introducing investors to the island to create "good jobs" in export-oriented factories that take advantage of Haiti's low labour costs. President Bill Clinton aggressively expanded the food aid program for this purpose during his term, but in 2010 he admitted to Congress that his policy had been a complete failure: *"The United States has followed a policy ... that we rich countries that produce a lot of food should sell it to poor countries and relieve them of the burden of producing their own food, so, thank*

---

[44] Kushner, J. (2016, Jul 26). Haitian farmers call on US to stop subsidizing its own. *The World.* https://theworld.org/stories/2016/07/31/haitian-farmers-call-us-stop-subsidizing-its-own

goodness, they can leap directly into the industrial era [...] It may have been good for some of my farmers in Arkansas, but it has not worked ... I have to live every day with the consequences of the lost capacity to produce a rice crop in Haiti to feed those people."[45] And yet, even after Clinton's surprisingly candid 2010 testimony, the U.S. continues to follow the exact same policy with things like subsidized clothing and construction projects, which are useful policies for the Americans who are contracted to deliver these philanthropic efforts to Haiti but are nevertheless having the same destructive impact on Haiti's local construction and clothing industries as these too are now being hollowed out by America's "generosity".

The Free Thought Project reports that after the 2010 earthquake, Hillary Clinton (as Obama's Secretary of State) organized a massive relief effort, but it later came out that the delivered shelters were laced with toxic chemicals and carcinogens and were bought from a company *"owned by longtime Clinton campaign donor Warren Buffett, [who] had already been sued by the US Federal Emergency Management Agency for providing unsafe trailers to the victims of Hurricane Katrina"*, and that *"millions of dollars in donated relief funds would go missing never to be seen again"*.[46] The Guardian reported in 2019 that of the $2.3 billion spent on Haiti by USAid, less than 3% passed through Haitian hands while 55% ended up in the hands of American companies located in and around Washington DC.[47]

One of the projects USAid was supposed to undertake in Haiti was the construction of a new port on the north side of the island — USAid was given the contract despite the fact that the agency hadn't built a port since the 1970s. The project went through endless revisions and delays. Five years into the two-and-a-half-year project there was still no estimate as to when construction would begin. Then it was discovered that none of the private companies that USAid hoped to attract to the new port had any

---

[45] Kushner, J. (2019, Oct 11). Haiti and the failed promise of US aid. *The Guardian*. https://www.theguardian.com/world/2019/oct/11/haiti-and-the-failed-promise-of-us-aid

[46] Via, D. Jr. (2024, Mar 14). Haiti is Among the First and Latest Example of US-Led Destabilization. *The Free Thought Project*. https://thefreethoughtproject.com/foreign-affairs/haiti-is-among-the-first-and-latest-examples-of-us-led-destabilization

[47] Kushner, J. (2019, Oct 11).

interest in relocating their operations to that location. So, to save face, USAid decided to expand a different existing port. It took another two years to discover that this project was also not economically viable. Seven years and $70 million dollars later, the project was abandoned altogether — the Guardian reports that the only thing that got constructed were two concrete electricity poles for the future port, but even those poles were never connected to the grid.[48] Likewise, a $300 million industrial park promoted by the Clinton Foundation only created 1,500 of the projected 65,000 jobs it had promised by 2012, even as the project evicted 366 families from their land in order to make room for the industrial park. However, a project that did get completed was a new Marriot hotel in Port-au-Prince, financed by a multinational company whose chairman was a friend of the Clintons.[49]

This deliberate ploy to push everyone into "good jobs" working for international mega-corporations is the essence of the neo-liberal post-WWII era — a global economic vision created by the unholy marriage between masterful transnational corporations and masterful globally interconnected governments. Hillary Clinton has called this vision "economic statecraft." The idea is that by combining foreign aid with diplomacy, broken countries can be transformed into attractive investment destinations for multinational companies, which in turn is meant to help the country in question attract foreign financing to build out its infrastructure even as their citizens are transformed into industrious worker bees toiling to make cheap stuff to sell into the global economy. It's a lovely theory if you're into that sort of vision of the future, but like all centrally planned ideas imposed from above, it hasn't even come close to working.

This economic model isn't even working for industrialized nations in their home countries anymore as our own nations slide deeper into conflict, dysfunction, and despair, and as citizens find their hands increasingly tied at every turn if they try to do anything outside of the confines of the corporate rat race, leaving them unable to do what's best for themselves, for their own local communities, and for their local ecosystems. In the 1960s, someone working a blue-collar job without a university education (i.e. a shoe salesman) could realistically expect to

---

[48] Kushner, J. (2019, Oct 11).

[49] Kushner, J. (2019, Oct 11).

own his own modest home and successfully raise a family on a single income, even if his wife was a stay-at-home homemaker. Sixty years later, despite scores of technological improvements, cheaper energy, a greater abundance of everything, and a stock market that soars to ever higher heights, this vision of the American dream is retreating ever further out of reach — neo-liberal globalism is even hollowing out the local economies of the countries that stood to gain the most by being in control of the liberal international rules-based system.

This hollowing out of local economies, rural regions, and small businesses everywhere in order to integrate them into the global, Wall-Street-led, export-dependent, corporate-controlled economy has provided a bonanza of windfalls to investors, politicians, and administrative hubs even as everyone else is gradually reduced to expendable (disposable) low-cost labour. And the bureaucratic decision-making process that dominates our political institutions ensures that even when useful insights about the failings of the current world order rise to the surface (like Bill Clinton's 2010 testimony in Congress about the failures of economic statecraft in Haiti), those insights are never translated into meaningful change as bureaucratic inertia, egos, political calculations, and conflicts of interests all work to perpetuate the status quo.

In a quote from Range Magazine, Allan Savory shared a story illustrating how long it takes for complex centralized political and corporate bureaucracies to incorporate new knowledge: "'*Captain James Lancaster sold four ships to India in 1601. He gave the crew on one ship limes, and they got no scurvy. The other three ships got no limes and 50 percent of the crew died. That was pretty convincing evidence but because the Royal Navy is a bureaucracy, led by brilliant officers, they discussed and argued about that for approximately 150 years at which point James Lind, a surgeon in the Navy repeated it, and got patients to recover. Then the Royal Navy argued another 49 years before they accepted it.' It took the Merchant Marine 70 more years to follow suit.*"[50] And in that story, there weren't even any financial gains to be had for defending the status quo. Just egos that needed protecting.

~

---

[50] Hadley, C.J. (Fall 1999). The Wild Life of Allan Savory. *Range Magazine.* https://www.rangemagazine.com/archives/stories/fall99/allan_savory.htm

In the parable of the Blind Men and the Elephant, at least the men groping about the elephant were blinded by no fault of their own. In our current system, we have rendered ourselves deaf, blind, and dumb through centuries of increasingly centralized decision-making, through the perverse incentives that emerge from such a centralized system, and through the compartmentalized, reductionist thinking that has taken over our political and scientific institutions. Such a top-down system is wholly incapable of grappling with complexity even when there aren't self-serving interests trying to game the system for their own benefit. Instead, complex issues will forever be reduced to simplistic single-cause explanations that play well to political audiences, the decision-makers will forever defer to whichever experts most impress them with their credentials, and "the facts" will invariably be form-fitted to suit the biases of the decision-makers.

Any system with a powerful central gatekeeper with the authority to make laws, extract taxes, and create winners and losers at the stroke of a pen inevitably turns into an elaborate circus in which much of society revolves around extracting favors from the gatekeeper, while truth-seeking, justice, and foundational principles all fall by the wayside. Consider the absurd picture of life at France's Palace of Versailles during the reign of Louis XIV (the Sun King) — arguably the quintessential example of centralized power during the era of monarchy, and not altogether different from the progressive circus that haunts our halls of power today. On any given day there were anywhere between 3,000 and 10,000 nobles of greater and lesser status living in the guest rooms at the Palace of Versailles, all vying for influence with the king. Since virtually everything depended on the whims of their king, practically the whole of French nobility permanently resided at the Palace of Versailles, out of necessity, in order earn (and maintain) the favor of their king. It was a never-ending game of jockeying played out in the courts, luxurious gardens, and between the sheets of Versailles' more than 2,300 rooms, where even a trip home to visit your estate could spell catastrophe if, in your absence, someone else managed to catch the king's eye and convince him to transfer your titles, privileges, or estates to someone more deserving. In refusing to grant favours asked for by some noble, Louis XIV habitually remarked, "*We never see him*", meaning that the hopeful claimant did not spend enough time playing the game at the Palace of Versailles. It's a snapshot of life in a golden birdcage, where the pomp, prestige, and never-ending theatrics are a thin veneer disguising lives trapped in a golden cage.

In essence, we have recreated some version of Louis XIV's Palace of Versailles within our liberal democracies, with all sorts of absurd counterfactual beliefs, word games, and "noble lies" emerging to justify the status quo and extract favors from the system, not least of which is the climate crusade. Our institutions, politicians, corporations, academics, and activists are the equivalent of the self-serving lesser nobility in Louis XIV's court, all vying for influence in an elaborate theatrical game to manipulate the gatekeepers, even as the gatekeepers' temporary perch at the top of the political food chain is equally dependent upon the approval of all those below who are playing manipulative games to earn the gatekeepers' favor. In such a system, what is true matters far less than what is *useful* as a tool for pulling strings, opening doors, securing advantages, and fending off competitive rivals. It's a wholly artificial political ecosystem in which everyone is permanently trapped in a never-ending game to defend their position in the food chain. The end justifies the means, irregardless of the cost to truth, decency, or the long-term viability of the nation. Those who refuse to play the game soon find themselves collecting dust on the sidelines.

The climate crusade, like so many other bizarre beliefs that define our current age (i.e., foreign interventionism, pandemic treaties, pronoun wars, victimhood culture, and the list goes on and on), is just one of countless absurd fictions that have become useful to a lot of people despite being utterly divorced from reality and despite being demonstrably harmful to those they claim to help. At least during the reign of the Sun King, there was a clear hierarchy to the throne. In the present system, everyone and yet no-one is truly in charge of the progressive monstrosity that rules over us. Whose proverbial head must roll to initiate the process of reform when even the president is merely a braindead puppet bobbling about on strings controlled by an impenetrable maze of institutions and influencers who have a stake in the outcome of the game? How to you reform this perverse, bloated, pseudo-democratic kleptocracy when almost everyone has a stake in preserving it in order to safeguard their own ill-begotten slice of the pie and when they all instinctually rally together to defend "the system" as soon as even a whisper of discontent emerges from among the long-suffering plebs from whom they extract their plunder?

What's remarkable about Haiti's tragic political and ecological saga is that, despite the fact that their centralized system of governance is a textbook example illustrating everything wrong with centralized power, the perverse incentives it creates, and the environmental degradation that inevitably follows on the heels of centralized top-down governance, every time there is a fresh effort to fix Haiti's broken system, whether from within or from without, the would-be reformers simply doubled down on the exact same top-down, progressive, centralized model of government (prop up a better dictator) and the exact same economic prescriptions (the plantation/factory economic model focused on exporting cheap stuff to world markets). The only thing that changed were the names of the people placed on the throne.

Perhaps none of the would-be reformers could imagine a more decentralized form of governance and a more localized economic system that would reduce the authority of those in power. Perhaps they didn't want to imagine it because the lure of centralized power is simply too great. Perhaps there was too much money at stake to want to imagine something else. And perhaps the institutions working in the background were too invested in preserving the status quo to allow any would-be reformers to introduce any meaningful reforms. As Tolkien's timeless epic *The Lord of the Rings* illustrated, even a humble hobbit who gets his hands on the One Ring will be consumed by its power and will be unable to let it go.

But perhaps the citizens themselves share the largest part of the blame, both within Haiti and within all the paternalistic "liberal" nations meddling in Haiti's affairs, because a citizenry that believes good governance depends on finding angels to place on a powerful throne are the enablers of the most dangerous and destructive noble lie of them all — the illusion of government as a benevolent shepherd.

> *"A sheep spends its whole life fearing the wolf, only to be eaten by its shepherd."*
>
> — African Proverb

Through our appetite to use the administrative state as a tool to engineer "progress," and ease the burdens of life, we share a great deal of the blame for creating the bloated political ecosystem that incentivizes a never-ending saga of corruption, manipulation, and conspiratorial scheming. The bigger the palace, the greater the opportunity for palace intrigues. Our appetites enabled the construction of the vast government trough, and now we're horrified that it is overrun by pigs. What is being done to us is merely the consequence of what we did to ourselves when we allowed ourselves to be tempted by the seductive lure of central planning.

The only way to break out of this destructive cycle and return as much autonomy as possible to individuals and local communities is to cut the size of the centralized political ecosystem down to an absolute bare minimum, as the Founding Fathers tried to do when they framed a constitution for a federal government that "governs best when it governs the least". Society's enthusiasm for a progressive administrative state shattered the limits imagined by the Founding Fathers as each successive generation expanded the reach of the government just a little further and, in doing so, gradually dismantled the restraints that once kept centralized government power from usurping individual and local autonomy. The powerful bureaucratic monstrosity that has emerged has long since taken on a life of its own, including creating its own myths (like the climate myth and the myth of benevolent liberal internationalism, to name but a few) to justify the continued expansion of centralized power and defend itself against would-be reformers.

The restoration of individual and local autonomy, the revival of evidence-base debate in science and politics, and the antidote to all the destructive consequences of central planning can only emerge from a culture that not only rejects the seductive fairy tale of the benevolent shepherd, but also builds a labyrinth of stories and values upon which to raise its children in order to inoculate them with an instinctual fear of the shepherd.

A constitution is the instruction manual written by past generations to tell future generations how to operate the political architecture they inherit from their forefathers — it is the effort to formalize the foundational principles that once limited government power to prevent it from exceeding its healthy limits. But in truth, a constitution is nothing more than a piece of dusty parchment — culture, tradition, and our knowledge of our own history are what give meaning to the words. The stories and myths we read to our children will determine how future generations interpret the words of the Constitution during their adult life and will influence whether they will resist or succumb to the temptations to step over the limits on power imposed by their forefathers. But during the Progressive push to expand the administrative state into every facet of our lives, society also unwisely outsourced its storytelling to state-funded educators and media corporations that work hand-in-glove with the administrative state. In so doing, we dismantled the last cultural antibodies that once gave new generations a measure of immunity to the seductive lure of central planning.

Every war, whether it is fought with sticks and clubs in a muddy field or with words in a political arena, begins and ends with the philosophical war for hearts and minds. By turning over the transmission of culture to the administrative state and to organizations that have cozy relationships with the government, we have in effect made the proverbial goat responsible to keep watch over the intergenerational garden. Now we reap what we have sown.

It does us no good to blame Woodrow Wilson, FDR, the Clintons, Bill Gates, George Soros, the activist-scientists, or any of the rotten institutions that are sprouting like mushrooms in the current system — they are responding to the incentives that we created for them through our appetite for a big meddlesome "benevolent" shepherd. Send them all on an extended vacation to Mars and the political ecosystem they leave behind would almost instantly be repopulated by a fresh crop of parasitic opportunists playing some version of the very same games. Until we cure ourselves of our appetite for a "benevolent" shepherd, the trajectory remains unchanged no matter who is put in charge of the throne.

It is not an accident that the mythological story of Icarus played out in three acts. The first act focused on the construction of the Labyrinth — the stories and myths designed to impose principles and values upon society's youth — symbolized by the effort to contain the destructive bull-headed appetites of the Minotaur with a maze too complex for anyone to be able to find their way through. This maze of restraint was so important as a means to safeguard future generations from themselves that King Minos even went to the extreme length of imprisoning the Labyrinth's architect (Daedalus) and his son in a high tower to try to prevent them from leaking the secrets of the Labyrinth to the rest of society.

The second act tells the story of how crafty Daedalus built wings for himself and his son to enable them to escape their past, which serves as the perfect metaphor for what happens when one generation transfers great power into the hands of the next generation in the hope of freeing future generations from the shackles of their history. Unless the passing of the torch is accompanied by a cultural inheritance of stories and myths to temper the next generation's youthful starry-eyed impulses, those that inherit the torch from their ancestors will be too innocent and too inexperienced to understand the strict limits imposed by their forefathers.

The third and final act ends in tragedy as Icarus, untethered from the ground and out of reach of his father's restraining hands, tries to use the god-like power of the wax wings given to him by his father to soar ever closer to the Sun. The story ends with the father burying his son even as he curses his hands for creating the very thing that led his son into lethal temptation.

The centralization of power and resources, whether under the guise of monarchy, dictatorship, democracy, or something else altogether, and whether it is justified by the imperatives of nation building, fighting "inequality", or fighting the "climate crisis", irregardless of the many forms it takes and the countless excuses that emerge to rationalize it, centralized power in all its forms across the span of history has always been utterly incompatible with truth-seeking, individual autonomy, thriving local communities, and healthy ecosystems. Lessons will be repeated until they are learned… or until all the soils are gone, the lights have been turned off on civilization, and the cycle resets anew. Such is the cyclical fate of empires. Nothing captures this cyclical aspect to human civilization better than 19th-century painter Thomas Cole's famous series of paintings, *The Course of Empires*. What's different this time is that, for the first time in our species' history, civilization has become global and so, while the ecological problems may be local, the political forces driving them have become deeply intertwined and international in scope.

Case Study #3: Haiti, a Long Descent into Hell

Figure 115. *The Course of Empire — The Savage State.*
Oil on Canvas, by Thomas Cole, 1834

Figure 116. *The Course of Empire — The Arcadian or Pastoral State.*
Oil on Canvas, by Thomas Cole, 1834

Figure 117. *The Course of Empire — The Consummation of Empire.*
Oil on Canvas, by Thomas Cole, 1836.

Figure 118. *The Course of Empire — Destruction.*
Oil on Canvas, by Thomas Cole, 1836.

Figure 119. *The Course of Empire — Desolation.*
Oil on Canvas, by Thomas Cole, 1836.

I'll leave the last words to President Herbert Hoover. He can hardly be described as a laissez-faire president — indeed, he was responsible for signing the infamous Smoot-Hawley Tariff Act into law after the Stock Market Crash of 1929, which triggered the trade wars that helped turn the stock market crash of '29 into a decade-long Great Depression and exported that Depression to the rest of the world. And yet, perhaps as a consequence of having experienced the realities of sitting on the presidential throne, as America launched itself on a new era of greatly accelerated bureaucratically-controlled economic planning under Roosevelt's New Deal (thus laying the groundwork for the modern bureaucrat-infested neo-liberal world order that we are still plagued with today), Hoover emerged as one of America's loudest voices trying to dissuade America from its blossoming love affair with central planning. In 1940, not long after the outbreak of WWII, he warned America about the insidious forces that unravel nations:[51]

---

[51] Hoover, H. (1941). *Addresses Upon the American Road.* Charles Scribner's Sons, New York.
https://hoover.archives.gov/sites/default/files/research/ebooks/b3v3_full.pdf

"Let us first examine the weakening of the structure of liberty in the United States. That is but part of a far larger war in the world today than a war of tanks and airplanes. It is even more dangerous to America. This is a war of hostile ideas, philosophies and systems of government. There are no neutrals in that war. That is where we have to fight.

[...] For more than a century and a half before the War of 1914 the whole continent of Europe up to the Russian border had been struggling upward toward liberty. These people in 20 races had attained a large degree of free government—free speech, free worship, orderly justice and free enterprise. After that war, liberty and peace seemed assured to the world.

And then in less than 20 years, hundreds of millions of these people surrendered freedom for bondage under totalitarian government. This abandonment of liberty has been the most gigantic revolution in history.

Two years ago, I was the invited guest of some twelve European countries [author's note: Hoover was the only US president to visit Nazi Germany and meet Adolf Hitler before the outbreak of WWII]. That gave to me a unique opportunity to inquire into some things that might help the American people.

I wanted to know more of what ideas and pressures had plunged these nations into dictatorships. There will flash into your minds that it was Communism, Fascism, or Nazism. That is not what I refer to. They were the effect. I was seeking the cause. Liberty had been weakened long before the dictators rose under those banners. There was a long poignant drama before the last act in this gigantic tragedy of civilization.

There were many disintegrating forces. But also in every single case before the rise of dictatorships there had been a period dominated by economic planners. Each of these nations had an era under starry-eyed men who believed that they could plan and force the economic life of the people. They believed that was the way to correct abuse or to meet emergencies in systems of free enterprise. They exalted the state as the solvent of all economic problems.

These men thought they were liberals. But they also thought they could have economic dictatorship by bureaucracy and at the same time preserve free speech, orderly justice and free government. They can be called the totalitarian "liberals." They were the spiritual fathers of the New Deal.

These men were not Communists or Fascists. But they mixed these ideas into free systems. [...] These so-called liberals shifted the relation of government to free enterprise from that of umpire to controller. Directly or indirectly they politically controlled credit, prices, and production of industry, farmer, and labor. They devalued, pump primed, and inflated. They controlled private business by government competition, by regulation and by taxes. They met every failure with demands for more and more power of control. And they employed that hand maiden of power, named "Gimme a Billion, Quick."

These leaders ignored the fact that the driving power of free economic life is the initiative and enterprise of men. When it was too late, they discovered that every time they stretched the arm of government into private enterprise, except to correct abuse, then somehow somewhere men's minds and judgments became confused. At once men became hesitant and fearful. Initiative slackened, production in industry slowed down.

Then came chronic unemployment and frantic government spending in an effort to support the unemployed. Government debts mounted. And finally, government credit was undermined. Out of the miseries of their people there grew pressure groups—business, labor, farmers, demanding relief or special privilege. Class hate poisoned co-operation.

Does this sound unfamiliar to you?

It was all these confusions which rang down the curtain upon liberty. Frustrated and despairing, these hundreds of millions of people voluntarily voted the powers of government to the man on horseback as the only way out. They did it in the hope of preserving themselves from want and poverty. They did it in hope of preserving their national independence.

[...] Let me say at once I am not interested in free enterprise because it is a property system. I am interested in it because intellectual and spiritual liberty can be sustained only by economic liberty. They are indissolubly bound in a common fate. [...] It was dynamic individual enterprise which raised the country from insignificance to greatness. It was not government that put these prairies under the plow. It was not government that flung these railways across a continent. It was not government that built these great factories or cities. It was not government that brought forth these great inventions. It was not governments that built these millions of churches, or added this wealth of music, art and literature.

There can be a free economy in America which releases the productive energies of men. Or there can be a dictated economy. These cannot be mixed. A free economy means sweat, turmoil, and competition — hard at times if you choose. It means ever new wrongs that must be righted, but it means also resistless growth and resistless progress.

The restoration of confidence does not require exploitation or monopoly. A free people can no more tolerate private economic power without checks and balances than we can tolerate political power without balances and checks. But in effecting great reforms there is a dividing line. Upon the right of this line is cure of abuse and solution of marginal problems. On the left of it are dictation and tyranny and discouragement of production.

[...] Does [Man] possess the right from the Creator to plan his own life, to dare his own adventure, to earn his own reward so long as he does no harm to his fellows? Or must he submerge his life, his liberties, and his independent personality in an omnipotent government?

If man is merely one of the herds, running with the pack, Stalin is right, Hitler is right, and, God help us for our follies and our greed, the New Deal is right.

But if man is an inviolable human soul, possessed of dignity, and endowed with unalienable rights, America is right. And this is a war that Americans dare not lose."

Chapter 28

Epilogue — Water Wars

"By means of ever more effective methods of mind-manipulation, the democracies will change their nature; the quaint old forms — elections, parliaments, Supreme Courts and all the rest — will remain...Democracy and freedom will be the theme of every broadcast and editorial...Meanwhile the ruling oligarchy and its highly trained elite of soldiers, policemen, thought-manufacturers and mind-manipulators will quietly run the show as they see fit."

— Aldous Huxley, Brave New World Revisited

As Kelowna battled the fires that swept through the city in the summer of 2023, another much less publicized battle was happening out of the public eye in the surrounding valleys as the government leveraged the crisis to extend its control ever deeper into farmers' lives. In essence, this story mirrors some version of what is happening all over the world as our democratic principles are relentlessly eroded away, slice by slice, one precedent at a time, under the guise of adapting to "climate change".

In mid August of 2023, the government of British Columbia sent armed Natural Resource Officers (bullet-proof vests and all!) to knock on farmers' doors along several watersheds in the interior of British Columbia, including our local watershed, ordering farmers to cease all irrigation, effective immediately. The legal cover for this order was a Fish Population Protection Order triggered by the drought conditions. Before I unpack the precedents set by this extraordinary directive, I first want to share the relevant excerpt from the official shutdown letter given to local farmers in order to provide context for where this story is going:

> **"BC Water Sustainability Act – s. 88 Fish Population Protection Order – Bessette Creek watershed**
>
> Due to continuing drought conditions this summer, persistent low stream flows are threatening the survival of endangered South Thompson River Chinook salmon in the Bessette Creek watershed. The enclosed Fish Population Protection Order (FPP

Order), made by the Minister of Forests under section 88 of the BC *Water Sustainability Act* (WSA) temporarily restricts the diversion and use of water for irrigation of forage crops from the sources in that watershed area, as described in the FPP Order, including Bessette Creek, its tributaries, and surrounding aquifers (groundwater wells), regardless of the precedent of rights under WSA.

You are receiving this letter because you are the holder of an irrigation licence or area transitioning groundwater user in the Bessette Creek watershed (Bessette Creek, tributaries and hydraulically connected aquifers). This FPP Order applies to irrigation licenses from both surface water (streams) and groundwater (wells) sources as well as those users that have submitted an existing use groundwater application within the transition period ending March 1, 2022. <u>This FPP Order applies to prevent the irrigation of forage crops only</u> and does <u>not</u> apply to water diverted and used for non-forage crops, such as market vegetables, stock watering, or domestic purposes..."

Following the order, these armed Natural Resource Officers patrolled the area daily looking to catch farmers still irrigating their crops. A three-part documentary called *Stolen Water* (posted to YouTube and Rumble)[1] documented the government's ham-fisted enforcement efforts in another local watershed in the nearby town of Westwold. The Westwold ranchers pointed out that water levels in their streams and wells did not seem particularly different from previous dry summers and demanded to see the scientific evidence underpinning the directive. The government flatly ignored their request.

So, the Westwold farmers began digging into whatever meagre data they could find on the government's official websites and stumbled upon the BC government's "groundwater level data interactive maps", available online, which clearly showed that at the time of the shutdown, groundwater levels at test wells in their area were hovering right around the average of the previous 10 years. Even in our watershed a few valleys

---

[1] GroundUp Pictures. (2024, Mar 27). *Stolen Water – Part 1*. [Video]. YouTube. https://www.youtube.com/watch?v=En-7bNw7pi4&ab_channel=GroundUpPictures

further to the east, the local test well showed the groundwater table sitting near, *but not below*, the bottom of the range of the prior ten years. I took screengrabs of the test well charts from both Westwold and Lumby not long after the directive was issued, shown in figure 120 below.[2]

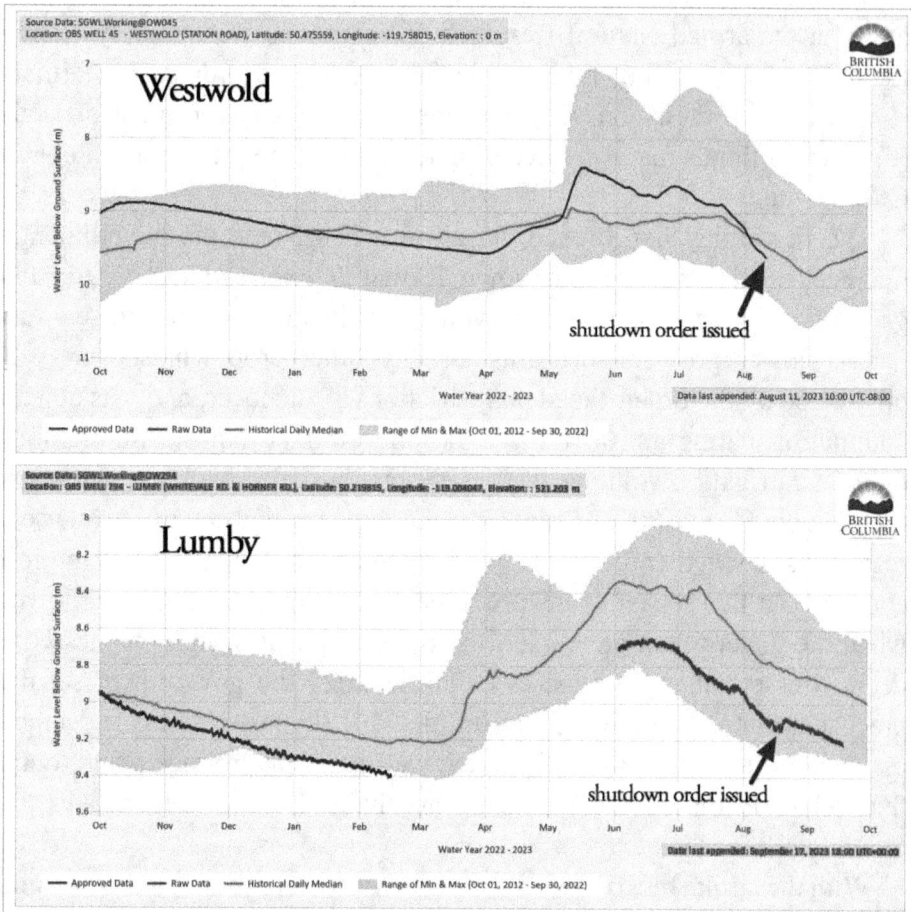

Figure 120. Groundwater data for observation wells in the Westwold and Lumby areas in 2023. The grey zone shows the minimum and maximum range of the past 10 years, the thin grey line shows the 10-year average, and the black line shows data for 2023. The arrow shows groundwater level on the date the unprecedented irrigation shutdown order was issued.

https://www2.gov.bc.ca/gov/content/environment/air-land-water/water/groundwater-wells-aquifers/groundwater-observation-well-network/groundwater-level-data-interactive-map

---

[2] "Groundwater Level Data Interactive Map." *Government of British Columbia*. Accessed 2023, Sep 17. https://www2.gov.bc.ca/gov/content/environment/air-land-water/water/groundwater-wells-aquifers/groundwater-observation-well-network/groundwater-level-data-interactive-map

The Westwold farmers organized a community meeting and invited government officials in the hope of clearing up the escalating tensions, but government officials categorically ignored their demands for dialogue and refused to attend the meeting. Armed Natural Resource Officers continued to patrol the area and advised frustrated farmers that, rather than getting angry at the armed Natural Resource Officers who were merely "doing their job" but were not responsible for issuing the orders, the farmers should take up their concerns with the province's government officials (the same officials that were avoiding the farmers' requests for dialogue). Check mate.

The first thing that jumped out at me when I read the government's unprecedented irrigation shutdown is that *it only applied to farmers growing forage for livestock.* Anyone else, including golf courses, turf farms, flower growers, orchards, or any other crop whatsoever, was unaffected by the order, faced no restrictions whatsoever, and was able to continue irrigating at 100% levels despite the fact that during previous droughts, irrigation volumes were traditionally scaled back in increments (i.e. reducing use to 80% of capacity) and were applied equally to all users regardless of whether they were growing forage, flowers, or grass for golf courses. But in a repeat of the precedent set during the Covid lockdowns when the government gave itself the right to decide which businesses it deemed essential versus non-essential, in 2023 the government set the precedent that it had the right to decide which types of farms it deemed essential vs non-essential. This time it was livestock producers that found themselves in the central planners' crosshairs. Each precedent lays the groundwork for the next.

With the stroke of a bureaucrat's pen, livestock production has become a riskier agricultural category in BC. Livestock producers who rely on irrigation will now find it harder and more expensive to get operating loans, equipment financing, and mortgages because the BC government has signaled that during dry years, livestock producers will be singled out in order to keep other farms irrigated. It also imposes a similar chilling effect on investment in any business that provides services for the livestock industry in BC. This is how individual sectors of the economy are quietly strangled without ever having to issue an outright ban via legislation. The message is clear — if you're planning to invest in agriculture in BC, invest in orchards and golf courses, not cows.

Pumping from groundwater or from rivers isn't the only source of irrigation water in the area. Local city irrigation districts have reservoirs (dammed lakes) up on the surrounding plateaus to capture and store water from snowmelt. They then sell water from those reservoirs to farmers via a network of pipes. This metered irrigation water costs a fortune, and it isn't available to farmers in the Bessette Creek watershed because the metered irrigation network does not reach most of the farms located in the Bessette Creek basin. And yet, ironically, this expensive metered irrigation water comes from storage reservoirs *located at the headwaters of the Bessette Creek watershed.* For example, Duteau Creek (the creek that runs through my parent's farm, which was part of the FPP Order that shut down groundwater wells and stream licenses in the Bessette Creek watershed) originates in three lakes — Aberdeen Lake, Haddo Lake, and Grizzly Lake — these same three lakes are the primary source for much of the metered irrigation water in the Greater Vernon Water district. But this expensive metered farm irrigation water was not restricted in any way during 2023 despite the irrigation shutdowns issued to farmers with water licenses to pump out of the creek flowing out of those very same three lakes. Farmers (including forage producers) who received metered irrigation water were allowed to continue watering at 100% capacity, leading to the bizarre situation where Vernon farmers paying for expensive metered water were watering at full capacity even as their next door neighbors just to the east (growing identical forage crops) were subject to fines (or worse) if they pumped even a single drop of water from their stream water licenses despite the fact that both sources of water ultimately come from the same lakes up on the Aberdeen Plateau. In effect, the government was robbing Peter to sell water to Paul.

Before 2023, water licenses on local streams and lakes were legally ranked by date of registration so that during a water shortage the oldest licenses have priority over more recently registered licenses. This has been the law since the founding of the province. Thus, any water license that predates the construction of the water reservoirs on the highlands had legal priority over metered irrigation water during a shortage (for example, the water license on my parents' farm on Duteau Creek predates the construction of these dammed reservoirs by over 40 years) — it's part of what determined the purchase value of that farmland. I can remember several instances during the dry 1970s and 1980s when the Irrigation District throttled the stream flow in Duteau Creek to such a degree that my father's irrigation pump was left high and dry. After complaining, the Irrigation District had to let more water through its spillway gates because it could not divert water for its metered irrigation supply at the expense of my father's much older irrigation license.

However, the 2023 Fish Population Protection Order overturned all that with the stroke of a pen. And it was done by bureaucrats reinterpreting the law, not by elected representatives passing new legislation in the provincial legislature. It's hard not to see this as a blatant theft of water and a cancellation of long-established water rights (a.k.a. property rights).

As the old saying goes, "never let a good crisis go to waste." Crisis is always the tried-and-true trigger for a massive expansion of government powers, and the drought of 2023 was no exception. A year earlier, in 2022, the BC government created a brand-new government ministry, called the BC Ministry of Water, Land, and Resource Stewardship. In the aftermath of the 2023 drought (and 2023 water wars), the authority and reach of this new Ministry were dramatically expanded as everything water-related was rolled into this new Ministry and as it was given vastly expanded enforcement powers. By early spring of 2024, the new Ministry's enthusiastic bureaucrats were hosting drought management meetings with local farmers — better late than never. I attended the local meeting in Lumby in April.

The first 15 minutes of the meeting were taken up by the usual paternalistic ramble about how to have productive and respectful two-way communications as part of the performative theatre that every government agency is now obligated to perform to keep woke political masters in distant halls of government happy. But far from being a two-way conversation in which farmers were given an opportunity to have input into the policies shaping their destiny, as the meeting progressed it became apparent that, in between the info-sessions about various new taxpayer-funded government grants available to farmers through this new Ministry (the carrot that accompanies the stick), the main purpose of the meeting was a one-way communication informing farmers of the bureaucrats' newly-expanded authority to enforce its directives and to educate farmers about the policy decisions reached by its industrious army of government-funded "experts". The main "two-way communication" that emerged from the meeting was that the Ministry asked farmers to communicate their email addresses to the Ministry to make it more efficient to notify farmers about future Ministry directives and shut-down notices. Any in-depth technical questions raised by farmers during the meeting were brushed off with some version of "it's complicated, we have experts working on these things, there's lots of mathematics, trust us, we know what's necessary." Trust the experts.

During the meeting the Ministry's bureaucrats proudly shared that they had been able to bypass the cumbersome legislative process by repurposing and "re-interpreting" older legislation (the Fish Population Protection Orders) to adapt them to present circumstances and the new Ministry's objectives, highlighting once again that much of the rule-making process in our current "democracy" is no longer done by elected representatives but by unelected and unaccountable bureaucrats and the "experts" working in those institutions.

As frustrating as this rule-by-bureaucracy is, until these government institutions are wholly stripped of much of their current powers in order to restore autonomy to individuals and local communities, perhaps it is a blessing in disguise that the centralized power of the government is currently concentrated in a diffuse collection of dysfunctional and self-interested bureaucratic agencies instead of the alternative in which power is concentrated in the unrestrained hands of any single leader. Better this bureaucratic nightmare than to empower a new Stalin, Mao, Hitler, Mussolini, Wilson, or Roosevelt. Can you imagine how much worse all this would be if the tyrannical instincts of so many of our braindead politicians weren't being restrained by bureaucratic inertia and institutional self-interest, and were instead given free rein to impose their hubris and their ideological whims on us by diktat? It's an important lesson to remember because, as the chaos and dysfunction of our era continue to grow, it is extremely tempting (as Haiti has done over and over again) to seek a way out of the chaos by empowering a strongman to sweep aside the red tape through the force of their personality and through their ruthless disregard of existing norms — "the man on horseback", as President Hoover called him — rather than taking the harder route of trying to cut government back down to size and restoring autonomy to individuals and local communities.

Even Rome's descent into imperial dictatorship was cheered by Rome's long-suffering citizens who saw Julius Caesar as the antidote to the stifling and corrupt rein of a senate-controlled republic. Be careful what you wish for... only three generations later Rome's citizens were suffering under the spurs of one of Rome's most cruel and tyrannical emperors (Caligula), who was allegedly so power-mad that he set about to appoint his favorite horse, Incitatus, as a consul as a gesture to humiliate the Roman Senate to show them that their authority had become so meaningless that they could be replaced by an animal at the emperor's whim.[3]

Despite the Ministry of Water's genuinely friendly demeanour at the water meeting and all the donuts, coffee, and free pens they handed out, the Ministry's hardnosed agenda as energetic central planner and controller was unmistakable. They announced that future enforcement efforts would be much more aggressive as these agencies had just given themselves the authority to increase fines from just over $1,000 in 2023 to a much higher rate of *$100,000 to $500,000 per infraction per day* going forward, with the option of escalating to criminal prosecution, jail time, and fines in excess of $2 million. One of the farmers pleaded whether, at the very least, could the Ministry please provide some kind of advanced warning a few weeks before a shutdown to give farmers the opportunity to prioritize their most vulnerable crops before losing access to water? One of the Ministry's bureaucrats responded that farmers should consider this meeting (in April, before the start of irrigation season) as their advance warning and that farmers should hereby consider themselves on notice for whatever water restrictions the Ministry might choose to impose, at its discretion and without further notice, in the months to come. And then one of the bureaucrats had the gall to complain about how hard the bureaucrat's job is, how emotionally difficult it was for them to have to turn off the farmers' water, and that these bureaucrats therefore had just as difficult of a summer in 2023 as the farmers who'd had their water shut off and their crops imperilled. It wasn't meant in a bad way — it was a poorly worded effort at expressing empathy for the farmers' plight — but it resonated like fingernails on a chalkboard nonetheless.

Another farmer brought up the issue I raised earlier that the Ministry of Water had essentially negated long-standing water rights by prioritizing metered water over much older water licenses. The bureaucrats responded that downstream water rights do not have a right to access stored water, only natural streamflow. The farmer responded that, by virtue of building dams, encouraging rampant logging, and altering the water flow inside the

---

[3] Nix, E. (2016, Jun 21). Did Caligula really make his horse a consul? *History.com*. https://www.history.com/news/did-caligula-really-make-his-horse-a-consul

watershed since the 1960s, the government had completely changed the water cycle in the upstream watershed and thus changed the natural behaviour of the streams, and thus had a responsibility towards farmers with pre-existing downstream water rights. The officials pressed on to answer a different question instead.

Another farmer pointed out that in view of the massive population increase in the area, which has quadrupled since the 1970s, why aren't more dams being built in the highlands to supply the growing population? After all, there's plenty of spring runoff wasted every year during the snowmelt that could be captured and stored. The representative from the Ministry of Water dismissed the idea outright, declaring that there would be no new dams built because it has become much too expensive (*"you just can't do the kinds of things you used to be able to do in the 50s, 60s"*). What was left unsaid is that with the bulk of the province's voting population concentrated in Vancouver and with so little local autonomy over the decision-making process required to build such dams, the noise made by urban environmental activists against new water reservoirs trumps the needs of smaller population centers and rural communities who would like to build out their local infrastructure to keep up with their growing populations and evolving circumstances. And with tax revenues being pooled in provincial treasuries, by the time provincial politicians finish divvying out cash for projects near and dear to the hearts of their largest voting populations, there just isn't enough money left in the piggy bank for anyone else. Without greater autonomy over local decision-making and greater local autonomy over how tax revenues are spent, distant political considerations will always trump the needs of local communities as career politicians work to appease those who pose the greatest threat to their chances of re-election.

At a follow-up meeting in June, a representative from the Regional District's water department (not from the provincial Ministry of Water) clarified that, contrary to what was said by the Ministry of Water at the April meeting, the local Regional District does in fact have a longstanding plan to enlarge the Aberdeen Lake reservoir, but that they have already been waiting more than four years (angry face) for the provincial government to issue the necessary permits to get started on the reservoir expansion. In the meantime, the projected cost, initially set at $20 million, has doubled (or more). What came through during the June meeting is that, with so many distant and more centralized layers of government superseding local authorities, local planners are just as helpless and frustrated as everyone else in their efforts to deal with local problems even though they clearly see both problem and solution.

She went on to advise the farmers, who by then were working on a series of proposals to build their own storage reservoirs to be able to store and manage the water guaranteed to them by their water licenses, that thanks to all the regulatory hurdles they should expect the approval process to take at least 10 years, or more, before they see any project reach completion, if they are lucky enough to get approval in the first place. Basically, at the glacial speed at which the regulatory process moves for anyone without teams of lawyers and well-greased political connections, you're building things for the benefit of your children and grandchildren. The local authorities are not the ones holding up the process — the bottleneck lies further up the political food chain within the heavily politicized provincial and federal layers of government and within the maze of competing bureaucracies that have authority over our lives.

The same local planner also pointed out that anything labeled as an "irrigation dam" is extremely unlikely to be granted approval by the provincial government in the current political climate. Moreover, applications by individual farmers would almost certainly be denied. She advised instead that applications need to be written as being a "community-led effort to protect stream flows for fish populations" (or some variation thereof), with the unspoken side benefit that as long as there is enough water for the fish, farmers will be able to continue to draw from their water licenses in the same watershed. In other words, thanks to politics, everything has to be couched in the right "newspeak" (as George Orwell called it) in order to navigate the political maze — if it doesn't have the right buzzwords and doesn't express empathy for all the right causes that are in vogue today, the approval (plus or minus any attached grants) will flow to someone else who knows how to play the game.

However, the net effect of being forced to cloak private interests as charitable communal projects designed to "protect the fish" is that, while it may temporarily alleviate current tensions, it ultimately means that the time, effort, and private money that these farmers are pouring into the construction and subsequent management of these charitable communal water storage projects does not actually give them an explicit guaranteed right to use that stored water, nor does it explicitly resurrect their right to use water previously guaranteed to them by their existing water licenses.

By the wording that's needed to jump through all the hoops, they are acting as philanthropists even as the extra water they add to summer stream flows nevertheless still belongs to the "community" and to the fish. The net result is that the red carpet is thus laid out for the government to change its mind sometime in the future as the water wars continue, leaving farmers as risk of being hung out to dry once again. It leaves the path wide open for future planners to redirect that extra stored water from those "community" reservoirs to some other purpose that they deem will better serve the community's collective interests in the future, whether for environmental reasons (pray that some environmental group doesn't discover some rare frog in need of water and a well-funded activist group to watch over it), or to supply water to the expanding urban population, or to supply irrigation water to the growing number of politically-connected agri-corporations moving into the region.

Without secure water rights, you do not have viable multi-generational farms, and every second and every dollar that farmers invest into their crops, infrastructure, and soil is a gamble. And yet here we are, watching political expediency transform the long-standing concept of ironclad water rights into whimsical water privileges. Haiti's example should provide a stern warning that without explicitly guaranteed property rights, you are forever at the whim of capricious gatekeepers and the parasites in their entourage who are perpetually lobbying them for favors.

The local planner also pointed out that despite the long-term planning being attempted by local planning authorities, in her experience approvals (and grants) from higher levels of government go through predictable cycles — during dry years only projects that focus on drought mitigation get approval (and are able to access grants) but then, as soon as the weather gets wetter, drought mitigation projects are categorically shelved, cancelled, or defunded as flood prevention projects suddenly become vogue. In short, grab as much money and secure as many approvals as you can from capricious planning committees while the mood is favorable because there's absolutely no long-term thinking or steady hand on the tiller (paraphrasing her words). And so, on and on it goes, like a chicken bouncing off walls as it runs about without a head. Such is the price we pay for centralization. And yet somehow, in our enthusiasm for a benevolent shepherd, we have given that headless chicken authority over almost every aspect of our lives.

During the April meeting, the government hydrologist from the new Ministry of Water presented charts showing the declining water levels in local streams and aquifers and stated that stream levels had *"fallen to the lowest levels in over 100 years, since records in the area began."* She followed up by stating that *"farmers just need to accept that they can only grow these kinds of crops in this area during 7 out of 10 years."* In other words, in her view this is the new normal that people simply need to learn to live with (paraphrasing once again). She then proceeded to point farmers towards a new government-run crop insurance program to cover the bad years. The path to government dependency is paved with good intensions.

Although I have no way of independently verifying the hydrologist's numbers, her claim about falling stream levels very much fits with what I described in previous chapters as to what happens to stream flows, aquifers, and groundwater tables as a consequence of deforestation and soil degradation — most notably the tragic story told about Lake Chad in the Sahel region where lake levels fell by 83% to 93% since the late 19th century. And so, I suspect her claim about stream and groundwater levels is likely true in the same way that the blind man holding on to the elephant's tail is 100% correct in understanding the nature of the elephant's tail while being wholly unable to see the bigger picture because, as always, the implication (subtext) is that these local climate changes are the consequence of global long-term irreversible changes to the climate caused by CO_2.

When looking at official annual precipitation records from across the province and from the neighboring province of Alberta (a sample of which are shown in figure 121 below[4]), you can see that there is <u>no</u> downward trend in precipitation over the past century as rainfall levels across the province continue to fluctuate within a normal historic range. Despite the unprecedented number of communities threatened by wildfires and despite water levels falling to historic lows in local streams and aquifers, 2023 was merely a dry year but still well within the normal variability of the local climate. With the exception of Kelowna (whose data does not reach back as far in time as the others), annual precipitation records do not show a downward trend in precipitation levels — the trend is clearly flat.

---

[4] "Canada Weather Stats." *Weatherstats.ca, with data from Environment and Climate Change Canada.* Accessed 31 Aug 2023. https://www.weatherstats.ca/

And so, in British Columbia we see stream flows, aquifers, and groundwater tables declining against a backdrop of *stable* precipitation trends. Drought without a reduction in rainfall.

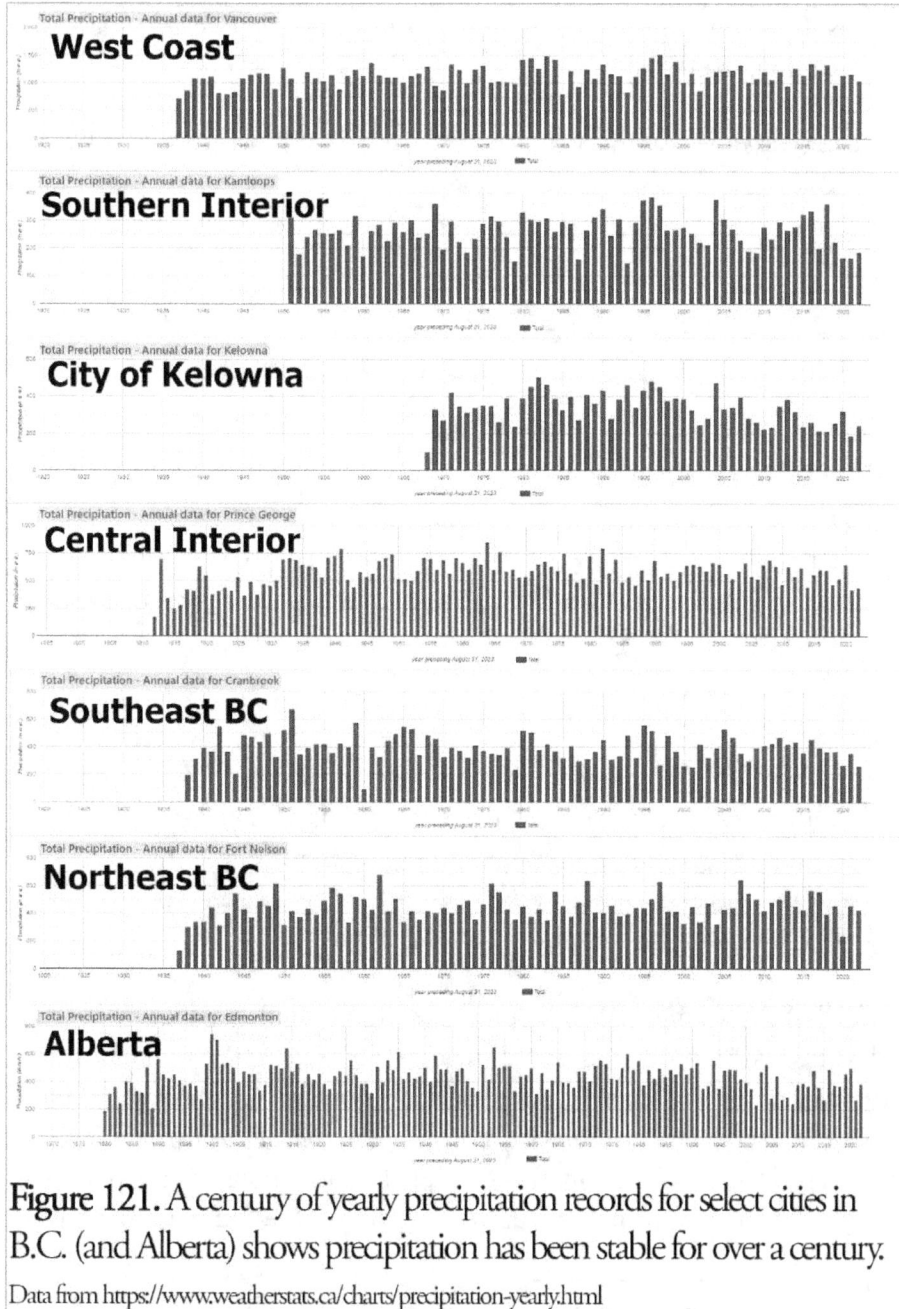

Figure 121. A century of yearly precipitation records for select cities in B.C. (and Alberta) shows precipitation has been stable for over a century.

Data from https://www.weatherstats.ca/charts/precipitation-yearly.html

Official government data even shows that when Canada is viewed as a whole, rainfall has *increased* across the country in each of the four seasons, as shown in figure 122 below,[5] as you would expect against the backdrop of a mildly warming climate that increases evaporation over the oceans. It would seem we are manufacturing a drought *despite* increasing precipitation. Well done!

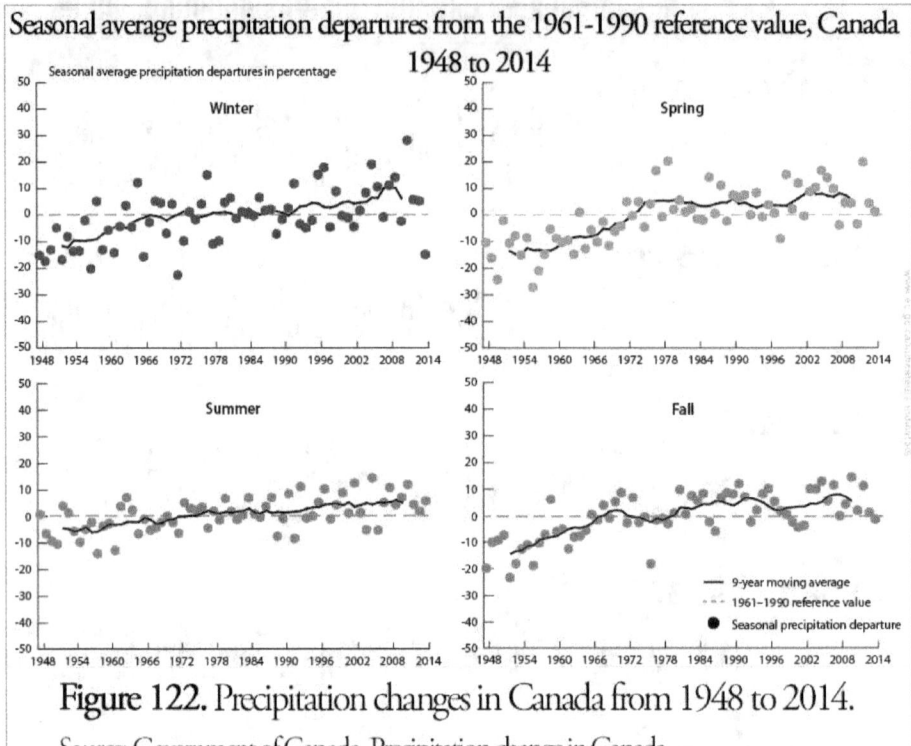

Figure 122. Precipitation changes in Canada from 1948 to 2014.
Source: Government of Canada, Precipitation change in Canada
https://www.canada.ca/en/environment-climate-change/services/environmental-indicators/precipitation-change.html

One of the farmers at the April meeting pointed out that as it stands today, over 40% of the Aberdeen Plateau (the source of much of the local irrigation water) has been clearcut, and that number rises to 55% when we include the acreage consumed by recent forest fires. The farmer's statistics are likely in the right ballpark — a quick comparison of air photos on Google Timelapse from 1984 until 2022 shows the dramatic increase in

---

[5] "Precipitation change in Canada." *Government of Canada.* 9 Jun 2016.
https://www.canada.ca/en/environment-climate-change/services/environmental-indicators/precipitation-change.html

bare clearcuts, which have not yet fully regrown, as shown in figure 123 below.[6] Clearly, cutting down so many trees has had a major effect on the water cycle in the region as evaporation from the soil increases, as soil erosion reduces the soil's moisture absorption capacity, and as bare deforested soils lose their snow faster in the spring.

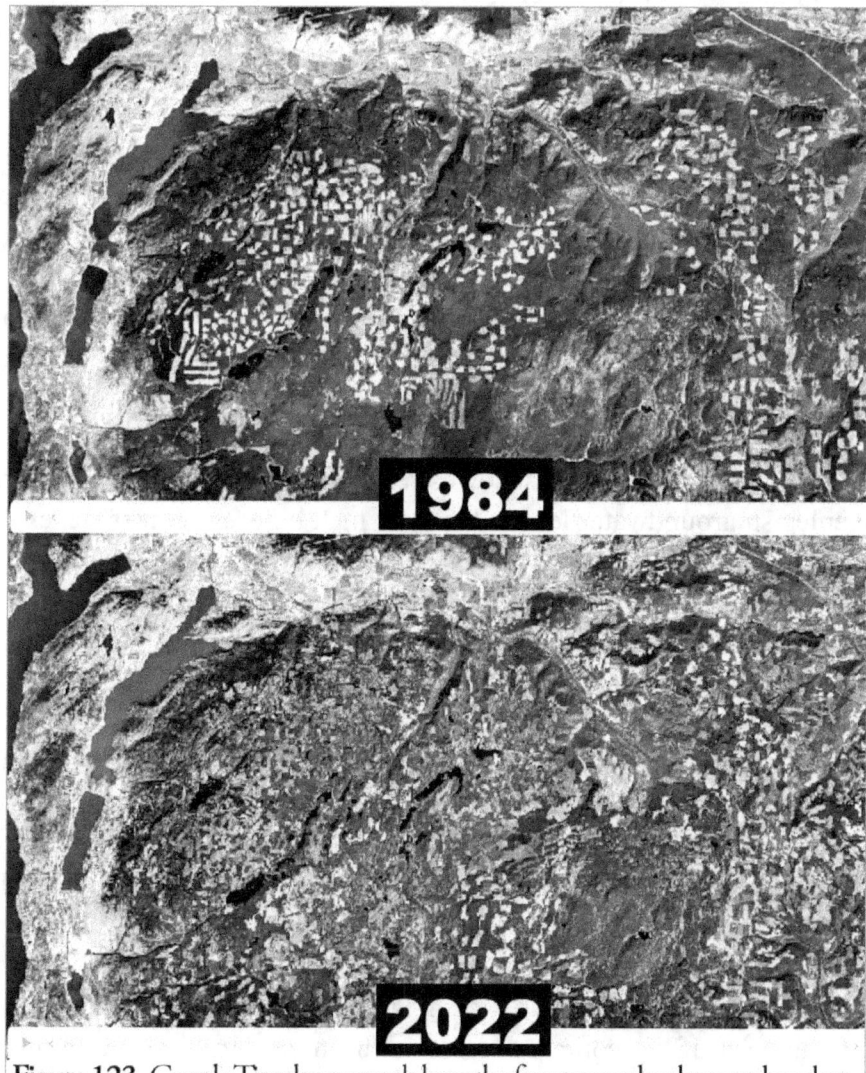

Figure 123. Google Timelapse reveals how the forest cover has been reduced on the Aberdeen Plateau, near Kelowna, B.C., from 1984 until 2022.

https://earthengine.google.com/timelapse/

---

[6] "Google Timelapse." *Google Earth Engine.* Accessed 8 Jun 2024.
https://earthengine.google.com/timelapse/

Just because these clearcuts will regrow in time doesn't change the fact that in the interim those bare patches have a dramatic impact on how much rainfall is lost to evaporation instead of being absorbed by the soil, how quickly water moves through the watershed and contributes to refilling local aquifers, and how local climate patterns will change as a result of the loss of tree cover. And new clearcuts continue to be added faster than old clearcuts regrow so the problem is only going to continue to get worse. Even BC's premier was recently forced to admit that "short-term thinking" has left BC's forests "exhausted".[7]

Much like John Wesley Powell's unsuccessful efforts a century and a half earlier to explain the causes of increased runoff from the hills in the Great American Desert, the farmer at the meeting attempted (unsuccessfully) to explain to the bureaucrats that aggressive logging in the surrounding hills has had a dramatic impact on how quickly the snow melt happens in spring because the loss of trees reduces the shade that once slowed snowmelt, leading to faster spring melts (and spring floods in the valleys), and that because of this accelerated snowmelt a greater portion of the meltwater is lost downstream instead of soaking into the soil to replenish groundwater levels. If snows melt before the frost leaves the ground, very little of the meltwater will infiltrate into the soil. And so, as summer advances, springs dry up, streams that once had healthy flow all summer slowly run dry, and aquifers decline because they are not being replenished with sufficient groundwater seeping off the surrounding hills.

When the farmer stated that our water woes will continue until we address the rampant deforestation of our watershed basin, the bureaucrats flatly responded that the forests in the watershed are the responsibility of BC's Ministry of Forests and are thus completely outside of the Ministry of Water's purview, and are most definitely out of the hands of both the municipal government and the Regional District's authority — such is the nature of the compartmentalized structure of our institutions and bureaucracies. Local hands are tied in knots.

In an echo of the complex forces set in motion in the lead-up to the Dust Bowl of the 1930s, the subtle changes in our land use seem negligible when considered in isolation, but added together they are gradually

---

[7] Owen, B. (2022, Dec 26). B.C. timber industry in throes of change, as premier warns of 'exhausted forests." *CBC News*. https://www.cbc.ca/news/canada/british-columbia/bc-timber-industry-change-exhausted-forests-1.6693769

turning our once thriving local ecosystem into a brittle ecosystem that is no longer able to cope with the weather extremes that make up the natural long-term cyclical nature of our local climate. The inconvenience of what would once have been nothing more than a dry year is transformed into a drought of existential proportions despite the fact that rainfall patterns are still well within the historic range of natural climate variability in the region and despite the fact that rainfall trends are actually increasing across the country. And then, during wet years, those same streams and rivers invariably run brown with soil stripped off the surrounding hills as they spill over their banks into the farms and villages below.

Meanwhile, deep below our feet the aquifers that once sustained our wells and replenished our streams and rivers are slowly drying up. It's a local story that is repeating itself all over the world. From the Ogallala Aquifer on the High Plains of Kansas, to the Paris Basin, to the Northern China Aquifer System, to the Central Valley Aquifer in California, to the Ganges-Brahmaputra Basin in India, to the Canning Basin in Australia, and countless aquifers in between, all over the world we see local planners struggling with aquifers that are drying out as they are exhausted not only by over-pumping, but because the land use above these aquifers has severely impaired the water cycle that refills those aquifers whenever it rains. According to data published by NASA and the California Institute of Technology, 21 of the world's 37 largest aquifers have already passed the tipping point on their way to being depleted.[8] The water wars are only going to intensify, but there is no cure as long as we continue to misdiagnose the cause. The fanatic crusade against greenhouse gases is slowly coming to bite us all because it is blinding us from being able to see the world as it truly is, and it is leading our governments to attack all the things that can fix this — livestock, fertilizer, low-intensity fires, property rights, and individual and local autonomy.

~

History is a graveyard of foolish ideas, misbegotten ideological crusades, and destructive phantasies. Each generation recognizes the

---

[8] "Map of Groundwater Storage Trends for Earth's 37 Largest Aquifers." *NASA Jet Propulsion Laboratory and the California Institute of Technology.* Accessed 27 Jun 2024. https://gracefo.jpl.nasa.gov/resources/48/map-of-groundwater-storage-trends-for-earths-37-largest-aquifers/

follies of past generations and thinks of itself as smarter, wiser, and better informed because it can see through the falsehoods that captured prior generations. And yet, that wisdom is an illusion. The falsehoods of yesteryear have become visible not because we are smarter or better informed, but because those earlier falsehoods have ceased to be useful. Society has merely moved on to new falsehoods to suit a new age, to legitimize new forms of plunder, and to express the sums of all its current hopes, dreams, fears, and vulnerabilities. And those new falsehoods are wholly invisible to us because, just like our ancestors, we too have wrapped ourselves in a dizzying maze of perverse incentives, "noble lies", and soothing fairy tales to prevent us from seeing the world for what it really is.

The climate crusade is merely a symptom — one of many — of what we became as a consequence of the incentives set in motion when we bamboozled ourselves with the lure of central planning as the solvent to all our problems. At the heart of the unravelling lies an ancient philosophical war about the purpose and limits of effective government.

The greatest tragedy of the myth of Icarus was not that his hubris led him to try to fly too close to the sun. That is the essence of human nature. The greatest tragedy was that even before Daedalus gave Icarus his wings, he recognized that his son was likely to test the limits of the tool placed within his grasp but did it anyway under the mistaken belief that he could rein in his son with a stern warning about the limits of his newfound power to fly.

Each generation inherits limits without the hard-earned experiences that created them. And so, much like the parchment guarantees of our rights and freedoms that our founding fathers used to place limits on our democratic systems of government, the constitutional limits on legislative power are nothing without the long, slow, laborious process of nurturing an enduring multigenerational culture willing to defend those limits at all costs, especially during difficult times when the urge is greatest to overstep those limits and trample on rights and freedoms for political expediency. Only culture can provide a counterforce to the universal constant of our human nature to test the limits of *any* tool that falls into our grasp. Of all the tools and technologies ever created by our species, none has more power, and none is in greater need of limits than the masterful administrative state, because the power to create is ultimately also the power to destroy.

The greater the power and resources that are placed in the hands of any gatekeeper, the more that society turns away from truth in order to play the great game of soliciting favors from the gatekeepers. And so, the incentives of a bloated centralized system soon take on a life of their own as they gradually erode away all the essential foundations that once allowed civilization to thrive — principles, values, constitutional limits, individual autonomy… and soil.

To break the invisible chains that enslave our minds, we urgently need to draw a line under the progressive legacy of Wilson, Roosevelt, and all the starry-eyed central planners who thought they could make the world a better place by empowering a masterful administrative state to elevate government from umpire to controller, only to inadvertently drive a stake through the heart of science, democracy, individual liberty, and the pursuit of truth. Everything is connected.

~

Three Rings for the Elven-kings under the sky,
Seven for the Dwarf-lords in their halls of stone,
Nine for Mortal Men doomed to die,
One for the Dark Lord on his dark throne
In the Land of Mordor where the Shadows lie.

One Ring to rule them all, One Ring to find them,
One Ring to bring them all, and in the darkness bind them
In the Land of Mordor where the Shadows lie.

— J.R.R Tolkien, Lord of the Rings

About the Author

Julius Ruechel has a BSc of Geology from the University of Alaska, Fairbanks. He is the author of *Grass-Fed Cattle: How to Produce and Market Natural Beef* (a how-to guide for farmers based on his experiences in the cattle industry) and the author of *Autopsy of a Pandemic: The Lies, the Gamble, and the Covid-Zero Con* (an investigation into the government's mishandling of the Covid pandemic). This is his third book.

You can read more of his essays and explore his independent investigative journalism on his website at *www.juliusruechel.com*. And you can read his articles about raising cattle on his educational cattle farming website at *www.grass-fed-solutions.com*.

www.ingramcontent.com/pod-product-compliance
Lightning Source LLC
Chambersburg PA
CBHW071909210526
45479CB00002B/345